Metallic and Molecular Interactions in Nanometer Layers, Pores and Particles
New Findings at the Yoctolitre Level

RSC Nanoscience & Nanotechnology

Series Editors:
Professor Paul O'Brien, *University of Manchester, UK*
Professor Sir Harry Kroto FRS, *University of Sussex, UK*
Professor Harold Craighead, *Cornell University, USA*

Titles in the Series:
1: Nanotubes and Nanowires
2: Fullerenes
3: Nanocharacterisation
4: Atom Resolved Surface Reactions: Nanocatalysis
5: Biomimetic Nanoceramics in Clinical Use: From Materials to Applications
6: Nanofluidics: Nanoscience and Nanotechnology
7: Bionanodesign: Following Nature's Touch
8: Nano-Society: Pushing the Boundaries of Technology
9: Polymer-based Nanostructures: Medical Applications
10: Metallic and Molecular Interactions in Nanometer Layers, Pores and Particles: New Findings at the Yoctolitre Level

How to obtain future titles on publication:
A standing order plan is available for this series. A standing order will bring delivery of each new volume immediately on publication.

For further information please contact:
Sales and Customer Care, Royal Society of Chemistry, Thomas Graham House, Science Park, Milton Road, Cambridge, CB4 0WF, UK
Telephone: +44 (0)1223 432360, Fax: +44 (0)1223 420247, Email: sales@rsc.org
Visit our website at http://www.rsc.org/Shop/Books/

Metallic and Molecular Interactions in Nanometer Layers, Pores and Particles
New Findings at the Yoctolitre Level

Jürgen Fuhrhop
Institute for Organic Chemistry, Freie Universitat Berlin, Berlin, Germany

Tianyu Wang
CAS Key Laboratory of Colloid and Interface Science, Institute of Chemistry, The Chinese Academy of Sciences, Beijing, PR China

RSC Nanoscience & Nanotechnology No 10

ISBN: 978-0-85404-166-4
ISSN: 1757-7136

A catalogue record for this book is available from the British Library

© Jürgen Fuhrhop and Tianyu Wang 2010

All rights reserved

Apart from fair dealing for the purposes of research for non-commercial purposes or for private study, criticism or review, as permitted under the Copyright, Designs and Patents Act 1988 and the Copyright and Related Rights Regulations 2003, this publication may not be reproduced, stored or transmitted, in any form or by any means, without the prior permission in writing of The Royal Society of Chemistry or the copyright owner, or in the case of reproduction in accordance with the terms of licences issued by the Copyright Licensing Agency in the UK, or in accordance with the terms of the licences issued by the appropriate Reproduction Rights Organization outside the UK. Enquiries concerning reproduction outside the terms stated here should be sent to The Royal Society of Chemistry at the address printed on this page.

Published by The Royal Society of Chemistry,
Thomas Graham House, Science Park, Milton Road,
Cambridge CB4 0WF, UK

Registered Charity Number 207890

For further information see our web site at www.rsc.org

Preface

Modern nanochemistry provides the most rewarding and promising opportunity today to re-discover the chemical elements by investigating all their material properties anew under the condition of nanometer confinement. A few chosen examples may help to visualize some of their magic attractions in the 21st century.

Hydrogen and lithium are the lightest reactive elements and serve as electricity sources in fuel cells and laptop batteries. Carbon forms covalent, non-metallic graphite wires, which shows the phenomenon of "ballistic" electron transport without heat. Heavy main group metals like Cs, Bi, Te, and Pb and their combinations with S and Se have a high electric and a low thermal conductivity, a property which leads to a similar cool electric current in thermoelectric nanowires. They are all dream materials for laptop wirings. Crystalline nanolayers of germanium antimony tellurides become amorphous under laser beam pulses within a nanosecond and recrystallize upon electric heating. This behavior became the basis of rewritable laser disks.

The separation of sodium-potassium ions dominates the geological and biological water chemistry. Hydrated sodium ions pile-up in the oceans and in the blood stream, whereas the equally abundant potassium ion is bound to phosphate and silicate nanoparticles of the soil and the protein fibers within muscles and neurons. Main group metal chalconides, provide diamagnetic semiconductors which separate electrons and electron holes and luminesce upon charge recombination in nanoparticles. Simple anodic oxidation of metallic aluminum produces regular patterns of nanometer holes coated with alumina and all kinds of brush-like nanowires assemblies may be formed in them and be set free by acid/base treatments.

The most peculiar and important property of transition metals is their variable soft or hard paramagnetism, which is most prominent in thin,

---

RSC Nanoscience & Nanotechnology No. 10
Metallic and Molecular Interactions in Nanometer Layers, Pores and Particles: New Findings at the Yoctolitre Level
By Jürgen Fuhrhop, Tianyu Wang
© Jürgen Fuhrhop and Tianyu Wang 2010
Published by the Royal Society of Chemistry, www.rsc.org

needle-like micro- or nanocrystals. Reversible and long-lived data storage of computers totally rely on optimized magnetic nanodomains.

All of the mentioned material properties strongly depend on confinement and ordering. Crystal surfaces react very differently from bulk crystals, the borderline between fluid layers is a much more favorable environment for nanoparticles than bulk media. Nanoparticles with several organic attachments (dyes, proteins, receptors, DNA *etc*) and asymmetric surface coatings may be constructed and become multifunctional like microorganisms An organic coating may also stabilize the most fragile nanocrystals like iron metal in air/water environments.

This book provides recent examples for these and many other peculiarities of the artificial nanoworld, which is ordered here by the periodic system of the elements. It helps chemists to think about the development of new material properties, which may improve electronic machines, geological, biological and medical reagents, which may, in the end, be applied worldwide not on the megaton, but on the kilogram scale.

Contents

Chapter 1	**Atoms, Molecules, Electrons, Light and Heat in Nanometre Confinement**	**1**
	1.1 Introduction	1
	1.2 Water, Toluene, Nanoparticles and Nanocrystals	3
	1.3 Nanolayers, Yoctowells and Receptors	9
	1.4 Nanorods, Nanotubes and Nanowires	17
	1.4.1 Appearance of Nanoparticles	18
	1.5 Electric Conductivity	21
	1.6 Magnetism	24
	1.6.1 Quantum Mechanics of the Fermi Sphere	27
	1.6.2 Thermodynamic–Kinetic Model	28
	1.7 Heat and Thermal Conductivity	35
	1.8 Luminescence	37
	1.8.1 Practical Applications	37
	1.8.2 Nanoparticle Luminescence	38
	1.9 Catalysis	43
	1.10 Density Functional Theory	46
	References	47
Chapter 2	**Organic Carbon**	**52**
	2.1 Introduction	52
	2.2 Nanodiamonds, Graphene Sheets and Carbon Nanotubes	53
	2.3 Hydrocarbons	72
	2.4 Alcohols, Ethers, Amines and Sulfides	82

RSC Nanoscience & Nanotechnology No. 10
Metallic and Molecular Interactions in Nanometer Layers, Pores and Particles: New Findings at the Yoctolitre Level
By Jürgen Fuhrhop, Tianyu Wang
© Jürgen Fuhrhop and Tianyu Wang 2010
Published by the Royal Society of Chemistry, www.rsc.org

	2.5	Aldehydes, Ketones, Acetals and Imides	94
	2.6	Carboxylate and Phosphate	94
	2.7	Esters and Metal–Organic Frameworks (MOFs)	101
	2.8	Polymers	104
	2.9	Dyes	130
	2.10	Cellulose	136
	2.11	Non-covalent Twisted Ribbons and Proteins	139
	2.12	DNA	143
	2.13	Fluoride	146
	References		147

Chapter 3 Main Group Elements 160

    3.1 Introduction   160
    3.2 Hydrogen   163
    3.3 Alkali Metals   164
        3.3.1 Lithium   164
        3.3.2 Sodium and Potassium   166
    3.4 Magnesium, Calcium and Barium   174
        3.4.1 Magnesium   174
        3.4.2 Calcium   174
        3.4.3 Barium   178
    3.5 Boron, Aluminium, Gallium, Indium and Thallium   179
        3.5.1 Boron   179
        3.5.2 Aluminium   181
        3.5.3 Gallium   184
        3.5.4 Indium   186
        3.5.5 Thallium   188
    3.6 Silicon, Germanium, Tin and Lead   189
        3.6.1 Silicon   189
        3.6.2 Germanium   196
        3.6.3 Tin   200
        3.6.4 Lead   201
    3.7 Nitrogen, Phosphorus, Arsenic, Antimony and Bismuth   202
        3.7.1 Nitrogen   202
        3.7.2 Phosphorus   209
        3.7.3 Arsenic   212
        3.7.4 Antimony   213
        3.7.5 Bismuth   213
    3.8 Sulfur, Selenium and Tellurium   214
        3.8.1 Sulfur   214
        3.8.2 Selenium   215
        3.8.3 Tellurium   217
    3.9 Fluorine   220
    3.10 Zinc, Cadmium and Mercury   221

		3.10.1	Zinc	222
		3.10.2	Cadmium	225
		3.10.3	Mercury	231
	References			232

Chapter 4 Iron **245**

	4.1	Introduction	245
	4.2	Steel Domains and Magnetism	246
	4.3	Layers on Steel Surfaces, Water and Corrosion	251
	4.4	Green Rust	257
	4.5	Fe(0) Nanoparticles	261
		4.5.1 Ferrocene	266
	4.6	Iron Oxide Nanoparticles	269
		4.6.1 Ferrites	284
	4.7	Photosynthesis and Catalysis on Iron Oxide Nanoparticles	286
		4.7.1 Catalytic Iron Nanocrystals in Industrial Processes	289
	4.8	A Brief Comparison of Carbon and Iron	291
	References		292

Chapter 5 Transition Metals **299**

	5.1	Introduction	299
	5.2	Titanium, Ziconium, and Hafnium	299
		5.2.1 Titanium	299
		5.2.2 Zirconium	317
		5.2.3 Hafnium	321
	5.3	Vanadium, Niobium and Tantalum	321
		5.3.1 Vanadium	321
		5.3.2 Niobium	325
		5.3.3 Tantalum	325
	5.4	Chromium, Molybdenum, Tungsten and Manganese	328
		5.4.1 Chromium	328
		5.4.2 Molybdenum	329
		5.4.3 Tungsten	334
		5.4.4 Manganese	335
	5.5	Technetium and Rhenium	341
	5.6	Ruthenium and Osmium	342
		5.6.1 Ruthenium	342
		5.6.2 Osmium	346
	5.7	Cobalt, Rhodium and Iridium	347
		5.7.1 Cobalt	347
		5.7.2 Rhodium	351
		5.7.3 Iridium	353

5.8	Nickel, Palladium and Platinum		353
	5.8.1	Nickel	353
	5.8.2	Palladium	355
	5.8.3	Platinum	357
5.9	Copper, Silver and Gold		361
	5.9.1	Copper	362
	5.9.2	Silver	365
	5.9.3	Gold	371
5.10	Scandium, Yttrium and the Lanthanides		381
	5.10.1	Scandium	382
	5.10.2	Yttrium	383
	5.10.3	Lanthanides	384
References			388

Subject Index   **403**

CHAPTER 1
Atoms, Molecules, Electrons, Light and Heat in Nanometre Confinement

1.1 Introduction

The chemistry within yoctolitre (10^{-24} L) holes or on the surface of particles with a yoctolitre volume is generally called "nanochemistry" because each dimension of these holes and particles measures a nanometre (nm): $1\,\text{yL} = 1\,\text{nm}^3$. In volumes of up to 1000 yL, *e.g.* $10 \times 10 \times 10$ nm, all distances between atoms and molecules on the particles or in the holes are within a few nanometres, which allows them to interact with each other and with ions and molecules that approach the surface. Holes left by missing single atoms, molecules or ions in yoctowell walls, and in the centres or surfaces of nanoparticles or nanocrystals, can easily migrate through the whole species.

Yoctolitre dimensions are practical for the estimation of constituent numbers and models based on them: 1 yL can contain ~ 100 metal atoms, 50 molecules of metal oxide (MO), 33 molecules of water or 25 molecules of metal dioxide (MO_2). Nanometre concepts are more appropriate if [111] surfaces have to be compared with [100] surfaces and where vectors, angles, and tenths of an Ångström become important.

The first special property of nanochemistry in or on yoctolitre-sized wells, pores, spheres or crystals depends on the fact that a large percentage of the atoms that form the holes or particles are surface atoms. Quite often the electrons on the hole or nanoparticle surfaces do not bind anything, but represent "dangling bonds" with special activities. The second dominating fact is the relevance of space-dependent physical properties of molecules, which

---

RSC Nanoscience & Nanotechnology No. 10
Metallic and Molecular Interactions in Nanometer Layers, Pores and Particles: New Findings at the Yoctolitre Level
By Jürgen Fuhrhop, Tianyu Wang
© Jürgen Fuhrhop and Tianyu Wang 2010
Published by the Royal Society of Chemistry, www.rsc.org

play hardly any role in bulk inorganic and organic chemistry. For example, glucose becomes water-insoluble in hydrophobic yoctowells, because the molecules have a hydrophobic edge, which sticks to the wall, and metallic Ag(0) nanocrystals seem to forget about their metallic electron cloud and fluoresce like covalent molecules, for reasons yet unknown.

The translation of stereochemistry into spin interactions of electrons and different kinds of magnetism, the chemical stabilization of electron–hole pairs in order to create luminescence of all colours by the variation of electric potentials, the change in metal crystal lattices on the surface of nanocrystals that renders them catalytically active or fluorescent, and the fixation of water-soluble carbohydrate edge amphiphiles in hydrophobic, water-filled yoctowells are subjects of current research, and chemists interested in the development of new properties should become familiar with them. Research in this field requires access to modern techniques such as rapid crystallization of vapour, laser ablation under water, preparation of atomically smooth surfaces, atomic force microscopy, electron transmission microscopy and electron scanning microscopy, among others.

Subjects covered in this book are, in order of increasing complexity:

- self-cleaning surfaces of nanometre roughness
- quantum dots (QDs), which provide everlasting colours and discrete energy levels instead of bands
- soft and hard magnetic particles for computers and other engineering uses
- minimization of functional AFM tips
- localization of many different single molecules in a very small aqueous space
- decomposition of chlorganic compounds in soils
- coupling of NMR signals with magnetic field steering
- fixation of proteins, DNA and cell surfaces on nanoparticles in water
- routine interconnection of nanoscale elements by nanowiring
- photolysis of water by sunlight
- synthesis of effective catalytic corners with different elements, particularly iron
- assembly of nanoparticles to form efficient thermoelectric elements
- minimization of computer hard disks.

Today's nanochemistry offers a unique chance to have a fresh look at the periodic system of the elements. Carbon chemistry, to name a popular example, is a huge field of classical and modern chemistry, but molecular coatings of nanoparticles that shrink in water on heating, graphite tubes showing ballistic electron transport or the dominance of dangling bonds in nanodiamonds, which makes them extremely reactive, are accessible only through nanochemistry and are of interest in industry and medicine. Magnetic data storage in computer or control devices, the conversion of sunlight into electric currents and of electric currents into light and refrigeration – all these processes depend on metallic nanoparticles and help to create or replace energy sources. Working with nanocrystals, nanowires and the inner walls of nanometre-wide wells and tubules in university labs will help young scientists to find rewarding problems for work in industry, which has to engineer, optimize, produce, sell and guarantee the nanosized constituents of computers, catalysts, light bulbs, solar converters, *etc.*

Most of the reference citations in this book date from 2005–2009 and are from easily accessible chemical journals. They provide a starting point for learning and research. Each important element of the periodic table has its own section, where relevant general properties are summarized and the element-specific ideas behind recent publications, together with selected results, are discussed. Altogether, the book is as "hands-on" as possible, and we have tried to provide the necessary physical and material chemical background in a descriptive manner. Nanochemists should know whether the products they develop in reaction flasks or on solid surfaces contain magnetic, luminescent or conductive layers, particles or wires.

Our aim is to make readers aware of current developments, while avoiding speculations about economic perspectives. The motivation for "going nano" should always be scientific success, the hope of finding something new and surprising.

1.2 Water, Toluene, Nanoparticles and Nanocrystals

The most useful nanoparticles have diameters between 1 and 10 nm, sometimes up to 20 nm, *i.e.* volumes between one and a few thousands of yoctolitres. The 1 nm species are best for structural and theoretical studies, but 10 nm guarantees relatively high stability and longevity; 5 nm is the standard compromise. Furthermore, all nanosized clusters of atoms or molecules that are of interest fulfil a function. The "functionality" may be as simple as being magnetic, or separation of sodium chloride into Na^+ and Cl^- ions, or as complex as the recognition and removal of two proteins from the bloodstream of a living organism.

We start with water, the lightest natural molecule on Earth with its molecular weight of 18. Its hexameric cluster is the lightest and most dynamic sub-nanoparticle, with a molecular weight of $6 \times 18 = 108$ Da. All molecules that are lighter than water, in particular hydrogen (H_2) and helium (He), eventually disperse into outer space after reaching the atmosphere. Water, however, forms droplets and ice crystallites in the atmosphere which could never escape the gravitational pull of the Earth, even in prebiotic times when the Earth's surface was volcanic and very hot. Our water molecules originally came here from outer space, but now they never leave the Earth, because they form nanoparticles, micrometre, and millimetre particles at heights of a few hundred metres. Liquid water is the only low-molecular-weight fluid that is stable over a 100 K temperature range. The formation of clusters and droplets has kept water on Earth for more than 4 billion years and allowed for the evolution of life here.

The oxygen atom of water is buried in the centre of a tetrahedron with two electron pairs and two protons at the corners. This polar ordering of the molecular surface dominates the behaviour of fluid and solid water. The oxygen atom holds the protons and electron pairs together, provides weight, and becomes important as molecular triplet oxygen as an end product of the water splitting in photosynthesis. Electrostatic hydrogen bridges between the electron pairs and protons of neighbouring water molecules then form a water pentamer with a molecular weight of 90, which is the unit of larger clusters and finally water droplets (Figure 1.1). But this is not all. In fluid water a sixth water molecule squeezes into the remaining empty space in the fluid pentamer clusters and forms extra hydrogen bridges, which are much less stable than the other four and correspondingly short-lived. It is this sixth molecule that accounts for the low viscosity of water.[1] It moves so quickly from one cluster to another that it keeps them mobile.

The major effect of hydrogen bonding is not viscosity, but high surface tension and boiling point. Bulk water attracts the surface water molecules strongly with one-sided hydrogen bonds, and the air on the upper side cannot compete. We know that 18 cm^3 of water contain 6×10^{23} molecules, so 1 nm^3 or 1 yL of water contains 33 molecules. The hexamer nanoparticle has a volume of

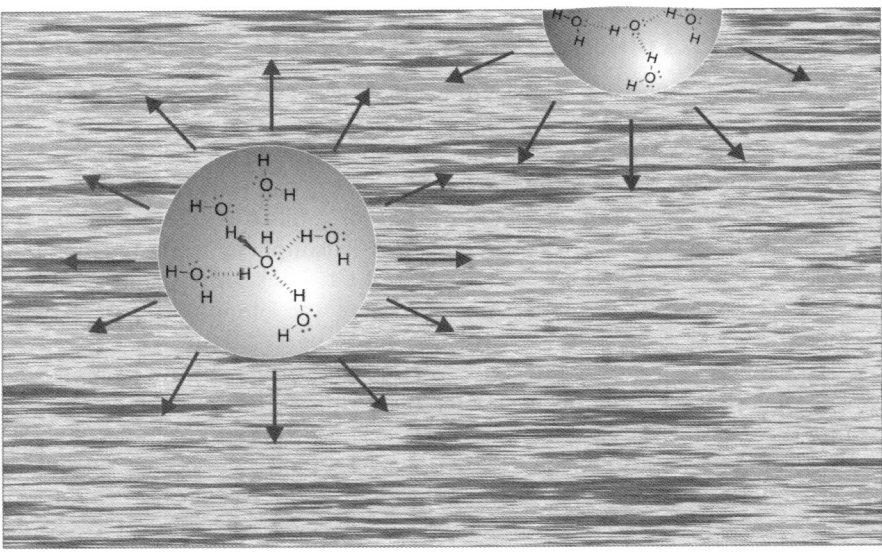

Figure 1.1 In liquid water a pentamer, $(H_2O)_5$, forms the dynamic building block of large clusters connected by linear hydrogen bonds. $-OH \cdots O$. Clustered liquids are usually viscous, *e.g.* glycerol, but water is not. It is fluidized dramatically by a sixth water molecule,[1a] which enters and leaves the pentamer very fast. Water clusters at the water–air interface are fixated by the internal water, and hardly interact with air. This effect causes surface tension and a high boiling point.[1b]

only one sixth of a yoctolitre. About 10 molecules of water form the inner hydration sphere of Na^+ and Cl^- ions, and each of the water molecules is part of a cluster. Only one or two NaCl units can thus be dissolved in a 1 yL well, and become centres of fluid water nanoparticles there.

Water is also the basis of life. More than 98% of the molecules of the human body are water, and it participates in each of the molecular actions of proteins, nucleic acids, ATP, *etc*. Its magical role in animal life is the formation of clusters around phosphate, K^+, Na^+, and H^+ ions, which permits strong electric potentials in nerves, muscles, brain and stomach.[1b] The pH of human blood and cellular fluid is always 7.4, which means there is less than 0.01 mg of protons in the whole body with exception of the stomach, with a pH of 0, where there is more than a gram of protons. Enzymes, nerves and muscles for example, would not be active at pH 5.

Soft nanoparticles and nanolayers made of carbon-containing molecules in water, *e.g.* micelles, vesicles and bilayer lipid membranes, are described in Chapter 2. Here we discuss only a few general points about the preparation and properties of metal and metal oxide nanoparticles and nanocrystals.

Toluene (=methylbenzene, $C_6H_5CH_3$) is a water-immiscible solvent. It floats on water, but both water and toluene form 1 nm long fingers, which protrude into each other at the interface for tens of picoseconds (10^{-12} s). Films of Au(0) NPs are formed if a hydrophobic gold(I) compound, namely triphenylphosphine gold chloride, $AuPPh_3Cl$, is dissolved in the upper toluene phase and a water soluble reductant, formaldehyde, CH_2O, in the form of the polar $P(CH_2OH)_4Cl$ salt, in the water below. Irregular Au_{55} clusters of uniform size (1.2 nm) were then formed at the toluene–water interface and seven such clusters formed a planar heptamer at the interface: one $Au(0)_{55}$ NP in the centre, six others around it. The lower side of this heptamer became hydrophilic by a water layer, the upper side was hydrophobic with a toluene solvation coat. Now came more gold particles from the water and toluene sides and three of them attach upside to the hydrophobic part, three others downside to the hydrophilic part. They occupied the gaps in the flat Au_{55} heptamer and reverted its polarity. The toluene side now had a hydrophilic surface, the water side a hydrophobic one. The polymeric metal consisted of 13 Au_{55} units. Growth stopped there because the short-lived, 1 nm long toluene fingers in water and water fingers in toluene did not allow for longer reaction times and larger assemblies. Only one NP layer on each side and the solvent caps stabilized the heptamer in the centre. The presence of the clusters and the associated organic layer hindered the progress of the "crystallization" unless the interface was disturbed by surface pressure and/or vibrations (Figure 1.2).[2]

The primary Au_{55} NPs were also formed in two steps: at first Au_{13} NPs were formed which attached immediately 42 more Au atoms in a second shell. Au_{55}, a "cuboctahedron" with a diameter of 1.2 nm is formed as relatively stable gold core. In the toluene/water case this Au_{55} was surrounded by a non-identified 1.1 nm organic shell, possibly $P(CH_2OH)_3$/toluene, and the coated Au_{55} NPs again clustered to form the described cluster of 13 Au NPs of 2.3 nm diameter

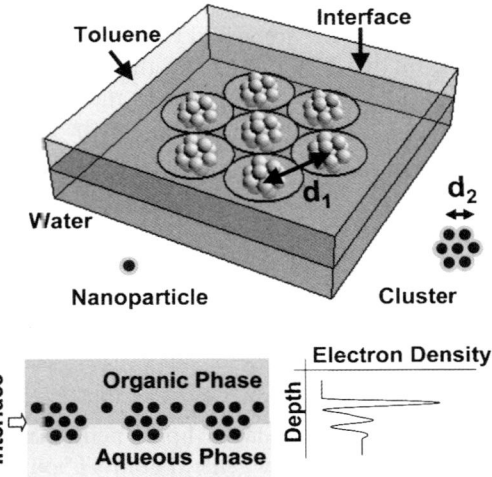

Figure 1.2 3D schematic of a 13-membered capped gold nanoparticle cluster and 2D model of three such clusters in water and toluene fingers.[2]

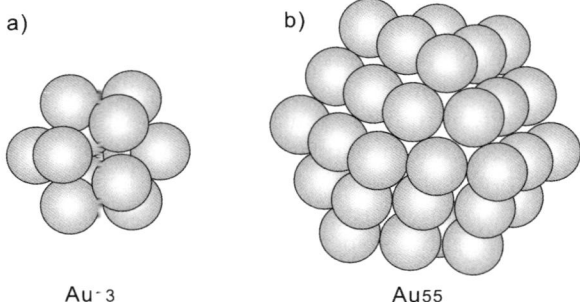

Figure 1.3 The primary nanocrystals of Au(0), namely Au_{13}, Au_{55} and $(Au_{55})_{13}$.[2] Attempts to purify and isolate Au_{55} failed.

each, one in the centre and 12 on the surface (Figure 1.3). Such interfacial assemblies are presumably standard intermediates in the formation of very small spherical NPs. Solidification of vapours on cold surfaces, laser ablation under water or rapid precipitation in acid/base reactions in uniform solvents probably produce the same intermediates but limitation of growth is caused there by short crystallization times, rather than by "fingers" between two phases.

The major crystal lattices of metals are summarized in a general structure proposed by Hume and Rothery (Figure 1.4). The possible positions of the atoms are identified by the letters a, b, c and d. If all positions are occupied by the same kind of atoms, e.g. Fe(0), the structure is cubic close packed (ccp). The Hume–Rothery figure contains eight of these ccp units and occurs in nature in

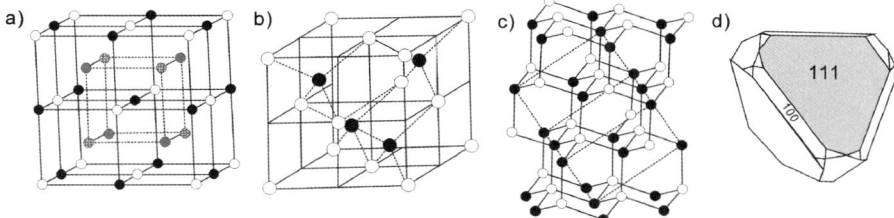

Figure 1.4 The Hume–Rothery phase with eight cubic centred cubic (ccc) cells (a) and diamond–zinc blende cells with missing corner atoms, first in a Hume–Rothery description(b), second in the form of C chair conformers (c). The (111) axis is perpendicular in the latter model. The actual crystal then contains surface planes connecting the planes of the Hume–Rothery model (d). In plane (111), for example, each next atom goes one step further in the x, y and z directions, corresponding to the diagonal plane of the Hume–Rothery model.[4]

zinc blende (ZnS), gallium arsenide (page 184 ff), zinc selenate (p. 212, 215, 231), and diamond. In zinc blende the a positions on alternating corners of the cubes are occupied by zinc, the b positions appear in alternating centres of the cubes and are occupied by sulfur; c and d are unoccupied. The usual drawing of the diamond structure with cyclohexane chairs is identical, but the Hume–Rothery cube is rotated by 45°, so that the (111) axis appears perpendicular. The (111) plane unit contains one atom; this plane grows rapidly and is the largest in the real crystal. The (100) plane unit contains two atoms, grows slowly and is much smaller in surface area.[3,4]

Metals crystallize in the lattice with the highest possible symmetry that is in agreement with the specific properties of the atoms. This leads mostly to simple cubic crystal lattices (with 8 neighbours) or hexagonal lattices (with 12 neighbours). Crystals of molecules are usually harder than those of amorphous materials, because the crystal is stabilized by various directed non-covalent forces, in particular crosslinking and hydrogen bonding (pages 11 and 14). Single crystals of metals behave differently: they are soft, but harden upon deformation. This is caused by sliding planes in the crystal, which are destroyed upon rearrangement. Corners are formed instead, which prevent any further movement (Figure 1.5).[4]

The most important metal on Earth is iron (see Chapter 4). It occurs naturally as an Fe(II) or Fe(III) oxide. In laboratory simulations of stream water containing Fe(II) chloride, 87% of the Fe(II) remained unoxidized after 24 h of aeration at pH 6.5. After addition of natural organic matter or organic carbon in the form of humic and fulvic acids from plant debris, 97% of the Fe(II) was converted to Fe(III) (hydr)oxides in the form of nanoparticles, which were adsorbed and stabilized by the organic matter. The primary FeOOH nanoparticles remained <200 nm in size and dispersed in water only in the presence of the organic matter. Their transport in stream water depended on this cooperation between phenol polymers and iron oxides. The most abundant

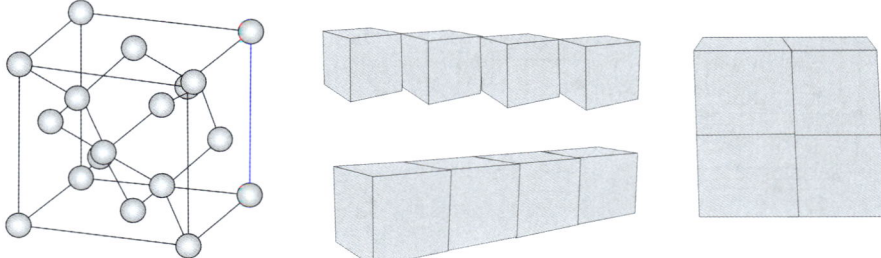

Figure 1.5 *Left:* Again the zinc blende (diamond, Si(0), Ge(0)) elementary cell (see Figure 1.4), this time in the regular cubic form. *Right:* a model of nanoparticles made of four such units with edge on, plane-on and two-planes-on attachments. The last cube corresponds roughly to a 1 nm nanocrystal. The surfaces are 24, 18 and 16 square units respectively; if the last cube is doubled, the surface will only be 24 squares instead of 32. Nanocrystal growth rapidly decreases the number of surface atoms and of (reactive) corners.

nanoparticles in nature appear at the interface between plant debris and iron salts.[5]

Nanocrystals of magnetite (Fe(II)Fe(III)$_2$O$_4$ or Fe$_3$O$_4$), are commonly used in medicine as carriers in the bloodstream, and may be steered to damaged tissues by an external magnetic field. The magnetite nanocrystals result from an initial short burst of nucleation, when Fe(II) and Fe(III) are coprecipitated with OH$^-$ ions. Later, when the concentration of the Fe(II/III) hydroxides reaches a critical supersaturation, the diffusion of solutes allows only a slow growth of the nuclei and leads to many different crystal surfaces. Small, monodisperse magnetite nanocrystals are therefore best isolated immediately after nucleation.

In such time-dependent crystallizations, the desired size and shape of the nanocrystals can always be tailored by the conditions, such as pH, ionic strength, temperature, choice of salts (perchlorates, chlorides, sulfates, nitrates, citrates), and the experimental optimization of the absolute and relative Fe(II)/Fe(III) concentrations. Immediate microscopic control of the result and preparation according to a detailed protocol is essential. The possible addition of any other molecule for stabilization should wait until one knows how to prepare the nanocrystals. Different anions often lead to different nanocrystals; variations in size and shape are then usually traced back to more or less negative charges on the surface of the nanoparticles. For example, if increasing amounts of citrate ions are added, the diameter of the nanocrystals will decrease by a factor of ~ 2. Chelation of the iron ions first retards the nucleation, and a citrate coat then inhibits further growth of the nuclei.[6]

Ferritin, the best-known Fe(III)OOH nanoparticle in animal biology, stores iron for haem synthesis in the liver and muscles. It is contained in a protein capsule (page 273 ff) and consist of a shell of eight subunits surrounding a hollow core in which the Fe(III) nanoparticle is deposited. Turkey liver, for example, contains ~1.1 mg ferritin per gram of liver tissue liver, and muscles ~50 mg/g. The Fe(III) oxide content is ~25% and it can be liberated by the reduction of Fe(III) to Fe(II) at pH 4.5. Fe(II) then migrates through 0.4 nm pores between the peptide units, and apoferritin remains. The ferritins are peculiar for three reasons, each of which makes them irresistible as research subjects:

1. They are the only metal oxide nanoparticle found in animals, and they are essential for the iron economy of our bodies. The oxidation state is Fe(III), although the body fluids including blood have a redox potential of 0 V, Fe(IIII)/Fe(II) of + 0.7 V. All the other iron in the blood and cells is Fe(II), but not in the main storage material. The reason is obvious: only Fe(III) forms insoluble salts, which are good for storage. Therefore there are six centres for the oxidation of incoming Fe(II) within the ferritin cavity, *i.e.* the oxidized tyrosine side chains of the protein, and the oxidant is ATP.[7] Rapid mineralization of ferritin coincides with the formation of a Fe(III)–tyrosinate complex. 50 iron atoms/ferritin molecule were rapidly oxidized in isolated ferritins and formed the Fe(III)–tyrosinate complex.
2. Fe(III) is a relatively strong Lewis acid at pH 7 and would destroy the tertiary structure of proteins if it came close to them. Somehow a basic hydroxide, Fe(III)OOH, is made within the ferritin cage and does not disrupt proteins.
3. The protein shell dissociates into several single peptides below pH 4 and re-forms at pH 7.[8]

Nanoparticles in the bloodstream must have diameters < 5 nm. The ferritin protein capsule derived from horse spleen is made up of 24 polypeptide subunits that fold to produce an 8.0 nm cage structure for up to 4500 iron oxide molecules: too large for general transport in the bloodstream. The *Listeria* ferritin cage, with an inner core diameter of 5.6 nm, accommodates only 500 atoms of iron or cobalt as oxides.[8,9] After the removal of the protein shell, iron oxyhydroxide nanoparticles with diameters of 2.5 and 6.0 nm could be produced from horse spleen ferritin, and particles in the 3.0 nm range from *Listeria* ferritin. These biological matrices are, however, much too expensive for any practical application. Synthetic yoctolitre vessels are needed (see, for example, MOFs, p. 104).

1.3 Nanolayers, Yoctowells and Receptors

Unpredictable things happen if one forces opposite charges together. For example, what happens to a water layer 1–4 nm thick between a pair of parallel hydrophobic plates when a perpendicular electric field, which introduces charges, is applied? In the absence of experimental results, computer modelling helps when extrapolation becomes difficult at nanometre dimensions. Open-ensemble molecular simulations indicate that the orientational asymmetry of

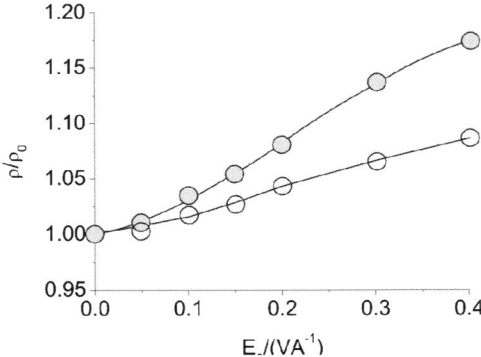

Figure 1.6 Relative increase in average liquid density as a function of applied electric field E_o in hydrocarbon-like confinements of width 1.64 nm for perpendicular (open) or parallel (filled symbols) direction of the field. The bulk reduced compressibility of water is 0.065. © American Chemical Society.[10]

water will be different at each wall (Janus effect) when a potential of at least 0.2 V/Å is applied. The positive wall should attract the oxygen atoms of water and be wetted, but the negative wall cannot deal with the protons and will stay dry (Figure 1.6). For a paraffin-like wall material, one elementary charge per 10 nm^2 suffices to render the positive surfaces hydrophilic, whereas the negative surfaces remain hydrophobic.[10] Such theoretical hypotheses are directly applicable in the evaluation of experimental results in membrane chemistry and neurochemistry, and should be taught in bioorganic chemistry courses.

Molecular monolayers made up of fatty acids standing upright on water are among the classics of nanochemistry. In 1891, Agnes Pockels had the brilliant idea of dissolving a few milligrams of fatty acid in ether, dropping some of it on a small area of clean water surface and watching it spread. Katherine Blodgett in 1934 then transferred the monolayers to solid surfaces by simply dipping small glass plates into the water. Langmuir–Blodgett monolayers thus started preparative nanochemistry.[11] If the fatty acids are replaced by double headed α,ω-diamido, -CONH-, groups, they may form hydrogen bond chains on the upper and lower surfaces of the layers, which become very stiff. The layers then become impermeable to amines and to water, which usually dissolve readily in membranes made of flexible alkyl chains. This impermeability occurs only if an even number of CH_2 groups lies between the two amide groups and if the chains are tilted against the solid base plane (Figure 1.7). Only if these conditions are fulfilled can both amide groups form parallel hydrogen bond chains, which render the monolayer crystalline and stiff.[12]

If one first binds porphyrin dyes or similar planar compounds flat on to an atomically smooth Au(I) or Si–R–NH$_2$ surface, the rigid, impermeable diamido monolayers build up form-stable yoctowells around them. The monolayer forms a rigid wall around the porphyrin base, which on stimulation with visible light produces a visible fluorescence spectrum at concentrations as low as 10^{-9} M.

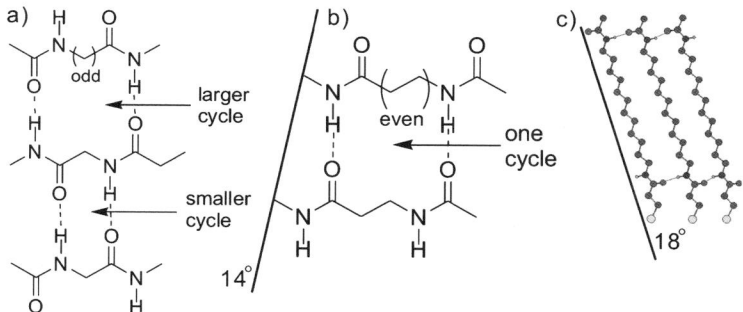

Figure 1.7 Odd–even and tilting effects on the formation of linear hydrogen bond chains in monolayers made of diamido bolaamphiphiles.[11]

This fluorescence is detectable by routine instruments and provides an easy way to analyse reactions within the yoctowells, which have a volume of $\sim 2 \times 2$ (porphyrin area) $\times 2$ (height of diamide wall) = 8 nm^3 = 8 yL, containing $8 \times 33 = 264$ water molecules or 44 water hexamers. Most of these are attached by their oxygen atoms to the hydrophobic walls by van der Waals forces.

Addition of a Mn(III) porphyrinate chloride quenches the fluorescence of the bottom molecule completely: it just disappears. If the Mn(III)porphyrin is the same size as the fluorescing porphyrin at the base of the yoctowell, it takes 30 min for the fluorescence to disappears completely (Figure 1.8). The 2 nm diffusion process becomes so slow because the incoming porphyrin is adsorbed again and again at the hydrophobic walls of the yoctowell, and its desorption and re-orientation to reach its final location at the bottom take a very long time. A Mn(III)porphyrinate that is slightly larger than the bottom porphyrin cannot enter the yoctowell at all, and therefore has no quenching effect.[13] Such selective analytical procedures must be developed for each new property of nanoparticles and yoctowells – unspecific bulk analysis methods are no help here.

Although yoctolitre volumes with hydrophobic walls are freely accessible to water and other solvents, the small volume obviously has peculiar properties. Small molecules move very fast, molecules that just fit move extremely slowly, and slightly larger molecules do not enter at all. Small water-soluble compounds with a hydrophilic and a hydrophobic edge, *e.g.* glucose and its linear dimer cellobiose, are so strongly adsorbed to the hydrophobic walls of the yoctowells that they crystallize irreversibly within the wells (Figure 1.9). Distilled water in the bulk medium above the well cannot dissolve the 2 nm cellobiose crystal, even after a year. The angled isomer maltose, which has no linear hydrophobic edge, does not crystallize at all in the well. Furthermore, the diffusion rates of molecules the size of the bottom porphyrin are 10^{12} times slower than the diffusion rates of molecules half the size. Adsorption to the hydrophobic walls slows down only the molecules that just fit.

Some rigid and acidic crystalline nanoparticles adsorb enzymes without affecting their catalytic activity. This has been demonstrated best for

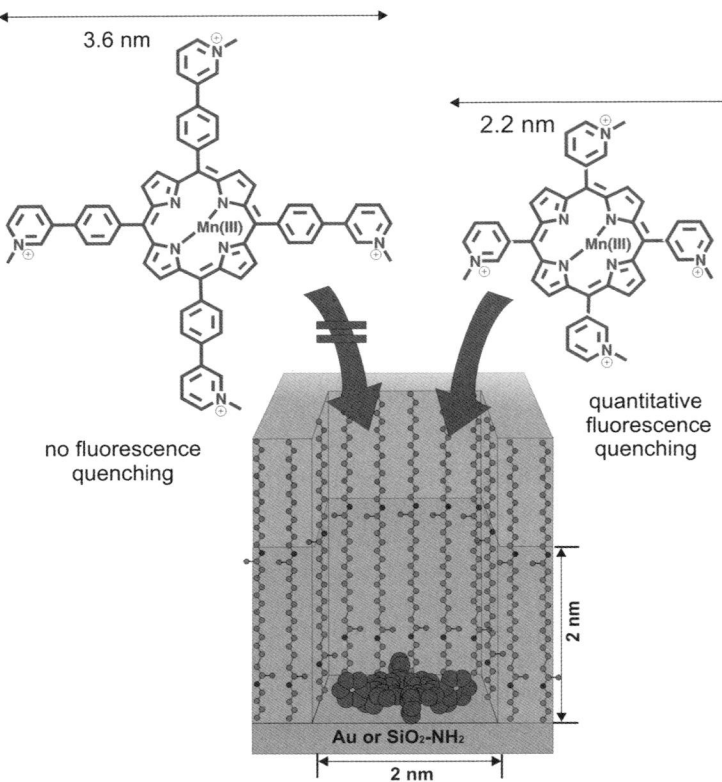

Figure 1.8 Mn(III) porphyrins of just the right size reach the bottom porphyrin within 30 min and quench its fluorescence quantitatively and irreversibly, as long as the yoctowells exist. Mn(III) porphyrins that are too large have no effect, as long as the yoctowells exist.[13] Heating above 70 °C or addition of dimethyl sulfoxide to the water solution destroys the amide hydrogen bonds, destroys the rigidity of the yoctowell walls and immediately leads to quenching of fluorescence by the Mn(III) porphyrin.

α-zirconiumphosphate, in which proteins can even be stored in fully hydrated and active form (see page 319).

Metal monolayers are also often stable, but usually appear as "weathered" multilayers. In haematite (Fe_2O_3), for example, the Fe^{3+} is octahedrally coordinated to O^{2-} and occupies two-thirds of the available octahedral sites for electrical neutrality (Figure 1.10). Neither hydroxyl ions nor water molecules are present within bulk haematite. In real geological environments, however, several layers of haematite are replaced with OH^- and H_2O, hydrated ferric oxides with dissolved NaCl (p. 5 and 171) are formed and release much more Fe(III) than dry haematite.[14]

The simplest atomically smooth layers are epitaxial metal films, usually silver or lead. A silver film with a minimum initial thickness usually of 2–3 nm is

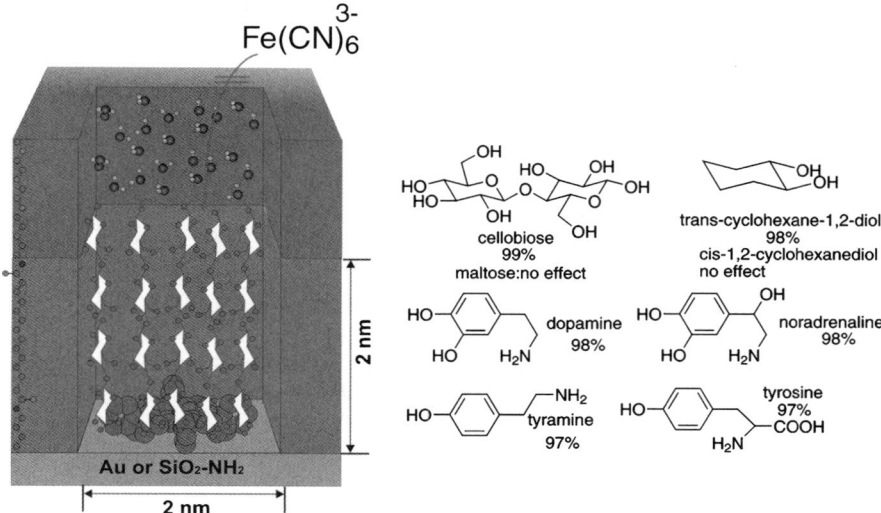

Figure 1.9 Glucose, cellobiose and many other water-soluble cyclic molecules with a hydrophobic edge (equatorial –CH–CH–CH– in glucose) are irreversibly attached to the hydrohobic yoctowell walls in contact with bulk water.[13]

obtained by the evaporation of silver at 135 K and condensation on smooth semiconductor crystals, *e.g.* silver or gallium arsenide. Nanoclusters with diameters of 2–3 nm are formed at first; the size is limited by the slow diffusion of the silver atoms at low temperature. When the spherical nanoparticles are warmed to room temperature they rearrange to a perfectly flat film 1.5 nm thick with small rectangular pits. These 1.5 nm deep pits form only in order to allow the overall critical layer thickness of 1.5 nm, corresponding to five silver atoms. A quantum well near the Fermi level, which is exclusively of sp character, is responsible for this five-atom limitation. Only then is the topmost quantum-well state unusually deep compared to those of the neighbours, leading to an overall lower energy of the layer. Similar observations were reported with lead (p. 201). The "magic" thicknesses of flat and atomically smooth metal and metal oxide layers are thus clearly dominated by electronic effects[15,16] (Figure 1.11).

Thin layers of similar smoothness can, however, also be produced routinely by the evaporation of carbon on to computer hard disks (p. 73 ff). Micron-thick layers of nanometre smoothness and laser-burnt nanometre grooves are standard products in disk technology (p. 36 and 80).

Natural geological materials are usually oxides of silicon and metals. The most common minerals are silicates and aluminates, made up of covalent silcate (SiO_4) and aluminate (AlO_4) tetrahedrons. These tetrahedrons are, however, not stable with four –OH groups but split off water and "condense" spontaneously to form polymeric layers, usually in connection with aluminium oxide layers (clays, feldspars). The aluminate layers neutralize the protons,

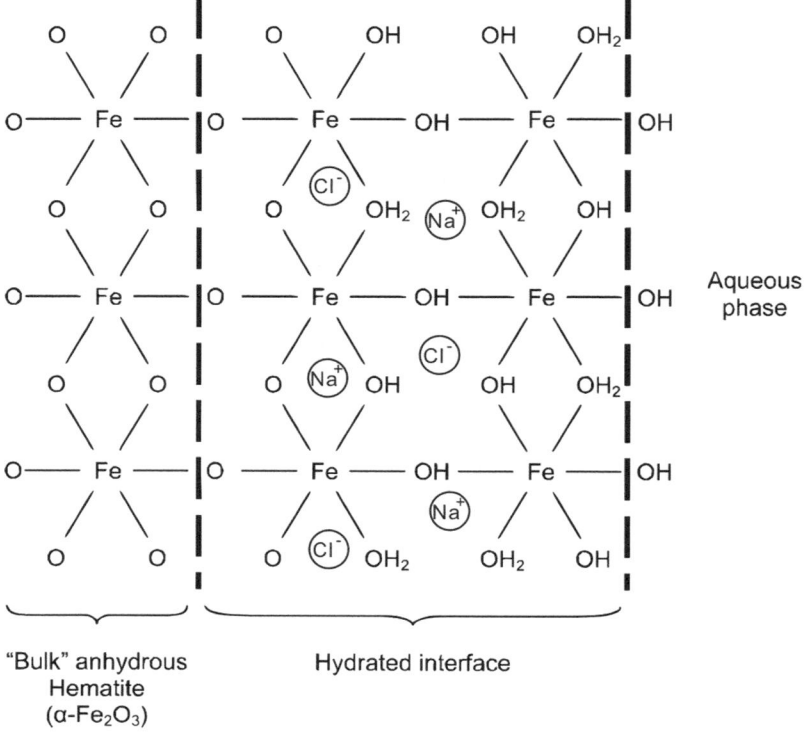

Figure 1.10 Models for the hydrate on haematite–goethite (FeOOH) surfaces.[14] Water penetrates into the crystal, forming a hydrated surface layer ~3 nm thick, and dissolves NaCl.

which are released at the silicate edges by their acidic –OH groups. Silicate layers, *e.g.* in mica, are important matrices for active particles (see page 183).

Zeolites are synthetic aluminosilicate polymers with cage-like units (Figure 1.12). These cages entrap ions, for example Na^+, which exchange spontaneously with bivalent ions, such as Ca^{2+} and Mg^{2+}. An atomically smooth surface that is routinely used and easily accessible is found in the pores of the zeolite L, which are hydrophilic, 0.75 nm wide and many microns long (page 195 ff).[17]

In order to detect and measure organic molecules or pairs of molecules molecular pairs by atomic force microscopy (AFM): (1) the molecules must be solidly fixed on the surface of the bulk carrier material, because a vibrating tip in the tapping mode transfers a lot of kinetic energy to the objects it approaches and pushes loose ones away; (2) the roughness of the carrier material must be <1 nm; and (3) the curvature of carrier particles must be low, otherwise the tip detects only objects that are on the most elevated point.

Sensitivity tapping mode AFM was tested on the smooth surface of an aminated silica particle with 1 nm high dots of a calixarene tetracarboxylate.

Atoms, Molecules, Electrons, Light and Heat in Nanometre Confinement 15

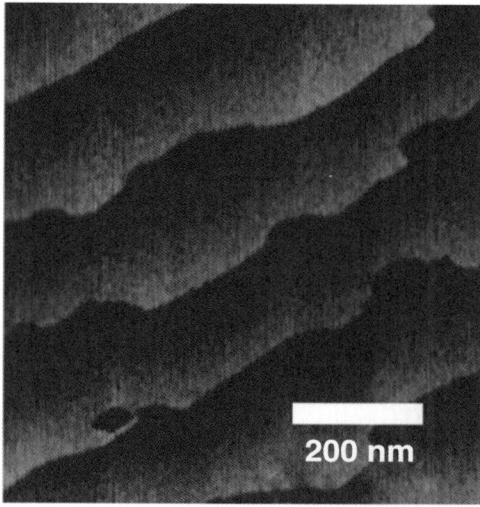

Figure 1.11 Atomic force nanograph of atomically flat Pb(110) surfaces of titania, separated by molecular steps.[15] Millions of dye molecules may be attached on each plane; the steps hardly disturb the order. © American Chemical Society.

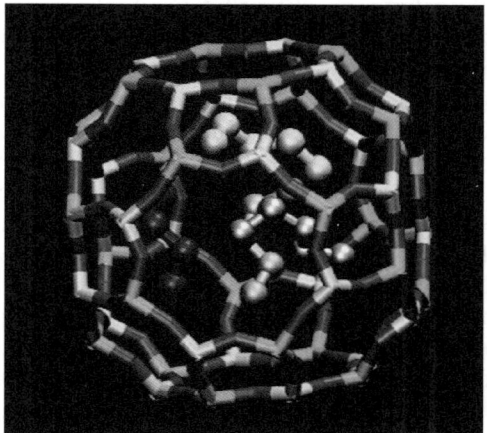

Figure 1.12 Model of a rigid zeolite cage and an entrapped flexible hydrocarbon chain. © American Chemical Society.[17]

A porphyrin standing on a tricarboxylate tripod substituent produced dots 2 nm high, and the reaction product dots were 3 nm high (Figure 1.13). The radius of the nanoparticle was >800 nm, *i.e.* its surface was practically planar, and 1 or 2 nm nanoparticles were clearly distinguishable. Scientifically, the most interesting result of this experiment was the finding that the calixarenes were completely occupied by the porphyrin; 2 nm porphyrin islands hardly occurred at all

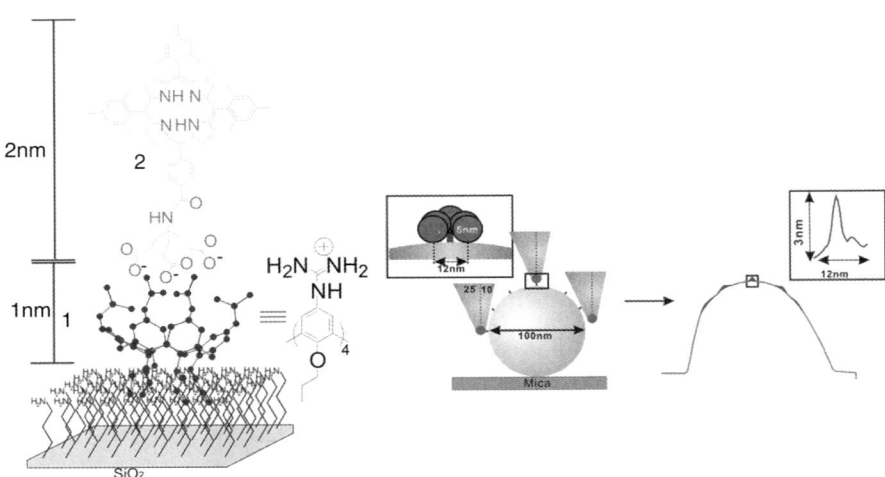

Figure 1.13 (a) Model of a calixarene–porophyrin dimer erected on a planar, aminated 100 nm silica nanoparticle surface. (b) The curved tip of an AFM detects the 1 nm high calixarene units as well as the 3 nm high dimer on the fixated, atomically smooth curved surface and produces the nanograph of the two-molecule tower (c).[18]

in the presence of the calixarene. The well-defined calixarene ligand is obviously a better acceptor for the porphyrin tricarboxylate than the much larger aminoammonium territory (see page 191) of the nanoparticle surface.[18]

Tailored nanopores are one of the most active areas of materials science. A silica called MCM-41 was the first commercial example of a material with pore channels in the 2–10 nm range. It was synthesized by a liquid crystal templating mechanism, and this technique was quickly extended to the fabrication of stable mesoporous transition metal oxides for catalysis.[17]

In the 1990s the soft, non-covalent micelles made of amphiphiles were replaced by equally soft but stable micelles made of linear copolymers with hydrophilic and hydrophobic segments, and dendrimer micelles with hydrophobic and hydrophilic cores and shells. These do not explode in a matter of milliseconds, nor do they degrade in solvents. Rather, they form different, well-separated hydrophilic and hydrophobic segments. For medical use they may be injected into the bloodstream, where they adsorb specifically to cells they encounter and deliver their drug payload only there (see p. 104 ff).[19]

Transport vehicles with yoctolitre-sized entrapped water or solvent volumes are just as important as smooth surfaces and exact nanometre pore sizes in bulk materials. Well-known examples for metal ions are crown ether ligands, which make K^+ ions membrane-soluble, cyclodextrins, and calixarenes, which transport hydrophobic perfumes and drugs into water. All three "cavitands" were discovered decades ago, but are still being investigated. Potassium–crown ethers with hydrophobic substituents have recently been used to kill cancer cells, by carrying away their potassium in rapid "flip-flops" through the cancer

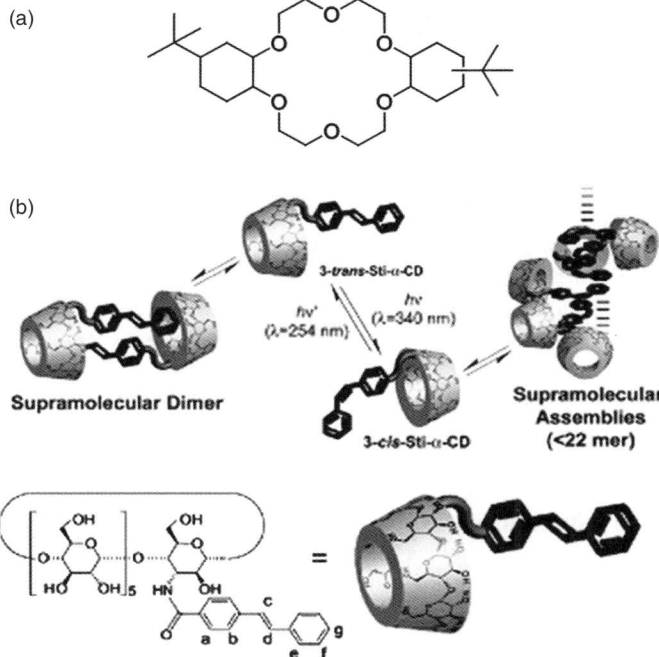

Figure 1.14 (a) Hydrophobic potassium–crown ether complex for "flip-flop" in cancer cell membranes leading to cell death by apoptosis.[19] (b) Structure of a cyclodextrine-*trans*-stilbene amide, its dimer in water and polymer after photoisomerization to *cis*. © American Chemical Society.

cell membranes (Figure 1.14a). Some cyclodextrins now dimerize under UV light to *trans*-stilbene-bridged nanoparticles (Figure 1.14b). Calixarenes with a hydrophobic alkyl group base can provide an almost cubic upper edge made up of heavily substituted aromatic rings and solubilize linear hydrocarbons such as octane and sodium dodecylsulfate in fatty tissues (p. 133).[20–22]

1.4 Nanorods, Nanotubes and Nanowires

There are 12 general properties of nanoparticles:

1. Nanoparticles with lengths and widths of 1–2 nm have almost 100% of their atoms on the surface, *i.e.* every atom is in contact with the environment. Covalent nanoparticles, *e.g.* nanodiamonds or high-valent metal oxides will not be saturated there; they will have "dangling bonds" facing their environment. Edges, corners, and the whole nanoparticle or crystalline nanocrystal, will not be as stable as bulk material or large crystals.

2. Nanoparticles may combine inorganic metals and metal oxides with organic sulfides and polymers, proteins or DNA. Highly complex chemistry can easily be concentrated in the space of a few yoctolitres.
3. Isolated polymer molecules on nanoparticle surfaces are much less sensitive to aggregation effects in water or solvents than free polymer molecules.
4. Suspensions of small nanoparticles are stable; precipitation occurs slowly or not at all. Nanoparticles may migrate from one solvent to another and can be efficiently solubilized in any medium by surface coatings.
5. Coloured nanoparticles, *e.g.* quantum dots (QDs), have discrete orbitals like molecules, and produce phonons from electrons and electron holes which are separated by ~ 5 nm, which is a large distance in a molecule. QDs interact strongly with sunlight in thin sheets or molecular layers of water.
6. Nanoparticles have a lower melting point than bulk material and sinter readily upon heating. This facilitates the removal of nanoparticles by melting them together. Coatings may keep the shape of the melt constant.
7. Nanoparticles are better catalysts than solid surfaces because of high reaction rates and large contact area per unit mass. In industrial use, nanoparticles are, nevertheless, stabilized by integration into solid matrices in order to prevent melting. Alumina (Al_2O_3) is the most commonly used matrix. In the environment, however, labile Fe(0) nanoparticles with anionic coatings in aqueous dispersions migrate freely even in soil and decompose when fulfilling their function of reducing industrial contamination.
8. Metal oxide nanoparticles are detectable in the bloodstream by low-intensity X-ray or magnetic resonance imaging (MRI) methods.
9. Magnetic metals and metal oxide nanoparticles can be steered to specific organs by external magnetic fields.
10. Nanometre-wide pores within silica or aluminium nanocrystals or poly(amide) sheets may contain isolated yoctolitre volumes of water or solvent. These provide models for biological systems such as blood vessels, and geological phenomena such as pores in sand grains and rocks.
11. Nanocrystals may self-organize in "superlattices" which lead to remarkable properties, such as the fluorescence of metallic Ag(0): for yet unknown reasons the electron gas of metals settles in fixed molecular orbitals.
12. Nanoparticles and nanocrystals are already very successful in computers, lighhting and toys and they are continuing to become cheaper and more reliable. From these applications we can then learn about their potential variability.

1.4.1 Appearance of Nanoparticles

We can generalize as follows:

- Nanoparticles are usually spheres: there is not much fine structure in 1–5 nm nanoparticles.

- Nanocrystals are usually cuboid in appearance and only one or two different crystal lattices appear on the surface. It is possible to selectively dissolve one crystal lattice: this process often leads to hollow cubes, whose properties are as yet almost unknown.
- Nanolayers are either fragile, non-covalent Langmuir–Blodgett monolayers (p. 93), which can be stabilized by amide hydrogen bonds, or complex organic copolymer landscapes with hydrophobic and hydrophobic domains (p. 111), or stiff metal oxide multilayers (p. 15).
- Nanorods and nanowires are usually crystalline and may be branching.

Nanorods and nanowires are often made in hollow matrices. The problems with their fabrication are discussed here in some detail, because they combine the syntheses of nanopores and nanoparticles.

Metal wires and tubules are formed in commercial porous membranes made of silicate or alumina, which are soluble in hydrogen fluoride, or in poly(carbonate) layers, which can be removed by solvents. Only silicate pores are narrow enough to be of interest for nanowires. The wider pores are useful for the formation of tubes with nanometre-thin walls or for segmented rods, especially barcode rods (see p. 183, 378).

The longest matrices for the formation of nanowires are "microstructured" silica optical fibres with pore diameters down to 10 nm and lengths of 30 cm or more. The main advantage of these templates is that the hole diameter can be designed with great flexibility and precision to repeat periodically or non-periodically, and to contain engineered cavities. The surface roughness is ~ 0.1 nm, without holes. Furthermore, the optical transparency of these templates makes them ideally suited for fabrication of photonic materials. Pores can be modulated by a glassblower, and the fibres are commercially available. The challenge for scientists lies in filling the optical fibre templates with functional materials. The relatively great length of the nanometre pores, with aspect ratios $\sim 10^6$, requires non-viscous fluids without any bubbles or precipitates and totally uniform solidification. So far the overall width of silicon wires grown in supercritical fluid carbon dioxide is of the order of 1 μm[23] (Figure 1.15)

Alumina pores have their attraction as a template mainly because of its low-cost processing. On the μm–nm borderline it has become an inexpensive technological material;[24] as a scientific tool on the nanoscale it needs further development. Anodic oxidation of thin aluminium electrodes leads to hexagonal patterns of 5–200 nm pores with more or less amorphous alumina walls. Phosphoric acid, for example, loads the alumina with negative charges down to pH 3.5, but oxalate has no such effect. Electro-osmosis of aqueous metal salt solutions fills the pores, and applied electric potentials precipitate metal tubules or wires on their surface. It does not speed up the filling of the pore very much, but it affects important details of the pore surface.

A 42 nm diameter iron nanowire grown in these pores was single-crystalline and had a coercivity of 1654 Oe. Increasing the pore diameter resulted in improved crystallinity of the nanowires, but their coercivity was reduced because of the decreased aspect ratio.

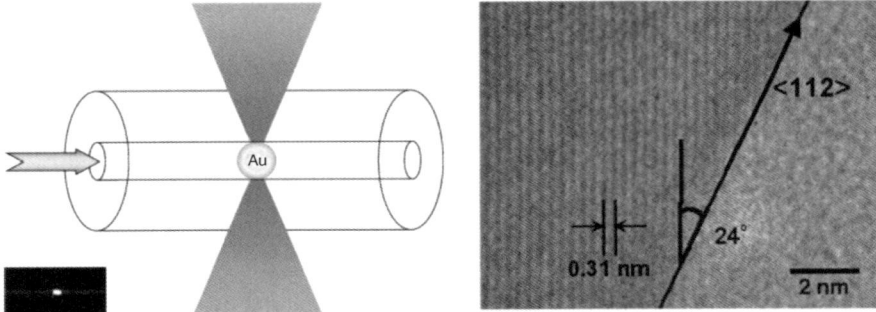

Figure 1.15 Model of an Au(0) nanoparticle deposited in a microstructured silica optical fibre. The nanoparticle was formed in a silicate pore by laser irradiation of an organometallic gold precursor dissolved in supercritical fluid carbon dioxide. It flowed down the template pores carrying a silane–helium (SiH_4–He) mixture, forming a 1 µm Si(0) wire which grew from the gold nanoparticle with a speed of 700 nm min^{-1} at 370 °C. High resolution bright-field electron micrographs showed an image with 111 lattice planes oriented vertically at 24° to the growth direction.[123]

A wide variety of nanorods and nanotubes has been formed within these pores, but their width is usually >50 nm. The base of the porous alumina is electrically isolated from the metal by a thin barrier layer of aluminium oxide. The most widely used underlayer is not aluminium itself, but a noble metal such as gold or platinum. This causes widening of the pores, however, because the metal base catalyses the electrolysis of water upon anodization or cathodization, and oxygen or hydrogen bubbles appear when the oxidative or reductive potential of the noble metal interface is reached.

The poly(carbonate) used for CDs and DVDs is hard when solid but has a very flexible melt. Arrays of micron-long nanowires were obtained by filling a commercial track-etched, porous poly(carbonate) template. It contained a large number of cylindrical holes with a narrow size distribution, but irregular diameters. Filling proceeded in solution by electrochemical deposition. Other useful commercial templates are anodized aluminium films or track-etched foils. At small pore diameters, the conductivity of poly(pyrrole) and poly(thiophene) fibres is higher than in bulk polymers. Copper- or gold-covered plates are also available, and supposedly lead to fewer defects in the wires, due to unidirectional growth starting from the attached metal.[24]

Before the prefixed poly(carbonate) foil ("membrane") is mounted in the electrochemical cell, it is ultrasonicated under water for 2 min. This step is crucial. If it is omitted, growth starts in hydrophilic pores and the number of pores in which growth proceeds increases with time, leading to an increasing current (potentiostatic mode). This current increase is an artefact unrelated to the effective pore diameter, which should be controlled by the current. Successful pore wetting is controlled by eye: the free membrane area should be fully covered by a water droplet after withdrawal. Electroplating occurs in a

Teflon cell with the foil (membrane) facing upward. There is no stirring or heating; a saturated calomel electrode (SCE) is used as reference for the applied potential and is placed 7 mm above the foil. Using a conventional potentiostat, the current is measured during electroplating at a fixed potential *vs.* SCE. The following aqueous electrolytes have been used:

- *for nickel:* 515 g/L Ni(H$_2$NSO$_3$)$_2 \cdot$ 4H$_2$O + 20 g/L NiCl$_2 \cdot$ 6H$_2$O + 20 g/L H$_3$BO$_3$
- *for cobalt:* 400 g/L CoSO$_4 \cdot$ 7H$_2$O + 40 g/L H$_3$BO$_3$
- *for poly(pyrrole):* 0.5 M pyrrole monomers (from a freshly distilled solution) + 0.1 M LiClO$_4$.

Typical deposition voltages were –1.2, –1.1 and + 0.8 V$_{SCE}$ for nickel, cobalt, and pyrrole, respectively. After electroplating, the membrane is inspected under an optical microscope. If the growth has been interrupted while the pores were not completely filled, the surface areas on the membranes covered with nanowires will appear black. If the pores were completely filled and growth has proceeded on the whole membrane surface, a 3D bulk film will have started to cover the membrane. In this case, the metal deposits appear reflective. These two simple observations test the degree of growth homogeneity on the aluminium foil. In order to measure the diameter of the wires by electron microscopy, the poly(carbonate) membrane is dissolved in dichloromethane at 40 °C.[24]

Highly concentrated electrolytes (1 M) are a prerequisite for monitoring the growth process with the aid of the *current–time* characteristic; dilute solutions give complex patterns. The wires resulting from 1 M solutions are, however, also oriented in a wide range of directions, as a result of the non-parallel alignment of the pores in commercial screen membranes. Initially poorly wetted pores led experimentally to similar results.

The wires were also observed to be wider in the middle than at the base or the top by a factor of up to 2.5, which amounted to a sixfold increase in area. It was shown that the poly(carbonate) pores were also not cylindrical with a constant cross-section, but rather cigar-shaped. Pores with a nominal diameter of 80 nm were as much as threefold wider in the middle section.[25]

1.5 Electric Conductivity

At least six types of conductivity are relevant to nanochemistry (more references and illustrations are given in the relevant sections later in the book):

1. transport of spinless charge carriers, *e.g.* polarons, in polyenes
2. ballistic electron transport in carbon nanotubes
3. p- and n-semiconductivity in silicon, germanium and various chalcogenides (sulfides, selenides)
4. pressure-induced piezoelectricity in quartz crystals
5. thermal conductivity in heavy metal chalcogenides
6. the common metallic conductivity of the electron gas.

If polyenes with conjugated C=C double bonds (carotenes, poly(acetylene); see p. 74 ff) or C=N–C double bonds (poly(aniline); see p. 123 ff) are mixed with a small proportion of oxidants, in particular iodine or protons, their conductivity rises to $\sim 10^{-3}\,\mathrm{S\,cm^{-1}}$, which is $\sim 10^{-3}$ times less than that of metallic copper and $\sim 10^{11}$ times greater than that of insulating hydrocarbons. The effect of oxidants and protons is not that of p-type or n-type doping, analogous to the generation of charge carriers in doped inorganic semiconductors. Polyacetylene, poly(*p*-phenylene) and polypyrrole were shown to display conductivity, which is not associated with unpaired electrons, but rather with low-energy spinless charge carriers, so called solitons and polarons (p. 76 ff) or bipolarons (p. 77). The polaron conductivity depends strongly on the potential of the environment, and the current is switched on and off depending on its oxidation potential or pH.

Ions, atoms and molecules are made up of slowly moving charges. Electrical forces dominate molecular interplay in general, especially in biological processes. All of our life processes, from thought and sight to the heartbeat and muscle movement, totally depend on currents of sodium and potassium in water and along porous lipid and protein fibres, and the phosphate groups that regulate them (p. 100, 166 ff).

Rolled-up graphite sheets (carbon nanotubes) are conjugated benzene planes. They become strongly conductive, if (i) they have the right stereochemistry, which allows the electrons or holes to run parallel along the tube axis and (ii) they are chemically p-doped with electron holes by oxygen or covalently bound nitro groups. Addition or adsorption of an electron-donating group may also provide up to 1 electron for every 1000 carbon atoms of the nanotube and and convert it to an n-type (n for negative) semiconductor, which is almost as stable in air as the p-type (p for positive). The conductivity is caused by *ballistic electron transport*, which means that the electron or the hole jumps hundreds of nanometres, touching only a few of the carbon atoms on the way and hardly heating up a carbon wire (p. 60 ff).

Semiconductors are common materials in electronic equipment. Later, we shall meet them in quantum dots (p. 225 ff) and all kinds of energy converters (p. 309 ff). The most common semiconductors in technological applications are crystalline and amorphous silicon. The crystals have a diamond-like structure and are usually n-doped with atoms of phosphorus, arsenic or indium (p. 189). When a phosphorus or arsenic atom enters the diamond lattice, only four of its electrons are bound. The fifth is pushed into the conduction band, an Si–Si bond is broken, and a non-bonding electron is introduced into the conduction band. In these n-type semiconductors the silicon lattice is negatively charged and the phosphorus or arsenic atoms in the lattice produce the positively charged counterions. Indium, on the other hand, with only three electrons, also disrupts an Si–Si bond, but leaves it positively charged. This produces a p-type semiconductor, which takes up electrons but does not release any. Both n- and p-type contacts open and close currents flowing in one direction.

The functional part of a semiconductor is the p–n junction, where excess electrons and electron holes come in contact. The electrons jump spontaneously into the holes and the junction becomes a more or less electroneutral barrier with high resistance. An applied voltage may, however, pull the electrons back into their old positions, so the resistance drops and a current flows. An opposite voltage, however, raises the barrier so that no current flows. The junction works as a rectifier.

Foreign atoms are usually implantated into semiconductor crystals on the surface of small crystals or thin slices of crystals by diffusion of the dopants at high temperatures or by using an accelerator to shoot particle beams at the semiconductor.

From the chemist's viewpoint, layers of crystalline titania (titanium oxide, TiO_2) are most promising research objects because they can be coupled with surface dyes and then used directly in a Grätzel cell (p. 309 ff) for artificial photosynthesis, *i.e.* the splitting of water into oxygen and hydrogen by sunlight (Figure 1.16). Atomistic state density simulations and experiments show clearly that the electron transfer from the dye to the titania conduction band is driven by vibrational motions, but the rate is hardly influenced by temperature, changing by only a factor of 2 between 50 and 350 K.[26a–c]

An electric voltage can also be used to deform piezoelectric crystals, such as thin slides or wires made of quartz (silica) or lead zirconate titanate ceramic. When charges are displaced in a piezoelectric crystal, its surfaces move by a few nanometres. This effect is applied, for example, in the active-head sliders of computer hard disk drives. The flying height of the head above the disk must be maintained at <10 nm with an accuracy of ~ 1 nm. This is achieved by the active-head slider, which is first kept at a distance of ~ 24 nm, set by the action of the air pressure above the rotating disk on an air bearing pad, and then regulated by the varying voltage piezo-electric at the end of the actuator.

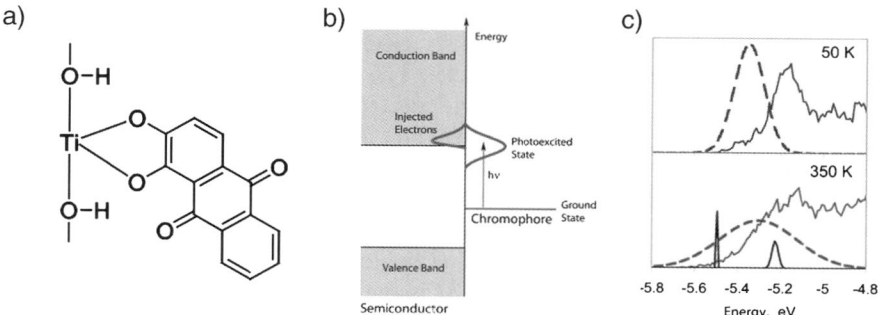

Figure 1.16 (a, b) Molecular model of the bis-phenol electron donor dye and an energy diagram of a dye–titania interface. (c) Thermal vibrational motions create a distribution of the photoelectron electronic-state energies of the dye molecules (dashed lines) and titania density of states (solid lines) for 50 and 300 K presented in arbitrary linear units as a function of energy. The dye-excited state is slightly homogeneously broadened (left peak) as a result of coupling to the semiconductor states, which are broadened much more (right peak) at 350 K. © American Chemical Society.[26]

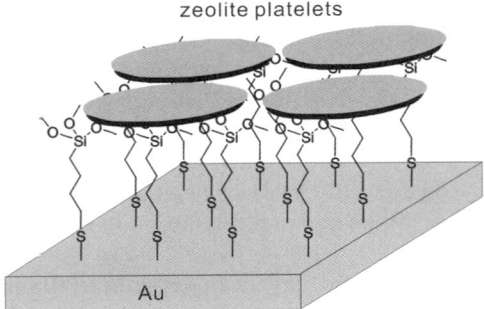

Figure 1.17 A gold electrode and zeolite nanoparticles are separated by three alkyl groups and a few heteroatoms. The potential of the electrode changes after addition of butane gas, because the distance and polarity change; it acts like a piezoelectric crystal, but is controlled by a gas. © American Chemical Society.[27]

SH-functionalized zeolite nanoparticles with a negative surface charge attached to bare gold surfaces change the potential of the electrode if a hydrocarbon molecule such as n- or iso-butane is added and dissolves in the membrane structure. The connecting alkyl chain stretches reversibly and mimics the movement of crystal planes.[27] (Figure 1.17)

The metallic conductivity of nanocrystals has not been studied so far, to the best of our knowledge; their thermoelectricity is introduced in section 1.7.

1.6 Magnetism[28,29]

The magnetism of matter can be traced back to tiny rings of electric current. The first, simple model is planetary: electron rotation with a speed of ~60 000 km/h around the atomic nuclei. The second, electron-inherent source of magnetism is the rotation of single electrons around their own axis, producing either a spin-up or a spin-down $2\mu_B$ magnetic moment.

The magnitude of magnetic fields and their spin-up or spin-down direction is determined by measuring changes in the magnetic fields of external DC currents flowing around the magnetic material. An electric current flowing around a given area produces a magnetic field H, which may induce or change a magnetization M in any material, where M is the total magnetic moment per unit volume. The specific magnetic moment for any material is denoted by μ; for free space it is μ_o. The magnetic response of the material is called magnetic induction or flux density B ($B = \mu_o H$). If the material becomes magnetized, the flux density is $B = \mu_o H + M$, if the magnetization is parallel to the external field, or $B = \mu_o H(1 + \chi) = \mu_o \mu_r H$, where χ is the magnetic susceptibility and $\mu_r = (1 + \chi)$ is the relative permeability (equal to 1 *in vacuo*). The material constant μ_r may be up to 10^5, which means that the magnetic field is 100 000 times stronger within the magnetic material than in free space.

Magnetic fields have practically no influence on chemical processes, either inorganic or organic. This is surprising, because paramagnetic oxygen behaves chemically very differently from diamagnetic singlet oxygen (see below). Generations of polar bears and penguins have lived in the strong magnetic fields of the Arctic and Antarctic without significant damage to their lung tissues or disorientation of the brain, although they produce ionic Na^+, K^+ currents in their brains and live with triplet oxygen. They have all breathed, metabolized and reacted normally, not in a "magnetized" way. (Although, of course, birds and fish use internal magnetic compasses, usually rows of haematite crystals, to find their way in the same manner as the captain of a ship or an aeroplane.)

Here we treat magnetism like any other chemical property of an atom or compound, with the conviction that such a simple descriptive knowledge of electron–spin interactions can be applied by chemists to invent and tailor magnetic materials in the same ingenious way in which atomic and molecular orbitals are applied in the planning of stereoselective syntheses and the development of dyes and cold light sources. Even a partial and descriptive knowledge of physics helps the spirit of chemical invention.

Magnetic materials are important, because the magnetic moments, south → north and north → south, are the 1 and 0 bits that make up the billions of digital pictures, numbers and letters of modern human society. Nothing can be switched so fast in such a small space, and remain so stable for so long, as the magnetic moment of a correctly arranged assembly of paramagnetic atoms or ions.

Magnetism is based on the spin of unpaired electrons in a binding orbital. A free electron in space or in the electron gas of a metal has a permanent magnetic moment of one millionth of a Bohr magneton, $10^{-6} \mu B$. It may be spin-up or spin-down, producing an elementary north or south pole. Without an external magnetic field, most magnetic moments cancel each other out. In the magnetic field H of an applied current, the spins become partly or totally oriented, *i.e.* they show *paramagnetism*. The f-orbital electrons of rare earth metals have localized moments. Their paramagnetism is roughly additive and depends strongly on temperature (*Langevin paramagnetism*).

Electrons in s,p,d-orbitals show quantum mechanical variability and have a permanent magnetic moment of $1 \mu B$ independent of temperature (*Pauli magnetism*). The experimental fact that an electron gas becomes paramagnetic in an applied field shows that it is susceptible: the number of spin-up electrons increases if a magnetic field pulls them in this direction. Reversible Pauli paramagnetism depends on this effect. Susceptibilities can only be measured by applying the magnetic field of a current or a ferromagnet. An elongated magnetic or diamagnetic sample, with one end in a strong magnetic field, is suspended from a sensitive balance. If the substance is ferro- or paramagnetic (or diamagnetic) the corresponding gain (or loss) measured by the balance indicates the force acting on the sample. For small samples, a magnetic field with a sharp gradient is applied along one axis in which the sample is free to move. The force acting on the sample is then again measured with a balance.

In general, the exchange integral between two electrons is positive if they are localized on the same magnetic atom or on covalently connected atoms. Paramagnetic *triplet states*, with both electrons having the same spin, are largely favoured there. The most important triplet state on Earth is, surprisingly, not a metal or metal oxide, but atmospheric molecular oxygen (O_2). This makes oxygen very slow in its chemical reactions, because a quantum mechanical rule forbids reactions of magnetic with diamagnetic compounds and spin reversals in molecules are very slow. If the oxygen molecules were diamagnetic, they would immediately oxidize and destroy organic molecules as quickly as molecular chlorine (Cl_2) does. Biological processes such as photosynthesis and respiration would be impossible. As it is, the paramagnetic oxygen molecule has first to be split into atoms by Fe(II) before it becomes aggressive, and this reaction can be controlled and handled by organisms with appropriate organic protein nanoparticles. The magnetic fields that occur on Earth cannot convert triplet to singlet oxygen: only UV light has enough energy to do that.

A pair of electrons on neighbouring magnetic atoms that are not covalently linked usually has a negative exchange integral and the electron pair is oriented spin-up–spin-down, which produces diamagnetism (*i.e.* the material is non-magnetic). Single electrons of radicals are usually paramagnetic, and line up only in the presence of external magnetic fields (Figure 1.18).

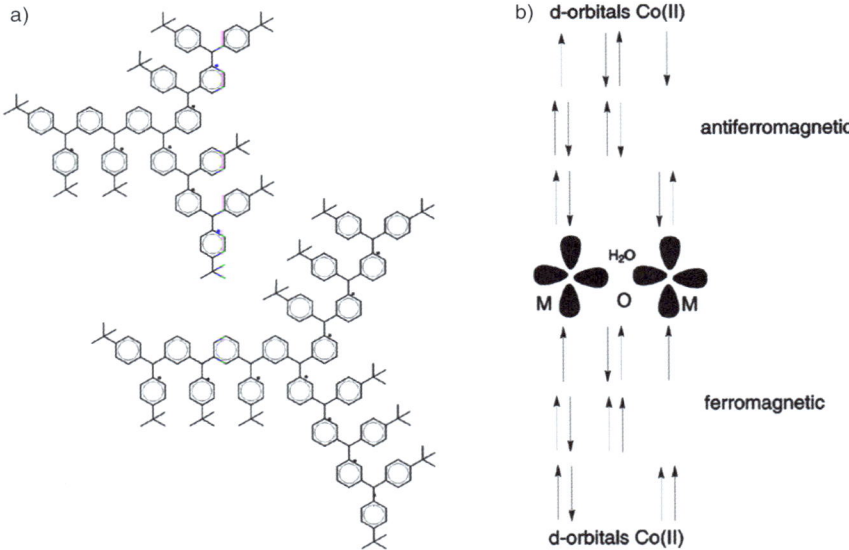

Figure 1.18 Left: Several unpaired electrons in organic polyradicals are paramagnetic. They (partially) line up in presence of external fields. Right: Paramagnetic cobalt (II) ions in hydrated and dehydrated crystals may line up antiferromagnetically (dehydrated, diamagnetic) or ferromagnetically (hydrated, paramagnetic).[29,30]

The magnetic properties of bulk metallic iron (p. 246 ff) depend to a great degree on the purity and previous history of the sample. A ferromagnet is made up of small magnetic domains, in which all electron spins are parallel (*saturated magnetization*). The total magnetization may nevertheless amount to zero when all the domains are oriented statistically with respect to each other. Magnetization can then be enforced by an external field H, and occurs parallel to the external field. At low temperatures, *e.g.* room temperature, M increases monotonically with H until saturation is reached. At high temperatures, above the so-called *Curie point* temperature, however, magnetization disappears completely. This bulk behaviour is independent of surface effects.

The movement of the grain walls may be reversible or irreversible. The total magnetization of a material aligns with a strong external field. The specific magnetization of iron is ~ 220 A cm^{-1}, and the Curie point ~ 760 °C. The specific magnetization of magnetite (Fe_3O_4) is ~ 55 and the Curie point lies between 500 and 620 °C.

The extraordinary permanent magnetism of iron, cobalt and nickel, which occurs without any applied field, needs an explanation. Why should only these three metals have a permanent magnetic moment, whereas all others are paramagnetic only and need an external magnetic field in order to become magnetic? One common theoretical explanation combines quantum mechanics with a classical thermodynamic–kinetic model.

1.6.1 Quantum Mechanics of the Fermi Sphere

In the case of solid metals, quantum mechanics localizes electrons within a fixed volume of atoms and lets the them move freely from atom to atom within this volume. This volume is called the *Fermi sphere or space* and it has radius k_F. The electronic states within the Fermi space are two planar waves projecting the sphere onto a circle with only two wave vectors, k_x and k_y in the projection of the sphere. The linear distance between the states is $2\pi/L$, where L is the energy difference between the planar electronic state waves. The highest occupied energy level possesses the Fermi energy E_F, and the density of states in the circle is given by the integral over Vk^2/π^2 dk. Each electron within this space can have either a spin-up or a spin-down state, which introduces a $g = 2$ factor as well as a step-like behaviour of the electrons energy in the quantum mechanical model. For metals, E_F on the surface of the Fermi circle is equal to the chemical potential μ of the metal, within 0.01%. In insulators or semiconductors the chemical potential μ corresponds to the band gaps between valence and conduction bands (see p. 124).

If the model is realistic, spontaneous spin splitting must now occur spontaneously in order to provide the observed permanent ferromagnetism. A majority of up-spins must produce a magnetic field in the Fermi sphere, and an electrical field in the Fermi field must induce a majority of up-spins. But which comes first? Does the molecular field determine the spin–spin interactions, or do the spin–spin interactions build up the molecular field? Furthermore, in metals the unpaired electrons are found only on the outer periphery of the

Fermi sphere. How do the electrons of neighbouring spheres interact with each other? What quantifies the electronic interactions of an electron gas on the surface of atoms? We have here a typical quantum mechanical problem and provide a concrete solution.

1.6.2 Thermodynamic–Kinetic Model

The chicken-and-egg dilemma and unlimited electronic interactions can be avoided by introducing and quantifying potential and kinetic energy terms for the electrons in the Fermi spheres of specific atoms. This approach reminds chemists of the kinetic *and* thermodynamic events in chemical reactions and immediately makes sense. Chemists know that the thermodynamics and kinetics, the speed and probability of a reaction between electrons, are largely independent of each other.

The "kinetics" of ferromagnetism are, of course, connected with the number of spin flips in the metallic region. Electrons with a high Fermi energy E_F will flip fastest, and the kinetic energy E_{kin} of the transformation will be proportional to $g(E_F)$. This is where the special situation of iron, cobalt and nickel comes in: positioned at the end of the first d-electron period of the elements, they have the most c-electrons in the smallest volume of all metals.

The potential energy E_{pot}, on the other hand, comes from the total magnetization M of the sample and may also be called atomic or molecular *field energy*. Two electrons interacting with each other produce a Coulomb energy proportional to $U = \mu_B^2 \lambda$, the product of their squared magnetic moment μ_B and the mean field factor λ. $\Delta E = \Delta E_{pot} + \Delta E_{kin}$ is proportional to $1 - Ug(E_F)$. Spontaneous spin splitting occurs if ΔE is positive or $Ug(E_F) > 1$ (Figure 1.19). It is this *Stoner criterion* that separates ferromagnetic from paramagnetic

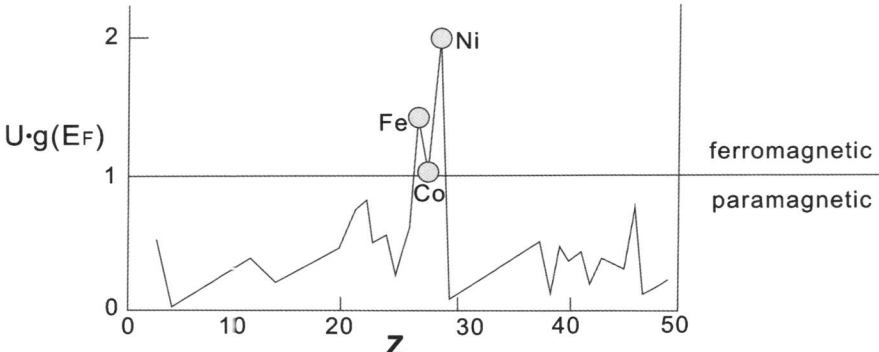

Figure 1.19 Qualitative explanation of the strong ferromagnetism observed only in iron, nickel and cobalt. These metals have most d electrons in the smallest volume, producing the highest Fermi energy E_F and Coulomb energy U. The product of both is >1, and a positive overall energy of the electron states leads to spontaneous spin splitting or ferromagnetism without an external field.[29]

atoms. Iron, cobalt and nickel are the smallest atoms in the first row of transition metals with a maximum of electrons in the d-orbitals. The density of (half-)occupied orbitals close to the Fermi level is exceptionally high, and ΔE_{kin} therefore has a maximum here. The overall magnetic field, ΔE_{pot}, characterized by λ, is only average. It is the large number of unpaired conducting electrons in a small space that leads to the observed intense electron–electron interaction and to the spontaneous spin splitting that leads to permanent ferromagnetism.

Fe(0) has one free electron in the 4s orbital, which is not oriented, and seven localized d-electrons, which produce a magnetic moment of $2.2\,\mu B$. Thus 4.6 electrons are spin-up, 2.4 are spin-down. In both spin bands there are $0.4\,\mu B$ energy holes, which means that the experimentally found exchange splitting does not use up the total energy difference between the Fermi energy E_F and the top of the d-band. The interaction between the unpaired electrons is strong, as explained above, but relatively weak for a ferromagnet. In nickel with 0.6 free electrons, 5 spin-up electrons and 4.4 spin-down electrons the spin-up band has 0.6 spin-down holes, but no spin-up holes. The exchange splitting of nickel is larger than the difference between Fermi and top d-band energy; it is therefore a strong ferromagnet.

The most descriptive material property corresponding to the elementary magnetic moment is *coercivity* or *coercive field* and is measured in oersteds ($1\,\text{Oe} = 1000/4\pi\,\text{A m}^{-1} = 79.6\,\text{A m}^{-1}$). 1 Oe corresponds to the magnetic field strength at a distance of 1 cm from a unit magnetic pole μ_B. As a material constant the coercivity defines the relative difficulty of erasing the magnetism of the material, e.g. cancelling out the magnetization of a ferromagnetic or paramagnetic material. The higher the coercivity of a material, the more difficult it is to erase its magnetism with external fields.

Natural magnetic materials orginate in outer space. Interstellar electron clouds, nuclei formed in the stars, and the cold of deep space produce para- and ferromagnetic elements and oxides. But what can chemists here on Earth do for the optimization, stabilization and fast interactions between magnets?

The weak metallic ferromagnet Fe(0) has a coercivity of 2; for Co(0) and Ni(0) it is of the order of 20–150. These values are modest when compared to AlNiCo, the permanent magnet commonly used in refrigerator doors, which has an "artificial" coercivity of ~ 2000 Oe; for the ceramic material $Nd_2Fe_{14}B$ it is as high as 4000 Oe. The non-metallic oxide paramagnetics haematite ($Fe(III)_2O_3$) and chromium dioxide ($Cr(IV)O_2$) have coercivities in the range 200–450 Oe. Chemists were able to increase this value by a factor of 100 by patient trial and error experiments. But how did they do this, and why is AlNiCo such a strong magnet?

The choice of nickel and cobalt is clear: they provide the unpaired d electrons and fulfill the Stoner criterion. But why 60% of diamagnetic aluminium, which is common to most commercial AlNiCos? First of all, Al(0) has a low melting point (660 °C) and forms alloys with cobalt and nickel. The original AlNiCo was an alloy with the approximate composition of $Al_{72}Ni_{15}Co_{13}$, but a large variety of other magnetic materials are also called "AlNiCo" and then distinguished by a number. AlNiCo 8, for example, with a coercivity of ~ 2000 Oe, is shipped and stored as non-magnetic alloy and rapidly magnetized by magnetic fields of 8000 Oe.

The magnetism of AlNiCo 8 is clearly based on elongated, needle-shaped (= acicular) cobalt–nickel nanocrystals with a diameter of 30–50 nm and an interparticle spacing of ~ 50 nm. Coercivity and remanence depend on the direction of magnetization of thin nickel needles of decagonal crystallites and their parallelism. The magnetic power of AlNiCo is completely and irreversibly lost at temperatures above 538 °C. It is probably simply the orientation of the nickel or nickel–cobalt needles precipitating from a cooling aluminium melt that leads to the high coercivity. When this orientation is lost, the high coercivity disappears. Magnetism is mainly a result of crystal ordering. At a high temperature, e.g. 538 °C, or under mechanical stress, these needles may break down to a powder and therefore can never re-orient. This special mechanism in the magnetization of metals is the just the reverse of domain formation, which is crucial for ferromagnetism to occur.[31–33]

Similar nanocrystalline grains with a width of 23 nm were found in FeNdB ribbon flakes by electron microscopy. The magnetization process is accomplished by the alignment of each single grain, and this process has even been visualized under a transmission electron microscope with sintered $Nd_2Fe_{14}B$ needles.[33]

One contribution of chemists is thus the preparation of thin, uniform, stable and immobile nanocrystals within an easy-to-handle matrix. The magnetic nanoparticles will always become smaller and cheaper, but they must not lose their orientation.

The magnetization of strong nickel magnets may even be nano-stereoselective. Disk-shaped nickel segments are permanently magnetized perpendicular to a rod axis, if the rod ordering becomes much longer than the disk diameter. In nanowires nickel is also magnetized parallel to the axis. Atomic force microscopes with a magnetic tip reproduce such alignments, and north–south pole interactions assemble the groups of rods lengthwise.[34a,b]

The deoxygenation of chromium(VI) peroxide (Cr_2O_5) to acicular chromimium dioxide (CrO_2) crystals was accomplished by heating an aqueous slurry of Cr_2O_5 under high-pressure oxygen or argon first to 200 °C for 1 h, then to 200–345 °C at 1700 at for a longer period of time. 0.5% of antimony oxide (Sb_2O_3) was added to the slurry to keep the particle size constant and raise the coercivity of the CrO_2 from 150 to 500 Oe. The reproducibility of the process was demonstrated in 5 lb (2 kg) batches, and scale-up by a factor of ~ 1000 was achieved without deterioration in the properties of the CrO_2. 20 g of CrO_2 was then dispersed in a vinyl copolymer formulation similar to that used in the production of commercial audio tapes (Figure 1.20). The volume loading of the magnetic material was 40%. A few mg of binder, formulation and solvents were loaded into a small sand mill, milled for 17 h and the magnetic dispersion was then filtered and coated ~ 330 μm thick on a polyester base. This procedure gave ~ 60 ft (20 m) of coated polymer tape.[35]

Magnetic Fe(0) nanocrystals are drawn to the region of highest magnetic field. Rotation of magnetic moments of individual nanoparticles occurs, for example, in paramagnetic metal chalcogenides like CrO_2 or Fe_2O_3. The most

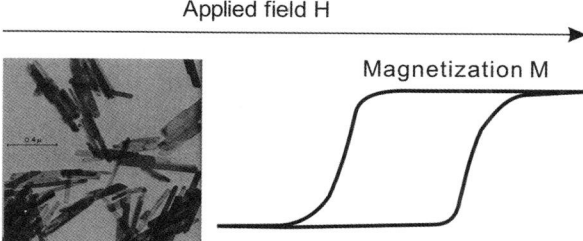

Figure 1.20 Acicular monocrystalline chromium dioxide (CrO_2) microparticles made for magnetization on polyester tapes are very "hard" as shown by the hysteresis (see Figure 4.6, p. 251). © American Chemical Society.[35]

common "nanocombinations" in commercial magnetic tapes and disks use metal oxides and organic polymers. At first Fe_2O_3 or CrO_2 particles with lengths between 100 and 1500 nm and length-to-width ratios in the range 5–10 were glued to polyester tapes.[35] The overall picture was similar to AlNiCo. It was immediately found that the magnetization of the CrO_2 particles and the tape noise depended strongly on the distribution of particle lengths and length-to-width ratios. The noisier samples had a wider range of particle lengths and length-to-width ratios. The crystallinity, the uniform orientation of the acicular axis of the CrO_2 particles (which was also the direction of easiest magnetization), and their perfect alignment in the polymer during coating reduced the signal/noise ratio of the tapes.

Thinner magnetic nanolayers, often without any detectable nanoparticles, are made by vapour deposition of metals or metal chalcogenides *in vacuo* on to cold substrates. The vapour is produced by an electron beam, by a glow discharge, or by the treatment with ion or laser beams. The energy beam is directed on to the piece of metal or metal chalcogenide, and the emitted vapour is then allowed to condense on a metal, glass or polymer surface. Laser beams are often used as they can be precisely targeted; they can ablate a molecular monolayer or multilayer from crystalline surfaces and deposit micron- or nanometre-thick sheets.

In a computer hard disk, thin layers of magnetic material rotate rapidly. Changes of current from a nanosized tip are used to read and write the magnetic nanoparticles in the rotating thin layers. Two magnetic layers, separated by a non-magnetic layer, are used. One layer rotates the spin-up, spin-down traces; the second, separated by the non-magnetic layer, always remains in one orientation. The resistance of the chromium or alumina separating layer then depends on the relative orientation of the spins in the two magnetic layers. A current that flows through all three layers meets a larger resistance if the magnetization in the immobile disk and the rotating disk are opposite to each other. When data are written to the disk, this weaker signal is translated to a molecular spin-up section or 0 byte. If both magnetizations are parallel, the stronger current marks a spin-down or 1 byte. The bulk property of the triple layer is called *anisotropic magnetoresistance* and it is the basis for both miniaturization and speed. The magnetization pattern of the disk is translated into electric current, and the pattern of the current encodes or decodes the digital text, pictures or music on the disk.

If a cobalt layer is evaporated on a gallium arsenide (GaAs) substrate, and another cobalt layer on metallic gold, these two cobalt layers possess different coercive fields and produce so-called *giant magnetoresistance* in the thin insulating, semiconducting or metallic layers. If a current runs perpendicularly through a cobalt/gold/cobalt triple layer, it will magnetize the two cobalt layers differently. At a current-induced field of –0.1 T, for example, both thin films are magnetized parallel in one direction ($\leftarrow$); at + 0.1 T parallel in the other direction ($\rightarrow$); at –0.02 T antiparallel in one ordering of the cobalt films ($\rightleftarrows$); at + 0.02 T in the antiparallel inverted ordering ($\leftrightarrows$) (Figure 1.21).[36a–c–40]

Both antiparallel arrangements of the spins in the two cobalt films will cause the current to decrease by $\sim 1\%$ and this change can be translated to a 1 bit. The low-resistance current through parallel magnetized layers becomes a 0 bit. The conducting gold spacer can be replaced by an insulating film of alumina which is only a few atoms or tenths of a nanometre thick. The current then tunnels through the thin barrier, and the giant magnetoresistance effect remains the same. It consists solely of the decreased resistivity of two ferromagnetic layers separated by a non-magnetic layer, when they rotate from antiparallel to parallel alignments.

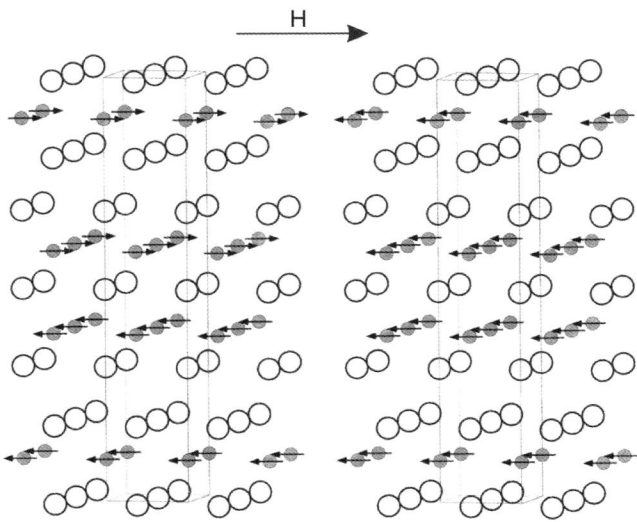

Figure 1.21 Magnetic structure of (left) the antiferromagnetic phase, (right) the ferromagnetic phase of $Sr_2LaMn_2O_7$. Mn sites are shaded and the Sr/La atoms are shown as unshaded circles. Arrows indicate the direction of the magnetic moments. © American Chemical Society.[36c]

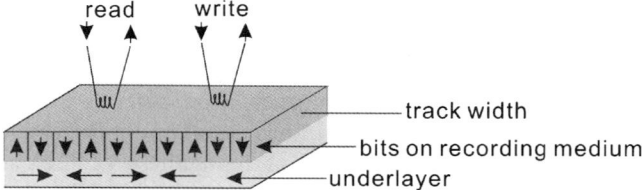

Figure 1.22 The heart of "magnetic tunnel junctions". The upper magnetic layer rotates bit by bit and interacts with changing external currents. The signal to the rotating magnetic layer is as much current (bit 0), or less current (bit 1) as produced by the giant magnetic resonance effect. The bits are recorded extremely fast, and later direct the electric current for reading and writing to the hard disk.[129]

Computer random access memory (RAM) is built on the basis of billions of such magnetic tunnel junction elements (Figure 1.22), which are connected by wires into which they discharge their 0 and 1 bits. The production of bits is extremely fast and needs no charging time. Most importantly, RAM retains its magnetic state when the computer is switched off.

In hard disks the material is not the most important consideration. They are always made of layers of iron, cobalt or nickel or a corresponding oxide, combined with almost any thin-layer material in between. It is the thinness and smoothness of the layers, as well as their stability to metal diffusion and layer deformation, that must be optimized.

Read–write elements for an alumina hard disk are made not of distinct nanoparticles, but of a continuous 1000 nm thick cobalt or haematite layer evaporated on to the aluminium alloy. This layer is protected by a $-(CH)_n-$ layer (p. 73 ff). A 3.5″ (9 cm) hard disk has a capacity of up to 1500 GB ($=1.5 \times 10^{12}$ B), and a magnetic head writes and reads on it in organized segments. A single magnetic bit on a 1000 GB disk is $\sim$1000 nm wide (track) and 10 nm long. Giant magnetoresonance sensors have become very complex mechanisms. The read/write heads are connected to sliders, which fly less than 10 nm above the smooth graphite surface that protects the metal layers of the disk from oxidation. The 10 nm distance is held constant by the air stream above the rotating disk, so air must be allowed into the hard disk chamber. However, the air must be kept absolutely clean for many years, as any dust immediately kills the magnetic contacts. The distances are getting smaller, and the speed is always increasing. Today the length resolution of a hard disk is only a factor of 100 greater than an atomic radius.

The atomic basis of paramagnetism is best understood in metallic nanoparticles and metal clusters within them. Only for such small clusters can electronic spin ordering be modelled and the results compared with measurements of nanoparticle magnetism. The beauty of metal clusters lies not only in the fact that their properties are size dependent, but that this size variation is non-monotonic. Some trends along groups or across periods of the periodic table are imprinted in clusters with as few as six atoms, if they are fixed in the octahedral symmetry which defines a common fragment of typical transition-metal crystal lattices.

The experimental cohesive energy in such a hexamer is 3.90 eV/atom for scandium, slightly more for titanium and has a maximum for vanadium. After that it decreases, showing a minimum for manganese (2.92 eV/atom), then increases again for elements to the right of manganese, having a broad maximum for iron, cobalt and nickel, and finally decreasing again for copper. The almost parabolic behaviour of the binding energy is a consequence of the change from bonding to antibonding character of the d orbitals as the d-band filling increases.[40]

The electronic structures of transition-metal clusters are also drastically altered by ligand shells. The magnetic moments of the surface atoms in large nickel clusters are completely quenched by carbonyl (CO) and ammonia (NH_3) ligands, because their σ lone pairs interact repulsively with the diffuse 4sp electrons of the metal to fill the 3d shell. When the number of NH_3 molecules is larger than the number of nickel atoms, however, the nickel–nitrogen bonds are stretched because of steric interaction between two NH_3 ligands and the nickel–nitrogen distances exceed the critical distance of 1.9 Å. Magnetism then reappears.[41]

The special properties of semiconducting metalloids (silicon and germanium) and metal chalcogenides (oxides, sulfides, arsenides, selenides and tellenides), as well as of metal nanoparticles, depend on electrons in the conducting bands or the Fermi level between valence and conducting bands and will be described in connection with the special elemental properties of nanoparticles.[41]

Reducing a bulk 3D system to two dimensions leads to nanolayers. Nanowires are effectively one dimensional and nanoparticles have zero dimensions. Below a diameter of ∼300 nm – and nearly all applicable nanoparticles are much smaller than this – and for single-domain nanoparticles only two opposite magnetization states are possible. Nanoparticles behave like paramagnetic domains; they show *superparamagnetic* behaviour. Fluctuations between the two states occur if the temperature rises above a critical *blocking temperature*, which represents the limit for stable magnetization. The fluctuation period below this temperature is reduced not only by rising temperatures, but also by a reduction of nanoparticle size. The 10 nm long sections of hard disks can always be taken as single domains.

The only ferromagnetic compound that does not contain iron, cobalt or nickel is Mn_2Ga_5, where a unique atomic arrangement and arrangement of bond angles leads to a high density of electronic states at the Fermi level in only one crystal plane. The saturated magnetic moment is 2.71 µB per formula unit and the Curie temperature is 450 K[42] (see p. 335 ff).

Recent examples of relatively large, egg-shaped micrometre spheres of haematite had a high coercivity of 3800 Oe at room temperature, but 55 nm nanoparticles reached only 52 Oe.[43,44] For soft magnets, which quickly gain and lose their magnetism, see p. 250 ff.

1.7 Heat and Thermal Conductivity

Heat is routinely used to self-assemble nanoparticles by techniques such as chemical vapour deposition or laser ablation. The temperatures applied to produce the vapours range from 300 to 1000 °C, the pathway of the vapour from a few millimetres to centimetres, and condensation occurs on smooth, heat-stable surfaces. Nanoparticles made from vapour or hot solutions (>300 °C) are always more crystalline than those precipitated from solution at room temperature. The latter procedure, standard for chemists, is often avoided by physicists, because interesting properties like stereochemically defined colours, conductivity in carbon nanotubes and strong fluorescence of silver nanoparticles result only from high-temperature preparation. So chemists have to buy a furnace and a pulse neodyme laser, and learn to handle them. Yields are usually low, but amazing properties depending on short-lived, frozen-out nanocrystals can be expected: silicon oxide (p. 190) and zinc gases, produced under water and condensed there to nanocrystals; magnesium oxide nanowire (p. 174); and fluorescing silver nanoparticles (p. 42).

An exception to the need for high temperatures is the pulsed Nd:YAG laser, which can also be used under water at room temperature and has already been used to produce carbon nanotubes from graphite (p. 57).

For fluorescing silver nanoparticles a better preparative method would be, of course, to start with stable and boring Ag(0) nanoparticles and use aqueous chemical treatments to convert them to fluorescent ones in gram quantities by controlled rearrangements. One has to research the differential surface chemistry of crystal layers, and it is predictable that each element or chalcogenide will behave differently.

One major aim of nanochemistry is to minimize *thermal conductivity* in thermoelectric materials that have high electrical conductivity. Electric currents flow through wires of such materials without warming them up or even cooling them down (p. 218 ff). Wires like this would make computers, batteries and refrigerators much more efficient. One difficulty here is the problem of obtaining reliable absolute values of thermal conductivities (in kJ) per temperature gradient. Distance and time dependencies are difficult to measure, because each probe is sensitive to all kinds of disturbances, in particular the temperature, porosity and crystallinity of the materials. Titania crystals, for example, have thermal conductivities of 21.9, 19.0, and 17.3 $Wm^{-1}K^{-1}$ at 300, 500 and 800°C respectively. For magnesium oxide at the same temperatures the values are 22.0, 16.2 and 10.6 $Wm^{-1}K^{-1}$. For quartz it is 38.2 $Wm^{-1}K^{-1}$ perpendicular to the axis and 72.2 $Wm^{-1}K^{-1}$ parallel to it. Glass is isotropic, with only 7.4 $Wm^{-1}K^{-1}$.

Relative heat conductivity, k, is a quantity that changes with the direction of an applied temperature gradient, even with a difference of just a few degrees. This property complicates measurements of k. In the classic Sénarmont method the material to be studied is covered with a thin layer of wax. a metal wire is inserted through a small hole in the wax, and then heated. The wax melts at a given temperature and a bulge of wax is pushed in front of the melting zone, which usually forms an ellipsoid on hexagonal crystals or a circle on glassy material or cubic crystals. The speed of the melting zone gives a relative measure of the heat conductivity of the material. The ratio of the large and small ellipsoid diameters corresponds to that of the k values, *e.g.* 4 for graphite along and perpendicular to the carbon planes. Some heat conductivities of thermeoelectric materials are given above.

A classic source of heat is the burning of carbon or hydrocarbons. This applies also on the micro- and nanoscale. Laser light with a wavelength of 640–760 nm, for example, is used to burn dyes on a compact disk (CD) in a desktop or laptop computer. The CD is made of a transparent poly(methacrylate) with a 180 nm surface layer of poly(carbonate). In commercial CD production the demanding step is the formation, melting and rapid solidification of a poly(carbonate) coating with a structured groove 500 nm wide and 5000 m long. The formation of this pattern requires an extremely fluid polymer melt, and the rapid reproduction of a pressed-on matrix of complex, micron-wide hole patterns. Storage of the CD, and playing it under a laser beam, requires an extremely form-stable solid, which keeps the printed pattern for long periods of time and does not deform in any way. Only irregular polymer chains made of rigid benzene units melt and solidify in this way, and require careful optimization of polymer weight distribution as well as the number and kind of crosslinks.

Commercially produced CDs or DVDs are pressed, but private products are burnt by a 50 mW laser beam (this is for an 8 × speed). The CD contains dye layers in the pressed grooves, which absorb the laser light and are burnt by it in the presence of atmospheric oxygen (Figure 1.23). The laser beam, shining through from the reverse side of the CD and following signals from the computer, either burns the pregroove dye layers to "yes" holes or leaves "no" pregrooves. When the CD is subsequently read by a laser beam, the reflection of the reading beam is completely quenched by the burnt holes. The laser must also, of course, distinguish clearly between the poly(carbonate) and the track.

Scanning electron micrographs of burnt pits show foamy spots. Since the distance between the reading head and the disc is of the order of 1 µm, 1000 times greater than the height of the foam bubbles, they do not cause a problem here although they would be fatal in the reading of hard disks, where high-density data must be read from a nanometre distance (see above).

Another material used in the classic CD is a 60 nm layer of silver. This conducts electrons and heat better than any other metal, is stable against oxygen and reflects light. The tiny fire on the CD will not harm it. Silver has only one binding electron, which interacts simultaneously with 12 neighbour silver atoms. On an average there is only one-sixth of a binding electron between two silver atoms, whereas carbon and silicon have two of them.

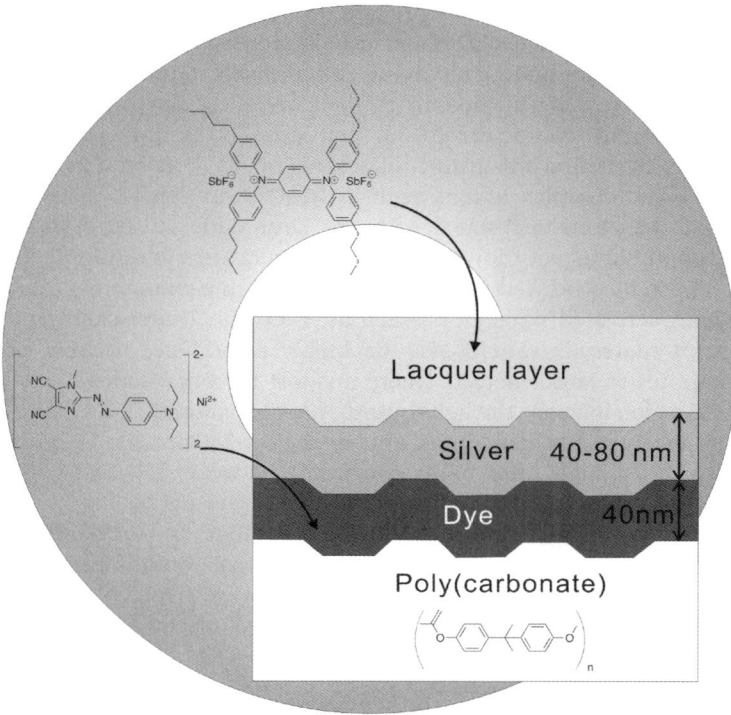

Figure 1.23 Simplified model of the triple layer on a CD before burning and SEM of four bits of burnt dye on the poly(carbonate) track. The low-molecular-weight poly(methacrylate) lacquer (see p. 105ff), cross-linked by UV light on the disk, contains a dye that protects the disk from the formation of singlet oxygen. A 150 nm layer of dye on silver is burnt by oxygen after being heated by a laser beam in the printing process.[45a,b]

1.8 Luminescence

1.8.1 Practical Applications

A flame directed on to a block of lime (calcium oxide, CaO) produces a bright white light. This property, which in the 18th and 19th centuries kept actors literally in the limelight, is known as *thermoluminescence*. Thorium(IV) oxide shows the same effect.

Photon emission by an excited material is called *photoluminescence* and is usually classified by the excitation source: plasma panel and field emission displays use high-energy UV light or low-voltage electrons; light emitting diodes (LEDs) and some fluorescent lamps work with near-UV or blue light. The ultimate goal, however, is *electroluminescence*, the direct excitation of semiconductors by electrical currents.

Today, most research goes into organic or inorganic phosphor materials for LEDs. Classical inorganic phosphors usually consist of a host lattice doped with activator ions containing low-lying energy levels. Lanthanide ions have 4f electrons that are well shielded from their chemical environment by outer electrons. Different, characteristic line emissions occur for each lanthanide. Furthermore, transition metal ions like Mn^{2+} or Ce^{3+} as well as s^2 ions like Pb^{2+} or Sb^{3+} and complex anions such as MoO_2^{4+} or NbO_3^{4+} show a strong coupling of the electronic states with the lanthanide lattice vibrations and produce broad bands with low quenching temperatures and a large variety of colours. The light produced is dim, however, and line emissions of lanthanide systems with activated phosphors, such as $Y_2O_3:Eu^{3+}$, are usually preferred. Luminescent materials should also be highly crystalline, because absorbed energy migrates to lattice defects where it is lost by radiationless deactivation. Sensitizers, which increase the light output, are also needed. Tb^{3+}, for example, luminesces nicely in the visible, but absorbs below 230 nm. Ce^{3+}, on the other hand, provides a high absorption coefficient between 250 and 350 nm and transfers the 254 nm light emitted by the mercury plasma to Tb^{3+}.

The blue and green phosphors used in the cathode ray tubes of classic TV screens were cheap zinc sulfides, with Cl^- ions as donor and Ag^+ ions as acceptor for blue and Cu^{2+}, gold and aluminium for green. Nowadays liquid crystal screens are used for small TVs, and the lanthanide phosphors have conquered the market for large, flat cathode ray tubes. Accelerating voltages are down to less than 5000 V as compared to 30 000 V for a cathode ray tube. Y_2O_2S is used as the red phosphor, but the green and blue zinc sulfide phosphors degrade surprisingly quickly. It turns out that the less energetic electrons are much more destructive, because they do not penetrate the 5 µm phosphor grain, and create only a high-density surface excitation. The green $Y_2SiO_5:Tb_3$ and the blue ZnS:Ag phosphor layers are therefore thinned out from micron thickness to 20 nm and a protective layer is established against the oxygen, which otherwise oxidizes the sulfur.

High-pressure mercury lamps operating at some 1000 Torr (133.332 kPa) produce white light with adequate colour rendering and efficiency. To produce either coloured displays or the white light of lamps it must be used efficiently by three narrow-band emitters with emission maxima at 450, 540, and 610 nm employing, for example, $BaMgAl_{10}O_{17}$:Eu, (Ce,Gd,Tb)MgB_5O_{10} and Y_2O_3:Eu. The invention of the blue LED based on gallium nitride (GaN) can be regarded as a breakthrough of material chemistry with respect to lighting. The variation of x and y in $(Y_{1-\bar{y}x}Gd_x)_3(Al_{1-\bar{y}y}Ga_y)_5O_{12}$:Ce (YAG:Ce) as a broad-band phosphor causes a shift of the YAG:Ce emission to values between 510 and 580 nm, the "centre" of white light.[46]

1.8.2 Nanoparticle Luminescence

Luminescence is not limited to semiconductors. Ice particles in outer space do it too. The photolysis of ice by UV light produces atomic O and H as well as $OH^{\cdot}$ radicals and the dimer H_2O_2. Jupiter's Galilean satellites – Europa,

Ganymede, and Callisto – and comets are largely comprised of "dirty" ice and they send emission signals that originate from ice. The ice lattice shifts and broadens, however; the emission lines (the g values in the electron paramagnetic resonance spectra) change and signal quenching occurs, all caused by hydrogen bonding of the OH˙ radical to H_2O molecules. Small- to medium-sized ice nanocrystals show completely different luminescence spectra.

Synchrotron UV irradiation of ice nanocrystals with up to 1500 water molecules yielded fluorescence excitation spectra similar to those of single water molecules. It is the excited OH* that emits its characteristic long-lived (~ 1 s) luminescence at 420 nm, originating from a spin-forbidden transition of the short-lived OH˙ radical (700 ns):

$$(H_2O^* \rightarrow OH^* + H)$$

sp^3-Hybridized semiconductors, such as elemental doped silicon or the chalcogenides indium phosphide, gallium arsenide, and cadmium selenide, are excited by visible light and promote an electron from the highest occupied molecular orbital of the valence band to the lowest unoccupied conduction band. This excited electron then moves rapidly in response to an applied electric field, because there is little lattice distortion accompanying its creation. Such lattice distortion remains vanishingly small around electron–hole pairs, which migrate through nanocrystals of these sp^3 excitable materials. Again the nanocrystals act like molecules, comparable to the Ag(0) and water clusters mentioned above. They interact with light *via* their electronic transition dipoles and, in the absence of defects, a semiconductor nanocrystal should exhibit a near unity fluorescence quantum yield because lattice distortions (*Franck–Condon factors*) are small and the electron and hole remain superimposed because of quantum confinement. This occurs when the radius of the nanocrystal becomes comparable to the bulk exciton Bohr radius, which is the natural separation between an electron in the conduction band and the hole that it leaves behind in the valence band. A typical value is 5.6 nm for cadmium selenide. A suitable model for the electronic structure of nanocrystals restricts electrons and holes by an infinitely high potential at the nanocrystal surface and replaces the continuous energy levels of solids by discrete nanocrystal molecular orbitals.

A *quantum dot* (QD) is a semiconducting metal compound (*e.g.* zinc oxide or a cadmium chalcogenide) with the diameter of the exciton Bohr radius or slightly larger, which means roughly 2–10 nm. Applications of QDs come from tunable band gaps, an energy difference of the order of 1–2 eV between the conduction and valence bands (corresponding to 96.5–193 s mol^{-1}, wavenumbers of 8000–16 000 cm^{-1} and wavelengths of light 660–1250 nm). "Tunable" means here that the band gap in QDs is always larger than the band gap in bulk material and depends on the size of the QD. The larger the dot, the more closely its band gap will approach the bulk value.[45]

Electrons cross the band gap and reach the conduction band only if they absorb radiation energy greater than or equal to the band gap difference. Subsequent emission of this energy, as the electrons fall back into the valence band, makes the QDs usable as light sources and indicators. The major effect of the small size of QDs is their discrete energy states, in contrast to the continuous energy states in bulk semiconductors. The energy levels in QDs are separated sufficiently that a few atoms or electrons more or less will change the band gap energy and the colour of emitted light.

In semiconductors, the width of the forbidden zone or gap is 3.3–4.0 eV, corresponding to light with a wavelength <400 nm. Semiconductors with a bandwidth >3.3 eV (*e.g.* zinc oxide) are colourless. Between 3.3 and 1.8 eV the colour changes from yellow through orange to red (zinc selenide, cadmium sulfide, gallium phosphide, zinc telluride, cadmium selenide). Band gaps <1.8 eV lead to black semiconductors with a metallic luster (cadmium telluride, germanium, silicon).

The electron–hole exchange in cadmium selenide is due to confinement of the electron and hole wave functions. The influence of the exchange term and nanocrystal asymmetry on the band edge electronic fine structure is shown in Figure 1.24. Electron and hole states in a 1S totally symmetric particle are eightfold degenerate, taking into account the electron spin and hole spatial degeneracies. In wurtzite nanocrystals the crystal field splits the $1S_{3/2}$ hole state into three $\pm 3/2$ and $\pm 1/2$ states, and exchange terms mix the electron and hole angular momentum, yielding five extra excitonic states.

For low-energy band edge excitations the lowest optically active $\pm 1^L$ state is predominantly populated, followed by radiationless relaxation to the ± 2 dark excitonic state. Emission from the latter, weakly emitting state accounts for the long (microsecond) luminescence lifetime. A small Stokes shift of the band edge excited luminescence is caused by the size-dependent exchange splitting between the ± 2 and $\pm 1^L$ states. Two experimental findings, the magnetic field dependence of the luminescence lifetimes and the low orbital phonon couplings, are in full agreement with this model. The microsecond decay seen at zero field decreases with increasing magnetic field, and the zero phonon line intensity increases dramatically. The luminescence behaviour of pure nanocrystals is completely molecular-like.[46]

Early nanocrystal models of an internal electron–hole pair assumed that one of the carriers is quickly localized in a shallow surface trap, but could not account for the strong magnetic field dependence of luminescence lifetimes. The luminescence of single cadmium selenide nanocrystals embedded in a thin polymer film turns on and off ("blinks") on a time scale of several seconds. The "on" period duration scales inversely with excitation intensity, but the "off" period is intensity independent. Non-radiative photoionization *via* two electron–hole pairs occurs: the ionized nanocrystal creates only a very low-emission quantum yield, the ionized nanocrystal is non-emitting. Eventually it is neutralized by a second photoionization event or the return of the ejected carriers, and the emission is restored.

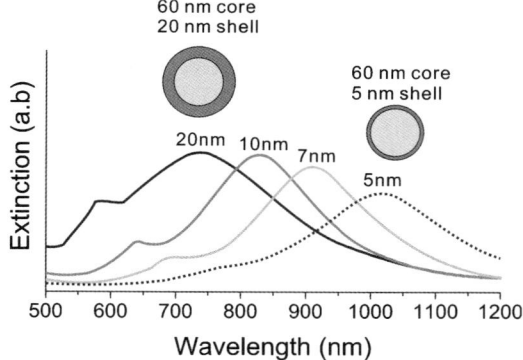

Figure 1.24 Optical (= electronic) spectra of 5–20 nm ZnO nanoparticles with a gold shell and a silica core. © American Chemical Society.[147]

The colour of metal nanoparticles cannot be caused by electron–hole pairs, which would never separate in a conductive metal, but the conduction electrons alone, as *plasmons*, absorb light energy. Surface plasmons are coherent oscillations of conduction electrons on a metal surface. They are excited by electromagnetic radiation at a metal–dielectric interface, *e.g.* gold nanoparticles in water (Figure 1.24). In the long-wavelength region (= electrostatic dipole regime) the intensity of the absorption band depends on dielectric functions of the metal nanoparticle and on the medium. In metallic nanorods, the absorption peak of the sphere splits into a strongly red-shifted long axis (large dipole) and slightly blue-shifted transverse mode peak (small dipole). In general, the optical, electronic, and catalytic properties of metal nanostructures are tuned by controlling their size, shape, and composition.[47]

Light emissions from lanthanide ions involve f-orbital electrons and are much narrower than those observed from QDs, metal nanoparticles or even organic molecules. The excitation within partially filled 4f orbitals is parity-forbidden and the transitions that occur produce only low molar absorption coefficients and slow emission rates. The f–f luminescent lifetimes are, however, very long, because the orbitals are shielded by the filled 5s and 5p orbitals. The quantum yield of lanthanide emission is high, and quenching or photobleaching effects, typical for fluorescent dyes, are not observed. The luminescence of lanthanide(III)-doped nanoparticles with inorganic cores and organic stabilizing shells is therefore used in protein labelling, long-lived light-emitting diodes or liquid-crystal displays.

The strong confinement of electronic wave functions in nanocrystals also leads to strong Coulomb interactions. The electron–hole and exciton–exciton interaction energies are inversely proportional to the natural exciton and biexciton radii. In group II–VI semiconductors the energy increases with decreasing nanocrystal size and a leakage of the electric field into the medium outside the nanocrystal may enhance the electron–hole interaction energy to

>100 meV, while the biexciton binding energy can be as large as tens of meV. Optical amplification in nanocrystals is due to multiexciton emission, but is also strongly affected by intrinsic non-radiative recombinations, which lead to very short (picosecond) lifetimes of optical gain. This latter fact is the major barrier to achieving lasing properties, but cadmium selenide and lead selenide nanocrystals provide tunable energy gaps and exciton–exciton interactions for the improved lasing performance of nanocrystals. Direct photogeneration of multiexcitons from single photons *via* carrier multiplication is possible. The absorption of a single photon can produce two or even three excitons, which results in internal yields of up to 300% for the conversion of light quanta into charge carriers.[48]

Another extremely sensitive analytical method, which allows quantitative characterizations of molecular complexes in 10^{-9} molar colloidal solutions, is *fluorescence quenching*. A 10^{-9} molar solution of a drug carried on nanoparticles in the bloodstream can easily be identified and quantified if the nanoparticle also carries a fluorescing dye.

One of the sensational findings of nanochemistry is the fluorescence of metallic silver: the electron gas of a metal behaving like electrons in the defined orbitals of covalent molecules. Nothing was really special in the preparation of the 2 nm Ag(0) nanocrystals or in their appearance under the electron microscope. The electron microscope beam resolved several planes with hexagonal atom patterns, but the overall form of the nanocrystals was more or less spherical. The major crystal planes of Ag(0) nanocrystals were detectable, but very little anisotropic growth appeared. The nanocrystals were formed by the thermal reduction of Ag^+ ions by a crystalline amino acid matrix, which also controlled the nucleation and migration of Ag(0) atoms and stabilized them in aqueous solution for more than 2 years at ambient conditions. Glycine (CH_2NH_2COOH) and silver nitrate ($Ag(I)NO_3$) were dissolved in water, the water evaporated, and the solid-phase mix heated at 445 K. The brown reduction product was resuspended in water, sonicated for 24 h and centrifuged. This process was repeated at least four times. The final solution contained glycine-coated silver nanoparticles with diameters ranging from 2 to 30 nm, as characterized by transmission electron microscopy (Figure 1.25). A few Ag(0) atoms fenced a few "metallic" electrons into a QD, and changed their gas-like independence to ordered electrons in a box with distinct energy levels instead of "bands". The nanocrystals, which are supposedly pure Ag(0), showed an intense red luminescence with a maximum close to 600 nm, and as yet no explanation for this is available.[49] No other Ag(0) or other metal nanocrystal has shown this phenomenon.[50]

As usual, the elemental analysis of the nanocrystals is not known. Is the glycine amino group present as an electron donor, or does the sulfonate together with the sulfide quench it? Such simple questions are never asked in nanocrystal chemistry, because they are impossible to answer with the nanocrystal "samples", which in this case were heated and sonicated and precipitated and dried, but never purified.

Atoms, Molecules, Electrons, Light and Heat in Nanometre Confinement 43

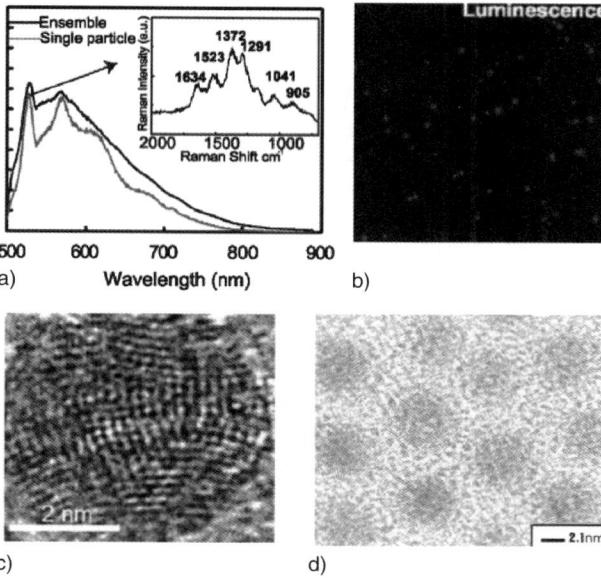

Figure 1.25 (a,b) Luminescence and (c) electron micrograph of silver–glycine nanocrystals;[49] (d) electron micrograph of hexagonal Ag(0)–dodecanethiol nanocrystals. © American Chemical Society.[50]

Bis(2-ethylhexyl)sulfosuccinate silver salt (AgAOT) and bis(2-ethylhexyl)sulfosuccinate sodium salt (NaAOT) in a ratio of 3:7 yielded crystalline Ag(0) nanoparticles with a 700-fold excess of hydrazine. They showed a strong plasmon absorption band at 400 nm, but no luminescence. Coating with a monolayer of dodecanethiol shifted the plasmon band to 430 nm, rendered the 2 nm nanocrystals highly soluble in hexane and produced perfect hexagonal arrangements, but again no luminescence.[50] The special nanocrystalline effect, which changes a metallic nanocrystal with a floating electron cloud to a molecule with fixed orbitals, is still unknown.

1.9 Catalysis

Heterogeneous catalytic reactions are not reversible; they run only in one direction. Transition states occur as dissipative structures on the surface of the catalyst, and oscillations and even chaos are observed.

The most practical starting point for chemist planning catalysis work with nanocrystals is probably Sabatier's *volcano curve* of heterogeneous catalysis (Figure 1.26).[52] The decomposition of formic acid to carbon monoxide and water (HCOOH → CO + H_2O), for example, is catalysed by transition metals and depends on the decomposition of transition metal formiates. $Au(HCOO)_3$, for example, has a binding enthalpy of -250 kJ mol^{-1}, $Pt(HCOO)_2$ of

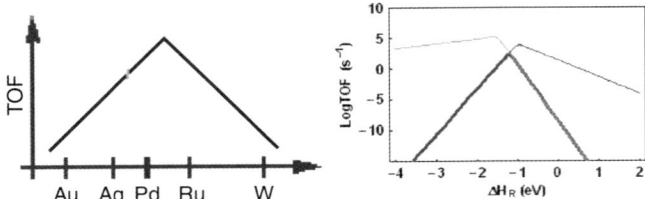

Figure 1.26 Sabatier's volcano curve of the heterogeneous catalysis for the dehydration of formic acid (HCOOH) to carbon monoxide (CO). The rate-determining steps are presumably adsorption of HCOOH and desorption of CO. A calculated plot of turnover frequency (TOF) against $\Delta H_{\text{(adsorption–desorption)}}$ gives the same curve. © American Chemical Society.[51]

350 kJ mol^{-1} and Fe(HCOO)$_3$ of 450 kJ mol^{-1}. Both the weakly binding Au(III) and the strongly binding Fe(III) formiates must be heated to 550 K to yield carbon monoxide, the intermediate Pt(II) only to 400 K. Strong binding is good for adsorption but bad for release; weak binding is fast in release but bad in activation and binding. Catalysis needs the compromise, namely Pt(II). Sabatier's volcano model has been in use for more than 100 years and as the basis for the selection of catalysts it is unsurpassed.[51]

Regular, long-lasting oscillatory kinetics are observed in oxidation reactions with molecular oxygen. Hydrogen, ammonia, hydrocarbons, methanol, and ethanol are the most common reducing agents used in industry. Another large group of reactions with rate oscillations is the reduction of nitrogen oxides (NO$_x$) in engine exhaust gases by carbon monoxide, hydrogen or ammonia, which is catalysed by noble (group VIII) metals. Both types of reactions are often carried out on the surface of nanocrystals embedded on "polycrystalline" surfaces, which have been mimicked with tips of various types of atomic force microscopes at pressures ranging from 10^{-9} mbar to 1 bar (10^{-4} Pa to 10 Pa).[52]

Density functional theory (see next section) is then used to calculate the energetics of catalytic surface reactions in detail under different conditions. One important feature of nanocrystals is that the concentration of low-coordinate corner atoms with respect to the other surface atoms is very large. If one just considers the reactivity of corner atoms, metal clusters containing 12 atoms are perfect models for the activation of carbon monoxide in its reaction with oxygen, for example (Figure 1.27). All the 12 atoms in the cluster are artificially fixed with the lattice constant of the bulk value to mimic the geometrically constrained corner of a larger 2–5 nm nanocrystal, which is then studied experimentally. It turns out that adsorption of carbon monoxide is considerably more exothermic on the 12-atom clusters than on the close-packed surfaces. The second elementary reaction, i.e. the coverage with molecular oxygen, may therefore become widespread enough to allow for an associative mechanism with two catalytically activated reactants (O$_2$* + CO* → CO$_2$ + O*). Platinum and palladium nanoparticles indeed possess very potent under-coordinated sites with optimal binding for carbon dioxide and oxygen under the right reaction conditions.[53]

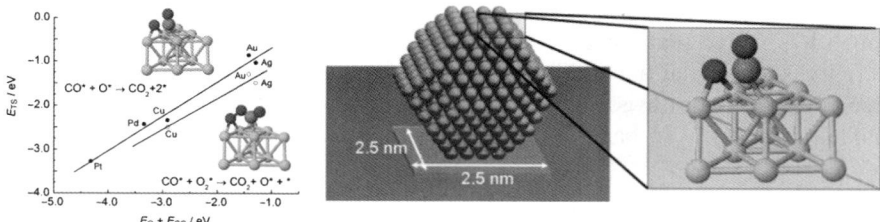

Figure 1.27 Model of the catalysis of the oxidation of carbon monoxide to carbon dioxide by oxygen on palladium or platinum clusters as models for nanocrystal corners as catalytic sites.[53,54]

More noble metals move to the reactivity peak of the volcano, when lower-coordinated metal atoms serve as active sites for the reaction. Reactive molecules such as oxygen or nitrogen monoxide are typically adsorbed by these sites. Platinum and palladium are more reactive (less noble) than gold and activate oxygen even in the surface plane, whereas gold becomes most active as a corner atom. For reactions involving less-reactive molecules, such as nitrogen, the best nanocrystal catalysts should be cobalt and nickel, the metals just to the right of the most active bulk metals in the periodic table, namely iron and ruthenium. It was also shown that by far the most active face of iron single crystals in ammonia synthesis at 20 bar (2 mPa) was the (111) plane, by more than two orders of magnitude in terms of area. This was attributed to special sites with a coordination number of 7.[54]

Turnover frequency (TOF) is always the best indicator for comparison of catalysts. TOF is measured by an isotope kinetic analysis in which the steady state of a catalytic reaction is running in a continuous reactor with complete mixing. A sudden switch of reactants, say from $^{14}N^{14}N + H_2$ to $^{15}N^{15}N + H_2$, followed by time-dependent measurements of the product composition with mass spectrometry, yields the TOF, which always depends on surface crystalline anisotropy, given by faces of a single crystal or clusters of a typical size of < 10 nm.[55]

The nano effect favours strong binding to low-coordinate corner atoms; metal atoms in the plane must be very active. Another advantage of catalysis by nanoparticles is the small amount of material lost in the process, which keeps costs low. Industrial catalysts are used in the form of foils, ribbons, wires, evaporated films, and supported catalysts, with silica or alumina being the support material in most cases. Palladium nanoparticles with a very narrow size distribution embedded in a zeolite matrix are usually most active, but cheaper alternatives often replace them in practice.

The electronic structure properties exhibited by 4d transition-metal sulfides in connection with hydrodesulfurization (HDS) catalytic activity indicate quite different behaviour, because sulfide is an extremely strong electron donor. High densities of non-bonding states below the Fermi level are decisive here for catalytic performance. The incomplete filling of the band gap at exposed transition metal atoms over the catalyst surface enhances the donation–back-donation mechanism from sulfur. In general 3d metals are much less active than 4d and 5d metals, for which the maximum in catalytic activity is reached at the middle of the transition metal period. For example, the PdS_2 surface is not a good desulfurization catalyst, because it does not have palladium d states to back-donate electrons.[56]

1.10 Density Functional Theory[57,58]

Nowadays density functional theory (DFT) provides the basis for the computation of energy, structure and properties of atoms, molecules and nanoparticles at much lower cost than the classical *ab initio* wave function techniques. Nearly all theory mentioned in this book is based on the DFT method. Two examples have already been given in the preceding sections, and the theory is briefly introduced here.

The basic idea concept of DFT is that at each point r the electron density $F(r)$ determines the ground-state properties of any atomic, molecular or nanoconfigured system, similar to the behaviour of the free electron gas of metals, which is controlled by the atomic nuclei and the ordered electrons. For molecules and the solid state, just one simple, one-parameter approximate exchange correlation function is used in the form of an exchange-only functional. Computational DFT is then an attractive alternative to wave function techniques, because the electron density of any nanosystem is the only quantity that is needed to describe the atomic, molecular and nanoparticulate ground states. Chemical concepts of the electron, *e.g.* electronegativity, redox potential and orbital geometry, are precisely defined in DFT, making it possible to use them quantitatively in calculations.

It has been shown that any finite domain of the ground-state electron density fully determines the ground state of the entire, boundary-less molecular system (the *holographic electron density theorem*). Each atom, each molecule, or the whole nanocrystal immediately changes its separate electron density function $F(r)$ when it is perturbed by a change in its number of electrons at a fixed external potential. Two types of quantities then emerge in the first-order derivatives: a global quantity that is a characteristic of the system as a whole (*hardness*, η) and a local quantity, $F(r)$, the value of which changes from point to point (*electronegativity*). Electronegativity is used in DFT as the major descriptor of relative reactivity. Hardness is the resistance of the chemical potential to changes in the number of electrons and is related to the energy gap between occupied and unoccupied orbitals, $\eta = 1/2(\varepsilon_{LUMO} - \varepsilon_{HOMO})$. The chemical hardness of atoms is mimicked in calculations by placing fractions of positive and negative charges on to neighbouring atoms and evaluating ionization energies or electron affinity.[59] More charge generally increases the hardness of atoms with respect to the isolated atoms.

References to a few applications of DFT are given in this book, but no attempt is made to explain any of the procedures, or to provide a catalogue of existing computer programs for specific problems.

DFT has been applied, for example, to give plausible models for the following experimental subjects and/or results:

- Yoctolitre-sized pores (*e.g.* in MOF-5) adsorb hydrogen molecules most strongly at their zinc oxide clusters rather than on the organic linkers (Chapter 2, p. 103).
- Pyrrolofulvalene and the viologen monocation radical do not change their planar conformation upon redox reactions, whereas the viologen dication tends towards a highly twisted C–C connection (p. 130).
- Zigzag boron nitride tubes produce various magnetic patterns at the boron and at the nitride edges (pg 98).
- A (5,5) boron nitride nanotube with a diameter of 0.69 nm and a length of 1.42 nm conducts water, but a corresponding (5,5) carbon nanotube does not (p. 206 ff).
- For the cubic structure of $GeTe-Sb_2Te_3$, a material for rewritable laser DVDs, the average number of valence electrons was determined (p. 199).
- Fe_3S_4 (greigite) nanosheets should behave like graphene (p. 286).
- Nitrogen species, in particular the interstitial NO^{2-} dianion, substituting for an oxygen in the titania matrix seem to be the only well-defined colourized dopants that make direct sunlight absorption feasible (p. 312).
- Charge compensation between a p-type (N) and an n-type (F) dopant in titania has a large stabilizing effect (p. 312).
- Niobium doping of molybdenum sulfide tubes leading to a density of electronic states close to the Fermi level (p. 333).
- The free energy of reversible atomic hydrogen bonding to various catalyst planes and edges (pg 370).
- The binding energy of different iridium clusters to NO_x (p. 352).

References

1. (a) F. Sciortino, A. Geiger and H. E. Stanley, Effects of defects on molecular mobility in liquid water, *Nature*, 1991, **354**, 218–221; (b) N. Smollin, A. Oleinikova, I. Brovchenko, A. Geiger and R. Winter, Properties of spanning water networks, *J. Phys. Chem. B*, 2005, **109**, 10995–11005; (c) J. Fuhrhop and T. Wang, *Sieben Moleküle*, Wiley-VCH, Weinheim, 2009, **19**, 1–63.
2. M. K. Sanyal, V. V. Agrawal, M. K. Bera, K. P. Kalyanikutty, J. Daillant, C. Blot, S. Kubowicz, O. Konovalov and C. N. R. Rao, Formation and ordering of gold nanoparticles at the toluene-water interface, *Phys. Chem. C Lett.*, 2008, **112**, 1739–1743.
3. W. Kleber, *Einführung in die Kristallographie*, VEB Verlag Technik, Berlin, 1971.
4. M. B. Bever (ed.), *Encyclopedia of Materials Science and Engineering*, Pergamon, Oxford, 1986.
5. J. W. Gaffney, K. N. White and S. Boult, Oxidation state and size of Fe controlled by organic matter in natural waters, *Environ. Sci. Technol.*, 2008, **42**, 3575–3581.

6. S. Laurent, D. Forge, M. Port, A. Roch, C. Robic, L. Vander Elst and R. N. Muller, Magnetic iron oxide nanoparticles: synthesis, stabilization, vectorization, physicochemical characterizations and biological applications, *Chem. Rev.*, 2008, **108**, 2064–2211.
7. R. K. Watt, R. B. Frankel and G. D. Watts, Redox reactions of apo mammalian ferritin? *Biochemistry*, 1992, **31**, 9673–9679.
8. G. S. Waldo and E. C. Theil, Formation of iron(III)-tyrosinate is the fastest reaction observed in ferritin? *Biochemistry*, 1993, **32**, 13262–13266.
9. H.-A. Hosein, D. R. Strongin, M. Allen and T. Douglas, Iron and cobalt oxide and metallic nanoparticles prepared from ferritin, *Langmuir*, 2004, **20**, 10283–10287
10. D. Bratko, C. D. Daub, K. Leung and A. Luzar, Effect of field direction on electrowetting in a nanopore, *J. Am. Chem. Soc.*, 2007, **129**, 2504–2510.
11. A. Ulman, *An Introduction to Ultrathin Organic Films from, Langmuir to Self-Assembly*, Academic Press, Boston, 1991.
12. J. Schneider, C. Messerschmidt, A. Schulz, M. Gnade, B. Schade, P. Luger, P. Bombicz, V. Hubert and J.-H. Fuhrhop, Odd-even effects in supramolecular assemblies of diamido bolaamphiphiles, *Langmuir*, 2000, **16**, 8575–8584.
13. S. Bhosale, S. Bhosale, T. Wang, G. Li, M. Kopascýnka and J.-H. Fuhrhop, Hydrophobic and hydrophilic nanowells, *Accts. Chem. Res.*, 2006, **39**, 498–508.
14. J. Hunjang, B. Dempsey and W. D Burgos, Solubility of hematite revisited: effects of hydration, *Environ. Sci. Technol.*, 2007, **41**, 7303–7308.
15. A. Imanishi, T. Okamura, N. Ohashi, R. Nakamura and Y. Nakato, Mechanism of water photooxidation reaction at atomically flat TiO_2 (rutile) (110) and (100) surfaces: dependence on solution pH, *J. Am. Chem. Soc.*, 2007, **129**, 11569–11578.
16. D.-A. Luh, T. Miller, J. J. Paggel, M. Y. Chou and T.-C. Chang, Quantum electronic stability of atomically uniform films, *Science*, 2001, **292**, 1131–1133.
17. I. Daems, G. V. Baron, S. Punnathanam, R. Q. Snurr and J. F. M. Denayer, Molecular cage nestling in the liquid-phase adsorption of n-alkanes in 5A zeolite, *J. Phys. Chem. C*, 2007, **111**, 2191–2197.
18. M. Kopaczynska, T. Wang, A. Schulz, M. Dudic, A. Casnati, F. Sansone, R. Ungaro and J.-H. Fuhrhop, Scanning force microscopy of upright-standing, isolated calixarene–porphyrin heterodimers, *Langmuir*, 2005, **21**, 8460–8465.
19. M. Marjanovic, M. Kralj, F. Supek, L. Frkanec, I. Piantanida, T. Šmuc and L. Tusek-Božic, Antitumor potential of crown ethers: structure-activity relationships, cell cycle disturbances, and cell death studies of a series of ionophores, *J. Med. Chem.*, 2007, **50**, 1007–1018.
20. K. Yamauchi, Y. Takashima, A. Hashidzume, H. Yamaguchi and A. Harada, Switching between supramolecular dimer and nonthreaded supramolecular self-assembly of stilbene amide-r-cyclodextrin by photo-irradiation, *J. Am. Chem. Soc.*, 2008, **130**, 5024–5025.

21. J. Rebek Jr, Contortions of encapsulated alkyl groups, *Chem. Commun.*, 2007, 2777–2789.
22. R. J. Hooley, S. M. Biros and J. Rebek Jr, Normal hydrocarbons tumble rapidly in a deep, water-soluble cavitand, *Chem. Commun.*, 2006, 509–510.
23. B. R. Jackson, P. J. A. Sazio and J. V. Badding, Single-crystal semiconductor wires integrated into microstructured optical fibres, *Adv. Mater.*, 2008, **20**, 1135–1140.
24. J. Oh and C. V. Thompson, Selective barrier perforation in porous alumina anodized on substrates, *Adv. Mater.*, 2008, **20**, 1–5.
25. C. Schoenenberger, B. M. I. van der Zande, L. G. J. Fokkink, M. Henny, C. Schmid, M. Krulger, A. Bachtold, R. Huber, H. Birk and U. Staufer, Template synthesis of nanowires in porous polycarbonate membranes electrochemistry and morphology, *J. Phys. Chem. B*, 1997, **101**, 5497–5505.
26. (a) W. R. Duncan and O. V. Prezhdo, Temperature independence of the photoinduced electron injection in dye-sensitized TiO_2 rationalized by *ab initio* time-domain density functional theory, *J. Am. Chem. Soc.*, 2008, **130**, 9756–9762; (b) W. R. Duncan, C. F. Craig and O. V. Prezhdo, Time-domain *ab initio* study of charge relaxation and recombination in dye-sensitized TiO_2, *J. Am. Chem. Soc.*, 2007, **129**, 8528–8543.
27. E. Biemmi and T. Bein, Assembly of nanozeolite monolayers on the gold substrates of piezoelectric sensors, *Langmuir*, 2008, **24**, 11196–11202.
28. U. E. Seiner and T. Ulrich, Magnetic field effects in chemical kinetics and related phenomena, *Chem. Rev.*, 1989, **89**, 51–147.
29. (a) M. Getzlaff, *Fundamentals of Magnetism*, Springer, Berlin, 2007; (b) D. Gignoux, Magnetic properties of metallic systems, in R. W. Cahn et al. (eds) Materials science and technology, Wiley-VCH 2005, **3A**, 367–457.
30. (a) S.-H. Kim, M. Bostrom and D.-K. Seo, Two-dimensional superdegeneracy and structure-magnetism correlations in strong ferromagnet, Mn_2Ga_5, *J. Am. Chem. Soc.*, 2008, **130**, 1384–1391; (b) Z. Duan, Y. Zhang, B. Zhang, D. Zhu, Crystal-to-Crystal transformation from antiferromagnetic to ferromagnetic diamondoid framework, *J. Am. Chem. Soc.*, 2009, **131**, 6934–6935.
31. E. Rotenberg, W. Theis, K. Horn and P. Gilles, Quasicrystalline valence bands in decagonal AlNiCo, *Nature*, 2000, **406**, 602–605.
32. C. L. Henley, M. Mihalkovic and M. Widom, Total-energy-based structure prediction for d(AlNiCo), *J. Alloys Comp.*, 2002, **342**, 221–227.
33. J. J. Kim, H. S. Park, D. Shindo, S. Hirosawa, M. Satoshi and H. Morimoto, *In situ* observations of magnetization process in Alnico magnets by electron holography and Lorentz microscopy, *Materials Trans.*, 2006, **47**, 907–912.
34. (a) A. R. Urbach, J. C. Love, M. G. Prentiss and G. M. Whitesides, Sub-100 nm confinement of magnetic nanoparticles using localized magnetic field gradients, *J. Am. Chem. Soc.*, 2003, **125**, 12704–12705; (b) J. C. Love, A. R. Urbach, M. G. Prentiss and G. M. Whitesides, Three dimensional

self-assembly of metallic rods with submicron diameters using magnetic interactions, *J. Am. Chem. Soc.*, 2003, **125**, 12696–12967.
35. J. P. Dismukes, D. F. Martin, L. Ekstrom, C. C. Wang and M. D. Coutts, Ferromagnetic chromium dioxide for magnetic tape, *Ind. Eng. Chem. Prod. Res. Develop.*, 1971, **10**, 319–329.
36. (a) G. Timp (ed.) *Nanotechnology*, Springer, Berlin, 2001; (b) W. R. Fahrner (ed.) *Nanotechnology and Nanoelectronics*, Springer, Berlin, 2004; (c) P. D. Battle, D. E. Cox, M. A. Green, J. E. Millburn, L. E. Spring, P. G. Radaelli, M. J. Rosseinsky and J. F. Vente, Anti-ferromagnetism, ferromagnetism, and phase separation in the GMR system Sr2-xLa$_{1+x}$Mn$_2$O$_7$, *Chem. Mater.*, 1997, **9**, 1042–1049.
37. P. Grünberg, R. Schreiber, Y. Pang, M. B. Brodsky and H. Sowers, Layered magnetic structures: evidence for antiferromagnetic coupling of Fe layers across Cr interlayers, *Phys. Rev. Lett.*, 1986, **57**, 2442–2445.
38. M. N. Babich, J. M. Broto, A. Fert, F. van Dau, F. Petroff, P. Etienne, G. Creuzet and A. Friederich. J. Chazelas, Giant Magnetoresistance of Fe/Cr Magnetic Super lattices, *Phys. Rev. Lett.*, 1988, **61**, 2472–2475.
39. A. Friederich and J. Chazelas, Giant magnetoresistance of (001)Fe/(001)Cr magnetic superlattices, *Phys. Rev. Lett.*, 1988, **61**, 2472–2475.
40. E. E. Fullerton and I. K. Schuller, The 2007 Nobel Prize in physics: magnetism and transport at the nanoscale, *ACS Nano*, 2007, **1**, 384–389.
41. J. A. Alonso, Electronic and atomic structure, and magnetism of transition-metal clusters, *Chem. Rev.*, 2000, **100**, 63–76.
42. S.-H. Kim, M. Bostrom and D.-K. Seo, Two-dimensional superdegeneracy and structure-magnetism correlations in strong ferromagnet, Mn$_2$Ga$_5$, *J. Am. Chem. Soc.*, 2008, **130**, 1384–1391.
43. D. Wang, C. Song, Y. Zhao and M. Yang, Synthesis and characterization of monodisperse iron oxides microspheres, *J. Phys. Chem. C.*, 2008, **112**, 12710–12715.
44. Y.-W. Jun, J.-W. Seo and J. Cheon, Nanoscaling laws of magnetic nanoparticles and their applicabilities in biomedical sciences, *Acc. Chem. Res.*, 2008, **41**, 179–189.
45. (a) K. Roth, Die Chemie der schillernden Scheiben, *Chem. unserer Zeit*, 2007, **41**, 334–345; (b) H. Mustroph, M. Stollenwerk and V. Bressau, Current developments in optical data storage with organic dyes, *Angew. Chem. Int. Ed.*, 2006, 2016–2035.
46. T. Jüstel, H. Nikol and C. Ronda, New developments in the field of luminescent materials for lighting and displays, *Angew. Chem. Int. Ed.*, 1998, **37**, 3084–3103.
47. M. Nirmal and L. Brus, Luminescence photophysics in nanocrystals, *Acc. Chem. Res.*, 1999, **32**, 407–414.
48. V. I. Klimov, Implications for lasing and solar energy conversion mechanisms for photogeneration and recombination of multiexcitons in semiconductor nanocrystals: implications for lasing and solar energy conversion, *J. Phys. Chem. B*, 2006, **110**, 16827–16845.

49. J. Zheng, Y. Ding, B. Tian, Z. L. Wang and X. Zhuang, Luminescent and Raman active silver nanoparticles with polycrystalline structure, *J. Am. Chem. Soc.*, 2008, **130**, 10472–10473.
50. A. Taleb, C. Petit and M. P. Pileni, Synthesis of highly monodisperse silver nanoparticles from AOT reverse micelles: a way to 2D and 3D self-organization, *Chem. Mater.*, 1997, **9**, 950–959.
51. J. Cheng, P. H. Ellis, S. French, G. Kelly and C. M. Lok, Brønsted–Evans–Polanyi relation of multistep reactions and volcano curve in heterogeneous catalysis, *J. Phys. Chem. C*, 2008, **112**, 1308–1311.
52. R. Imbihl and G. Ertl, Oscillatory kinetics in heterogeneous catalysis, *Chem. Rev.*, 1995, **95**, 697–733.
53. H. Falsig, B. Hvolbæk, I. S. Kristensen, T. Jiang, T. Bligaard, C. H. Christensen and J. K. Nørskov, Trends in the catalytic CO oxidation activity of nanoparticles, *Angew. Chem.*, 2008, **120**, 4913–4917.
54. L. C. Grabow and M. Mavrikakis, Nanocatalysis beyond the gold-rush era, *Angew. Chem. Int. Ed.*, 2008, **47**, 7390–7392.
55. M. Boudart, Turnover rates in heterogeneous catalysis, *Chem. Rev.*, 1995, **95**, 661–666.
56. R. Oviedo-Roa, J. M. Martínez-Magadan and F. Illas, Correlation between electronic properties and hydrodesulfurization activity of 4d-transition-metal sulfides, *J. Phys. Chem. B*, 2006, **110**, 7951–7966.
57. P. Geerlings, F. De Proft and W. Langenaeker, Conceptual Density Functional Theory, *Chem. Rev.*, 2003, **103**, 1793–1874.
58. W. Koch and M. C. Holthausen, *A Chemist's Guide to Density Functional Theory*, Wiley-VCH, Weinheim, 2000.
59. K. D. Sen and C. K. Jørgenson (ed.) *Electronegativity; Structure and Bonding*, 1987, **66**, Springer, Berlin.

CHAPTER 2
Organic Carbon

2.1 Introduction

The word "organic" in the chapter title refers to a typical department in the brains of chemists. It has little to do with organs and does not now mean "of biological origin". Organic chemistry is simply the chemistry of covalent carbon compounds, and the term might therefore be considered superfluous. On the other hand, the carbon chemicals made or modified by industry stem from "natural" coal, oil and gas, originating from photosynthesis. This fact is buried deep in the consciousness of chemists and justifies the use of the adjective "organic" for carbon chemistry alone. All biological processes involve carbon chemistry, and the nanochemistry carbon is also dominated by the element's unique covalent bonding and its stereochemistry.

Three fundamental non-covalent interactions dominate the chemistry of life: (1) hydrophobic interactions between hydrocarbon chains in membranes and in enzyme and receptor gaps; (2) hydrogen bonds, which regulate the behaviour of biopolymers (carbohydrates, proteins, DNA and RNA); (3) charge interactions, in particular between phosphate and sodium or potassium ions, which give rise to the ion currents in nerve and muscle networks. The same three binding effects also dominate the chemistry of artificial organic polymers in nanochemistry.

Carbon is the only chemical element that forms covalent chains with itself and equally stable bonds to hydrogen, nitrogen, oxygen, sulfur and halides (Figure 2.1). Millions of different covalent molecules with a carbon skeleton are stable in the terrestrial climate created by fluid water, oxygen biradicals, molecular oxygen and sunlight. The highest binding enthalpy in carbon single bonds is found in the CH bond with 99 kcal/mol; for the other elements found

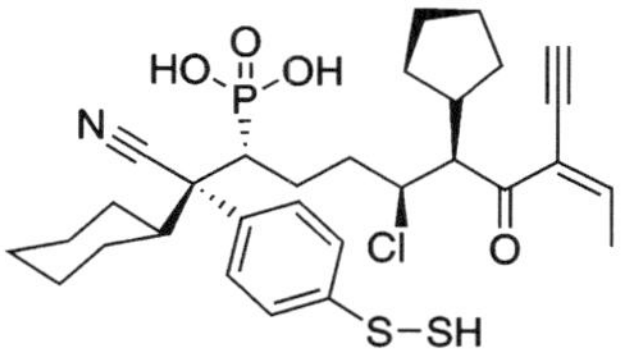

Organic Carbon 53

Figure 2.1 This molecule exemplifies the unique complexity of stable bonds to carbon and its stereochemistry. It contains covalent C–H, C–C, C–N, C–P, C–S single, double and triple bonds, rigid benzene and cyclohexane rings, a flexible (pseudorotation) cyclopentane ring, a *cis*-configured C=C double bond (CO > C≡C) and various chiral carbon atoms. The compound is extremely complex, but accessible by synthesis and stable in water, air and sunlight. The large variety of heterocycles is not exemplified here. Carbon is the only element that can take up so many elements in stable compounds. In the long run it will be digested by microorganisms living in water.

in biochemical compounds it is 80 ± 10 kcal/mol. Double and triple bonds range from 140 to 220 kcal/mol.

Hydrocarbons are stable to moderate heat and light, water and oxygen, but oxidize explosively when heated. C=O and C=N double bonds add and release water, alcohols, phosphate and amines, to form or degrade esters and amides. In nature and in industry, catalysts steer the oxidations and condensations. The best catalysts for all these reactions are again carbon compounds, *e.g.* enzymes, organic phosphate esters or the walls of nanosized pores in cell membranes. Carbon compounds are perfectly adapted to change chemically under the sunny and rainy conditions of the Earth's atmosphere. No other element performs in this way; only carbon is "alive".

Most biological carbon compounds are made by plants from carbon dioxide with the help of sunlight. Wood and all other plant material is eventually either recycled to carbon dioxide or converted first to phenolic earth humus, then to coal or hydrocarbons – the major resource for human survival in cold climates, for fast transport and for the synthetic chemical industry.

The carbon nanomaterials of central interest in 2005–2009 were elementary carbon (nanodiamonds and nanotubes); electron-conducting, particle-enwrapping and layered polymers; and fluorescent dyes.

2.2 Nanodiamonds, Graphene Sheets and Carbon Nanotubes

Colourless diamonds are gigantic C_n molecules made exclusively of covalently connected carbon tetrahedrons. The crystal structure of diamonds is that of zinc blende (p. 8), which is also found in metal nanocrystals. Diamonds are formed from other carbon compounds, in particular from graphite, at high temperatures and under high pressure, *e.g.* at 200 °C and 6000 bar (608 kpa).

About one-third of the carbon of the universe is probably diamond carbon, the rest graphite and carbides. Meteors, for example, contain diamond carbon nanocrystals of diameters between 0.2 and 10 nm, which are in general the most stable values at high temperatures. In smaller and larger carbon nanocrystals, graphite structures prevail.

The simplest synthesis of diamond nanocrystals is by detonation of graphite containing explosives. The chaos of an explosion is thus not only destructive but also creative – on a nanoscale. In a typical 3 nm detonation diamond $\sim$30% of the carbon atoms are on the surface, where only three covalent single bonds are possible. Such nanodiamonds therefore possess many "dangling bonds" on their surface. The stability and hardness of macroscopic diamonds disappear in nanodiamonds, which are reactive and form partial graphite or carboxylate coatings ("bucky diamonds"). Some of the unsaturation and unpredictable reactivity never disappears, but always remains dormant under the surface.[1]

Irradiation of diamond powder with high-energy electrons and annealing at 800 °C in the presence of air or silicon generate nitrogen or silicon vacancy defects. These provide colour centres in diamond nanocrystals and the most stable quantum emitters at room temperature. Charged or neutral nitrogen emits fluorescence in the visible (575 nm) or IR; silicon vacancies emit at 737 nm. Such single defect centres are observable and provide the most stable photoactive structures on a microscope tip, able to release single light quanta separately.[2,3b]

Chemical reactions are abundant on nanodiamond surfaces:

- Commercial 3.2 nm diamond nanocrystals as obtained by a detonation process were oxidized at their sp^2 and sp^3 defects by HNO_3/H_2SO_4 to give surface carboxyl groups.
- Air at 420 °C gave also surprisingly clean nanodiamonds with oxidized surfaces. These oxidized carbon nanoparticles are chemically more stable than any nanoparticles based on metal oxides and they are harmless, if they are kept small enough.[2–5]
- Single- and two-component molecular patterns were photochemically grafted on to hydrogen-terminated diamond surfaces using sub-band gap 254 nm UV light and 1-aminodec-1-ene or 1-dodecene as reagents.[6] N-doped nanodiamonds were coupled with gold nanospheres and served as single photon (575 or 700 nm) source.[3b]
- Polymers and biotin were attached covalently, and water- and chloroform-dispersible nanodiamonds were thus obtained.[7]
- Surface oxidized anionic 2–8 nm nanodiamonds were also assembled in a poly-L-lysine multilayer. The biocompatibility of the diamond film was then examined for cellular gene expression by a real-time polymerase chain reaction as well as by the knockdown of a lipopolysaccharide-mediated inflammation by macrophages facing the multilayer.[8]

Conventional diamond films contain micron-sized grains, leading to rough surfaces. Ultrasonication of nanocrystal slurries in water is a reliable method of depositing the diamond nanocrystals on silicon and other smooth substrates. When the density of diamond nuclei is not sufficient, large clusters are formed preferentially and produce numerous voids. Pre-coating with carbon or metallic layers then improves the formation of closed layers of diamond nanocrystals. Tungsten-coated silicon wafers (WSi) led, for example, to a fast accumulation of the carbon in the tungsten layer, because of the limited space for inward diffusion of carbon. Once the tungsten carbide layer completely covered the tungsten full coverage with diamond nuclei resulted and uniform, well-adherent films were formed.[9]

Nanowires made of metals and metal chalcogenides lack chemical stability in electrolyte solutions and are therefore difficult to integrate into complex systems. Diamond has an outstanding chemical stability, is ultrahard and gives a low background current as well as a wide working window. The dangling bonds on the surface can be hydrogenated or functionalized by oxidation. Transparent boron-doped diamond thin film on undoped silicon was produced by microwave-assisted chemical vapour deposition of a 100 nm diameter diamond powder. The thin-film electrode possessed a transparency of 40–60% in the mid and far IR regions, was operated in the four-wire resistance measurement mode (p. 129) and exhibited good electrochemical activity for a number of aqueous and non-aqueous redox systems. IR spectra can thus be measured at any potential.[10]

Vertically aligned diamond nanowires with optimized geometrical dimensions of 10 nm length and an average separation of 11 nm were obtained from a film of diamond carbon nanocrystals by reactive ion etching with O_2 (97%)/CF_4 (3%) gas. The diamond etching rate was ~ 1.0 nm s^{-1} and self-aligned nanocrystals caused the formation of vertically aligned diamond nanowires. Longer etching times led to ill-defined mixtures of nanowires. The primary diamond paste on macroscopic diamond plates was made from a pseudostable suspension of 8–10 nm diamond nanocrystals in water by ultrasonication (200 W, 20 kHz, 12 h). The diamond plate was immersed into this suspension and sonicated (100 W, 10 min) to seed diamond nanocrystals. The desired dense layer of diamond wires consisting of nanocrystals formed within 10 min and had an oxidized surface, which produced an anodic and a cathodic peak 868 mV apart. This enormous potential separation was mainly a result of a low boron doping (7×10^{19} cm^{-3}) of the nanodiamonds. After phenyl attachment, the redox peak currents decreased by 22% and shifted to an 1160 mV peak separation.[11]

Ionic carbides contain either sp^3 carbon (*e.g.* Al_4C_3), acetylenic sp carbon (*e.g.* CaC_2) or allenic sp^2 carbon (*e.g.* Mg_2C_3). They are not useful as materials, because they decompose in both water and air. The most important carbides are unreactive non-stoichiometric interstitial compounds of transition metals, in which carbon atoms occupy tetrahedral gaps in the cubic or octahedral metal lattices. Many of these carbides are exceedingly hard. Steel, for example, is an alloy of <2.6% of carbon in iron. This alloy is in fact a carbon carbide named cementite (Fe_3C), which forms the border of domains in solid iron, but crystallizes at higher concentrations of carbon and then destroys the malleability of

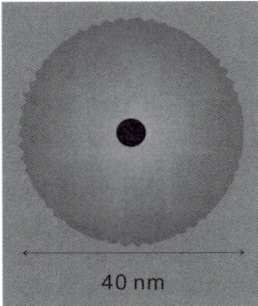

Figure 2.2 Schematic drawing of an electron micrograph of aggregated soot particles with a diameter of 40 nm as produced in incomplete combustion. The black dot indicates a primary nanocrystal. Heat and flame shape of dictate the size of the loose graphene net, which is rather uniform in a given flame.

steel. Along with boron carbide (B_4C; p. 180) or silicon carbide (SiC), artificial nanodiamonds also form the best commercial abrasive material.

The standard structure of elementary carbon on Earth consists of planar benzene layers in hexagonal arrangements, which appear in crystalline graphite as well as in isolated graphene monolayers (see Figure 2.3), and in the disrupted and rolled-up particles of carbon black or soot (Figure 2.2) with typical diameters of 20–100 nm and loosely packed condensation nuclei in the centre. The typical density of soot is 1.85 g cm^{-3}, as compared to 2.26 g cm^{-3} for graphite. The outer surface area of soot particles is around 100 m^2 g^{-1}, but heating may expand it to 1000 m^2 g^{-1}. The particles are soft and black, like pencil streaks.

Graphene sheets are two-dimensional crystals one atomic layer thick, or sp^2-hybridized carbon atom layers, which consist exclusively of hexagons. The name "graphene" comes from graphite and -ene, indicating irregular patterns of double bonds. The planarity of the graphene sheets dictates rigid benzene rings as units, the benzene rings causing the planarity of graphene. In graphite these sheets are planar and quite regular, but if isolated graphene sheets are peeled off they demonstrate an irresistible tendency towards irregular stacking and rippling. This is predicted by theory, which does not allow perfect crystals in two-dimensional space. Only bending and stretching movements can in principle stabilize atomically thin membranes through their deformation in the third dimension. Isolated graphite stacks are not flexible enough to recover from re-stacking to re-form graphite.[12] Only chemical oxidation leads to stable sheets (p. 84).

Single graphene layers become stable if they are rolled up into nanotubes, where the reactive edges are reduced in length or completely disappear. Preparation of these nanotubes involves chemical vapour deposition of carbon, heating of a hydrocarbon gas in a furnace tube reactor, or laser ablation from graphite. All these procedures lead to colloidal nanotubes rather than sheets. Suspensions of nanotubes are prepared by sonication of raw soot in water with

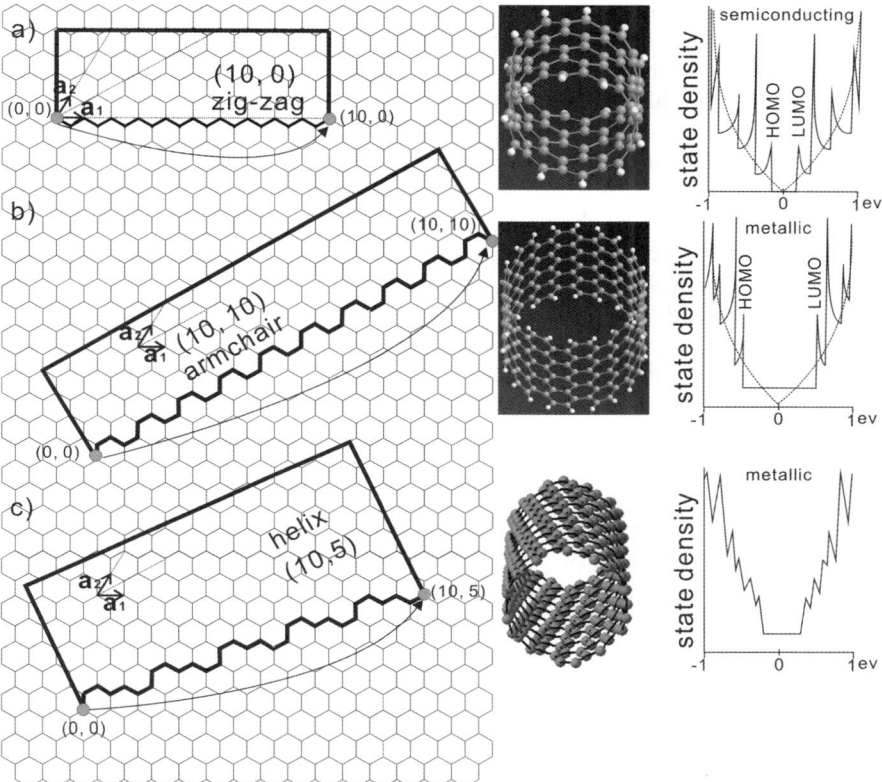

Figure 2.3 *Left:* (a) Honeycomb structure of a graphene sheet with the zigzag, (b) with the armchair and (c) with the helical line at the bottom. The lattice vectors a_1 and a_2 are indicated in the left hand corner and the sheets for the formation of 10,10 and 10,0 and 10,5 tubes are framed. Single-walled carbon nanotubes are then formed by folding the sheet along the lattice vectors. The molecular motif of the outer edge of the 10,10 armchair tubule, for example, is a planar 20-atom carbon ring with 10 C–C bonds. This ring repeats again and again along a_1 or a_2 and builds up the entire tubule. The diameter of the nanotube is ~1.2 nm. *Centre:* Contour of the a_1 and a_2 valence bands and of armchair carbon nanotubes without irregularities in the vicinity of the Fermi level eF. They cross at a Fermi level eF = 0.0 e; the 10,10 nanotube is metallic. *Right:* Electronic density states for metallic and semiconducting carbon nanotubes. The states close to the Fermi level correspond to the highest occupied and lowest unoccupied orbitals. In metallic nanotubes there is no energy gap between them. The difference in conductivity between stereoisomers is significant only in narrow (diameter <3 nm) single-wall nanotubes.[1]

the addition of dodecyl sulfate, other amphiphiles or block polymers, which wrap micellar coatings around the nanotubes.[13–15]

Formally, single-walled carbon nanotubes are formed by folding the graphene sheet along the lattice vectors a_1 and a_2, connecting the *meta* positions of benzene at a position which is defined as 0,0 on any carbon atom of a graphene sheet. Any other carbon atom then has two coordinates: m in the a_1 direction and n in the a_2 direction. The diagonal line connecting the *para* positions gives the $m=n$ line. Folding of the graphene sheet, connecting (0,0) with (m,n), e.g. (10,10), and rolling the sheet up to form a cylinder of uniform diameter completes the creation of the carbon nanotube. The wall always runs strictly parallel to an axis; the carbon nanotubes are made of undisturbed planar sheets, which are just rolled up around a central axis. The benzene units of tubes with zigzag or armchair edges are all oriented along this axis; in helical tubes the benzene rings twist around this axis (Figure 2.3).[1]

The indices n and m not only describe the connection point to (0,0) in the first step of tube formation, but also fix the circumference C of a nanotube: $C = a(n^2 + nm + m^2)^{1/2}$. The constant a is equal to the length of the a_1 or a_2 vectors, the distance between the *meta* positions in benzene or 0.246 nm. The diameter of a carbon nanotube is C/π. A 10,10 nanotube therefore has a circumference of $0.246 \times 300^{1/2}$ nm = 4.26 nm and a diameter of 4.26/3.14 = 1.36 nm. The circumference of a 20,20 nanotube is $0.246 \times 1200^{1/2} = 0.246 \times 34.6 = 8.51$ nm and its diameter is 2.71 nm; a 10,0 nanotube has $C = 2.46$ nm and $d = 0.78$ nm.

The special n,m vectors of graphene nanotubes thus settle the diameter, the conformation of the edges and the stereochemistry of the benzene pattern. The geometrical structure of the rolled-up polymer units is defined by their chiral vector $na_1 + ma_2$. In order to avoid double counting of one type of tube, n must always be larger than m. Zigzag tubes are characterized by $m=0$, armchair tubes by $m=n$; different values of m and n indicate chiral helices. Rolling up along a_1 or the bisector gives relatively short zigzag or armchair unit cells; rolling up at an arbitrary angle θ produces chiral tubular units of any length. Graphene tubes thus provide a new type of curvature stereochemistry, which, as we shall see, determines colour, luminescence and electron conductivity.

The many conjugated π electrons in the tube might produce the black colour of graphite or have different colours depending on their diameter. The latter assumption proved to be correct. Separation of carbon nanotubes by density gradient ultracentrifugation in aqueous sodium dodecyl sulfate–sodium cholate mixtures drove the nanotubes to areas where the tube density matched the density of the medium, and films of different colours were seen. This technique was applied to commercial single-wall nanotubes with diameters between 0.7 and 1.7 nm and of different synthetic origins: from high-pressure carbon monoxide conversion, laser ablation or electric arc discharge. The ultracentrifuge achieved high-resolution sorting. Carbon nanotubes with a diameter of ~1.0 nm showed no broad absorption bands beyond 700 nm, but gave visible bands with peaks between 450 and 680 nm. Those with diameters of 1.4–1.6 nm absorbed around 550 nm (Figure 2.4), where the human eye is most sensitive and where visible solar radiation is most intense.[14]

Single-wall nanotubes whose chiral indices (n,m) are related by $(n-m)q(3) = \pm 1$ have a direct band gap and produce strong luminescence spectra. The 981 nm absorption band ("$E11$ exciton transition") of these non-metallic single-wall nanotubes comes to 40% from the narrowest isolated semiconducting

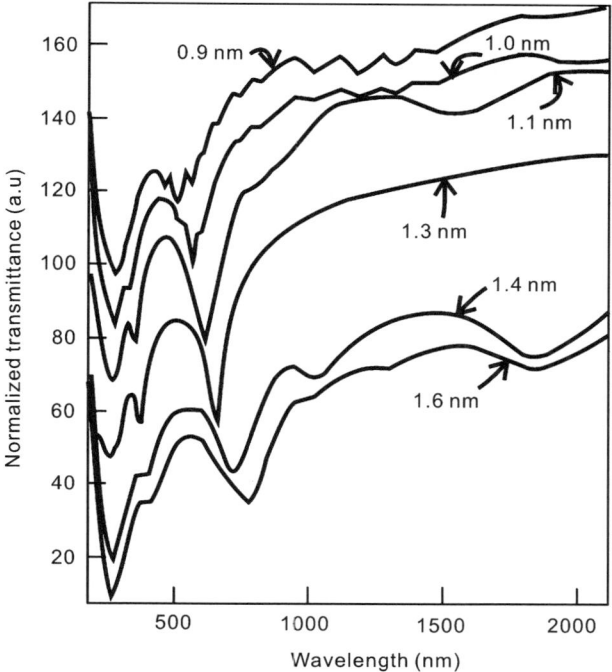

Figure 2.4 The transmittance of sorted metallic single-wall carbon nanotube films as obtained from laser-ablation material, normalized to the π-plasmon at ~270 nm. The films were produced from carbon nanotubes with the indicated dominant diameters. Their long-wavelength metallic transitions shifted from 509 to 778 nm. © American Chemical Society.[14]

(6,5) tube, which also produces the strongest photoluminescence upon excitation of the 571 nm absorption band (Figure 2.5). The measured quantum yield of the isolated fraction is 1.2% for this nanotube, but in aggregates it drops by a factor of ~1000.[14]

Cyclic voltammetry of multiwalled nanotubes enhanced the charging current after 1000 cycles and the effective electrochemical area in the tubules increased. Three oxidations occurred at 0.40, 0.51 and 1.7 V, followed by two reductions at –0.71 and –1.68 V. The redox sequences produced defects in the sidewalls of the tubes, similar to the surface functionalitiy of porous charcoal, as indicated by the appearance of an intense blue luminescence. An aromatic amine was chosen as a "solvent" for the tubule cluster and it introduced some nitrogen, analogously to the h-BN discussed on p. 202 ff. A nitrogen to carbon electron transfer at emissive trap sites and a radiative recombination of excitons from carbon nanotubes of different sizes are the probable sources of the emission. Black carbon begins to glow at room temperature.[15]

The most fundamental electronic transport property is electrical conductivity. We would expect carbon nanotubes to exhibit this property if the

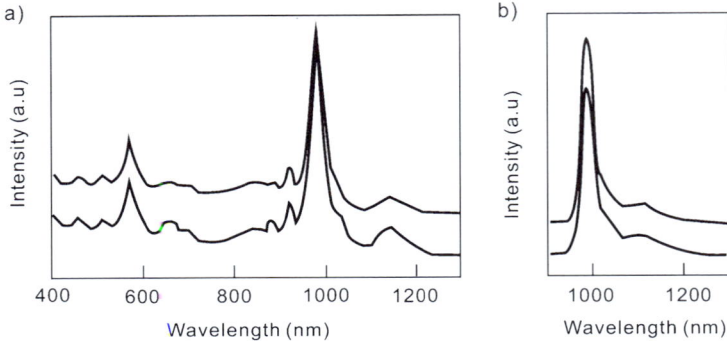

Figure 2.5 (a) Absorption and (b) photoluminescence (excitation at 571 nm) spectra of the semiconducting thinnest known [5,6] nanotube. © American Chemical Society.[15]

energy of the π electrons approaches the Fermi level. Carbon nanotubes are considered by theory as one-dimensional (1D) structures along a fibre axis, because the single atom layer is ultrathin. A two-dimensional (2D) energy dispersion of the π bands should occur, however, if the chromophore is twisted and becomes a helix around the axis. Density of states (DOS) calculations for the 1D wires indicated that $2n + m = 3q$ for single-wall carbon nanotubes, with q as an integral small number 1, 2, 3, 4, ..., yielded an overlap of the valence and conducting bands. The electrons of the high-energy valence bands (= the highest occupied molecular orbitals) should be metallic, and the carbon nanotubes in general become zero-gap semiconductors. Experimentally it was then found that armchair structures and some zigzag tubes, where the benzene rings lie parallel or perpendicular to the tube axis, do indeed show metallic electron conduction (cf. coronene tubes, where the rings are perpendicular to the axis). The chiral $m \neq n$ tubes are semiconductors; multiwall nanotubes are usually less conductive and difficult to differentiate. In single-wall nanotube populations with diameters between 0.7 and 2.7 nm about one-third is metallic and two-thirds are semiconducting. In summary, the carbon nanotubes behave like graphite-type wires, which is not really a surprise, and the conductivity depends on the stereochemistry of the tubes, which is of interest for the synthesis and and separation of carbon nanotubes.[16]

The surprising feature of *ballistic electron transport* in carbon nanotubes came from a deceptively simple experiment. "Ballistic" means that the resistance is independent of the length of the wire or tube; the electron jumps over the carbon nanotube up to micrometre distances at voltages < 100 mV, at room temperature. The most straightforward explanation for the effect is that the "wavelength" of the two conducting bands of the outermost conducting layer dominate electron transport and allow for no scattering.[16–17]

The carbon nanotubes were produced in a carbon arc in a 1 mm gap between a 7 mm diameter graphite anode and a 5 cm diameter graphite cathode. Multiwall carbon nanotubes 5–25 nm thick and 1–10 μm long, together with

Organic Carbon

amorphous graphite flakes, were found on the anode. A fibre was separated from the deposit and attached to the modified probe of a scanning probe microscope. The fibre was then lowered into contact with a liquid metal surface, typically mercury or molten gallium under silicon oil. A voltage of 100 mV was applied to the tip and the current measured using an oscilloscope, which also recorded the position of the nanotube with respect to the liquid metal surface. Contact between the carbon nanotube and the liquid metal surface resulted in a jump in the conductance. The successive steps in a trace result from several tubes coming successively into contact with the mercury. Several thousands of cycles were run, the tip always being raised and lowered by 1–10 µm. Initially the steps in $G(x)$ were poorly defined because of the rough fibre surface. After hundreds of cycles the fibre was displaced to a fresh area of the metal and a stable pattern of steps was finally established: the electron microscopic pictures showed smooth fibre surfaces. The mercury phase had washed away the graphite nanoparticles[16] (Figure 2.6).

The resistance of the carbon nanotubes was measured as a function of the contact length. It should be low for wires which are in contact with mercury over their whole length, and very high if the carbon nanotube only touches the mercury and stands alone between the microscope and the metal. This was not found. The resistance of the carbon nanotube dropped upon contact and did not change within the mercury phase. The conductance of all wires was approximately the same. It was quantized with the quantum of conductance $G_0 = e^2/h = 1/13 \text{ k}\Omega^{-1}$, not only independent of the length of

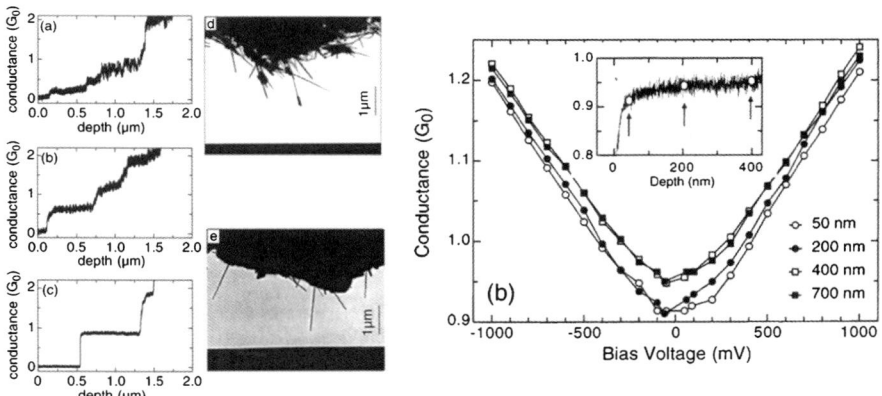

Figure 2.6 *Left:* (d,e) TEM image of a multiwalled carbon nanotube fibre tip opposing a mercury surface, the lowering of one, two and three tips into the liquid metal and measuring the conductance as a function of the position. In (a) and (b) the conductance per tip is about 0.6 G_o, in (c) the cleaned carbon nanotubes yield >0.9 G_o. *Right:* A full cycle–conductance trace, where the fibre is first lowered to the mercury and subsequently withdrawn.[16] *Inset:* Example of cone-shaped meniscus attached to the tip of the nanotube which occurs when the nanotube is pulled out of the (non-wetting) liquid just before contact is broken. © American Chemical Society.

carbon nanotube–mercury contact but also independent of the diameter of the carbon nanotube. The conductance exactly doubled when two wires instead of one touched the mercury. Furthermore, the carbon nanotubes behave as one-dimensional conductors, electronic transport occurs only on the outer layer, and current densities are greater than in any other material at room temperature.

The important message is that the intrinsic resistance of carbon nanotubes is low, of the order of a few hundred $\Omega/\mu m$ if the carbon layers are intact (see Figure 2.6 for diffusive electron transport). The scattering in the metallic sub-bands of metallic carbon nanotubes is much smaller than in the doped or thermally populated semiconducting sub-bands of the same nanotubes, and most electrons traverse the nanotube without scattering. Nanotubes shorter than 200 μm are ballistic conductors at room temperature over their entire length. Surfactants in connection with water greatly increase the resistance of carbon nanotubes by doping. Currents greater than 1 mA become diffusive and destroy the tubes.

Theory indicates that the thermal conductivity of defect-free carbon nanotubes is also ballistic. Distances of up to 10 μm are overcome in a single jump at room temperature, and heat scattering vanishes for its long-wavelength phonons.[17]

Molecular dynamics simulations indicate that water molecules confined in infinitely long 1.1 nm carbon nanotubes should diffuse through a fast ballistic motion mechanism for up to 500 ps at room temperature. Experiments with 7 nm carbon nanotubes did indeed give extremely high flow velocities. The flow of water through a freshly fabricated solid polystyrene film with aligned and crossing 7 nm multiwalled carbon nanotubes ($5 \times 10^{10} cm^{-2}$), was measured at ~1 atm (1 MPa) applied pressure. Macroscopic defects of the membrane were excluded. The flow rates were four to five orders of magnitude faster than conventional fluid flow would predict through pores of 7 nm diameter. Contrary to predictions based on hydrodynamics, the flow rate did not decrease with increased viscosity; more hydrophilic fluids flowed faster. The flow of hydrogen-bonded fluids decreased after a few minutes, but this does not happen with aqueous solutions of potassium chloride. The wall of the carbon nanotube is a frictionless surface with remarkably long "slip lengths" of 3–70 μm. The slip length in a resting tube is defined as the distance inside the wall at which the extrapolated fluid velocity would be zero. Since the observed slip lengths are much longer than the pore radius of 3.5 nm, the wall must be practically frictionless. The slip length decreased as solvents became more hydrophobic, which indicates a stronger interaction with the wall of the nanotube.[18–20]

The electronic ground state of carbon nanotubes is fourfold degenerate – at each energy level there are two spin states (spin-up and spin-down) and two orbital states (clockwise and anticlockwise) with respect to an applied magnetic field parallel to the tube's axis. Single conducting electrons were confined in a dot of the nanotubes by an electric "Coulomb blockade" and these four states were indeed shown to separate in a magnetic field and to form a distinct pair in

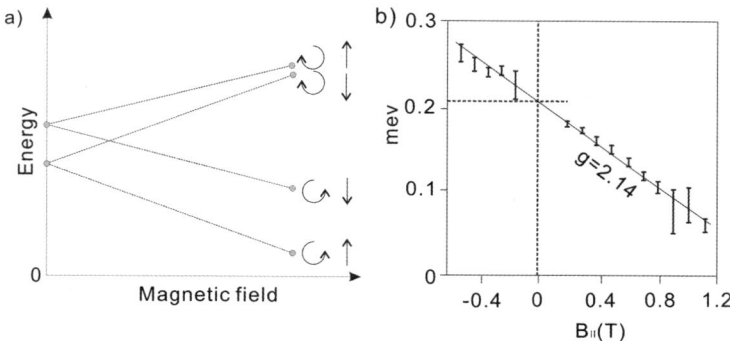

Figure 2.7 (a) Scheme of an electron revolving around the nanotube with a spin parallel to the nanotube axis circumference. The carbon pz orbitals are perpendicular to the surface and rotate around the spin. (b) Both the usual energy splitting by the Zeeman effect with $g = 2.14$ and the zero-field spin-orbit splitting of $\Delta = 0.21$ meV occur in EPR spectra of thin carbon nanotube radicals.[21,22]

the absence of the magnetic field (Figure 2.7). This suggests that the spin and orbital movement of electrons in carbon 1D nanotubes are about as strongly coupled as in 2D graphene sheets.[21,22]

Carbon nanotubes are insoluble in water or solvents, but they spontaneously exfoliate on dilution of dispersions in N-methyl poly(pyrrolidone). A dynamic equilibrium is established there, characterized by significant populations of individual nanotubes and small bundles dispersed in the polymer. This occurs because the surface energy of the pyrrolidone polymer (p. 113 ff) matches that of the graphitic surface. A zero enthalpy of mixing occurs and leads to long-lived dispersions, with a measurable negative enthalpy of mixing. Such fine dispersions should be useful in the manipulation of carbon nanotubes within nanoscopic architectures.

The application of as-produced carbon nanotubes is limited by their unavoidable polydispersity, resulting in mixed chemical, optical and electric behaviour. In analysed mixtures with a relatively narrow distribution of 1–2 nm wide single-wall nanotubes, for example, only about one-third was metallic. The rest were semiconducting, with band gaps that varied inversely with nanotube diameter and broad absorption bands ranging from 400 to 1600 nm.

Narrow-diameter (1–2 nm) single-walled carbon nanotubes are highly curved and chemically more active than planar graphite or wide nanotubes because of strain and loss of hybridization. Large-diameter nanotubes are chemically quite inert if no structural defects are present. It is also difficult to incorporate large-diameter, multi-walled nanotubes into polymeric, metallic or ceramic matrices because of their weak adhesion and the slow formation of sp^3-hybridized carbon atoms between the tube walls and the matrix. The tubes slip within the matrix. The best chemical connection was found to result from fluorination of the nanotubes and defluorination *via* heating to $>600\,°C$ *in vacuo*. Modelling suggested that as a result of this procedure the graphene structure is torn and becomes locally very flexible.[23,24]

Single and multiwall carbon nanotubes also separate at an oil–water interface. The highly curved single-wall tubes with small surface areas have less interaction with the aqueous phase and increased interaction with the oil phase, and therefore accumulate at the interface.[125] Semiconducting single-wall nanotubes are less ordered along the axis and less polar than metallic ones. They adsorb flat organic dyes on their hydrophobic surface, whereas metallic tubes do not. Long, solubilizing alkyl substituents on the adsorbed dye, *e.g.* *meso*-tetraphenolether-porphyrins, then selectively dissolve the semiconducting tubes in organic media, so the metallic nanotubes can be filtered off (Figure 2.8). A chiral biporphyrin even separated chiral single-wall nanotubes into right- and left-handed populations. The selection of specific nanotube diameters and chirality has been improved by wrapping the nanotube mixture with polyfluorene "hairy rods", where the aromatic rods align along nanotube helices and the oligomethylene chains takes care of the solubilization of the assembly. Sapphyrins without "hairy" substituents, but having a flexible chromophore, give similar effects.[26a,b]

Another useful phenomenon in carbon nanotubes is the easy field emission of electrons. Multiwall carbon nanotubes are mechanically more stable than single-wall tubes and are therefore more suitable as electron-conducting tips for AFM. The tips must be as sharp as possible for good resolution of the micrographs, and sharpening was achieved by electrically driven vaporization (Figure 2.9).

This should, however, be impossible if ballistic transport of electrons uses only carbon atoms which are more than 1000 nm or 8000 carbon atoms apart. Ultrasonication, surfactant stabilization and thermal annealing at 450 °C to burn out carbonaceous particles had presumably partly destroyed the surface carbon layers. The resistance of the carbon nanotubes rose from $10\,\Omega/\mu m$ for

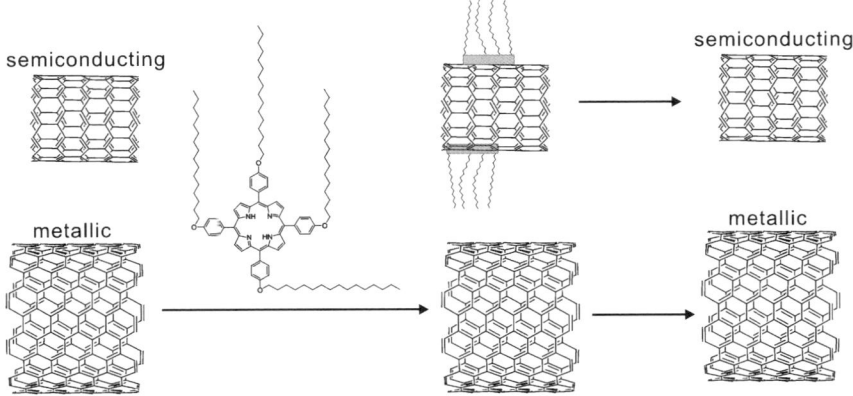

Figure 2.8 Hairy porphyrins and other flat chromophores bind exclusively to semiconducting, less ordered nanotubes. © American Chemical Society.[26]

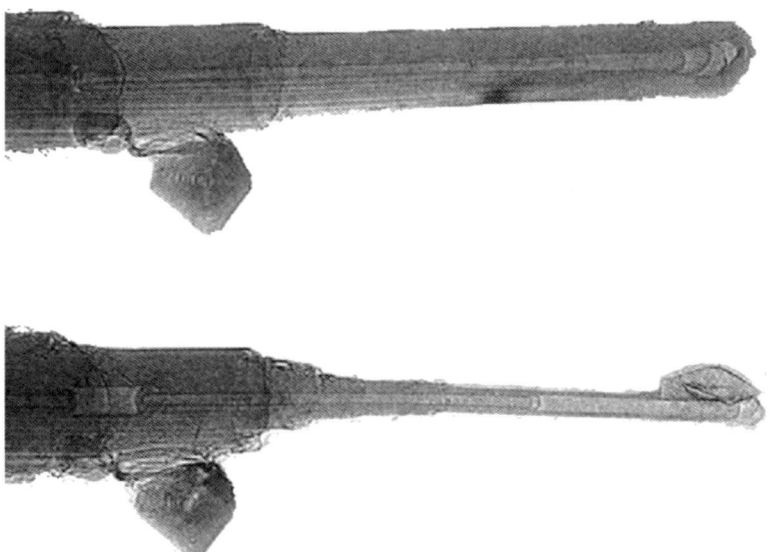

Figure 2.9 Self-sharpening of a 37-wall nanotube to a 2-wall tip with a diameter of 2.1 nm.[27]

ballistic electron transport to 100 kΩ/μm for diffusive transport and the outer walls were evaporated first near the end of the nanotube, leaving the core walls intact. The peeling and sharpening process continued until the innermost tube with the smallest diameter protruded, often with a tip almost as thin as a single-wall nanotube. The tip may, for example, remain as a three-walled, electrically conducting nanotube with a radius of 2.5 nm. The observation that the outer layers break down first indicates that the diffuse electrical current in multiwall nanotubes also flows mainly in the outer carbon layers of the tube, which agrees with the finding that thicker single-wall tubes conduct better. The effect may be the consequence of charge repulsion in an electron current: the less space there is, the more difficult it becomes to load the tubule with charges.[27]

Carbon nanotubes with an internal diameter of 1.1 nm have been filled with continuous rows of single C_{60}-fullerenes, which were 0.2 nm away from the walls (Figure 2.10). The outside π-orbital of the fullerene is presumably much more extended than the inside counterpart of the nanotube. In 2.2 nm tubules, zigzag patterns of the entrapped fullerene molecules were observable by transmission electron microscopy. Conducting nanotubes filled with dyes and electrons should be useful as antennae for sunlight, but probably contain too many molecules in a row to be applicable for photosynthetic charge-separation processes.[28]

To fill the carbon nanotubes with fullerenes, they were first heated in dry air at 500 °C for 30 min to open the tube ends. The encapsulation of C_{60} into the open-ended single-wall nanotubes was achieved through a gas-phase reaction. Aqueous micellar solutions of C_{60}-filled carbon nanotubes ("nanopeapods")

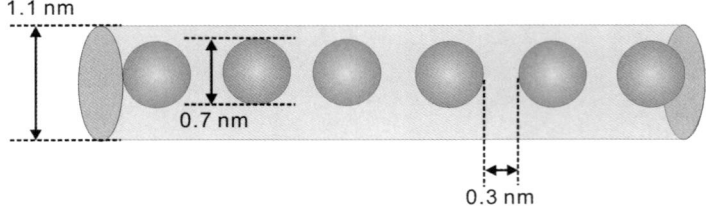

Figure 2.10 Model of the fullerene spheres in a 1.1 nm carbon nanotube.[28]

for photoluminescence measurements were prepared with dodecylbenzene sulfonate and sodium cholate as surfactants.

A 2D photoluminescence contour plot of empty 1.2–1.4 nm carbon nanotubes was assigned to individual nanotube diameters by using empirical relations. All peak positions were slightly and uniformly red-shifted after the tube opening, because water entered the interior space and caused a reduction of the electron–electron repulsions and the exciton binding energy within the walls of the tubes. The photoluminescence of the small tubes, diameter <1.25 nm, was assigned to (11,6), (15,1), (10,8), (12,5), and (11,7) tubes in the spectra, with and without added fullerene spheres.[29] These tubes were too small to entrap C_{60}. Photoluminescence of nanotubes with diameters >1.25 nm, such as (13,5), (12,7), and (10,9), appeared at different positions from those of the unfilled single-wall nanotubes. The strongest photoluminescence peak caused by the (13,5) tube in Figure 2.11A presumably moved to the new strongest peak in Figure 2.11B, as the filling never significantly affected the quantum efficiency. Likewise, the second and third strongest peaks in Figure 2.11B can be assigned to the (12,7) and (10,9) tubes.[29]

The benzene surface of single- and multiwall carbon nanotubes is, of course, reactive towards strong electrophiles and oxidants, which at first tend to produce small accumulations of defects on the sidewalls and then add to the network, which they disrupt locally. "Substitution" is only possible at the dangling bonds on the edges. Semiconducting carbon nanotubes become doped with electron holes in the presence of molecular oxygen or other oxidants, and then act as p-type field effect transistors between two electrodes, usually made of titanium. Adsorption of a reducing agent, *e.g.* a polyamine such as polyethyleneimine, on the other hand, donated ~1 electron to every 1000 carbon atoms in the nanotube and converted its p-type semiconductor character to that of an n-type field effect transistor (Figure 2.12). The nitrated or oxidized carbon nanotubes do not take up charges from an anode; the aminated carbon nanotubes do not accept electrons from a cathode.[30,31]

An orthogonal voltage on the carbon nanotubes will polarize them and transfer them to an On state, and the conductance along the fibre axis will rise by ~10%. An electric On state should also be more polar and therefore more reactive in electrophilic substitutions. The metallic tubules have higher electron densities near the Fermi level and should also be more reactive. Both

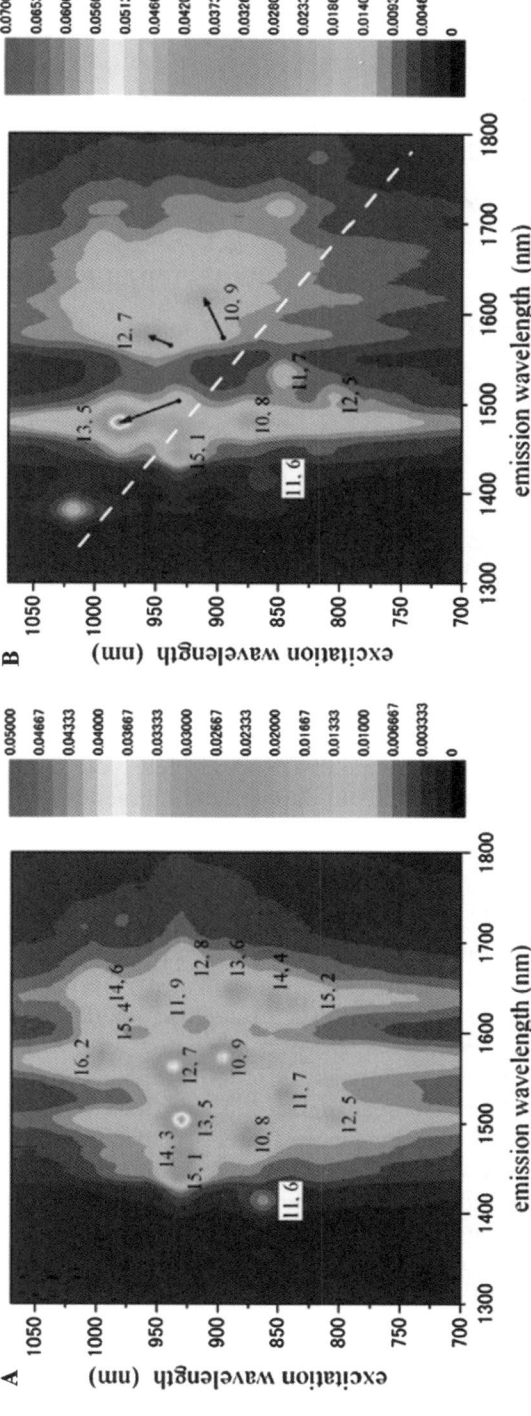

Figure 2.11 2D photoluminescence contour plots of (A) single-wall carbon nanotubes and (B) the corresponding C60 nanopeapods with entrapped fullerene 60. The arrows indicate the emission shifts upon fullerene entrapments for (13,5), (12,7), and (10,9) nanotubes with diameters >1.25 nm. Several signals for the small tubes did not shift, but some of the weaker signals are not detectable. The fullerene does not emit any photoluminescence signal. © American Chemical Society.[29]

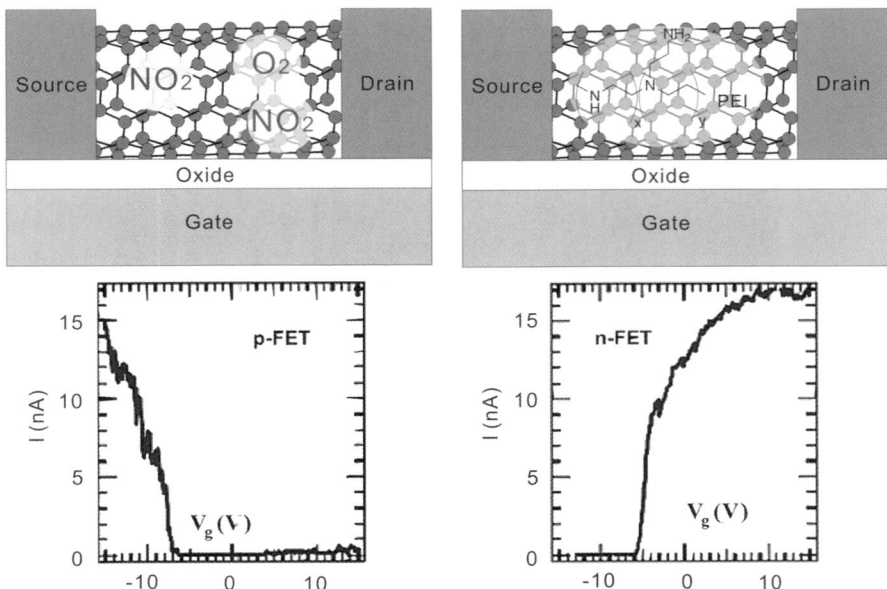

Figure 2.12 NO$_2$ and O$_2$ bind to an nanotube and withdraw electrons, NH$_3$ and amines donate electrons. The carbon nanotubes react with extreme sensitivity, switching from p-type to n-type. © American Chemical Society.[31]

predictions were verified experimentally by the yields of diazotation (see below): it is higher in the nanotube bundles containing metallic species and in the On state. The thermal back reaction at 500 °C is also more efficient in the On state (Figure 2.13).[32]

Small-diameter single-wall nanotubes are particularly reactive towards radical attack because of large curvature strain. A most efficient reaction sequence is the reductive nitrogen elimination of azides (made from aniline and nitrite) to form aryl radicals followed by "click chemistry" on alkynes, namely the Cu(I)-catalysed [2 + 2] cycloaddition of another azide. A well-defined polystyrene azide was prepared for this purpose by an atom transfer radical polymerization, terminal bromination and reaction with NaN$_3$ in DMF (Figure 2.14).[33,34] The first and most crucial indication that the "click" coupling of the polymer had created something useful was the fact that the nanotubes became soluble in organic solvents. In fact, the new conjugate gave clear, dark brown solutions in tetrahydrofuran that remained stable for at least 3 weeks.[34] Raman spectra showed the characteristic radial "breathing" band at 250 cm^{-1}. "Tangential" (1590 cm^{-1}) bands of the polymer chain and a "disorder" band at 1290 cm^{-1} indicated the sp^3 hybridized carbons within the graphene framework of the nanotube. This band increased dramatically, when more of the aniline was added, again indicating that the amine added readily to the electron-rich n-doped tubules.

Organic Carbon 69

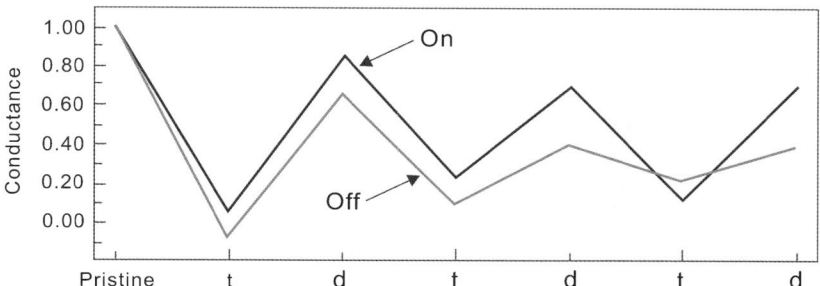

Figure 2.13 Conductance variations of single-wall nanotube transistors during diazotation/elimination cycles in the On and in the Off states. A voltage of 1 V was applied between the source and the drain. The On state gives more current and more elimination in the defunctionalization region. © American Chemical Society.[32]

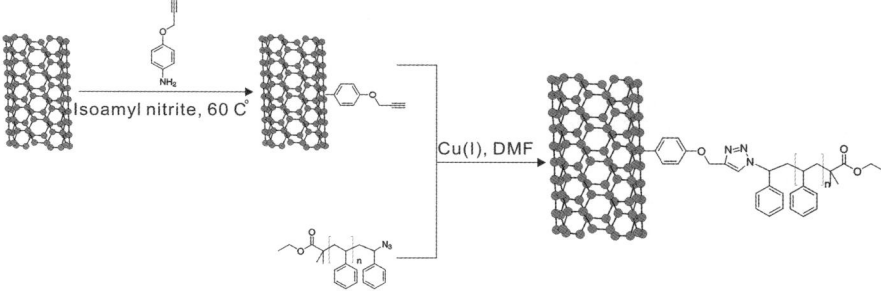

Figure 2.14 Typical synthesis of a polymer-wrapped soluble carbon nanotube by click reactions. © American Chemical Society.[33,34]

The main remaining problem with ultrathin carbon nanotubes is the poor yield of all known preparation methods. An elegant attempt to produce uniform nanotubes started with a cobalt-filled protein capsule as radius-determining matrix. Biochemistry took care of uniformity. Thermal chemical vapour deposition with methane was performed at 500 Torr (67 kPa) for 5 min to grow single-wall carbon nanotubes between two silicon pillars in a correspondingly structured landscape. This surface was covered with a large number of apoferritin grains (p. 273 ff) filled with 4.7 ± 0.7 nm Co(II)nanoparticles. Cobalt was chosen instead of the original Fe(II) because it was held more tightly within the protein cage and had a lower diffusion coefficient. The temperatures chosen were 900 and 1000 °C. Both the methane and the apoferritin quickly pyrolysed. The resulting carbon vapour and the Co_{40} units combined to form carbon hanging uniformly between the silicon pillars, which offered the best attachment points in the environment. The average tube diameter decreased from 1.53 ± 0.37 to 1.17 ± 0.27 nm by lowering the growth temperature from 1000 to 900 °C, and the electron micrographs were magnificent (Figure 2.15).[35]

Figure 2.15 Electron micrographs of carbon nanotubes suspended between silicon pillars grown at 1000 °C and a histogram showing the diameter distribution of the same tubes but grown on flat surface at 900 and 1000 °C. © American Chemical Society.[35]

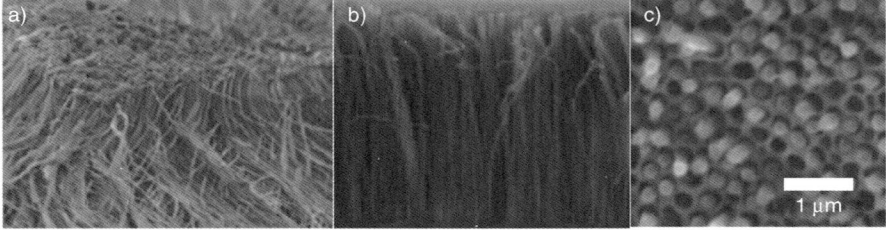

Figure 2.16 Transmission electron microscopic side views of (a) the entangled carbon fibres after dissolution of the alumina membrane and air drying without CO_2 and (b) after washing with water and ethanol and standing in liquid CO_2 in a steel cell, at 60 °C and 150 bar for 1 h, decreasing the pressure and opening the steel cell at 60 °C. (c) Top view of (b). © American Chemical Society.[37]

Wide nanofibres and carbon nanotubes of uniform diameters can be made in aluminium oxide wells with a silicon bottom (p. 182 ff), which act as matrices. Such aluminium foils with thousands of well-defined alumina-coated pores are easily accessible, but they are not really useful on the nanoscale.

A well-aligned array of free-standing carbon nanorods on silicon wafers was prepared by a calcination process of a soluble, low-molecular-weight bakelite (phenol + formaldehyde) mixed with a commercial poly(oxyethylene)–poly(oxypropylene) triblock copolymer (MW = 12 600). The mixed polymer sol infiltrated the wells in an aluminium oxide membrane and aged there to become a gel, which firmly adhered both to the alumina-coated holes and to the silicon wafer carrier at the bottom. Heating to 600 °C in a nitrogen atmosphere for 3 h decomposed the polymers and carbonized the metal oxide walls. The alumina membranes were then removed with 6 M HCl, generating 200 nm wide carbon nanofibres with an average length of 60 μm. Such a release of 1D nanostructures is usually followed by their entanglement. This was prevented by treatment with supercritical carbon dioxide, which filled up the liquid–gas interface during the drying process. Hexagonal patterns of 5 nm pores at the places where the alumina had been before, at an average distance of 12 nm, became visible on the surface (Figure 2.16).[36,37]

Organic Carbon 71

The commercial "structure-directing agent" Pluronic F127 is a symmetrical triblock copolymer, poly(ethyleneoxide$_{106}$propyleneoxide$_{70}$ethyleneoxide),[106] MW 12 600. In alumina pores it produced a few 100 nm thick carbon rods with a complex internal stereochemistry and put them in rows on a computer chip.

Alumina has a high surface energy, fibre precursor sols a much lower one. The sols spread out evenly over the alumina pore surface, forming a meso-ring upon cooling driven by the reduction in surface energy. A confined self-assembly process towards the pore centre follows. Concentric rings with decreasing radii are formed during the ageing periods with the formation of carbon fibres upon heating. Six or seven such rings build up the carbon fibre. Circular mesochannels are formed, concentrically coiled around the axis of the carbon fibre. The confining effect of the alumina walls then decreases; no more rings are formed parallel to the alumina wall, but columnar-oriented mesochannels form at the centre of the carbon fibre, because the carbonization does not provide enough material to fill the fibre completely.[36]

The diameters of individual fibres in electron micrographs were between 150 and 220 nm, corresponding to the alumina pore dimension. In isolated fibres electron microscopy showed porous channels in hexagonal arrangements at the edges. The mean diameter of the mesopores was ~ 5 nm, the pore-to-pore distance ~ 12 nm. Top-view images of the fibres, still in alumina pores, revealed patterns of 6–7 rings running parallel to the alumina or fibre surfaces. They are presumably formed by these hexagonally ordered mesopore channels. Some of these fibres also contain columnar channels at the centre, running parallel to the axis (Figure 2.17).[36]

Silica nano test tubes have also been isolated from alumina pores, and reversible corking was achieved. Amino silane 3-aminopropyltrimethoxysilane produced a cationic amine-ammonium rim at the end of the nano test tubes

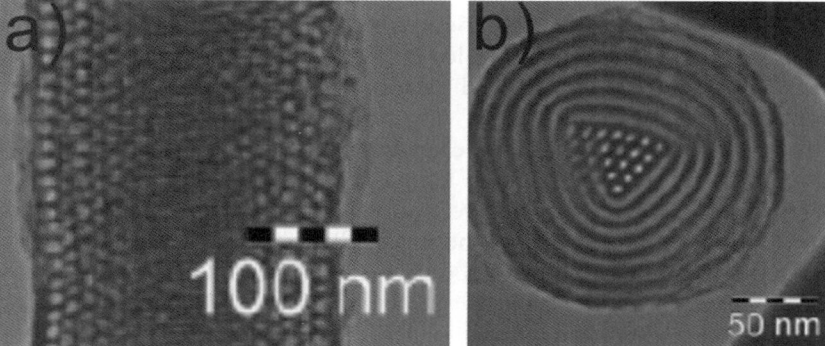

Figure 2.17 (a) Side-view electron micrograph of a short part of an individual mesoporous carbon nanofibre of Figure 2.14 and (b) top view of a fibre still in the alumina pore with the carbon rings making up the fibre, mesochannels between them and some columnar channels in the centre.

Figure 2.18 *Left:* Model of the heat-driven carbon nanotube bearing a nanosheet over a thinner carbon nanotube axis. *Right:* two electron nanographs.[42]

while they were still embedded within the pores of the alumina template, and anionic polystyrene aldehydes attached themselves as latex nanoparticle corks. The length of the nano test tubes was determined by the thickness of the alumina template (500 nm), and the outside diameter by the diameter of the pores in the template (80 nm). The walls of the nano test tubes were 3 nm thick and may also be loaded.[37–41]

An electric motor based on carbon nanotubes for the transport of nanobjects was realized between two gold electrodes. The current burnt away the outer walls, with the exception of the metal contact areas. A 500 nm nanotube remained, which rotated around another fixated 2 mm long nanotube that functioned as an axis. A current running through this axis moved an attached platelet over a distance of ~ 1 mm with a speed of $\sim 1\,\mu m\,s^{-1}$; the driving force was a heat gradient (Figure 2.18).[42]

2.3 Hydrocarbons

The membranes of plant, animal and human cells consist mainly of two linear $(CH_2)_n$ chains ($n = 7 - 19$), which are hydrophobic and arrange in water as molecular bilayers (p. 103), with water-soluble head groups. How these chains were first formed on Earth is not known: artificial "prebiotic" syntheses without enzymes always lead to small, branched hydrocarbons, never to extended chains.

In the 1950s experimental investigations began with methane, ammonia and water vapour blown into an electric arc between two carbon electrodes, imitating the atmosphere above the slowly cooling volcanic surface of the early Earth. The products of the radical reactions in the arc were then isolated by simply pumping the "atmosphere" into cold water and isolating the reaction product. The result of this prebiotic synthesis procedure were small carboxylic acids and amino acids formed from the primary product hydrocyanic acid (HCN). Its sensational successes were amino acids and nucleic bases; its main failure was the total absence of the linear alkyl chains of fatty acids. Membranes, and thus biology, remained out of reach. The long-term interactions of carbon intermediates with silicate surface matrices, which presumably must have occurred in evolution, were missing in the lab experiments.[43]

The primitive conditions of prebiotic syntheses, which start with methane and use the gas phase as medium and destructive heat as driving force, have contributed to nevertheless sophisticated machinery, namely the hard disks of computers. The number of data bits stored on a 10 cm computer disk increases

Organic Carbon

if the size of the magnetic read/write head and its distance from the rotating disk surface can be made smaller. The demand for faster computers has shrunk the disk–head distance to a few nanometres and increased the speed of the rotating disk to 10 000 revolutions per minute or more. Today the read/write head flies in the air stream above the disk at a typical distance of 3 nm and the ultrasmooth coating of the 100 μm thick magnetic layer must be as thin as 10 nm, without being centrifuged away from the fast-rotating disk. In addition, it should perform as a lubricant if the read/write system touches it accidentally, but on the other hand it should not smear on to the head surface a few nanometres away. Last but not least, it must remain unchanged for years under a strong current of air. (a-CH_x) and (a-CN_x) polymer layers with graphite-type structures fulfil this task as rugged and atomically smooth surface coatings, which rotate rapidly a few nanometres below the read/write magnets. The most exciting "nanohydrocarbon" is the statistically branched and crosslinked amorphous polymer called (a-$CH)_x$, which gives the optimal nanometre smoothness to the disk surface coupled with a guaranteed lifetime of several years even in currents of warm air. These 10 nm thick hydrocarbon or nitride layers can easily be hardened or softened and they are inexpensive.

How can an insoluble, crosslinked hydrocarbon be attached to aluminium or its alumina surface containing magnetic iron oxide particles, the usual surface of a hard disk? The answer is to take a mixture of hydrocarbon gases and argon/hydrogen, ionize it by means of a carbon arc or any other fast heating system and than blow it onto the rotating disk (Figure 2.19). Methylidyne (CH) is the stable dissociation product from 300–450 K, but above 470 K it decomposes to adsorbed carbon and gaseous hydrogen. The electron density is predominantly localized on the distorted CH fragment and a three-centre bond between the carbon, hydrogen, and one of the ridge atoms of the platinum

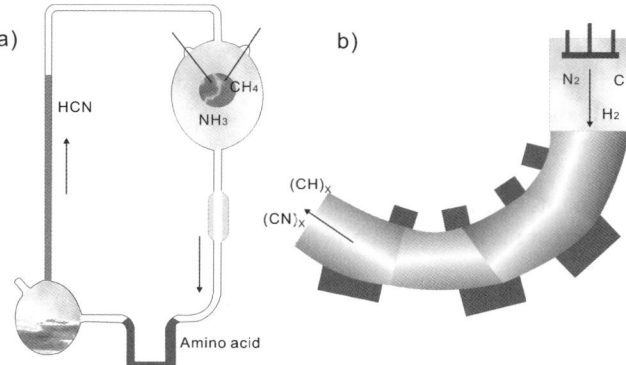

Figure 2.19 Classic "prebiotic synthesis" in a carbon arc[43] and modern nanometre coating of a computer hard disk in a pulsed high current arc.[44] In the latter case fine particles in the beam are filtered off in a curved tube with magnets, which only lead the light gaseous ions, whereas particles and droplets are precipitated or filtered off.

catalyst stabilizes it. The dominant contribution to the formation of the covalent CH–Pd bond, in the case of palladium-catalysed hydrocarbon dehydrogenation, originates from the degenerate 1π orbitals of the CH fragment mixing with the palladium states.[44–46] Amorphous, statistically branched and crosslinked CH–hydrocarbons will then form and the intermediate radicals will bind to the OH-groups on the disk surface. (a-CH_x), the standard overcoat material of hard disks, is formed within 20 ms from hydrocarbons and argon/hydrogen. If the resulting ion beam from the pulsed arc is accelerated toward the alumina/iron oxide surface by magnets, as in a mass spectrometer, the desired 8 nm thick, dense (a-CH_x) or 2 nm carbon layer films are formed and integrated into the surface. Addition of argon/nitrogen to the gas mixture leads instead to amorphous nitrogenated carbon, (a-CN_x), whereas ammonia as nitrogen source produces hydrogen cyanide (HCN) and the chemically more complex, ill-defined mixtures of prebiotic synthesis. Nevertheless, if the surface properties and lifetime of the product are superior, one might even think of HCN polymers as a starting point for computer hard disks.

The CH–polymer is not inert towards air, with which it will be always be in contact on the rotating disk. The surfaces of a-CHx overcoats are relatively inert to oxidation in the sense that the dissociative sticking coefficient of oxygen is $\sim 10^{-6}$ and the saturation coverage of oxygen during exposure to air at atmospheric pressure is 6%. This value is relatively low, and surprisingly much lower than that obtained during exposure to oxygen gas or water vapour at $< 10^{-3}$ Torr (1.3 Pa). There must be some very slow oxidation process at normal air pressure, which etches the a-CHx overcoats, remove the oxygen adsorbed at low exposures in the form of CO or CO_2 and leaves a surface that is inert to further oxidation.

The amorphous hydrocarbon network, without any uniform structural motif, solved a major problem in nanotechnology and contributed to a billion dollar market. For the final attachment of a fluorinated lubricant on top of the (a-CH_x), see p. 147.

Polyenes, in particular the carotenoids and polyacetylene, are the archetypes of conjugated carbon compounds, which become conductive if doped with reducing agents or oxidants. They will probably never form practical "wires", because they are too sensitive to oxidants, acids and light. Their excited states, their NMR spectra and their mechanisms of electron conduction are, however, theoretically understood and they serve as models for the more practical conductive materials, such as polypyrrole, polythiophene and, above all, polyaniline.

Polymerizations of functionalized 1-alkynes are effected by tungsten-based catalysts (p. 334), which need a phosphine co-catalyst and yield diarylacetylenes. Unsubstituted polyacetylene is notoriously unstable and poorly soluble, but conjugated benzene units make it stable to air and heat and render it soluble, thanks to the suppression of backbone interactions by the pendant R groups. Coupling of the polyacetylenes to carbon nanotubes or helical polyene structures produced soluble and conductive nanowires (Figure 2.20).[47]

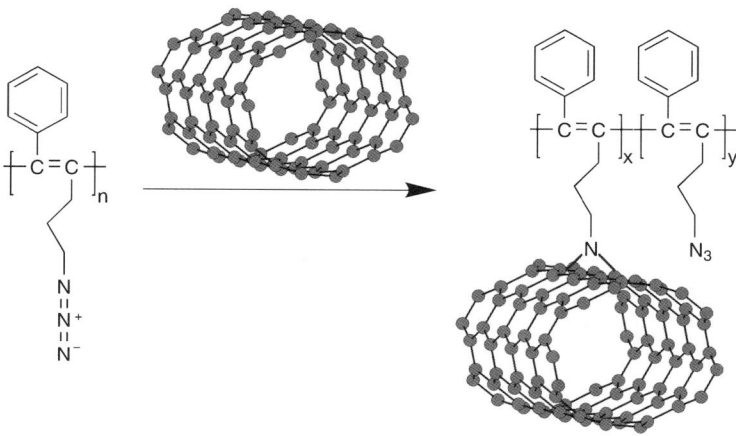

Figure 2.20 Soluble and conductive carbon nanotubes.[47]

Typical charge carriers in polyacetylene treated with electron donors ("n-doped") are resonance-stabilized polyenyl anions of ~29–31 CH units in length, with highest amplitude at the centre of the defect. It turned out, however, that long conjugation lengths are unnecessary for significant conductivity. In reality the experimental data are consistent with a rate-limiting interchain transport, as most of the time the electrons jump from one chain to another ("intersoliton hopping"). In order to avoid slow reduction steps, this charge transfer between carbanions often involves a disproportionation equilibrium between a neutral dianion pair (ion triplet) and a radical anion pair. The introduction of positive charges by oxidation of polyacetylene, "p-doping", has similar effects and also produces metallic electrical conductivity.[48]

The solitons were characterized in protonated carotenoids by ^{13}C-NMR spectroscopy, which shows chemical shifts corresponding to a charge-density wave with the positive charge located on every other carbon atom. In carotene monocations it was located at the even-numbered carbon atoms in the centre of the molecule, and in dications, where charge repulsion prevails, at the periphery[49] (Figure 2.21). At the same time there was a long wavelength shift of the visible/near IR absorption bands from ~500 nm of the neutral polyene to 800 nm for the monocation and to 1000 nm for the dication.

Most interesting, however is a massive increase of electric conductivity of the cations in comparison to the neutral dye. From first principles, one might think that a polyene of infinite length should be metallic, because the HOMO–LUMO energy difference, or band gap, vanishes and the population of the conduction band requires no energy. This is not the case. The conduction bands in linear polyenes are half-occupied by electrons (or electron holes). This leads to two inequivalent ground states, R and L, because each chain may begin with a hole or an electron; alternating bond lengths occur in both (Figure 2.22). Polyenes are thus insulating materials, because there is an energy gap between

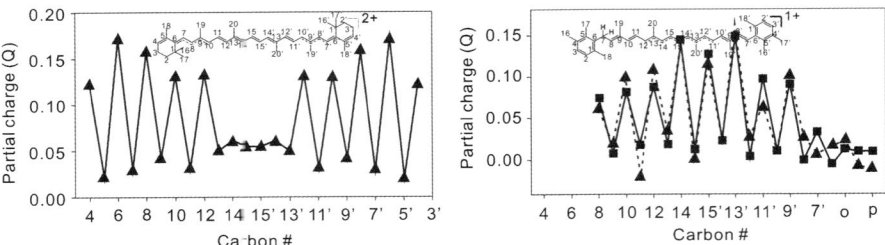

Figure 2.21 π-electron distribution (soliton model) of a protonated carotene monocation (*left*) and of an oxidized dication (*right*). © American Chemical Society.[49]

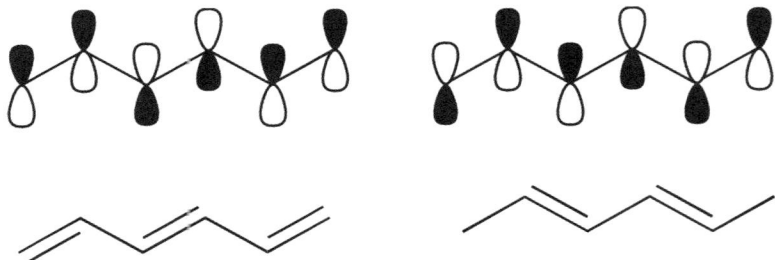

Figure 2.22 The Peierls distortion in polyenes. In terms of organic chemistry it replaces a double bond by two unpaired electrons, a biradical, or by the negative and positive charge of the excited state. A carbene on both ends of L would imply an extra electron pair, and is therefore not possible. The theory of the Peierls state, however, favours a metallic electron gas.

Figure 2.23 The soliton (= carbene) conduction in polyacetylene.[49,50]

these two forms, referred to as Peierls distortion. It must be averaged out before the electrons become metallic.[50]

This "averaging" happens upon doping with strong oxidants or reducing agents. The band gap is cut in half ("mid-gap") without any increase in the magnetic susceptibility. There are no triplet states or radicals, but the two equivalent polyene chains R and L are interconverted through the intervention of a mobile charge carrier, a soliton, which is described as a mobile charged or neutral defect ("kink") in the polyacetylene chain that propagates down the chain and thus reduces the barrier for single–double bond interconversion (Figure 2.23).[47,48]

In the intrachain charge transport the soliton has a finite width. When the length of the polymer chain exceeds it, the soliton becomes mobile, since the soliton may localize at any arbitrary site on the polyene chain with no change in energy.[47-52] This statement is not in accordance with conventional resonance theory, which describes valence bond structures as single entities. The latter view is, however, only realistic with small, resonance-delocalized ions; it is not correct for long conjugated systems. ^{13}C-NMR chemical shifts of diphenylpolyenyl anions indicate this very clearly. A plot of these shifts vs. $1/N$, where N is the number of polyene carbon atoms, shows a linear dependence of the aromatic and α-carbon chemical shifts for all anions of length >0.5 nm. The chemical shift differences vanish, however, at $N = 4.3–4.5$, corresponding to a polyene length of ~ 30 C–C bonds. This provides direct verification of the soliton width and illustrates that resonance delocalization represents a compromise between resonance energy and bond localization.[48]

Surprisingly, the formation of two solitons (called a bipolaron in the solid state) or two charges in one polyene molecule is also favourable, since ion pairing can be followed immediately by single electron transfer to a neutral polyene to produce two radical anions (polarons). This bipolaron hopping or radical anion disproportionation mechanism closely matches the observed charge transport energetics. Hexatriene, for example, is readily reduced to its dianion. The crystal structure of its dilithium salt has been determined and the most conductive form of potassium-doped polyacetylene has the general formula C_6H_8, which corresponds to a hexatriene dianion and a neutral hexatriene, both in equilibrium with two hexatriene radical anions.

Central to conventional microelectronics are interfaces between regions with different doping types (n vs. p) or densities. Although thermodynamically unstable, interfaces between dissimilarly doped regions of silicon (p. 189 ff) are kinetically stable due to the immobility of the dopant atoms. In contrast, the dopant ions of conventionally doped conjugated polyacetylenes can be mobile, which presents a potential problem in the fabrication of interfaces between dissimilarly doped conjugated polymers, because bulk chemical reactions occur and destroy the system. Such bulk reactions can, however, be prevented either thermodynamically or kinetically. An example of thermodynamic control by large differences in redox potentials is the polypyrrole–polythiophene pair (see p. 122 ff); kinetic control over interfacial reactivity was achieved by preventing counterion diffusion using polyenes with covalently fixated, charged substituents.[49-51]

Internally compensated n-type and p-type forms of polyacetylene were provided by poly-(tetramethylammonium 2-cyclooctatetraenylethane sulfonate) substituents with tetramethyl-ammonium counter-cations and poly[(2-cyclooctatetraenylethyl)-trimethyl-ammonium] with trifluoromethanesulfonate counter-anions. Cyclic voltammetry of 1.5 μm thick films of both polymers on glassy carbon electrodes showed single electron oxidation peaks for the cation radical formation (p-dopant) for both molecules, but a reductive potential (n-dopant) only for the cationic polymer. The voltammetric waves of polyenes are not symmetric: they differ, for example, by 100 mV for the

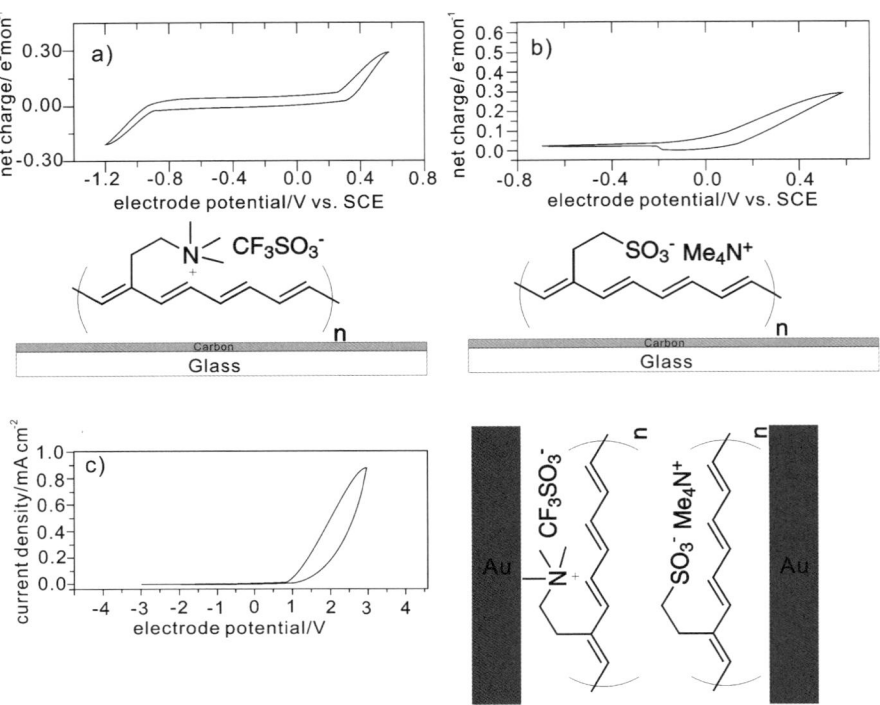

Figure 2.24 Single-electrode voltammetry data for films of (a) P_C, (b) P_A and (c) $P_C + P_A$ on glassy carbon electrodes in 0.075 M Me_4NBF_4. © American Chemical Society.[51,52]

p-doping of P_A and P_C polymers (Figure 2.24). The (P_+/P_C) wave produced 0.3 electrons per monomer or needed 12 C=C double bonds for one electron, whereas 16 C=C double bonds for one electron were utilized in the n-doping wave. In case of the sulfonium polymer an n-doping wave was not found. It did not take up electrons, because the sulfonate polymer could not get close enough to the electrode under the negative scan conditions, so an effective electron-transfer process did not occur. A unidirectional electronic transport was observed in a device featuring the interface between the two ionically functionalized, undoped polyacetylenes sandwiched between two gold electrodes, Au–P_C–P_A–Au.[51,52] The origin of the current relied on the asymmetry of ionic processes with respect to the sign of applied bias. The current is clearly electronic, since gold electrodes cannot support steady-state ionic currents.

The adsorbed dyes that reflect laser light in CDs and DVDs must have high refractory and extinction coefficients. Reflection is needed for reading, absorption for burning. CDs are written with 780 nm lasers, DVDs with 650 nm lasers. Low refraction of the dye layer would lead to low reading speeds; low absorption brings longer burning times and more heat. Absorption and refraction depend on wavelength – usually it is observed that the refractory

Organic Carbon

index has a minimum at the absorption peak. Furthermore, the dye should be heat stable up to 200 °C and then burn down quickly at a sharp degradation point. This implies that the dye should contain as much hydrocarbon as possible and as few heteroatoms or metal ions as possible. The dye is dissolved in a low-boiling solvent and spin-coated on the poly(carbonate) substrate with preformed land and pre-groove spirals. The layer thickness is ~150 nm in the pre-groove and 50 nm on land. Homogeneity of the dye layer is mandatory; no particles, not even dimers, are allowed (ref. 45a,b in Chapter 1).

The requirements for a practicable CD/DVD dye are therefore:

- The extinction coefficient must be >200 000 at 780 nm for a CD, or >100 000 at 650 nm for a DVD.
- The refractory index must differentiate strongly, *e.g.* 50%, between the 150 nm dye layer in the grove and the 50 nm layer on the land.
- The dye must not crystallize or oxidize upon storage.

After extensive quality controls and drying, the reflective layer of silver (for CDs) or aluminium (for DVDs) is sputtered on in an argon beam. The edge of the disk is then thoroughly washed to remove any surplus dye from the spin-coating procedure. The rest of the disk is protected from the washing solvent by the silver layer. The last step in CD production is spin-coating with lacquer, and control.

Hydrophobic dye surfaces were observed to be more stable than hydrophilic ones. Adsorbed water molecules presumably trigger crystallization. Conjugated polyenes with hydrophobic benzene end groups were long lived. Two nitrogen atoms as electron pair donors and acceptors established high extinction coefficients and refractory index, the triene with electron donating and accepting aniline ends achieving the long-wavelength absorption. Dimethyl and butyl substituents disturb the planarity and crystallization of the polyene-pyrrolenine chromophore; perchlorate or hexafluorophosphate counterions for the ammonium salt guarantee insolubility in water and prevent recrystallization. The trimethine was optimal for the 650 nm lasers of DVD recorders, the pentamethine for the 780 nm lasers of CD recorders. Both differentiate perfectly between land and groove (Figure 2.25). The positive charge of the imine prevents autoxidation by oxygen during storage in the dark. The hydrocarbons in the centre and the periphery take care of fast burning at intermediate temperatures and ensure there is no slow degradation at low temperatures (ref. 45a,b in Chapter 1).

A polyacetylene core with dendrimer and benzylamine substituents (see p. 120) was also realized. AFM indicated that the addition of a dendron to the periphery did not change the spherical conformation of the polyacetylene connected *via* a soft propionate spacer. Complexation, decomplexation and recomplexation of the polymers occurred several times, without any apparent change of conformation. The polyacetylene backbone changed conformation and degraded slowly in cycles of acid–base additions. Upon acidification it opened to stretched, wormlike structures and also aggregated. The dendrimeric

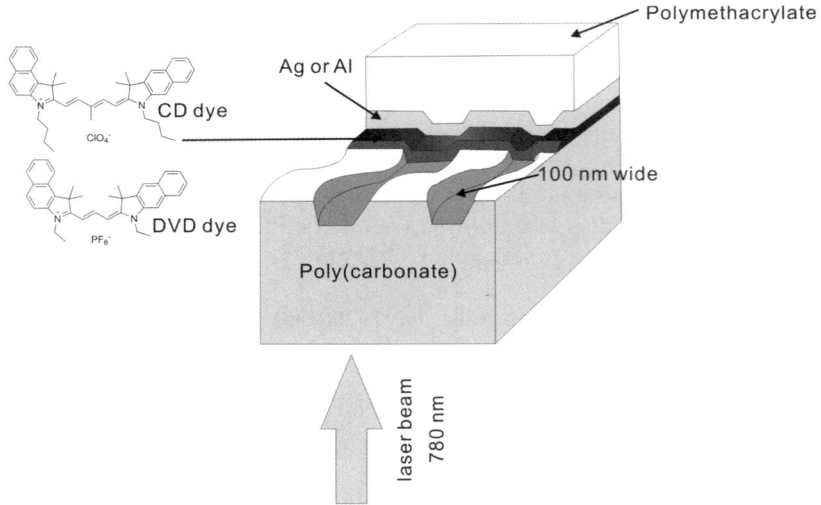

Figure 2.25 Placement of the dye layer in CDs and DVDs and structure of two appropriate dyes: they have electron donors and acceptors at the end for high extinction coefficients, charges and counterions for diffraction, long alkyl side chains for solubility and small alkyl groups at the chromophore for prevention of dimerization (ref. 45 in Ch. 1).[53]

worm then folded back to micelles when the pH was raised above 10, where the benzylamine lost its positive charge[55] (Figure 2.26).

The conjugated C=C double bonds of the styrene polyene poly(p-phenylenevinylene) emit light when submitted to a constant direct current at 5 V. The benzene–alkene polyene is not stable in air, however, and it is too difficult to process to be used for lightning. The polymer was then introduced into a suitable-sized zeolite pore (p. 195 ff) and mixed with caesium chloride and a conducting poly(acrylate-acrylamide) copolymer, both serving as electrolytes to promote charge migration. Entrapment in the silicate nanopores protected the polymer from the oxidation by air, electricity was conducted by the salty environment and the chemiluminescence became efficient and long-lived enough to be used in the first generation of organic light-emitting diode (OLED) lamps.[53,54,56] Doping of the olefinic with various electron and hole injection dopants, *e.g.* fullerenes and viologens, increased the efficiency of the cell (Figure 2.27), which thus came into use for the small displays of mobile phones. These dyes have not yet been used for television screens, because different colours still fade at different rates and carefully established "natural" shades develop colour distortion within a year.

Crystalline diacetylenes polymerize upon UV irradiation or thermal stimuli to anisotropic polydiacetylenes again in the crystalline state. The translational distance between the diacetylene units must be 0.49 nm in the monomer crystal, with an angle of inclination of 45° between the unit axes of the combining molecules. Only this specific arrangement leads to the desired linear

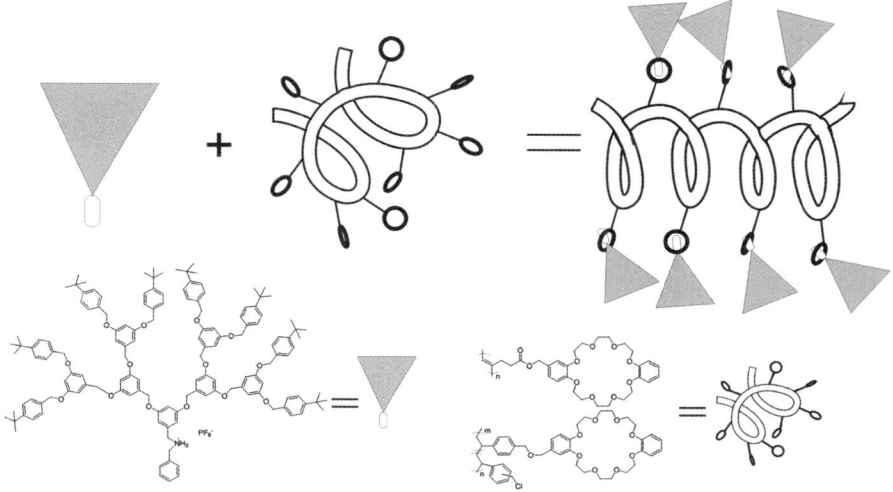

Figure 2.26 Folding and stretching of a polyene-connected dendrimer assembly. Each dendrimer substituent of the chain or helix constitutes a 2 nm carbon nanoparticle. © American Chemical Society.[55]

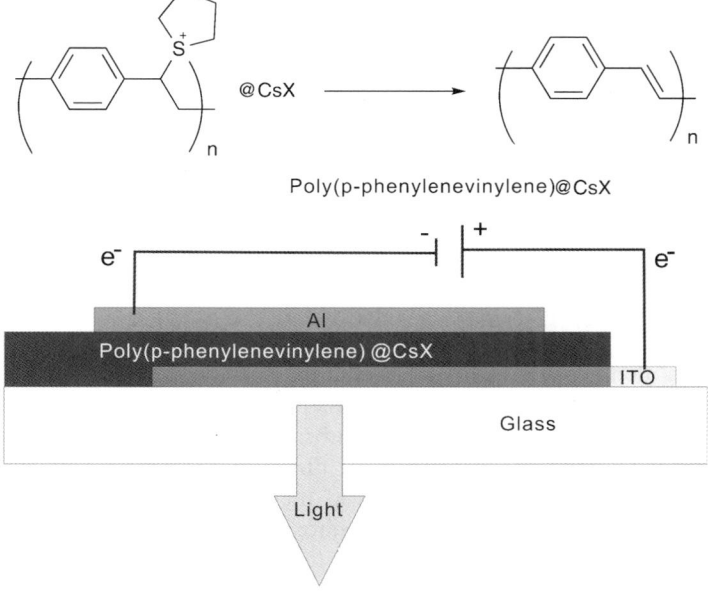

Figure 2.27 Poly(p-phenylenevinylene) layers in zeolite pores emit light when submitted to a constant current at 5 V. Dissolved caesium chloride promotes charge transport. © American Chemical Society.[56]

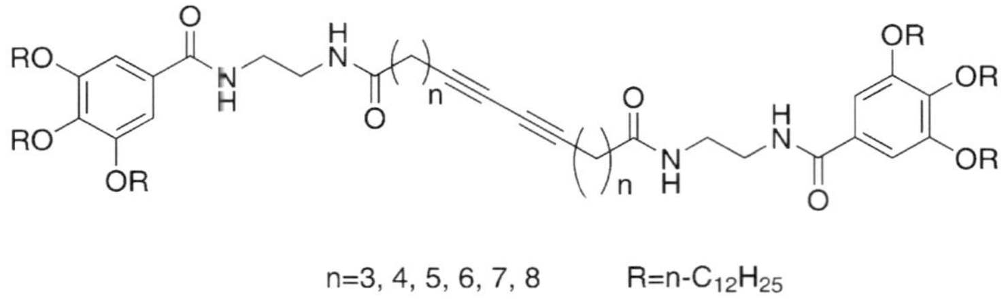

Figure 2.28 A candidate for chemically more stable light-emitting layers.[55]

stereochemistry and polymer length. A suitable orientation of crystal lattices dominates the covalent polymer synthesis, since the effective length of the conjugation systems is strongly affected by the planarity of the main chain. A 20-mer should be red, a 30-mer blue or green. This was verified experimentally, but linker effects came as a surprise: odd numbers of alkyl linkers (3, 5, and 7 CH_2 units) changed the red colour to orange (520, 526 and 530 nm), even numbers to blue (610, 615 and 625 nm).[57] Furthermore, fibres with odd numbers of CH_2 linkers were more readily bleached and conducted electricity less efficiently than the even-numbered analogues. Even-numbered linkers obviously did not disturb the linearity of neighbouring amide hydrogen bonds and, correspondingly, the linearity of the diacetylene triple bonds. Odd-numbered linkers, on the other hand, destroyed the order of the hydrogen bond chain, which led to a twist of the diacetylene units and short-wavelength shifts (Figure 2.28).[55]

The fact that 3 + 3 independent methylene groups do not produce an even-numbered linker indicates that the mobile conformation of the diacetylene unit renders the amide units on the left-hand side independent of those on the right-hand side. The environment of diene or diyne polymers in the crystalline state should be symmetrical in crystalline assemblies, otherwise long-wavelength absorption, high-extinction coefficients and electron conductivity diminish.

2.4 Alcohols, Ethers, Amines and Sulfides

Short-chain alcohols, in particular methanol (CH_3OH, bp 65 °C) and ethanol (C_2H_5OH, bp 78 °C), are occasionally used as water-miscible solvents for the destruction of micelles and vesicle membranes. Their linear hydrogen bond chains, instead of the 3D water cluster, diminish their dielectric constant by a factor of 3 as compared to water, and they do not separate metal ions or protons efficiently from carboxylates or phosphates. 10% of ethanol lowers the dielectric constant by 25%, although a water molecule solvated in pure ethanol loses only 14% of its hydrogen bonding relative to pure liquid water. Furthermore, their alkyl groups solvate the alkyl chains of the membrane core. The combination of both effects then leads to dissolution of spherical assemblies at low concentrations, typically $< 10^{-4}$, and to crystallization and precipitation at high concentrations, typically $>10^{-3}$ M. Water–alcohol mixtures are, however, common solvents for copolymers that tend to form micelles, *e.g.* the popular Pluronic F127 (p. 70).

Water–ethanol mixtures contract, the effect being largest at approximately equal volumes (Figure 2.29). For example, 520 mL of ethanol and 480 mL of water yield only 963 mL. The ethanol–water azeotrope with 95.6% of ethanol and 3.4% of water boils at 78.2 °C, pure ethanol at 78.32 °C. Nevertheless ethanol can easily be accommodated in the hydrogen-bonded network of water molecules without altering its structure. Calculations for the aggregates in the gas phase showed that cyclic patterns with two hydrogen bonds are the most stable and they were though to be responsible for the ethanol–water azeotrope.

Graphitic oxide is usually prepared from powdered flake graphite first by oxidation with sodium nitrate followed by potassium permanganate in aqueous suspension for 30 min. The mixture gradually thickens, becoming pasty and brownish grey in colour. Hydrogen peroxide is then added to reduce the residual permanganate and manganese dioxide to manganese sulfate. The graphitic oxide suspension turns bright yellow, and a yellowish-brown filter cake is filtered off, washed and dried. A carbon:oxygen atomic ratio between 2.1 and 2.9 is found for bright yellow samples; higher carbon:oxygen ratios have a green to black hue. A wide range of oxygen functional groups on both sides of the graphite plane and its edges is present, mostly epoxide and hydroxide on the surfaces and carboxylate on the edges (Figure 2.30).

The oxidized surface layers readily exfoliate and yield well-dispersed solutions of individual graphene oxide sheets in both water and organic solvents. Metal^{2+} ions bind tightly to carboxylic acid groups at the edges of individual graphene oxide sheets, and Lewis acidic metal ions also induce the ring opening of epoxides. Intercalation of metal ions between the graphene oxide sheets produces the mechanical stiffness and strength observed in M-modified graphene oxide paper.[60]

Chemical oxidation of graphite to graphene oxide, conversion of a majority of the oxygen groups to dangling bonds by hydrazine, and aryl grafting with an aryl diazonium salt for 1 h at room temperature in aqueous sodium dodecylbenzenesulfonate micellar solution gave a hydrophobic

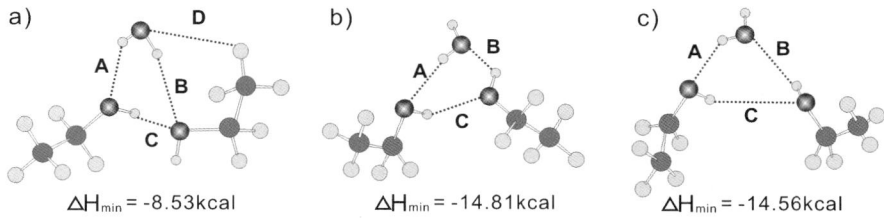

Figure 2.29 Cyclic hydrogen bridges in (EtOH)$_2$–H$_2$O trimers and calculated enthalpies. Such assemblies are the most stable ones and possibly responsible for the formation of the azeotrope. © American Chemical Society.[58,59]

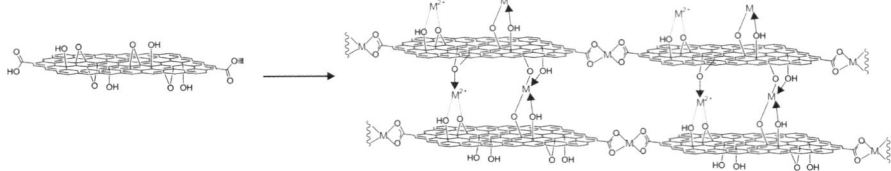

Figure 2.30 Graphene hydroxide stabilization by hydrophilic spacers.[60]

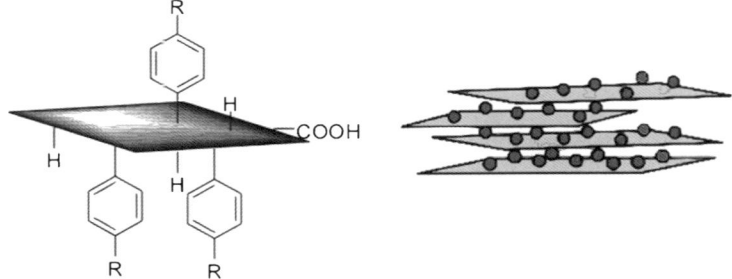

Figure 2.31 Benzene rings render oxidized graphene sheets soluble in 1-methyl-2-pyrrolidinone; Pt(0) nanocrystals keep them apart in the dry state. © American Chemical Society.[62]

graphene with 1 phenyl ring per 55 graphene carbon atoms (Figure 2.31). This product, with substituted phenyl spacers, was soluble in N,N'-dimethylformamide and 1-methyl-2-pyrrolidinone.[61] Face-to-face aggregation of graphene sheets was also efficiently inhibited by 3–4 nm fcc Pt(0) nanocrystals as spacers. These nanocrystals were produced in an aqueous suspension of oxidized graphene sheets at pH 9 and precipitated by acidification on the graphene surfaces. Nitrogen gas adsorption showed a surface of $862\,\text{m}\,\text{g}^{-1}$, which was 20 times greater than the accessible surface area in condensed graphene sheets without the platinum nanocrystals. Face-to-face stacking was effectively limited to ~3 aggregated sheets per stack.[62] Another promising single sheet polymer was made of cross-linked, long-chain dendrimers (p. 120 ff).

Hydrazine-reduced graphene oxide sheets also showed room-temperature ferromagnetism, which was not detectable before reductive treatments. A hysteresis was observed with a superconducting quantum interference device in the field range between -2500 Oe and $+2500$ Oe and the magnetism was attributed to long-range coupling of extended spin units existing as defects, which were set free in the annealing process.[63]

Chemical vapour deposition (CVD) produces gram quantities of highly crystalline graphene ribbons 20–30 µm long and 2–40 layers thick (Figure 2.32). Ethanol solutions containing ferrocene and thiophene (10:1 w/w) were carried

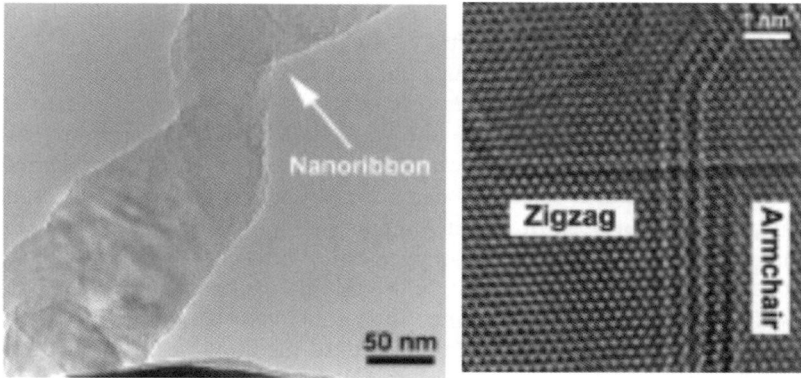

Figure 2.32 Electron micrographs of a graphene nanoribbon and its zigzag and armchair edges. © American Chemical Society.[64]

as an aerosol by an argon current into a quartz tube located inside a two-furnace system heated to 1223 K. The resulting black powder in the tube did not contain any of the ill-defined carbon nanotubes, but exclusively nanoribbons, which behaved like a metal. Lithium, potassium, sulfuric acid, and ferric chloride could be intercalated.[64]

Diethylether $((C_2H_5)_2O)$ is a low-boiling (36 °C) liquid, which is immiscible with water and is an important solvent for the extraction and separation of hydrophobic molecules. Tetrahydrofuran $((CH_2)_4O)$, its cyclic ether analogue with two hydrogen atoms fewer, behaves completely differently. Although this molecule possesses two hydrophobic sides, it mixes with water in any proportion, because fast vibrations of all its five ring atoms ("pseudorotation") prevent any aggregation, making tetrahydrofuran a very good solvent for carbon compounds and metallorganic reagents. Ether and tetrahydrofuran are not popular as solvents in nanochemistry, because they do not dissolve either polymeric coat molecules or nanoparticles very well.

Poly(ethyleneglycol) $(HO-(CH_2CH_2O)_{n-1}-CH_2CH_2OH)$, on the other hand, has the useful property of being equally soluble in water and benzene or many other organic solvents. It is therefore an extremely potent dispersing agent. Oligo(ethyleneglycols) combined with an alkyl chain are used to combine oil, in particular diesel, with water. This is useful, because at the high temperatures of the diesel engine the admixed water dissociates into OH• radicals, which oxidize hydrocarbon chains faster than air that has to be mixed in from outside. OH• oxidation ensures complete combustion and no soot. But how can one get 20% of water into diesel oil? The problem of mixing a lot of water into oil with as little detergent as possible is best solved with n-alkyl-oligo(ethyleneglycol) ethers, *e.g.* $H(CH_2)_{10}-(OCH_2CH_2)_5OH$. Just 3% of this amphiphile, or a cheap technical mixture of the same kind, will combine 20% of water with diesel oil in the form of a long-term stable microemulsion, which looks like a sponge under the electron microscope. The "pores" of the sponge are water droplets with a diameter of ~5 nm, and the "matrix" is fluid oil. The viscosity of the microemulsion is comparable to that of the oil.[65]

Oligo(ethyleneglycol) also provides the only known hydrophilic surface that does not adsorb proteins in aqueous environments. The ordered water structure on the surface behaves too much like water itself, and the proteins find no anchor (Figure 2.33). Protein–receptor interactions are not inactivated by this effect, however. This selective behaviour can be very helpful in medicine, in preventing the adsorption of blood proteins by nanoparticles.[66]

Ethers are the most water-stable oxygen-containing carbon compounds, even in strongly acidic or basic solutions. This stability was presumably the basis of the evolution of the cell membranes of archaebacteria, which are stable in the boiling acids of volcanic lakes, the standard medium of life during the first billion years of evolution. Their cell membranes are made up of water-insoluble, methylated alkyl chains, the water-soluble head groups on both sides of the chain consisting of acid-stable ethers. The archaic building blocks of natural monolayered membranes are called *bolaamphiphiles*, because their molecular structure is similar in shape to a South American sling called a *bola*,

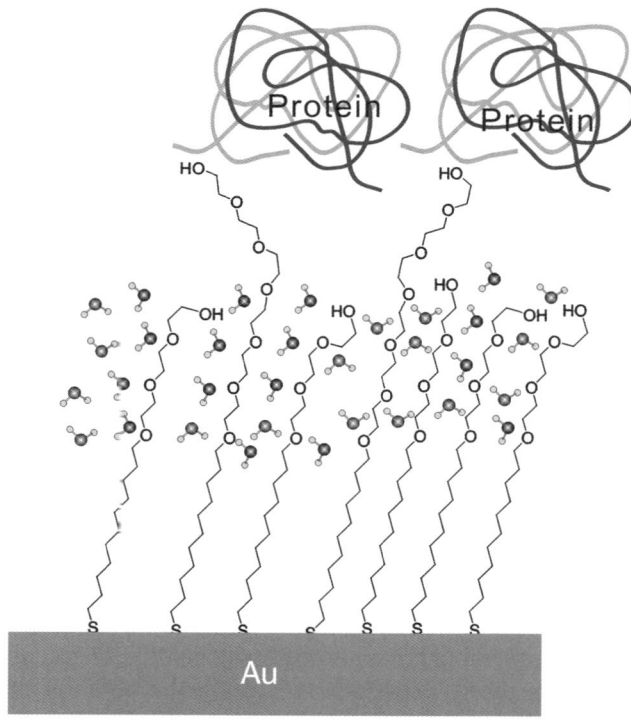

Figure 2.33 The water layer on oligoethyleneoxide monolayers rejects proteins, but receptor molecules on their surface remain active.[66]

Organic Carbon 87

a rope with a wooden ball on each end (Figure 2.34).[67] An archaebacterial membrane surface, with its ether coating, also rejects proteins.

A water droplet on a hydrophilic poly(ethyleneglycol) surface of isotropic roughness is almost spherical, but flattens out to have a large contact area on the bottom. The immobilizing effect of this flattening can be diminished by grooves that entrap air. The droplets still flatten in the same way, but the small contact area now makes them very mobile. They roll along under the influence of gravity and clean the substrate surface as they go. Such a surface was prepared by fixation of ethyleneglycol dendrimers on smooth oxide surfaces[68] (Figure 2.35).

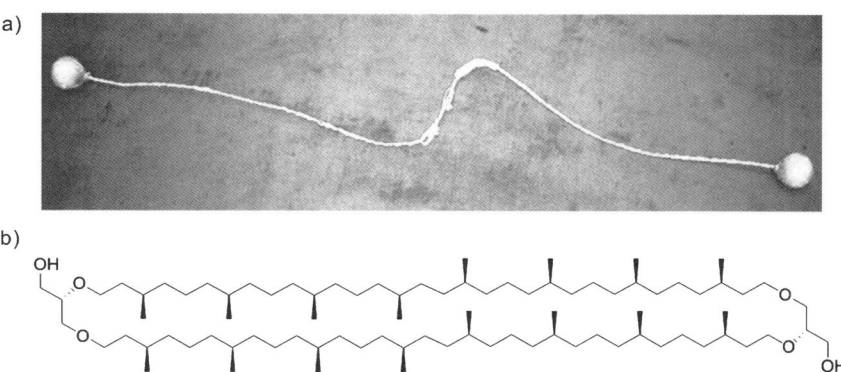

Figure 2.34 A South American bola (a), and the acid-stable tetraether bola of an archaebacterium (b).[67]

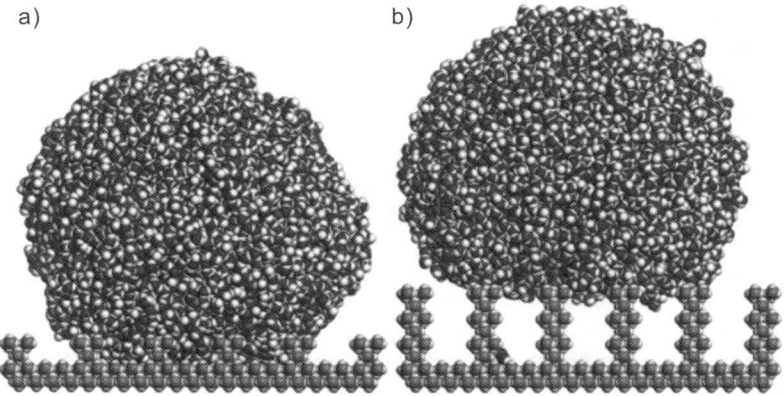

Figure 2.35 Configuration of a large water droplet (a) on a smooth hydrophilic poly(ethyleneglycol) surface with wetted and sticky contact and (b) on the hydrophilic, rough surface with the composite and loose contact. © American Chemical Society.[68]

Amines are basic, with a pK_a close to 10; at pH 7 they are almost completely protonated. Nitrogen is the only non-metallic element that carries positive charges in non-acidic water.

The protons of ammonium salts also form very strong hydrogen bridges with -OH groups and stable, hydrogen-bridged salts with carboxylates and phosphates. The effects of the positive charges of ammonium salts are much less pronounced. Ammonium chlorides, nitrates or acetates as well as neutral amines are often highly soluble in water and in many organic solvents. Perchlorates and other large anions, the same size as ammonium, are usually much less soluble, and tend to crystallize.

Trimethylammonium salts are common in neurotransmitters (acetylcholine) and form less stable salts than their protonated counterparts. R–NH$_3^+$ molecules are, however, readily deprotonated on the surface of negatively charged membranes, and the neutral amine, R–NH$_2$, then readily passes through the hydrophobic membrane. This is a major difference from neutral carboxylic acids – their hydrogen bond connected dimers are very polar and practically insoluble in membranes, and never pass through them. Phosphates, with a pK_a of 12.5, are even more hydrophilic and never lose their last negative charge under physiological conditions. Monovalent metal salts of fatty acids ("soaps") and long-chain phosphates form preferably spherical micelles, and this does not change if the micelles takes up solutes (p. 98).

Ammonium counterions also lead to spherical micelles first, but they may carry phenyl side chains, which often stabilize fibres. Cetyltrimethylammonium bromide forms micelles in water, which assemble to fibres in presence of salicylate counterions. These fibres are fluid, dissolve massive amounts of small, hydrophobic molecules, *e.g.* benzene, and show the effect of viscoelasticity in aqueous suspensions. The replacement of salicylic acid by chiral adenosine monophosphate anions, however, did not lead to fluid fibres. Binding interactions between the adenine units presumably led to stacking and solidification of the fibres.[69]

Micellar rods are also useful as templates for pores in silicas and manganese oxides, alumina, zirconia and other transition metal oxides. Tetramethyl ammonium micelles were, for example, heated with sodium aluminate or silicate gels to 150 °C, and later removed by thermal or chemical treatment. The organic templates must be uniformly incorporated in a dense embedding matrix without aggregation in order to form well-defined pores in the "ceramics" (Figure 2.36).[70]

Multishell architectures with hyperbranched poly(ethylene imine) in the centre, a hydrophobic spacer and an amphoteric oligoethyleneglycol shell encapsulate non-polar and ionic molecules in organic or aqueous environments. The amino core may become positively charged in acidic media, but the amide bonds do not hydrolyse. The poly(ethylene imine) core also reduces Au(III) ions to Au(0) nanoparticles and many organic or aqueous solutions are stable for days (the exception is hydrocarbons; Figure 2.37).[71,72]

The most important biological effect of the positive ammonium charge is the transfer of nerve impulses by non-chiral (acetylcholine) and chiral stimulants

Organic Carbon 89

Figure 2.36 Transmission electron micrograph of MCM-41 material based on ammonium amphiphile templating and having a pore size of 2 nm. © American Chemical Society.[70]

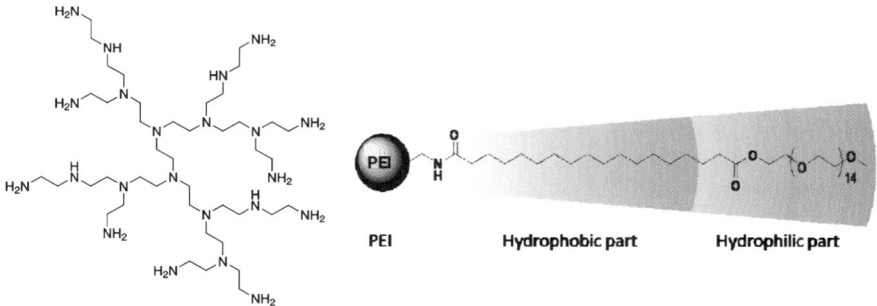

Figure 2.37 Basic poly(ethylene imine) (PEI) nanoparticles with a hydrophobic protein-repellent coating.[71,72]

and β-blockers (amphetamine, ephedrine, metoprolol). Such molecules may be linked to nanoparticle surfaces and work there as anchors at the corresponding membrane receptors in biological membranes (Figure 2.38). These molecules interact selectively with the helices of membrane proteins and membrane surfaces. Ephedrinium as well as the diastereomeric pseudoephedrinium myristate fibres dissolved magnesium octaetylporphyrin, a porphyrin, which is oxidized by Fe(III) to cation radicals similar to those of chlorophyll in photosynthesis. Circular dichroism spectra indicated that the dye was integrated into the myrisrate fibres (see pg 97) next to the dissolved chiral ephedrine on the surface.

Addition of an axial pyridine ligand pushed the dye out of pseudo-ephedrinium myristate, but not out of ephedrinium myristate fibres, because the phenyl and pyridine groups rejected each other only in the first case,

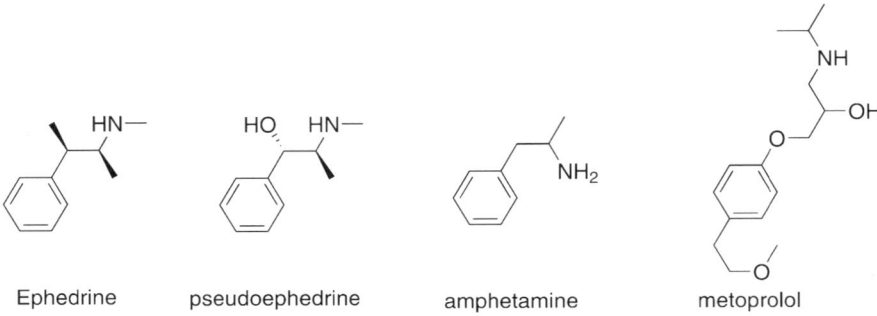

Figure 2.38 Typical structures of amino neurotransmitter molecules.

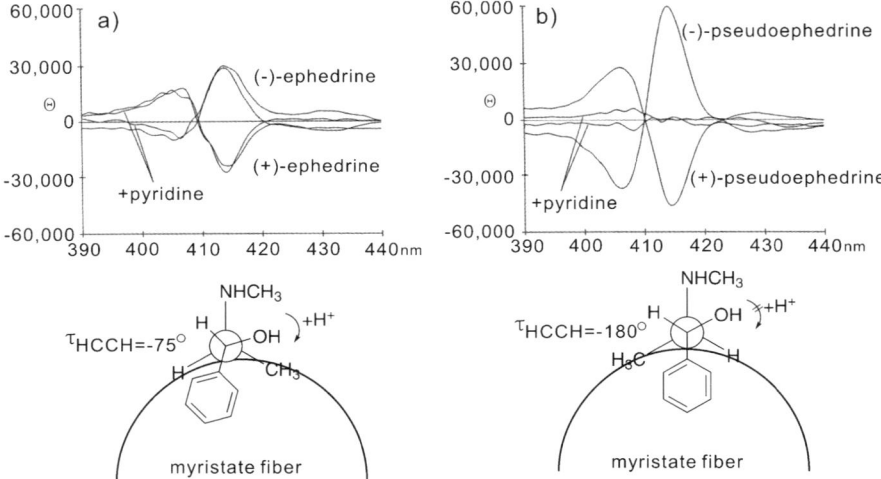

Figure 2.39 Circular dichroism spectra of viscoelastic solutions of (a) ephedrinium and (b) diasteromeric pseudoephedrinium myristate with 0.1 mM magnesium octaethylporphyrin dissolved in the fibres, before and after addition of 10 mM pyridine. Pyridine displaces the ephedrine cation and cancels the circular dichroism, but has little effect on pseudoephedrine.[73]

due to perpendicular or parallel orientations of pyridine and phenyl rings (Figure 2.39).[73]

Uniform and smooth silica particles with a dense amine coating (see atomic force micrograph on p. 16) are obtained from tetraethoxysilane (1.5 mL) and concentrated ammonia (3 mL, 28%), which are dissolved and stirred in anhydrous ethanol at room temperature for 24 h in the dark. Another 8 h reflux in water (400 ml) with the aminated silica gave ultrasmooth nanoparticles with a uniform diameter of 100 ± 10 nm. These nanoparticles were water-soluble

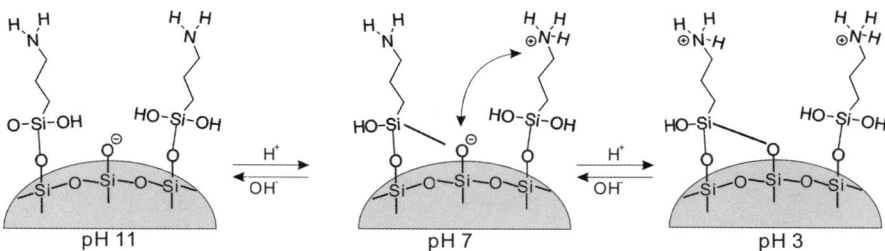

Figure 2.40 Hypothetical scheme of the surface of aminated silica nanoparticles in water at different pH values (see text).[74]

at pH 11, precipitated quantitatively at pH 2 and were then centrifuged for purification. Quantitative redissolution occurred at pH 11 (addition of NaOH) (Figure 2.40). Acidification with HCl to pH 7–8 caused no precipitation; aggregation was slow, as indicated by electron microscopy. Addition of 50% ethanol kept the 100 nm nanoparticles in solution at pH 7 or 11 for at least 6 months.[74]

The dissolution at pH 11, where all amino groups should be electroneutral, was tentatively traced back to deprotonation of some Si–OH groups. Repulsive interactions between these hidden negative charges then prevented coagulation. The extra stabilization by ethanol is probably caused by solvation of the propyl linkers of the amines. At low pH, the silicate layer is probably dehydrated to form Si–Si bonds with silicate neighbours of the surface layer and loses its repulsive character, whereas the ammonium polyelectrolytes stick together *via* ion pairs and hydrogen bonding. This dehydration of the monolayer is fully reversible on acidification.

Hydrosulfides (R–SH) are acidic with a pK_a of ~ 8. The deprotonated sulfide anions are rapidly oxidized to electroneutral radicals by molecular oxygen and dimerize to water-insoluble disulfides. This dimerization is the most frequent crosslinking reaction in animal biochemistry and is responsible for the insolubility of surface tissues (skin, hair, finger and toe nails, hooves). In nanochemistry the covalent addition of RS– to the Au(I) surface of Au(0) solid electrodes or colloidal particles is most important. The S–S bridge of lipoic acid also adds spontaneously to gold colloids in the presence of citric acid, which presumably acts as weak reducing agent. The energy associated with Au–S chemisorption is 28 kcal/mol (thiolate on gold) and has been traced back to the formation of Au(I)-thiolate on the surface. Protons have been proposed as oxidants for Au(0) (RSH + Au(0) → RS-Au(I)Au(0) + ½H$_2$).[75,76]

Lipoate-coated gold nanoparticles with a diameter of ~ 200 nm are red like ruby glass. At pH 11, the lipoate-coated gold particles showed the same 520 nm plasmon absorption band as citrate gold (p. 374) at pH 9. Upon titration with HCl, the colour of the sol changed rapidly from red to blue. Shifts and shapes of the absorption bands were essentially the same as observed for coagulating

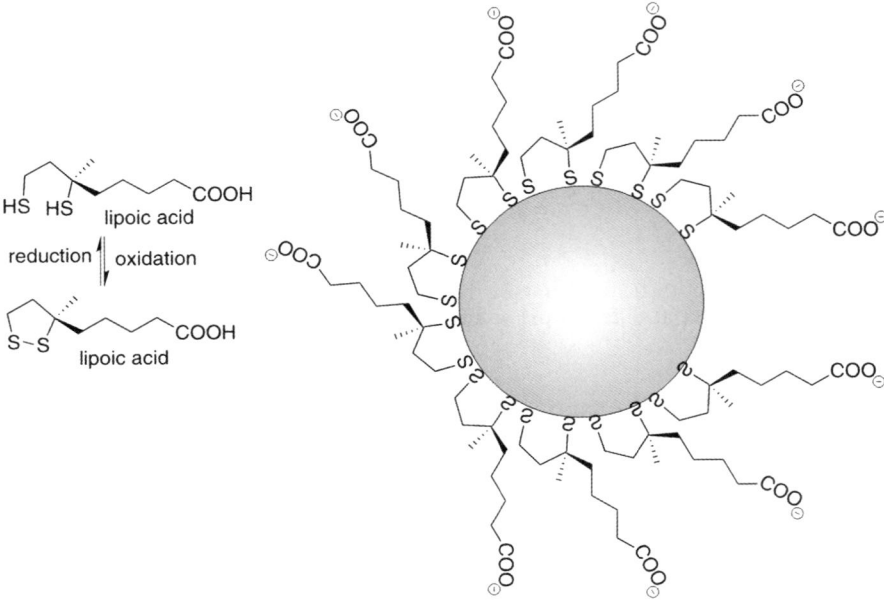

Figure 2.41 Typical lipoic acid coating of gold, silver and other metal nanoparticles.

gold particles coated with ω-thiol fatty acids (Figure 2.41). After several minutes, the blue colloidal solution became turbid and grey precipitates became detectable by the naked eye. Upon addition of sodium hydroxide, however, these precipitates redissolved completely at pH 11. The 520 nm band showed no change in either width or height after several cycles The Au(I)–S bond is thus much more stable to acid treatment than that of the citrate. To the best of our knowledge, no other totally reversible coagulation of medium-sized gold colloids is known (Figure 2.42). Treatment with sodium borohydride again led to a quantitative disruption of the Au(I)–S bond, and the metal precipitated as black Au(0) powder[76] (Figure 2.42).

Sulfides are perfect electron donors and therefore potential electron conductors. They even provoke a weak magnetism in diamagnetic zinc oxide nanoparticles (page 223 ff). The most useful electric sulfur materials are based on tetrathiafulvalene, a cross-conjugated triene containing four sulfur atoms as bridges for electron donation and transport (Figure 2.43). Crossed-wire conductivity measurements in biphenyl and related molecular assemblies embedded in *n*-alkanethiol monolayers showed that electron transport occurred through individual molecules and did not involve intermolecular hopping. Tetrathiafulvalene islands embedded in decanethiolate monolayers on gold coupled with each other to form non-covalent dye chains and also conducted electrons.[77]

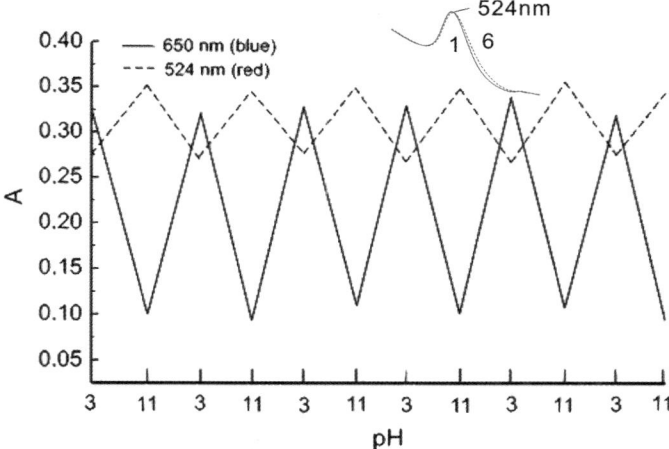

Figure 2.42 Completely reversible aggregation by change of pH. *Inset:* gold spectrum (pH 11; 524 mm) before the first addition of HCl and after the sixth addition of NaOH. © American Chemical Society.[76]

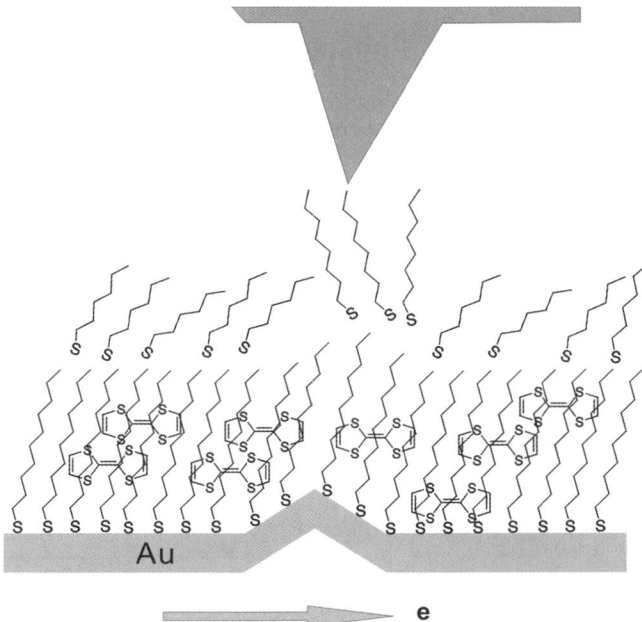

Figure 2.43 Schematic model of the tetrathiofulvalene intermolecular electron conduction path on a gold surface.[77]

2.5 Aldehydes, Ketones, Acetals and Imides

Aldehydes and ketones are the most important synthons in animal metabolism, e.g. in the citric acid cyle and glycolysis. They play only minor roles in nanomaterials, because they are not stable enough in the atmosphere or in the bloodstream. Addition of one drop of an acid or base solution to a solution of an aldehyde or a ketone produces an enol or enolate and it will form C–C bonds with the carbonyl carbon; aldehyde derivatives are sensitive to oxygen; and both aldehydes and ketones quickly add water, amines and sulfides. Uncontrolled polymerization and addition reactions are not desirable on nanoparticles or in yoctowells, and carbonyl compounds are not much used here. Exceptions are the internally protected aldehydes of carbohydrates, e.g. cellobiose chain polymers, the circular maltose oligomers of cyclodextrins and various reduced aldehydes and ketones, the polyols.

As reactants, aldehydes and ketones are almost exclusively applied as crosslinkers and spacers for artificial polymers, proteins and nucleic acids. Serum albumin, for example, was loaded onto a porous poly(acrylamide) hydrogel, in which benzaldehyde connected two hydroxyl end groups to an acetal within the polymer chain. At pH 7.4 no albumin was set free, but at pH 5.0 the gel was split into water-soluble polymers, and the encapsulated albumin was completely released within 2 h[78] (Figure 2.44).

2.6 Carboxylate and Phosphate

The carboxylate group of acetic acid (CH_3COOH) has a pK_a of 4.8, which means that at pH 4.8 half of it is protonated and there is a strong hydrogen bond connection between COO– and HOOC. At pH 7, >99% of the carboxyl groups are negatively charged and hydrogen can be replaced by ammonium or metal ions with up to three positive charges. Higher charges are usually

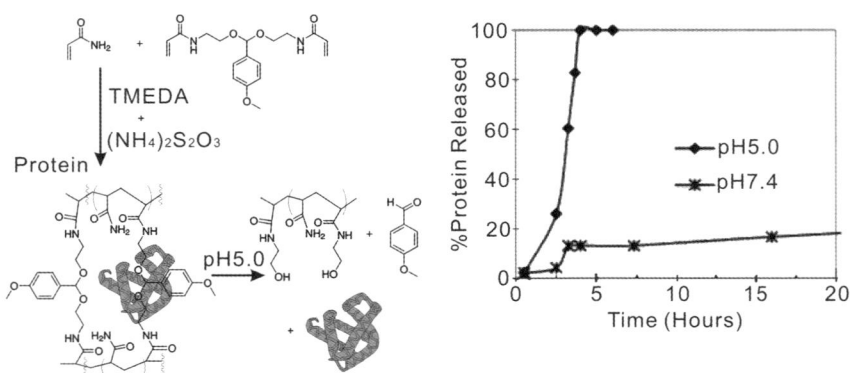

Figure 2.44 pH-dependent release of albumin from acid-degradable acetal hydrogels. © American Chemical Society.[78]

Organic Carbon

neutralized by the OH– or =O of water. α-Ammonium acids have a pK_a of 2.5. They are >100 times more acidic than acetic acid, because the positive charge of the neighbouring ammonium group rejects protons. Bi- or trivalent metal salts of carboxylates and phosphates are much less soluble than monovalent ones, and the "soft", polarizable mercury and cadmium salts are most insoluble.

In the environment of metal ions the water clusters become organized in the form of hydrates: 4–8 electron-rich oxygen atoms are attracted to a greater or lesser extent by the positive charges, and the water shells determine the interactions of the metal ions with the environment. The small Li^+ ion (diameter 146 pm) binds up to eight water molecules in crystalline salts, Na^+ (232 pm) binds six water molecules tightly and does not lose them in contact with basic carbonyl groups, whereas the larger K^+ (304 pm) usually gives water-free crystals (p. 168) and releases the water even in aqueous solution, if it comes into contact with the carbonyl groups of protein amides (p. 169). With the two oxygen atoms of carboxylate groups each bearing half a negative charge, metal ions with only one positive charge, such as Li^+, Na^+, K^+, Ag^+, and Au^+ may, depending on solvent, crystal structure, and type of metal ion, bind either to none, one, both or one-half of the oxygen atoms leaving a carboxylate anion, neutralizing it or reversing its charge to +1 (Figure 2.45).

The pK_a of the carboxyl group changes by several orders of magnitude in molecular assemblies of amphiphilic fatty acids. Their long hydrophobic alkyl chains cause the formation of large spherical, tubular or rodlike micelles and vesicles, or planar layers in the form of surface monolayers or liquid crystals. Sodium palmitate and myristate, for example, form fibres of micron width at pH 13, which rearrange to planar crystals of a 1:1 acid/carboxylate mixture at pH 9. At the higher pH these soap gels or "curds" are stable in dilute aqueous solutions, and can be isolated in the dry state and spun. High-resolution TEM and laser optical diffraction revealed a carboxylate group distance of 0.44 nm in

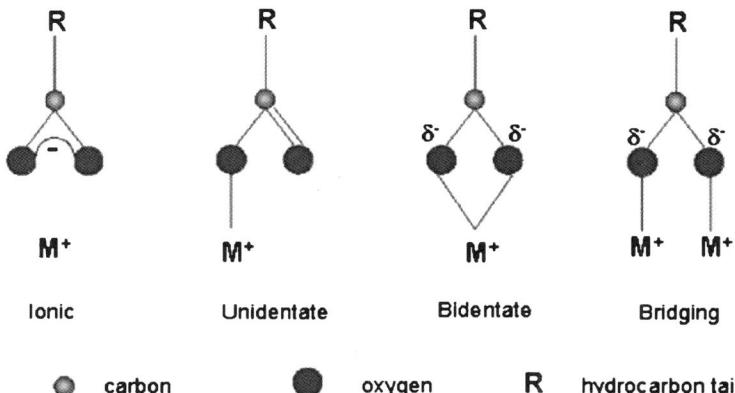

Figure 2.45 Possible reactions of monovalent metal cations, M^+, with the carboxylate ion of fatty acids.

a hexagonal arrangement. Combustion analysis of the isolated pH 9 crystals showed one Na^+ for each two myristic acid molecules. The IR spectrum of the 3D crystals indicated the formation of a hydrogen-bonded 1:1 compound between myristic acid and its sodium salt ("acid soap"): the carboxyl band occurred at $1725\,cm^{-1}$ and there was no carboxylate band at $1568\,cm^{-1}$. A strong band at $1450\,cm^{-1}$ presumably indicated the COO– group, whose IR absorption was displaced by hydrogen bonding. However, if 5×10^{-3} M NaCl was added at pH 9.8 the curds and the gel rearranged to thin crystals with one Na^+ for each myristic acid, showing the expected carboxylate IR band at $1568\,cm^{-1}$ only. A twofold excess of sodium in the form of hydroxide (pH 11.5) finally produced exclusively bundles of 3.8 nm micellar fibres, again with the $1568\,cm^{-1}$ carboxylate band together with a strong water absorption at $3433\,cm^{-1}$, which was absent in the other two preparations.

A comparison of eight fatty acid sodium salts $CH_3(CH_2)_n$-COONa with $n = 8$, 10, 11, 12, 13, 14, 16, and 18 under identical conditions under the light microscope showed that the shorter fatty acid salts remained dissolved and produced 3D crystals upon concentration while the longer fatty acids gave only very short fibres. Only myristic and palmitic acids grew to long fibres and produced gels with sodium hydroxide. Substituting lithium, potassium, ammonium, or ethylene diammonium ions for sodium also destroyed the fibres. Potassium always led to stable clear solutions, and the other ions always yielded platelike precipitates (Figure 2.46). Under the given conditions fibres with a circular cross-section were limited to sodium myristates above a pH of 9.8.[73]

The softest natural materials are the hydrocarbon chains of fatty acids, in particular the polyunsaturated compounds with *cis*-configured C=C double bonds connected by freely rotating methylene groups (CH_2). The most prominent examples are the ω–3 acids from fish and linseed oils, *e.g.* linolenic and docosahexaenoic acids, which are responsible for fast responses of neurons to changes in the aqueous environment as well as for the capability of red blood cells to move freely through capillaries with a width of only a few nanometres.

^{13}C relaxation rates in NMR spectra provided a detailed description of chain dynamics which showed that (1) the acyl chains of docosahexaenoic acids explore their full conformational space in tens of nanoseconds, and (2) proteins do not measurably alter rates of methylene bond movement, but only affect collective modes of lipid motion with longer correlation times (Figure 2.47).[79] The saturated inner part of the fatty acid moves much more slowly than the CH_2 groups neighbouring double bonds, and the terminal groups move even faster. For unsaturated fatty acids on steel, see p. 251 ff.

The simplest soft nanoparticle is the soap micelle. Soaps are mostly the sodium salts of fatty acids or the corresponding sulfonic acids (RSO_3Na), which do not precipitate in hard water containing calcium ions. The alkyl chains may be *all-trans* configured and form reefs, or they may bend until the core is completely filled to form a hydrophobic droplet. Common to both models is a large distance between the hydrated head groups. The typical diameter of a soap micelle is 3–5 nm and each one may dissolve one "big" molecule, *e.g.* one porphyrin dye, or two small ones, *e.g.* a nucleic acid base pair.

Organic Carbon

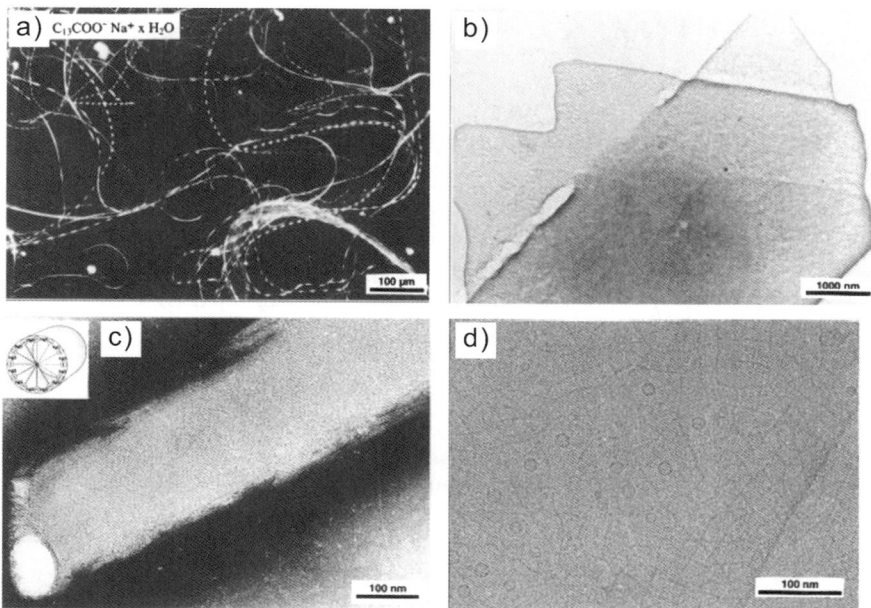

Figure 2.46 Light and electron micrographs of sodium myristate crystals and fibres: (a) acid soap, producing a hexagonal pattern of head groups; (b) soap, showing a smooth surface; (c) soap monohydrate crystals made of 3.8 nm micellar fibres; (d) cryoelectron micrograph of a viscoelastic solution of ephedrinium myristate acid showing entanglement of thin micellar fibres (see text and Fig. 2.39).[73]

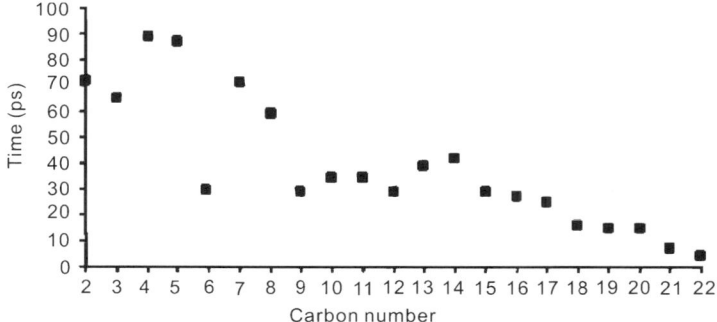

Figure 2.47 ^{13}C relaxation rates of the free C_{22} chain of docosahexaenoic acid with *cis* double bonds on C3, 4, 7, 8, 10, 11, 13, 14, 16, 17, 19, 20. These rates hardly changed when the acid was adsorbed on a protein surface. © American Chemical Society.[79]

The lifetime of a micelle depends on the length of the chains and the charge of the head groups. The half-life of an anionic sodium dodecylsulfate micelle is far less than a millisecond, but electroneutral C_{24} micelles live for several seconds. After these half-lives both types of micelle explode. Above a certain

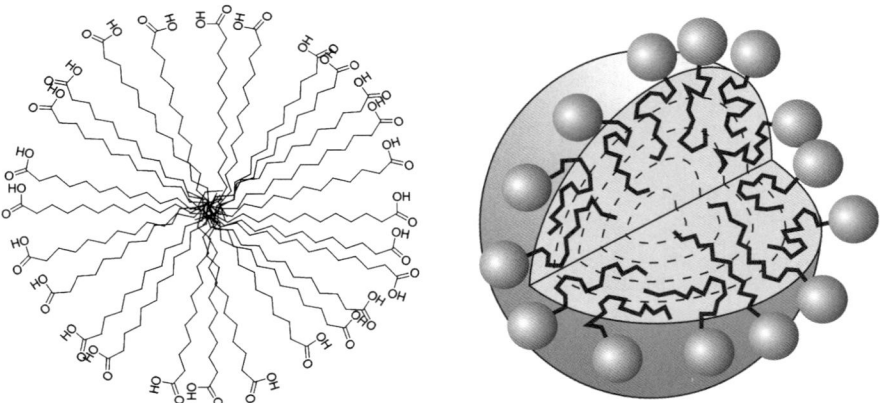

Figure 2.48 Two models of fatty acid micelle structures with water-soluble head groups at the water front and insoluble alkyl chains in the centre. *Left: all-anti* conformer, reef model; *right:* coil conformations, droplet model.[80–83]

temperature (Kraft temperature) and concentration (critical micellar concentration) they reorganize. Below the Kraft temperature micellar solutions begin to crystallize and become turbid; below the critical micellar concentration the micelles dissociate into ill-defined aggregates and monomers. The bulk micellar solution may, however, be stable for years (Figure 2.48).

The micelles are formed because the interaction between solvent molecules (usually water) is stronger than that between the solute and water. Micelle formation instead of crystallization is a result of solvation of the head groups and non-solvation of the hydrophobic core. The insolubility of hydrocarbons in water is due to entropy. ΔG for the transfer of butane into water, for example, can be split up into $\Delta H = -4.2\,\text{kJ/mol}$ and $T\Delta S = +28.7\,\text{kJ/mol}$. The entropic character of the dissolution of non-polar molecules in aqueous media is known as the *hydrophobic effect*. The rejection between water clusters and hydrocarbons causes the strong adherence between hydrocarbons in water, which is in contrast to the weak van der Waals interactions between hydrocarbons in air.

Storage of light energy in redox couples with one partner in the micelles and the other in bulk water has been repeatedly verified. A straightforward example is sodium dodecylsulfate vesicles with a dissolved zinc porphyrinate and Cu(II) counterions. The irradiated porphyrinate ejected electrons, which first reduced the Cu(II) to Cu(I); the ferricyanide, separated from the micelle surface by its negative charge, then pulled the electron into the bulk water.[80] Electron transfer was fast and quasi-irreversible, but the photolysis of water or other applications was never realized. None of the non-covalent membrane systems investigated was really effective in the photolysis of water, for example.

Although micelles have not proved to be useful in chemical reaction chains, they have of course been commercially successful for more than a century as washing powders, as detergents in water–oil emulsions for the floating of oil on minerals, and as ingredients in aerosol sprays.

If soaps are dispersed in toluene or other apolar solvents in the presence of a small percentage of water, they invert: the polar groups entrap a small water volume and the hydrophobic chains are solvated by the organic solvent. Both normal micelles and these inverse micelles are useful for dissolving a molecule or two in hostile environments, but they do not expand to the nanometre size needed for the formation of nanoparticles.[80–83]

Morphology-controlled synthesis of colloidal nanocrystals also uses fatty acids, usually the flexible, non-precipitating oleic acid, octadecylamine or tetradecylphosphonic acid, often diluted by non-coordinating, viscous solvents such as 1-octadecene. These solvents can be heated to high temperatures in closed reaction vessels (*solvothermal procedure*) and remain viscous enough to guarantee relatively slow crystallization processes. Metal oleate ammonium salts or tetradecylphosphonate are formed and build up inverse micelles with the metal in the centre.

If a chalcogene, *e.g.* selenium, is needed for the formation of semiconducting quantum dots (QDs), trioctylphosphine (TOP) is useful. It forms a soluble complex, *e.g.* TOP/Se(0), and reacts with the entrapped metal ion, *e.g.* Pb^{2+}, first to form spherical quantum dots, *e.g.* lead selenide nanocrystals. These may then rearrange to cubes, stars, flowers, hexapods, and tetrapods, with diameters of a few nanometres, depending on solvent mixture, growth temperature and time. Addition of carbon nanotubes usually induces longer rods and tubes as well as octahedrons. Recrystallization of nanocrystals formed in acidic media in long-chain amines often improved crystallinity.

The lead selenide was always synthesized by heating lead oxide with selenium/TOP. Oleic acid gave spheres or cubes, whereas the branched tetradecylphosphonate induced branching. Addition of carbon nanotubes had little effect in oleic acid; the nanoparticles only became somewhat smaller. The presence of the branched, more strongly coordinating phosphonate ligands, however, enhanced the kinetically controlled [100] growth of lead selenide nanocrystals along the axis of the carbon nanotubes, and octahedrons and strong side arms at low temperature were the result. The typical width of these nanocrystals was 20 nm and could be reduced with oleic acid in non-viscous high-boiling solvents such as toluene. Smaller nanocrystals, *e.g.* uniform 4.7 nm lead sulfide QDs, were then formed (Figure 2.49).[84]

Polymer-capped and luminescent lead sulfide nanoparticles were made by replacing oleic acid with poly(acrylate) or other polymers in tetrahydrofuran, methanol, or water. Spherical micelles made from amphiphilic block copolymers also encapsulated hydrophobic drug molecules[85] and, most importantly, in contrast to non-covalent soap micelles, they did not destroy cell membranes.

Even the environment of charged, monodisperse microspheres has been quantitatively characterized in the case of crosslinked polystyrene sulfonate and tetra-alkylphosphonium nanoparticles. Their mobile sodium or chloride counterions were transferred to electroneutral, partly oxidized silicon or glass surfaces until one elementary charge per 2000 nm^2 was established

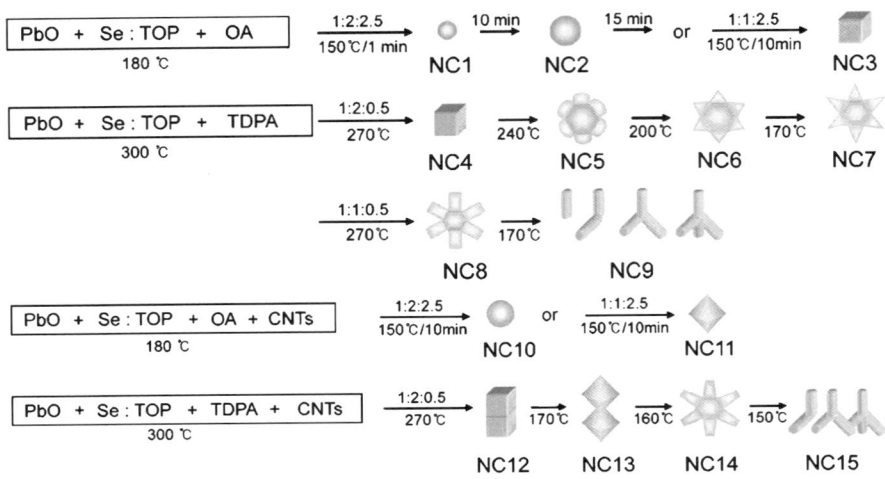

Figure 2.49 Models of inversed micelles with central metal oxide-selenium particles and hydrophobic phosphine and fatty acid coatings, which had also been used as solvents at high temperatures. Lead oxide, PbO; selenium, Se; powdered trioctylphosphine, TOP; oleic acid, OA; tetradecylphosphonic acid, TDPA; and 1-octadecene, ODE, were commercial products. 1 mg of carbon nanotubes was added to ~300 mg Pb–Se mixture in the second procedure. The ratios are indicated in the scheme. © American Chemical Society.[84]

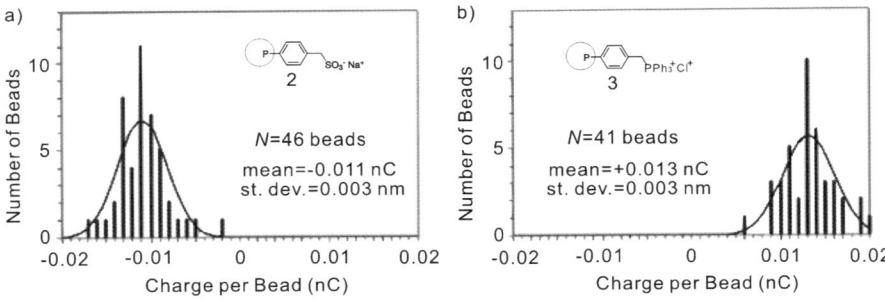

Figure 2.50 The surface charge of silicon nanoparticles may be changed at will with sulfonate or phosphonium salts on the surface. © American Chemical Society.[86]

(Figure 2.50). This limited transfer corresponded to a loss of 0.5% of the mobile counterions of the charged nanoparticles.[86]

Phosphate groups collect K^+ more efficiently than the strongly hydrated Na^+ in water, forming insoluble salts with multivalent metal ions. Ca(II) phosphates occur in our bodies as bones and teeth (p. 175 ff): long-chain phosphonates, with one C–P bond instead of P–O, stick tightly to Zr(IV) oxide nanoparticles. Insoluble, multivalent metal phosphates are difficult to crystallize, because they precipitate too fast as irregular crystallites. The same is true of the

Organic Carbon

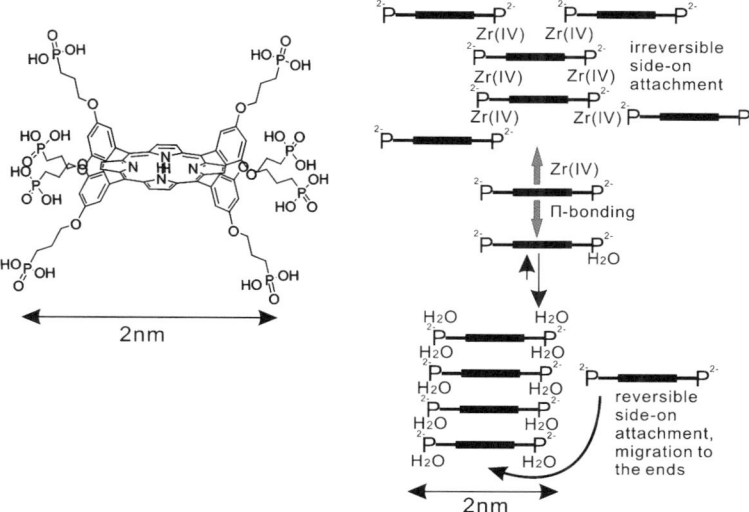

Figure 2.51 End-on growth of porphyrin-octaphosphonate fibres by addition of Zr^{4+} ions led exclusively to single-molecule nanorods.[87]

self-organization of linear phosphate rods. Only the slowly crystallizing porphyrin octaphosphonate, with four phosphonate groups on each side of the porphyrin plane and a molecular weight of ~ 1000, arranged to micron-long rods with a single molecular diameter of 2 nm upon protonation. Addition of Zr(IV) salts led preferentially to irreversible side-on addition and fibrous nanotubes of micron length (Figure 2.51).[87]

2.7 Esters and Metal–Organic Frameworks (MOFs)

Carboxylic and phosphate esters of glycerol (fats, lecithins) occur naturally in the membrane lipids of nerve and muscle tissues and in egg yolk (a cheap commercial source). Fats and oils made of a large variety of fatty acid esters are used as food; as skin protectors and cosmetics; and as lubricants for steel pistons, or for protein chains that rub against each other (in leather shoes, for example).

Fatty acid esters are useful only in double-headed bolaamphiphiles,[88] which were first found in the tetraethers of archaebacteria (p. 87) (Figure 2.52). In nanochemistry they have been replaced by the more easily accessible tetraesters, which are useful as coatings and which themselves form soft nanoparticles in water, namely spherical, monolayered vesicle membranes. The macrocyclic tetraesters are easily accessible at the 100 g scale and macrocyclization yields are $> 70\%$. Such monolayered vesicles provide water droplets 200 nm in diameter within a functional asymmetric membrane 2 nm thick. Unlike micelles, many vesicles have no measurable critical concentration ($< 10^{-10}$ m), and they are

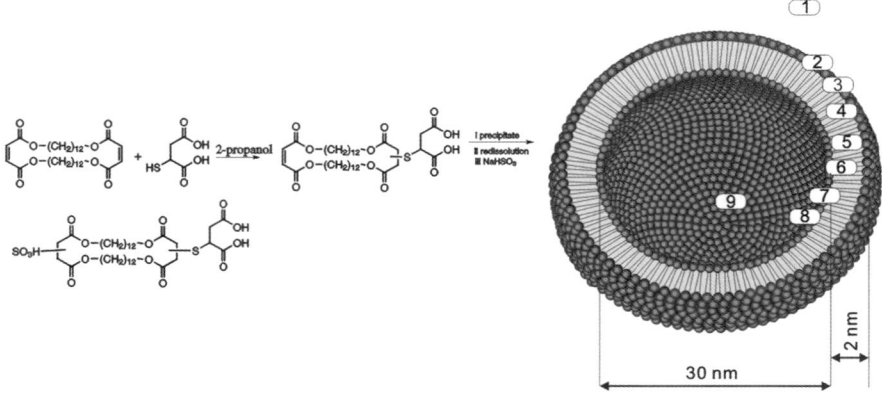

Figure 2.52 Molecular structure of a typical asymmetric bolaamphiphilic tetraester and model of the different areas of vesicles and the 2 nm monolayer membrane.[89]

stable entities in water. They therefore provide two strictly separated water volumes. However, their nine imagined reaction centres have only been prepared and characterized, not as yet really applied in long-term charge separation or any useful reaction chain.[65,89,90]

Voltage-gated channels through bilayer lipid membranes were also made from polarizable esters with oligoethylene glycol units. The most successful synthetic ion pore was made not of an elegant macrocyclic edge amphiphile, but of a trivial polyester, where all O–CO bonds pointed in the same direction (Figure 2.53).[90]

Phosphoric acid has three acidic OH groups, with pK_as of 2.15, 7.1 and 12.4. The diester is quite stable to water, bases, and nucleophiles, because the remaining negative charge ($pK_a = 2.15$) prevents attacks on the phosphorus atom. Monoesters and triester are readily hydrolysed.

The usefulness of phosphate chemistry is, however, connected with the short-lived monoesters formed by ATP with the serine side chains of proteins. These dianionic monoesters are responsible for the active transport of sodium and potassium in nerves and muscles and thereby dominate the life of animals (including humans). Phosphate binds K^+, without a hydration shell, much tighter than Na^+, which never lets go of its hydration sphere in aqueous environments. The preferred addition of K^+ to phosphate esters and anhydrides as compared to $Na^+(H_2O)_6$ is the dominant reaction of animal nerve tissues. Without phosphate esters we would have no thoughts, no feelings. Phosphate is the only known single anion which strongly differentiates between these two ions (p. 100); carboxylate and silicate do so only cooperatively in polyelectrolytes. This is the reason for the much higher sodium content of the oceans, which dissolve sodium as a hexahydrate, whereas potassium is bound to the polyelectrolytes of sediments. But that is really all we know about the

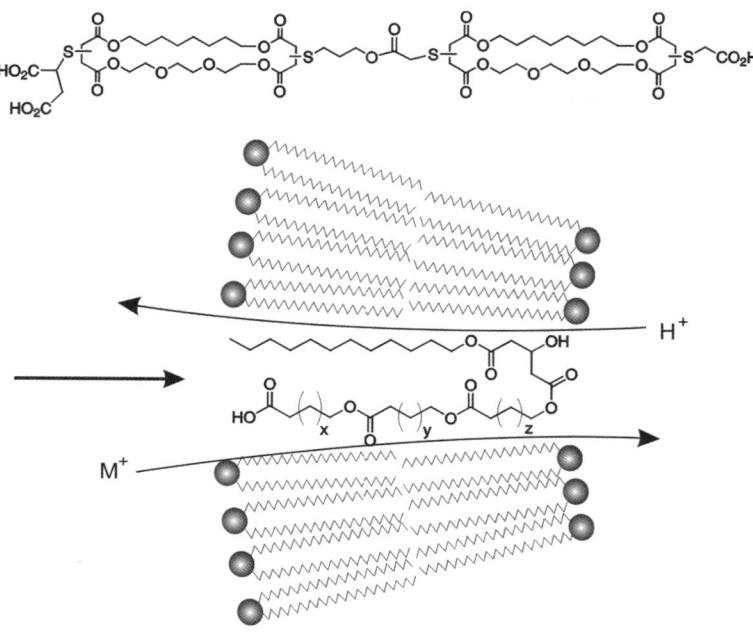

Figure 2.53 Molecular structures and a model of a voltage-gated channel in lipid membranes.[90]

chemistry of phosphates in membrane pores: the mechanism of active transport of sodium is not known.

Metal–organic frameworks (MOFs) are nanoporous coordination nanocrystals with an extremely high surface area of up to 4500 m g^{-1} and interchangeable organic ligands. They may be suitable for practical hydrogen storage. MOF-5, the main commercial MOF, has a highly symmetric crystal structure with 106 atoms per unit cell. 1,4-benzenedicarboxylate units link Zn$_4$O clusters together. The resulting framework with formula unit Zn$_4$O(benzenedicarboxylate)$_3$ consists of cubic 1.3 nm pores, the face and/or the edges of the benzene rings pointing toward the centre of the pore. The adsorption of molecular hydrogen on the pore walls creates an attractive potential well for hydrogen in the centre of the pore. MOF-5 stores 1.3 wt% of molecular hydrogen at 77 K and 1 atm ($\sim$1 MPa). The first binding centre is surrounded by the faces of the benzene rings, the second by the edges. There are five distinct adsorption sites on the edges of the pores (Figure 2.54). Density functional theory calculations show that hydrogen molecules adsorb most strongly at the metal oxide cluster rather than on the organic linker.[91]

A total of 20 different adsorption sites is available on each Zn$_4$O cluster, but adsorption of hydrogen falls off sharply after 16 of the sites are populated. The finding that hydrogen adsorbs preferentially on the metal clusters suggests looking for nano metal oxides, rather than organic metal complexes,

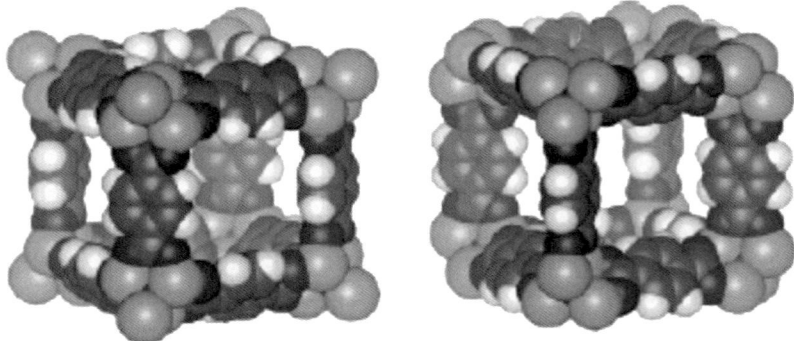

Figure 2.54 The two types of pore in MOF-5. © American Chemical Society.[93]

as potential hydrogen storage materials. Only the presence of accessible unsaturated metal centres and the existence of pores and channels in a size range well suited to the H_2 molecule seems to be necessary. Introduction of triazines instead of benzene ligands increased the storage of molecular hydrogen to 1.9 wt%.[92] The material with the greatest surface area was obtained by dissolving a 3:1 mixture of $Zn(NO_3)_2.6H_2O$ and benzene dicarboxylate in diethylformamide and heating it at 80 °C without stirring under nitrogen. The colourless cube-shaped crystals were washed with anhydrous dimethylformamide and anhydrous methylenedichloride. Water-sensitive $Zn_3(OH)_2$(benzenedicarboxylate)$_2$.2-diethylformamide was isolated and adsorbed 7.1 excess wt% H_2 at 77 K and 1 bar (100 kPa). The commercial minimum is 6 wt% at room temperature.[93]

MOF-5 is stable at very low water content, and water adsorption is reversible then, but it becomes unstable when exposed to >4% water. The weak interaction between zinc ions and carboxylate oxygens in MOF-5 allows attack by water molecules: $((Zn_4O)(benzenedicarboxylate)_3 + 4H_2O \rightarrow [(Zn_4O)(H_2O)_4(benzenedicarboxylate)_2]^{2+} + benzenedicarboxylate^{2-})$. The surface area decreases from 900 to 45 $m^2 g^{-1}$ upon exposure to water.[94] 100 nm MOF nanoparticles were produced in colloidal p-perfluoromethylbenzenecarboxylate ($CF_3C_6H_4COOH$) which bound to the surface of the particles as a cap on Zn_4O sites. Various methods of obtaining single crystals with different cavities have been summarized.[95,96]

2.8 Polymers

Microemulsions, micelles and vesicles made of small amphiphilic molecules that assemble in water or organic solvents disappear from research papers in the early 21st century. It looks as if their potential is now known. In recent nanochemistry experiments these non-covalent monolayers have largely been replaced by polymers, block polymers and dendrimers.

Electron-conducting block copolymers have been mentioned in the section on polyenes (p. 76 ff). Their saturated analogues are much more accessible and stable. They do not conduct electrons, but are chemically stable and allow massive surface crosslinking (shell crosslinked nanoparticles) as well as further functionalization.

The most common micellar copolymer is poly(acrylic acid)-b-polystyrene with block lengths from ~30 to 150 repeat units often elongated by poly(ethyleneglycol) units. These copolymers are made by radical chain reactions, which are tolerant of functional groups and impurities. Living radical polymerizations lead to polymers of defined lengths and to copolymers with defined segments, because a brake stops the polymerization after a predetermined time and because the number of polymer strains and radicals growing at a time is very limited, *i.e.* it is "controlled". This is achieved by the establishment of a rapid dynamic equilibrium between a minute amount of growing free radicals and a large majority of the dormant species. Atom-transfer radical polymerization (ATRP) will not occur, or will occur very slowly, if the equilibrium constant becomes too small. The monomers form radicals reluctantly, because the initiator has a slowly transferred halogen or nitroxide group. The concentrations of catalyst and additive, which steer radical formation and halogen transfer, are optimized in small-scale test reactions. Solvent and temperature also influence the equilibrium constant K_{ATRP} and are also optimized by trial and error.[97]

In the most successful ATRP reactions the dormant chains are alkyl halides and the monomers are typically styrenes, (meth)acrylates, (meth)acrylamides, or acrylonitrile. The active radical is formed by redox reactions between a Cu(I) complex, *e.g.* CuBr(bipy)$_3$, and a halide, most often a bromide or chloride, activated by small amounts of pyridine, by more or less polar solvents and by the temperature. An equilibrium constant $K_{ATRP} = k_{act}/k_{deact}$ for living polymerization is established, where the deactivation reactions, namely radical termination by coupling or disproportionation, stop the chain growth. Water and liquid carbon dioxide solutions, suspensions, emulsions, mini-emulsions, and dispersions have all been used in ATRP. Polar solvents often help to dissolve the monomeric catalysts; Cu(I)Br(bipy)$_3$, for example, was most active in ethylene carbonate or in non-polar solvents containing 10% v/v dimethyl formamide.

The alkyl halide initiator determines the number of growing polymer chains. If initiation is fast and transfer and termination negligible, then the number of growing chains is constant and equal to the initial initiator concentration. The halide must rapidly and selectively migrate between the growing chain and the transition-metal complex: bromide or chloride is best. Initiation should be fast and quantitative. Alkyl halides with aryl, carbonyl, or allyl groups, are potential candidates and were modelled by *ab initio* calculations.[98]

The most important component of ATRP is the catalyst. The metal centre must have at least two readily accessible oxidation states separated by one electron, bind to halides, and have an expandable coordination sphere that upon oxidation selectively accommodates the halide and a non-fugitive ligand. Copper catalysts are good in terms of versatility and cost.

Polymerization is terminated by radical scavengers (*e.g.* TEMPO, tetramethyl-piperidine-N-oxide). The polymerization is tolerant to a variety of functional groups, such as -OH and -NH$_2$, and insensitive to H$_2$O, CH$_3$OH, and CH$_3$CN. Regio- and chemoselectivities are similar to those in conventional free-radical polymerizations. The common copper bipyridine catalyst is probably a mixture of activating tetrahedral Cu(I)(bipy)$_2$ and deactivating bipyramidal BrCu(II)(bpy)$_2$ (Figure 2.55).[97,98]

Initiator-modified gold nanoparticles were produced by the reduction of aurochlorate (AuCl$_4^-$) with sodium borohydride (NaBH$_4$) in the presence of an α-hydrosulfide-ω-bromide (Figure 2.56), which then worked as initiator in the Cu(I)-catalysed ATRP of *n*-butyl acrylate (CH$_2$=C(CH$_3$)COOC$_4$H$_9$). When the Au–S bonds were not stable enough to keep all of the polymer coating, it was crosslinked by simple addition of dimethacrylate (CH$_2$=C(CH$_3$)COCH$_2$CH$_2$COOC(CH$_3$)=CH$_2$) to the reaction mixture. Long-lived crosslinked shell and tethered poly(butylacrylate) brushes on colourized gold nanoparticles were thus obtained in one step. The stabilized gold nanoparticles with initiating sites on the surface further served as macroinitiators for polymerization of the remaining *n*-butyl acrylate monomers, and formed the brush with dangling poly(butylacrylate) chains.[99]

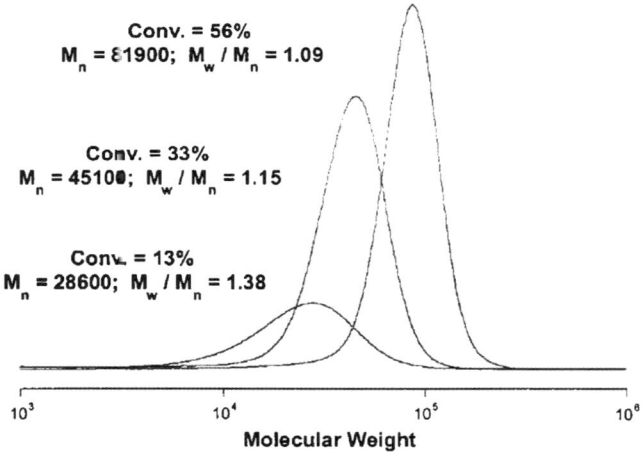

Figure 2.55 Evolution of the molecular weight and polydispersity in the ATRP of methyl acrylate in presence of methyl 2-bromopropionate, Cu(I) and 4,4'-di-*tert*-butyl-2,2'-bipyridine at 90 °C. The polymer growth shows a typical linear increase of the molecular weights with conversion in the ATRP of methyl acrylate. The rate constants of propagation for acrylates are relatively large, and high polydispersity was observed initially, because several monomer units were added after each activation step. As the reaction progressed, the chains become uniform due to continuous exchange reactions and the polydispersity dropped with conversion. © American Chemical Society.[97,98]

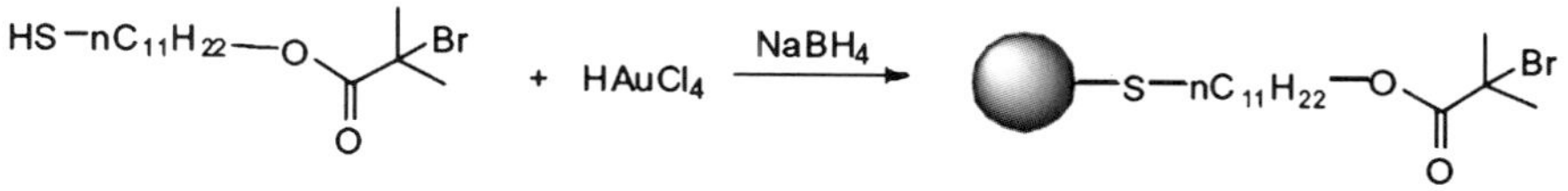

Organic Carbon 107

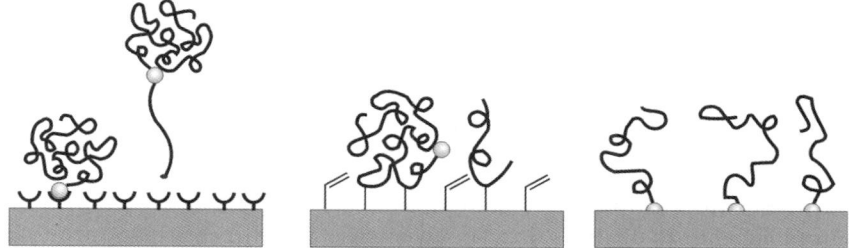

Figure 2.56 Model of a gold nanoparticle with a surface-bound bromide with spacer for surface-initiated ATRP.[99]

Figure 2.57 Surface-attached polymers, here in the presence of water, may appear (a) as "mushrooms" with the highest density of material in a distance of a few nanometres above the solid; (b) as regular brushes made of coiled or stretched polymers with block-like or parabolic density profiles; and (c) as flat, irregular monolayers, so-called "pancakes" with a height of <1 nm.[100]

Such a surface-initiated polymerization (SIP) leads to high polymer densities on the nanoparticles, since the end groups of polymers are already densely planted on the nanoparticle surface and offer plenty of binding sites for more monomers.

Grafting to particles or solids is, of course, not limited to alkylhalides and ATRP. Reactive OH groups on silicate, amino groups of aminated nanoparticles, Au(I) or Au(I)SH on Au(0) nanoparticles, *etc.* react directly with a carboxyl or SH end groups or with several carboxyl or SH substituents on proteins or synthetic polymers. Brush-like, dense coatings with more or less stretched polymers appear if the polymer chains, *e.g.* poly(ethyleneglycol) or poly(ethylene), are located in a good solvent and if only one end group binds to the surface. However, the polymer chains fold to form "mushrooms" if they come in touch with a bad solvent (Figure 2.57). The occupation of a solid comes to a halt relatively early, because both the brush and the mushroom layers block the diffusion of incoming chains and have no affinity to the solid either. "Pancake-type", flat and irregular coatings <1 nm in height result from polyfunctional polymers reacting with densely functionalized surfaces. All three types of polymer layers readily dissolve small molecules.[100a–c]

The amphiphilic block acrylate–styrene copolymers were, for example, used to construct covalent, crosslinked and functional micelles in order to entrap and transport ∼400 positron-emitting ^{64}Cu atoms per 10 nm particle. Crosslinking throughout the amphiphilic shell layer occurred with 2,2-(ethylenedioxy)-bis(ethylamine), functionalization with the macrocyclic chelator 1,4,7,10-tetra-azocyclododecane-tetraacetic acid *via* an extra diamine or lysine linker. Quantitative determination of the ^{64}Cu content showed that the diamino spacer was less efficient than the shorter lysine. Furthermore, large numbers of negative charges on the micellar surface as introduced by sulfonation were also favourable for ^{64}Cu binding and for injection into the bloodstream in medical applications such as positron emission tomography (PET) and radiotherapy.[101a] The hairy, shell crosslinked nanoparticles appeared as stable, non-aggregating spheres with diameters <10 nm in electron micrographs (Figure 2.58).[101a–c]

The poly(styrene)$_x$–poly(acrylate)$_y$ copolymers also allowed syntheses of macroscopic foils with hydrophobic 20 nm wide pores, which may provide matrices for the synthesis of well-defined metal nanotubes or even nanowires. The hydrophopic part of the copolymer phase-separated upon heating above a "glass temperature" of $\sim 70\,^{\circ}$C and annealing, yielding a hydrophobic and a hydrophilic end. If this was done in an electric field between two metal electrodes, *e.g.* an aluminium cathode and a gold anode at a distance of a few microns, and if the hydrophobic chain length was much larger than the hydrophilic one, then the minor acrylate component formed polar channels in a poly(styrene) block. Hydrolysis with hydrochloric acid cut out the polyacrylate and also dissolved the aluminium electrode, leaving a porous poly(styrene) foil with 20 nm wide hydrophobic wells on a gold foil. Commercial poly(imide) foils (*e.g.* Kapton H, see Figure 2.59) were used as carriers for the gold foil and the stiff, hydrophobic holes in the polystyrene membrane were filled with metals or metal salts to form a macroscopic array of conducting 20 nm nanowires on a gold electrode supported by the Kapton.[102]

Kapton H or poly[N,N'-(oxydiphenylene)pyromellitimide] is a form-stable polymer resistant to flame and high temperatures and with a low dielectric

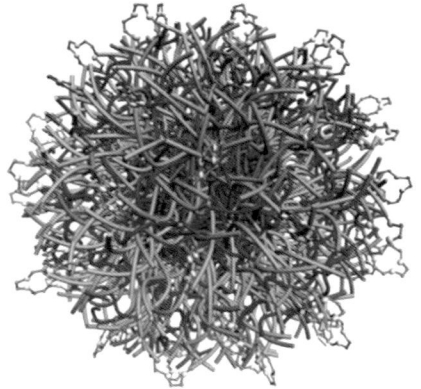 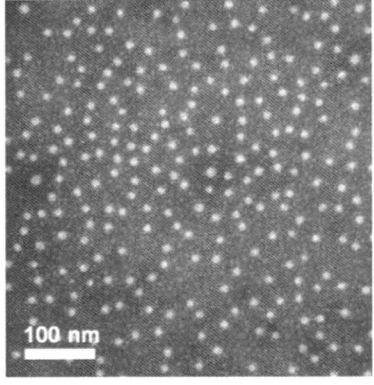

Figure 2.58 Polymer micelles with ligands for the formation of internal ^{64}Cu complexes for positron emission. © American Chemical Society.[101]

Organic Carbon

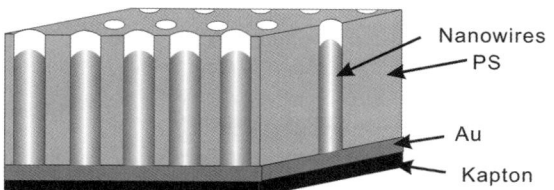

Figure 2.59 Model of the poly(styrene) block on Au(0)–Kapton H with 20 nm pores and nanowires. © American Chemical Society.[102]

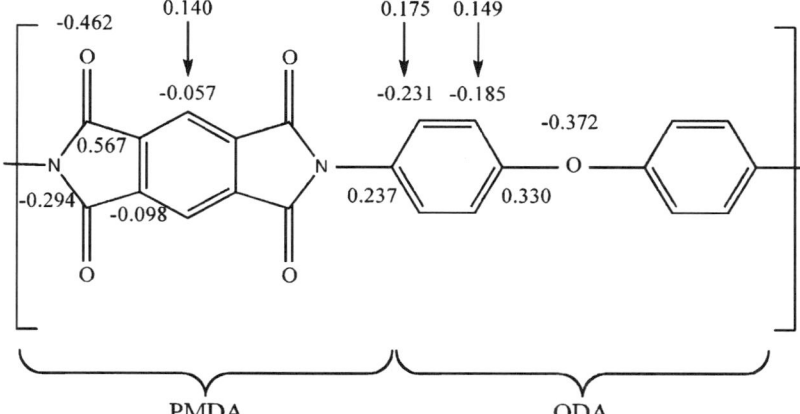

Figure 2.60 Kapton (PMDA-ODA) monomer with partial charges. Arrows point to charges carried by hydrogen atoms. PMDA, pyromellitic dianhydride; ODA, oxydiphenylenediamine.[103]

constant, with two succinimides connecting hydrophobic benzene units (Figure 2.60). Kapton films were developed for the microelectronics of space technology, where they work as an insulating basis for printed circuit boards. Molecular dynamics simulations of interactions between Kapton and water clusters, as well as the corresponding NMR and IR spectra, all indicated chain-like, continuous water clusters around the polymer, similar to those in poly(vinylpyrrolidone).[102,103]

Such crosslinked polystyrene micelles carrying charged groups on the surface also transfer a small portion of their mobile sodium counterions on to electroneutral surfaces upon contact (Figure 2.61). This process is proportional to surface area and the magnitude of charge. Only a small fraction (<0.5%) of the mobile counterions on the surface is removed from their covalently bound partners, and the contacting surface takes up one elementary charge per 2000 nm^2. This value is similar for both positively charged and negatively charged counterions.[104]

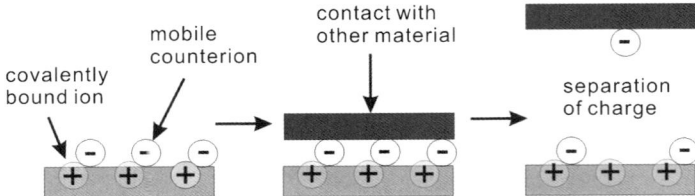

Figure 2.61 "Electret" character of electroneutral surfaces: they allow small-scale mechanical charge separation.[104] (An electret or electride is an ionic compound in which an electron is the anion.)

Triblock copolymers are also accessible from living polymerizations, *e.g.* poly(ethyleneglycol)$_{45}$–poly(styrene)$_{150}$–poly(acrylate)$_{108}$. Such copolymers allow the ordering of nanoparticles either on the surface or the inside of inverse micelles. The linear ethyleneglycols are as soluble in water as in toluene, and they bind only weakly to metal ions. The acrylates are soluble in water, but precipitate with bivalent metal ions, *e.g.* Cd^{2+}. Styrene dissolves in organic solvents but not in water, and also has no interaction with metal ions. Cd^{2+} thus goes selectively to the poly(acrylate) block to form a cadmium-acrylate core of nanoparticles, which are then surrounded first by poly(styrene), and above that by poly(ethyleneglycol). Such inverted micelles with an acidic core then polymerize to rods in tetrahydrofuran with a little water by polyether–water hydrogen bonding. More water splits these rods into spheres. The insoluble cadmium-acrylate cores remain untouched, but the polystyrenes clump together under an ethyleneglycol shell. Addition of hydrogen sulfide then forms an inner coat of cadmium sulfide, which covers the acrylic acid micelles in the polystyrene block.

If the hydrogen sulfide is added already to the tetrahydrofuran solution of the triblock polymer with Cd-acrylate, however, then the rods are destroyed and the hydrophobic styrene comes to the surface, forming inverse micelles with a CdS-coated polyacrylate in the centre. Addition of excess water, which is miscible with tetrahydrofuran, first brings the rods back, but then the poly(ethyleneglycol) migrates to the surfaces and carries the inverse micelles with the hydrophobic cadmium sulfide coating along with it. A small variation of the synthetic steps thus either keeps the luminescent nanoparticles in the hydrophobic interior of a micelle or brings them to the surface of a micelle in water (Figure 2.62).[105]

Single crystals of poly(ethyleneglycol)-hydrosulfide(PEO-SH) were used to immobilize 6 nm gold nanoparticles and to half-coat them with PEO-SH ("grafting from"). 11-Mercaptoundecyl-2-bromo-2-methylpropionate was then fixated on the free surface of the nanoparticle and ATRP with poly(methyl methacrylate) and a Cu(I) catalyst was performed ("grafting to"). After dissolution of the PEO-SH single crystal, very small Janus gold nanoparticles were obtained, with a hydrophilic poly(ethyleneglycol) coating on one half and hydrophobic poly(methyl methacrylate) on the other.[106]

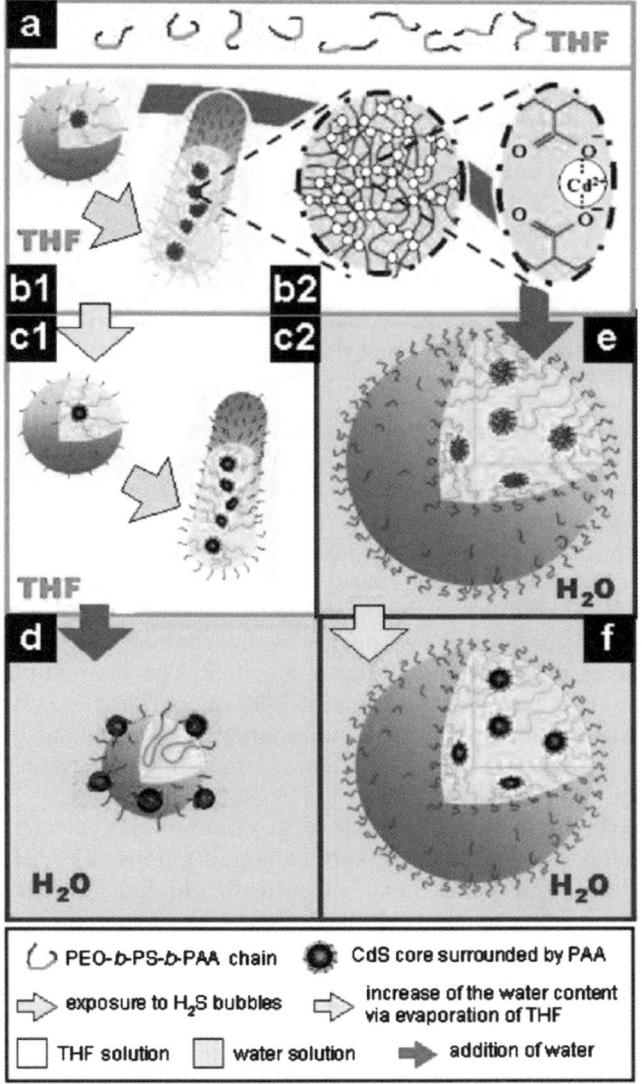

Figure 2.62 Models of the nanoparticles made from PEO(45)-*b*-PS(150)-*b*-PAA. (a) Single triblock copolymer molecules in tetrahydrofuran (THF). (b) Ionically crosslinked triblock micelles. (b1) primary spherical inverse micelles (b2) wormlike micelles at higher water content. (c) Triblock copolymer structures with CdS quantum dots: (c1) spheres in THF; (c2) rods in water-rich solutions. (d) core micelles in water, surrounded by CdS nanoparticles. (e) Multicore cadmium acrylate supermicelle (SM) structures, following the change of the solvent to water of the PSIMs shown in (b). (f) Water-soluble SM triblocks with CdS cores. © American Chemical Society.[105]

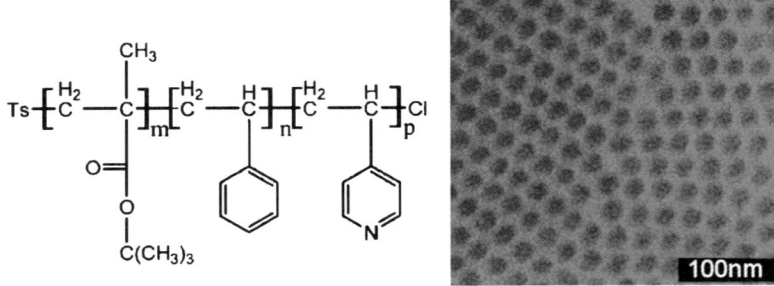

Figure 2.63 Molecular structure and electron micrograph of an acryl ester–pyridine pore. © American Chemical Society.[107]

Poly(*tert*-butyl methacrylate-*b*-styrene) copolymers with a glass temperature of ~100 °C were polymerized by ATRP with 4-vinylpyridine to yield poly(*tert*-butyl methacrylate-*b*-styrene-*b*-4-vinylpyridine). After iodine staining the pyridine phase appeared dark (Py^+ I^-) under the electron microscope, and hexagonally ordered cylinders appeared at a spacing of 23 nm (Figure 2.63). The methacrylate and pyridine phases were separated from each other by the styrene phase. A charged iodonium-pyridine mosaic had been formed, and membranes with long-range homogeneous, hydrophilic and functional pores should be accessible within this glassy polymer. The biamphiphilic triblock copolymer poly(ethyleneglycol-*b*-styrene-*b*-4-vinylpyridine) leads to soluble micelles in water, which wrap metal nanoparticles, in particular of Pd(0), in a monomolecular pyridine layer. Such nanoparticles may be useful for catalysis in aqueous environments.[107a,b]

Dry poly(vinylpyrrolidone) is the most common polymer protectant of transition metal nanoclusters, but surface-bound anions (BF_4^-, HPO_4^{2-}, polyacrylate or dodecylsulfonate) and a solvent with a high dielectric constant, *e.g.* propylene carbonate, cyclic -O-CHCH$_3$CH$_2$-O-CO-, are equally helpful. In the presence of water poly(vinylpyrrolidone) is less efficient in the stabilization of nanoparticles, but then attaches chains of water along the partly positively charged nitrogen and negatively charged oxygen of the cyclic amide (lactam) groups.[108] These water chains then lead to striking differences between H_2O, HDO and D_2O in microgels.[109] Furthermore, poly(vinylpyrrolidone) is a prospective material for use as serum in artificial blood preparations, because it is biocompatible and remains well hydrated at body heat.

Complexes of poly(acrylic acid) and poly(vinylpyrrolidone) (PVP) are stable at low pH values, and dissociate above pH 5 due to deprotonation of the acid. Such a complex was adsorbed to silica nanoparticles and crosslinked with poly(ethylene glycol)-α,ω-diacrylate. The silica core was dissolved with hydrogen fluoride and stable polymer vesicles were thus obtained. An excess of PVP over poly(acrylic acid) guaranteed water solubility of micron-sized vesicles (Figure 2.64).[110]

Pyrrole reduces Ag(I) ions to Ag(0) metal atoms, which first assemble to Ag(0)–polypyrrole nanosheets and then rearrange to Ag(0)–polypyrrole

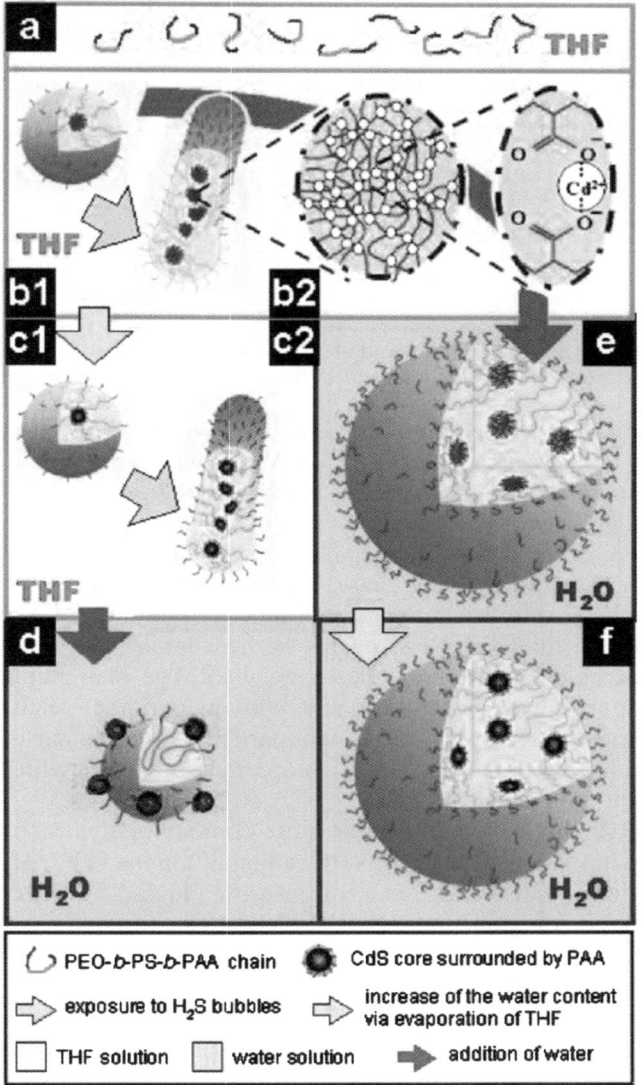

Figure 2.62 Models of the nanoparticles made from PEO(45)-*b*-PS(150)-*b*-PAA. (a) Single triblock copolymer molecules in tetrahydrofuran (THF). (b) Ionically crosslinked triblock micelles. (b1) primary spherical inverse micelles (b2) wormlike micelles at higher water content. (c) Triblock copolymer structures with CdS quantum dots: (c1) spheres in THF; (c2) rods in water-rich solutions. (d) core micelles in water, surrounded by CdS nanoparticles. (e) Multicore cadmium acrylate supermicelle (SM) structures, following the change of the solvent to water of the PSIMs shown in (b). (f) Water-soluble SM triblocks with CdS cores. © American Chemical Society.[105]

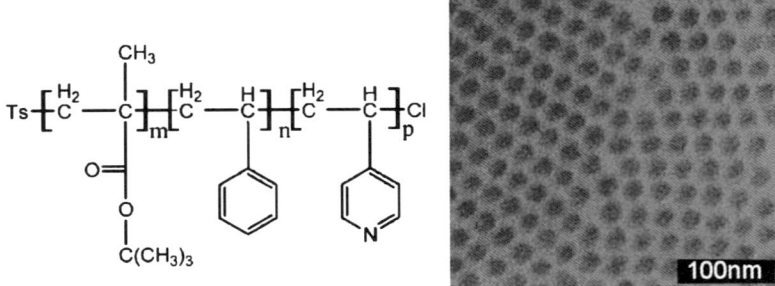

Figure 2.63 Molecular structure and electron micrograph of an acryl ester–pyridine pore. © American Chemical Society.[107]

Poly(*tert*-butyl methacrylate-*b*-styrene) copolymers with a glass temperature of ~100 °C were polymerized by ATRP with 4-vinylpyridine to yield poly(*tert*-butyl methacrylate-*b*-styrene-*b*-4-vinylpyridine). After iodine staining the pyridine phase appeared dark (Py^+ I^-) under the electron microscope, and hexagonally ordered cylinders appeared at a spacing of 23 nm (Figure 2.63). The methacrylate and pyridine phases were separated from each other by the styrene phase. A charged iodonium-pyridine mosaic had been formed, and membranes with long-range homogeneous, hydrophilic and functional pores should be accessible within this glassy polymer. The biamphiphilic triblock copolymer poly(ethyleneglycol-*b*-styrene-*b*-4-vinylpyridine) leads to soluble micelles in water, which wrap metal nanoparticles, in particular of Pd(0), in a monomolecular pyridine layer. Such nanoparticles may be useful for catalysis in aqueous environments.[107a,b]

Dry poly(vinylpyrrolidone) is the most common polymer protectant of transition metal nanoclusters, but surface-bound anions (BF_4^-, HPO_4^{2-}, polyacrylate or dodecylsulfonate) and a solvent with a high dielectric constant, *e.g.* propylene carbonate, cyclic -O-CHCH$_3$CH$_2$-O-CO-, are equally helpful. In the presence of water poly(vinylpyrrolidone) is less efficient in the stabilization of nanoparticles, but then attaches chains of water along the partly positively charged nitrogen and negatively charged oxygen of the cyclic amide (lactam) groups.[108] These water chains then lead to striking differences between H_2O, HDO and D_2O in microgels.[109] Furthermore, poly(vinylpyrrolidone) is a prospective material for use as serum in artificial blood preparations, because it is biocompatible and remains well hydrated at body heat.

Complexes of poly(acrylic acid) and poly(vinylpyrrolidone) (PVP) are stable at low pH values, and dissociate above pH 5 due to deprotonation of the acid. Such a complex was adsorbed to silica nanoparticles and crosslinked with poly(ethylene glycol)-α,ω-diacrylate. The silica core was dissolved with hydrogen fluoride and stable polymer vesicles were thus obtained. An excess of PVP over poly(acrylic acid) guaranteed water solubility of micron-sized vesicles (Figure 2.64).[110]

Pyrrole reduces Ag(I) ions to Ag(0) metal atoms, which first assemble to Ag(0)–polypyrrole nanosheets and then rearrange to Ag(0)–polypyrrole

Organic Carbon

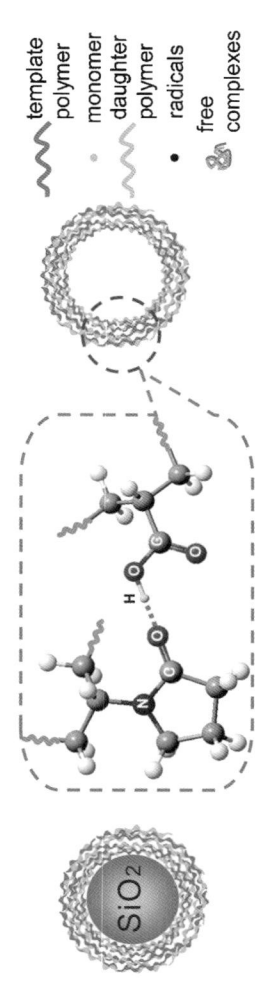

Figure 2.64 Model of the poly(acrylic acid) and poly(vinylpyrrolidone) complex, its coat around the silica nanoparticle and the removal of the silica core by HF.[110] The silica nanoparticle functions as a template: 3-trimethoxypropylate introduces reactive double bonds on its surface, poly(vinylpyrrolidone) stabilizes the aqueous suspension and the indicated complex with poly(acrylic acid) forms a polymerizing shell. "Grafting to" and "etching out" processes with vinyl monomers and water then produce water-soluble polymer cages. © American Chemical Society.

nanocables (Figure 2.65).[111] PVP acts as a capping agent to form silver nanowires, and steers pyrrole monomers on the [111] and [100] facets of silver nanosheets. The interaction between PVP and the [111] facets is much weaker than that between PVP and the [100] facets. Therefore, PVP is strongly adsorbed on the [100] facets, and pyrrole monomers distribute on a surface where a large amount of PVP solvent exists. At the same time, silver atoms easily diffuse on the [111] facets, and will be adsorbed there. The polypyrrole stays on the [100] facets, whereas the [111] Ag(0) facets grow. As the reaction continues, Ag(I) ion and pyrrole concentrations will both decrease, the Ag(0) nanosheets will become smaller and smaller, and the Ag(0) nanowires with thin polypyrrole coatings grow.[112–114]

Grafting of monomeric, liquid vinylpyrrolidone and benzoyl peroxide directly on to raw carbon nanotubes in the presence of Pt(0) produced a Pt–PVP–carbon nanocomposite with an extraordinarily high electrical conductivity of 31 S cm^{-1} (Figure 2.66).[115]

Thiol-terminated PVP was synthesized *via* reversible addition-fragmentation chain transfer (RAFT) polymerization, in which a terminal unit of an inert

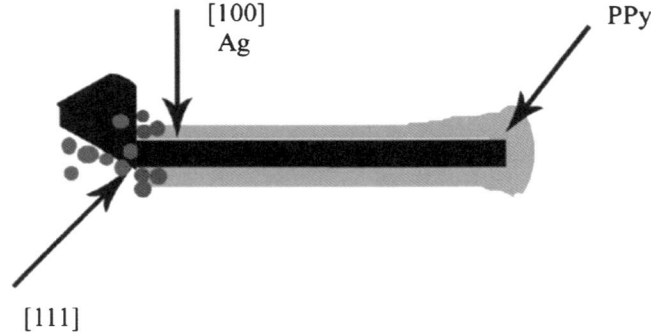

Figure 2.65 Model of the growing Ag(0)-poly(vinylpyrrolidone) nanocable from an Ag(0)-poly(vinylpyrrolidone) nanosheet. Pyrrole monomers block the sheet, Ag(I) ions are reduced and added as Ag(0) to the nanowire. © American Chemical Society.[111]

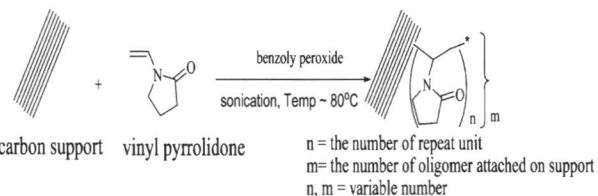

Figure 2.66 Model of vinylpyrrolidone grafting and polymerization on a raw carbon nanotube material. © American Chemical Society.[115]

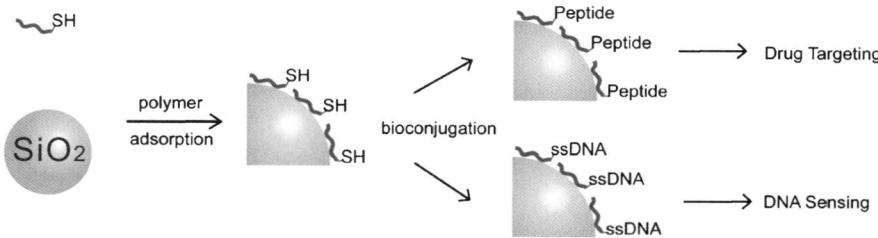

Figure 2.67 Model of an SH-terminated PVP, which is adsorbed on silicate nanoparticles without using the SH group, followed by addition of peptide or DNA and sensing activity. © American Chemical Society.[116]

polymer is functionalized. In this case a thiocarbonylthio group, SCOSH, was introduced and reduced by sodium borohydride to a thiol. Adsorption of the polymer to colloidal silica particles did not involve the SH group, which was then used to bind peptides or oligonucleotides. Both remained functional in their new environment in molecular recognition assays, and the polymer-coated silica nanoparticle (Figure 2.67) worked as a water-soluble, non-aggregating drug carrier.[116]

The three blocks of poly(vinylphenol), methyl methacrylate and vinylpyrrolidone were all miscible through hydrogen bond or dipole–dipole interactions, but the phenol–pyrrolidone hydrogen bond was dominant. This internal binding pattern led to various stable self-assembly morphologies such as hexagonal and lamellar phases.[117] Dyes dissolved in different regions of 245 nm, water-soluble poly(styrene-*block*-vinylpyrrolidone) nanobeads indicate pH, oxygen, temperature and other environmental changes by rapid colour changes.[118]

Nanoparticles based on poly(*N*-vinylcaprolactam) are temperature- and pH-sensitive and are used as switchable or stimulus-responsive materials. Chemically this constitutes the rare case of a useful statistical mixture of ill-defined polymers. Neither the sequence nor the block formation matters. It is their chemical composition which (1) determines the location of functional groups within swollen crosslinked parts of microgel nanoparticles and (2) allows targeted adsorption as well as covalent grafting of small molecules, polymers or proteins on their surface. No special polymerization procedure is necessary: the polar groups migrate automatically to the surface in water and the hydrophobic parts form the core. In the swollen coil state solvent–chain interactions dominate; in the collapsed state chain–chain interactions are more important. It does not matter where the vinylcaprolactam is located in the polymer chain: all copolymers with acetoacetoxy methacrylate and vinylimidazole form "microgels" in water, which are temperature- and pH-sensitive and bind to metal ions or biomolecules.

The same mixed polymer with a high percentage of caprolactam and very little imidazole also gave pH-dependent microgels. Scanning field emission electron microscopic images showed a corona-like shell around the particles, which was not present in the absence of vinylimidazole. As usual, small

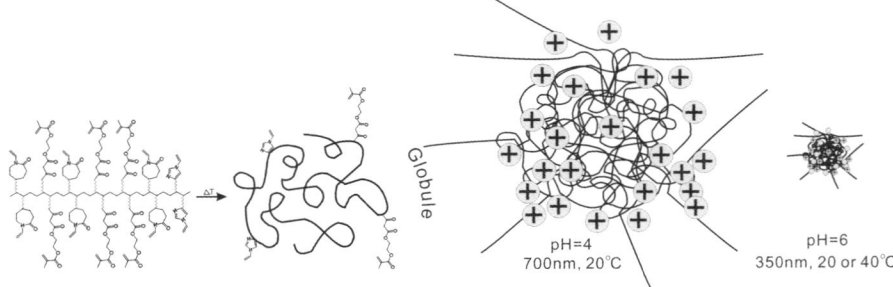

Figure 2.68 The statistical polymer with caprolactam, acetoacetoxy methacrylate and vinylimidazole side chains formed large coils in water at room temperature, which became smaller upon heating, when the inner hydration water was partly removed. Both hydrophilic substituents remained on the surface. The nanoparticles also shrank when the pH was raised to 6. A rise in temperature then had no effect, because the inner hydration water had already gone. © American Chemical Society.[119]

amounts of hydrophilic units in a hydrophobic particle strongly increase its hydrophilicity and allow for much greater swelling in water, especially if the units can add protons. Protons tend to move into hydrophobic environments and they take up a lot of water. This swelling reached a maximum at pH 4, and was caused by the strong electrostatic repulsion between the charged imidazolium groups. The diameter of particles containing 3% of imidazole doubled (Figure 2.68).

These variations of the microgel dimensions did not cause particle aggregation. The colloidal stability was monitored with an analytical centrifuge and a linear increase of the sedimentation constant by a factor of 4 was observed, when the pH was increased from 4 to 10. After that it levelled off. There are imidazolium groups up to pH 10, and it is not the size but the density of nanoparticles that dominates sedimentation. Surprisingly, the electrophoretic mobility was zero even at pH 7.5 and 10 °C, whereas at 40 °C practically no mobility change occurred from pH 4 to pH 10. Heating allowed the polymer network to collapse, shifted the charged imidazolium units to the outer layer and increased their pK_a. In the swollen coil state, water–chain interactions allowed the entrance of salts into the polymer network; in the collapsed state, chain–chain interactions dominate and the cationic counterions are forced into the water.[119]

The same melting behaviour of polymer gels can be adjusted to occur around the human body temperature of 37 °C. Water-soluble poly(N-isopropylacrylamide) exists in water in an expanded conformation up to 32 °C. At higher temperatures it becomes compact and its phases separate: the gel fluidizes and the polymer precipitates. The reasons for the phase transition is cooperative dehydration of the polymer chains. Relaxation frequencies of the chains depend on electric dipoles and were measured at different temperatures. There is a steep decrease between 30 and 40 °C, depending on the detailed polymer structure. The hydration number was determined to be 11 water molecules per monomer unit below 32 °C in aqueous solution, and diminished drastically slightly above this temperature irrespective of the molar mass of the polymer. The water content could not be determined in the final precipitate, but IR spectra indicated strong NH . . . OC hydrogen bonds between the isopropylamide groups, so the solid was probably totally dehydrated. It is the cooperative destruction of hydrogen-bonded chains at 32 °C that makes the difference. Similar shrinking effects in proteins presumably also occur in humans at "fever" temperatures above 40 °C. Dielectric spectroscopy gave 11 water molecules per monomer, other spectroscopic methods only 2–3. Hydrogen bond bridges between the surface water molecules thus perhaps provide eight extra water molecules, which are, however, quickly lost at ∼40 °C. This picture agrees with the instability of the bulk water network at this temperature (Figure 2.69).

In the swollen coil state water dissolves many proteins and other amides. At higher temperatures water is expelled and intramolecular hydrogen bond formation leads to a collapse of the polyamides.[120] A microgel nanoparticle is also an optimal medium for the rapid formation of uniform apatite nanoparticles from Ca^{2+} and PO_4^{3-}.[121]

Linear block copolymers are usually applied as multilayer coatings of solid surfaces, or as a monolayer coating of nanoparticles. The coverage of surfaces with high curvature and/or many edges is often most successful if it starts with the covalent addition of monomers followed by living polymerization. Classic anionic living polymerization, which requires thermally labile blocks, rigorous exclusion of oxygen and water, and ultrapure reagents cannot be used here, but free radical polymerization overcomes problems with functional nanoparticles.

The thermal fragmentation of nitroxides provided the first reliable method. The carbon–oxygen bond of alkoxyamines is "dormant" at room temperature, but "wakes up" upon heating to 125 °C, which is acceptable for almost any nanoparticle. Thermal fragmentation provides a stable nitroxide free radical, which reacts at near-diffusion-controlled rates with carbon-centred radicals. The polymeric radical undergoes chain extension with the radical monomer to yield another polymeric radical, which then recombines with the nitroxide, and the cycle of homolysis–monomer addition–recombination is repeated (Figure 2.70).

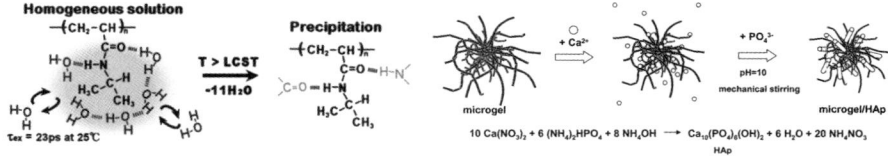

Figure 2.69 poly(N-isopropylacrylamide) loses water on the periphery and shrinks in aqueous solutions between 30 and 40 °C – a "fever" temperature at which water clusters also break down in bulk water. At pH 10 Ca^{2+} ions agglomerate in these polymer nanoparticles and uniform apatite nanocrystals are formed within the polymer. © American Chemical Society.[120,121]

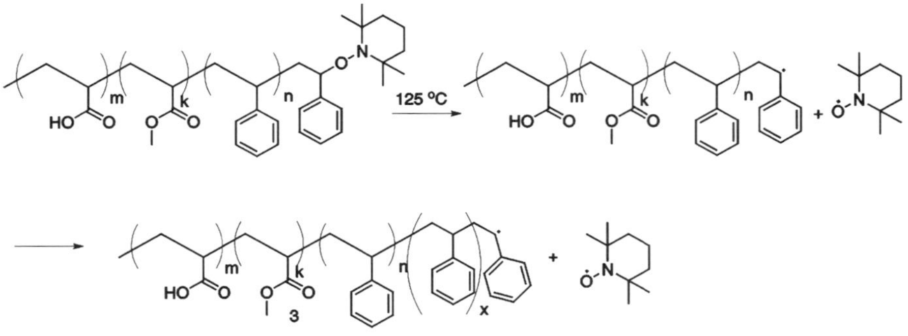

Figure 2.70 Transformation of a dormant nitroxide to a reactive radical and addition of a styrene molecule to form a new radical for the chain reaction and a typical, unrealistic formula of a triblock polymer from living polymerization.[122]

The stable nitroxide may be replaced by free radical transition metal species to obtain a variety of copper-, nickel-, or ruthenium-mediated "living" free radical systems. All of these systems depend on reversible homolysis of a covalent, inactive species followed by monomer insertion and reversible recombination. The overall concentration of radical chain ends is any time is low and the polymer chain grows in a controlled, or pseudo-living, process. "Living" free radical polymerizations are conducted under non-demanding reaction conditions, e.g. at 125 °C under a nitrogen atmosphere without any purification of reagents or solvents. Linear "biopolymers" of natural or synthetic origin, synthetic spherical dendrimers typically made of four dendrons (tree-like, branched fragments) and synthetic block polymers obtained by "living" or "controlled" polymerization constitute the most common attachments and coatings of nanoparticles. A typical copolymer made from acrylic acid, its methyl ester and styrene is shown in Figure 2.70, although concrete numbers for k, m, n must not be taken seriously. A synthetic block polymer will never approach a biosynthetic protein in purity. A variation of ±5 for each number may be realistic, and it is close to impossible to verify realistic proportions of impurities. Products of "living" polymerization under careful control provide nicely reproducible material properties that's all.[122]

Slow radical chain polymerizations of alkenes are often used in a related polymerization technique called atom transfer radical polymerization. Methylmethacrylate, for example, adds carbon tetrachloride in the presence of redox active Ru(II), and a sterically hindered aluminium base polarizes the carbonyl group adjacent to the C–Cl bond. Added methacrylate then substitutes the chloride, the resulting cation radical is again inactivated by chlorine radical and another metacrylate attacks. Highly monodisperse polyesters are thus formed after a given time, and may be coupled with anionic polyacrylate or polystyrene under practically identical conditions. Cu(I)/Cu(II) mixtures often replace the carbon tetrachloride as a mild radical source.

A free radical initiator was then first silanized and and then immobilized on silicon wafers. This surface initiated the formation of styrene brushes and bound them directly on the surface. Friction force measurements were then performed by AFM using a 5 μm spherical silica tip, which touched the polymer layer, and different solvents to demonstrate the dependence of brush lubricity on the solvation of the brush's interior. The measured coefficients of friction for the silica tip in poly(styrene) brushes were 1000 times lower in toluene than in 2-propanol or *n*-butanol. The elastic character of the polystyrene depends on isolated, solvated chains. Without solvents (or softeners) the benzene rings stick together and friction with external materials increases.[123]

The "grafting from" method of binding polymers on to nanoparticles, as just described, usually binds an optimum amount of polymer, because the monomers occupy the particle surface completely, as long as there are reactive groups. "Grafting to" nanoparticle surfaces that already carry polymers (Figure 2.71) is less efficient, because the attached polymer chains act as efficient barriers for incoming polymer chains, whereas they dissolve monomers readily.

Synthetic polymer chemistry also provides perfect covalent spheres with different large spaces in the centre and densely packed, almost inaccessible surfaces. These molecules, which look like nanocrystals, are regular dendrimers with tree-like branching. They can be synthesized by a repetitive multistep reaction at the 100 g scale, and are water and acid stable if the monomers are connected by ether linkages. The most common synthesis uses phenolate anions and phenyl bromides as educts. Simple bromination–substitution cycles of disubstituted catechols thus produce up to five generations of double-branched, tree-like dendrimers in each cycle. In the fifth generation the surface is fully occupied, so no further branching to produce a sixth generation is possible.

Two typical examples for the unique possibilities of such dendrimer syntheses, combining benzene chemistry with nucleophilic substitutions, are shown in Figure 2.72:

- An alkyne with two second-generation dendrimers is cyclized with a cobalt catalyst to give a flat nanoparticle 4 nm wide, in one step.

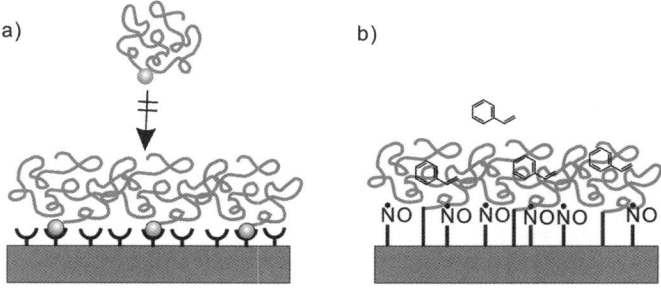

Figure 2.71 Models of "grafting to" and "grafting from" nanoparticle surfaces. © American Chemical Society.[123]

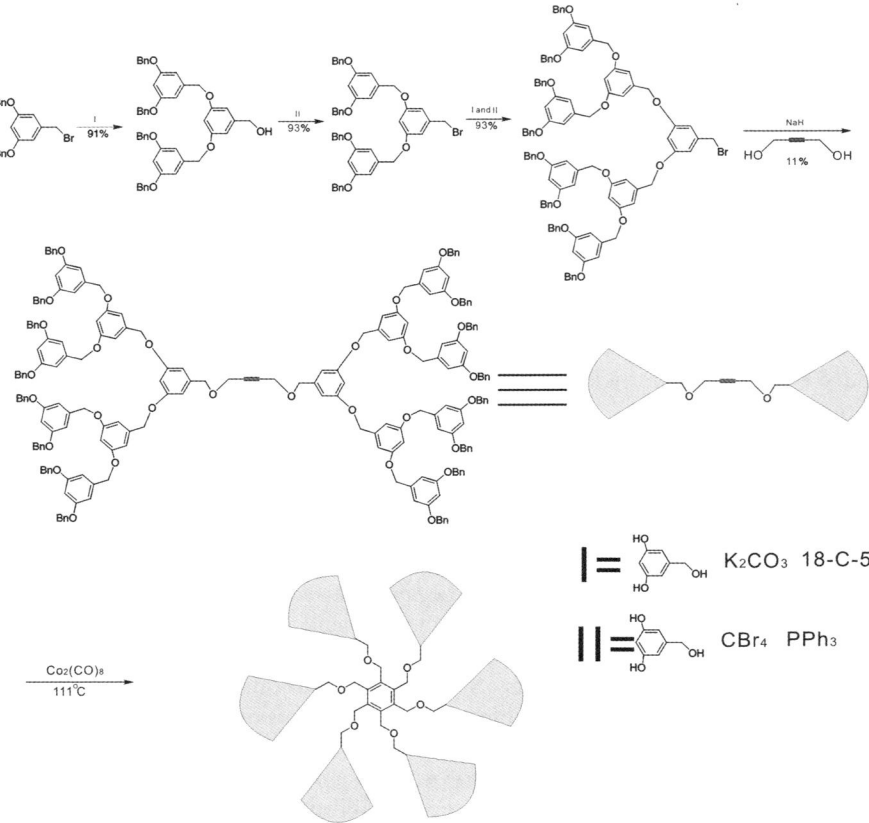

Figure 2.72 Typical dendrimer synthesis scheme.[124]

- From a biphenyl-4,4′ diol, simple step-by-step bromination–phenol coupling produced a perfect Janus-type micelle, which turned benzene surfaces polar and water surfaces apolar.[124,125]

If first-, second- or third-generation dendrons are attached as substituents of linear polymers at appropriate distances from each other, the original polymer coil is stretched into a basically linear chain wrapped in a dendritic layer. A filled, rigid cylinder is thus created from a mobile thread (Figure 2.73). Such molecules, with diameters up to 10 nm and plenty of empty space in the centre, have been cross-linked *via* lateral anthracene units and formed then stable monolayer films over 45×45 μm-sized holes.[126a,b] Such interconnected dendrimer chains provide today the only stable, single layer polymer sheets. However, they contain nanoholes, and are expensive. But they do neither fold nor aggregate.

The core of each dendrimeric sphere has empty space for the entrapment of water-insoluble drugs. In the fourth or fifth generation the end groups are tightly

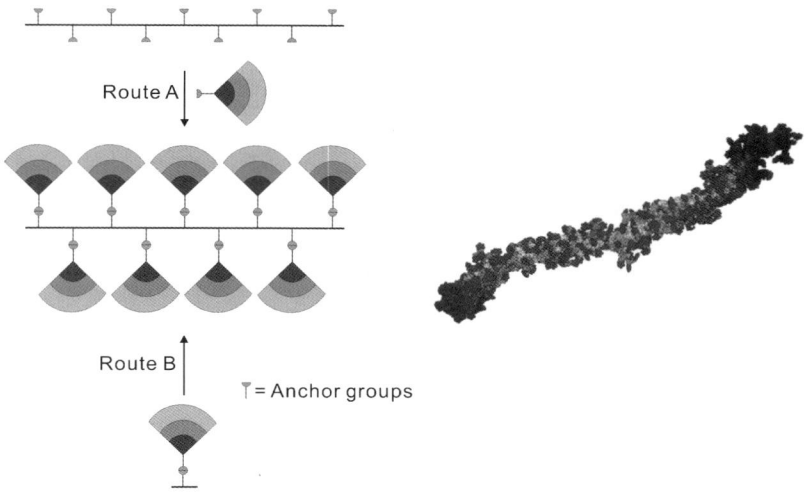

Figure 2.73 Models of synthetic chains of dendrimers.[126]

packed and function as a protective coating or shell. The dense periphery and the spacious centre thus allow for selective liberation of guests from the "dendritic box". Furthermore, the shell may also be hydrophilized by terminal sulfonation. A problematic aspect of dendrimers is that end groups of low generations tend to fold back on themselves, which removes the density gradient. Furthermore, synthetic pathways to fourth- or fifth-generation dendrimers are long, but the smaller, more readily accessible units are often not useful.

A "water-like" dendrimer, which might be developed to form perfectly camouflaged carrier molecules in blood, is the highly branched polyglycerol (Figure 2.74) with an average molecular weight of 5000. The terminal diols can be masked selectively with acetone, leaving isolated OH groups on the periphery that may then be esterified. One successful carrier based on this hyperbranched polyglycerol contained functional pyrrole-phenyl units in its core for the solubilization of hydrophobic drugs and fluorescent dyes. A *para*-iodide was Suzuki-coupled with a substituted *p*-phenylboronate or *via* Cu(I) catalysis to introduce the second phenyl group with useful substituents, *e.g.* the acetylene needed for click reactions.

The diameter of the dendrimers is usually 2–3 nm. They are perfectly covalent nanoparticles and are detectable by routine AFM. They may serve the same carrier purposes as metal oxide based nanoparticles, except that magnetism is missing. Pyrrole units on dendrimer spheres lead to conducting, aromatic and heterocyclic polymers.[127]

Single-chain poly(phenylenes), poly(pyrroles) and poly(thiophenes) are not graphite-like networks, but classical, relatively flexible chain polymers. Each unit contains six conjugated π electrons like benzene, and each unit can be oxidized to a π radical or be doped as a radical semiconductor. The only non-aromatic polymer is poly(acetylene), which consists of ethylene units. The triple

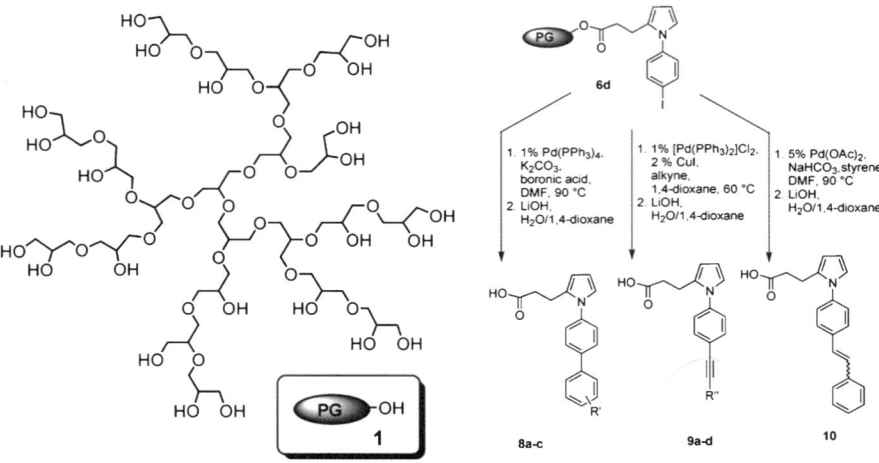

Figure 2.74 Structure of a polyglycerol molecule. PG-OH indicates the same molecule, where all diol groups are ketalized by acetone and cannot be esterified. Only one OH group reacts with the bulky acid on the right: the inner OH groups are not accessible.[127]

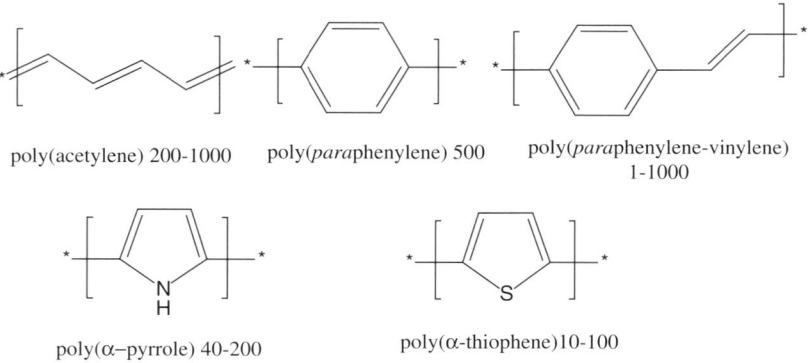

Figure 2.75 Structure and conductivity (in $\Omega\,cm^{-1}$) of some polymers.[128]

bond electrons of the acetylene monomers are used up to form the extra single bond (Figure 2.75).[128]

Poly(pyrrole) layers were used as an electrical-stimulating medium to grow differentiated bone marrow and nerve cells,[129] but they are difficult to prepare in quantity and with reproducible, stable electron conductivity.

Apart from medical applications, poly(thiophenes) are nowadays more popular then poly(pyrroles), because they are more stable and many pure thiophenes are commercially available. α-Connection alone between the thiophene cycles is not satisfactory, because the C–C single bonds are too flexible

Figure 2.76 Typical molecular structure of a thiophene conducting material.[130]

and electron transport becomes difficult over zigzag pathways. Increasing the rigidity leads to better stacking and better charge transport properties, but decreases solubility. The first synthetic solution of this problem is usually the introduction of flexible alkyl substituents. Substituents improve solubility and do not interfere with redox chemistry. Secondly, the α-connected parts are mixed with rigid "fused units", *e.g.* thiophene tetramers connected through α- as well as β-carbon atoms (Figure 2.76). These fused oligomers are also known for their very good environmental stability owing to their relatively large band gaps, and the non-planarity of only α-connected parts is largely wiped out in crystalline layers by the rigid neighbours. The ideal copolymer is reasonably flexible in solution and rigid in conductive layers. The field-effect hole mobility of the copolymer, $0.33\,\mathrm{cm^2/V\,s}$, makes it a practical semiconductor.[130]

Perhaps the most practical linear polymeric, electron-conducting dye is the aniline black of shoe polish. Take aniline, add some base, oxidize with air and mix with a fat. The paste thus obtained, which was one of the original products of IG Farben in 1850, has uses beyond colouring and cleaning black shoes. This shoe polish is nowadays on its way to becoming a high-tech electronic material on the surface of chips and nanowires. The reason for this is that the black and insulating imine-enamine chain becomes a colourless and electrically conductive poly(enamine), if mechanically blended with a solid organic acid. The electrical conductivity of the original black base is of the order of $10^{-9}\,\mathrm{S\,cm^{-1}}$, and equimolar mixtures of the base with camphorsulfonic or picric acid have a conductivity of $10^{-2}\,\mathrm{S\,cm^{-1}}$. Heating to 85 °C increases conductivity by a factor of 1.4 for the blend with camphorsulfonic acid and as much as 8.9 for mixtures with the electron-donating phenolate of picric acid, which presumably also formed charge transfer complexes.[131]

The proton from the acid goes to the imine nitrogen of the base and the protonated quinone diimine becomes benzenoid by transfer of an electron from nitrogen to the neighbouring ring. The remaining unpaired electron on the nitrogen, which represents the electron hole in the valence band, is responsible for the electrical conductivity. The current in polyaniline does not run over chains of positive charges – solitons, as described for polyenes on p. 77 – but over localized radical ions. They are localized on the nitrogen bridges, but travel along the polyaniline backbone, because the benzene rings next to the nitrogen have become aromatic 6π systems after N-protonation (Figure 2.77). This leads to an upward shift of the highest occupied molecular orbital (HOMO; valence band) and a downward shift of the lowest unoccupied molecular orbital (LUMO; conduction band; see Figure 2.78). The ionization creates a hole (unfilled level) in the HOMO, which can take up electrons on top of the valence band. The metallic character is established at the lowel level; no

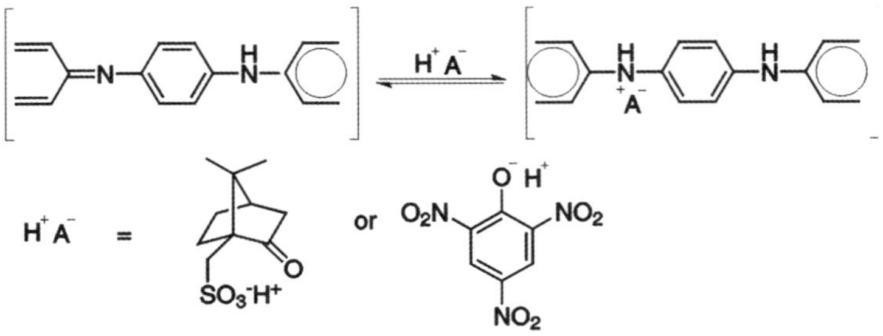

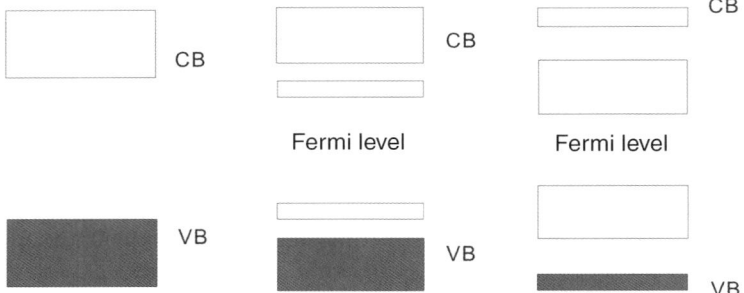

Figure 2.77 Activation of a poly(aniline) electron-conducting chain.

Figure 2.78 From left to right: Schematic arrangement of valence and conducting band in insulators, semiconductors and metals. The Fermi level E_F begins at the upper edge of the valence band.

lattice distortion takes place on the polymer chain, as the positive charge is delocalized over the whole chain.[132–134]

This situation corresponds to the conduction mechanism in doped organic polymer materials. Such a charge localization process is called a polaron in condensed-matter physics, but in chemical terms it is a radical ion. The polaron binding energy is of the order of 0.05 eV, the valence band remains full and the conductivity band empty. There is no appearance of metallic character since the half-occupied level is localized in the gap.

When a second electron is removed from the polymer chain a bipolaron is formed, which consists of two electrons coupled through a lattice vibration or a phonon. The electronic states appearing in the gap for a bipolaron are further away from the edges of the valence and conducting bands than for a polaron, and Coulomb repulsion is screened by the presence of counterions ("dopants"). The polymer is still insulating, because everything happens in localized gaps.

Conductivity only arises when the bipolarons recombine at high doping levels to form spinless bipolarons. The density of states at the Fermi level is still vanishingly small (<0.03 states eV^{-1} per monomer), but the massive overlap between the bipolaron states leads to the formation of two ~0.4 eV wide bipolaron bands within the gap. The band gap has widened from 3.2 eV in the neutral state to 3.6 eV in the 33% doped state, but the bipolaron states coming from the valence band and conducting band edges render the spinless bipolarons with their two charges mobile at high dopant concentrations, where Coulomb attractions are largely screened by the counterions. The bipolarons are spinless charge carriers as suggested by the lack of any ESR signal in highly conducting polymer films. All valence or conducting bands are either totally filled or empty, and mobile bipolarons, not electrons, transport the current. At very high p-doping levels, about one dopant per every other monomer unit, the lower and upper bipolaron bands could merge with the valence and conductivity bands, which would lead to metallic behaviour.

Such doping levels may occur in polyaniline, but are impossible to produce experimentally in the hydrocarbons, polypyrrole and polythiophene. Here, hopping among soliton states is a more likely mechanism of electron conduction. At low doping levels, charged solitons are formed either directly from existing neutral solitons or by recombination among polarons. Upon increasing the doping level, soliton states at midgap start overlapping and form a soliton band. The conduction process may then be based directly on mobile, charged solitons or on disorder-induced band tailing effects, which would allow electrons at the top of the valence band to hop among soliton states.

In polymeric materials, the charge carriers not only move along the chains but also jump from chain to chain. In macroscopic conductivity this process is probably rate-limiting, and models again propose the hopping of solitons or bipolarons.

In spite of its salt character, polyaniline is hydrophobic; nevertheless it changes its electric properties in the presence of water. In the dry state it conducts through 1D bands, provided it is at least semicrystalline in one direction. Hydration of the NH groups disrupts the parallel alignment of the chains and decreases conduction. Furthermore, when dissolved, polyaniline behaves as a normal polymeric redox couple, without special properties. If the strands are crosslinked and hydrated, however, electron-conducting redox hydrogels are formed. Here electrons diffuse in all directions through electron-transferring collisions between hydrated reducible and oxidizable polymer segments. Water then lowers the local viscosity and reduces attractive Coulombic interactions. As a result, segmental mobility and electron diffusion increase together with the permeability of water-soluble ions and molecules. Water thus introduces all kinds of short-cuts for the conduction of electrons in poly(aniline) gels. The Fermi level between the valence and conducting bands gets filled up (Figure 2.78).[131]

Figure 2.79 shows visible–near-IR absorption spectra of polyaniline at potentials of 0.65 V, where polyaniline is neutral (reduced) and insulating; at +0.35 V, where polyaniline has its maximum conductivity; and at +1.35 V, where the oxidized polyaniline again becomes very insulating.

Electrically conducting products are the most conclusive proof of the protonation of polyaniline, but supporting evidence was also found in the IR spectra (Figure 2.80). The 1593 and 1498 cm^{-1} bands of the free base are

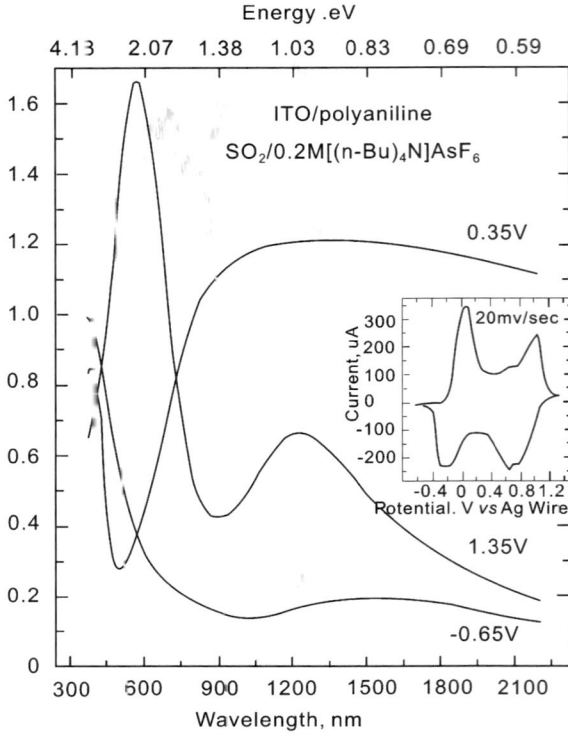

Figure 2.79 Visible-near IR spectrum of polyaniline on indium–tin oxide electrodes (p. 186) at the indicated potentials at $-70\,°C$. The integral area of the inset cyclic voltammogram indicates 3.4×10^{-7} mol or 1.8×10^{16} repeat units per cm^2 of polymer. This corresponds roughly to $10^{16}\,nm^2$ or $10^{-2}\,cm^2$. One out of 100 aniline units carries an electron hole in the valence band. © American Chemical Society.[131]

consistent with quinone and benzene ring deformations. They show a red shift to 1569 and 1480 cm^{-1} in the protonated sulfate. The 1308 cm^{-1} band of the free base and the protonated species is assigned to the C–N stretch of secondary aromatic amine, the 832 cm^{-1} band to an aromatic C–H out-of-plane bending vibration. In the region around 1100 cm^{-1} aromatic C–H in-plane bending modes are also observed in all spectra. The most characteristic band of the conducting protonated form is, however, observed at $1238 \pm 1\,cm^{-1}$ and originates from a $C-N^+$ stretching vibration of the bipolaron structure, which is responsible for electron conduction in polyaniline.[131]

Polyaniline is poorly soluble and needs a carrier to dissolve it. Uniform latex nanoparticles are commercially available and still dissolve in organic solvent when heavily coated with polyaniline. Here X-ray photoelectron spectroscopy (XPS) provides information about the chemical properties of the shells around nanoparticles down to a depth of $\sim 5\,nm$. The asymmetrical profile of an N(1s) spectrum of conducting polyaniline on latex was fitted to three components at

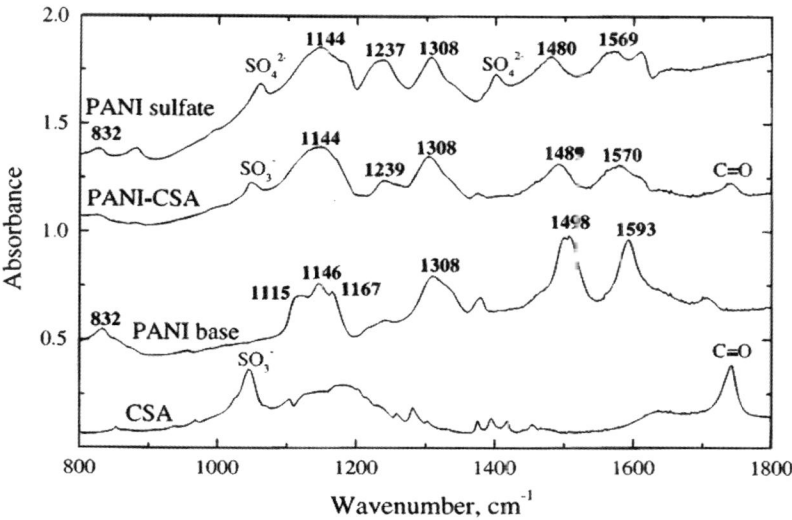

Figure 2.80 IR spectra of conducting poly(anilines). See text. © American Chemical Society.[131]

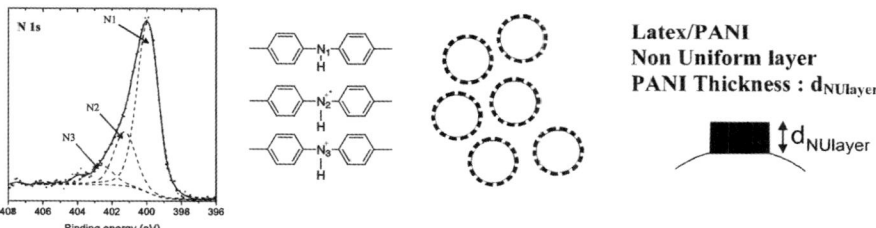

Figure 2.81 Scheme of aniline-coated latex nanoparticles. © American Chemical Society.[132]

400.0, 401.2 and 402.6 eV, which were assigned to the three kinds of nitrogen shown in Figure 2.81, namely neutral amine (1), cationic radical nitrogen atoms (2) with a delocalized positive charge, and the most positively charged nitrogen cation (3). There is no distinct border between these different types of nitrogen: they are interchangeable. The positive charges and the single electron are delocalized over the aromatic rings and the whole polymeric chain.[132]

The same XPS spectrum was reported for a polyaniline powder doped with a phosphate monoester with a short hydrophilic ethylene glycol (OEG) segment (Figure 2.82).[132] The visible–near IR spectra and cyclovoltammograms of this mixture were measured at pH values from 0 to 10 and always showed metallic behaviour, whereas with hydrochloric acid as dopant, the voltammograms and near-IR bands disappear, together with conductivity, above pH. The OEG–phosphate counterion dissolved the polyaniline in water, inducing an electric

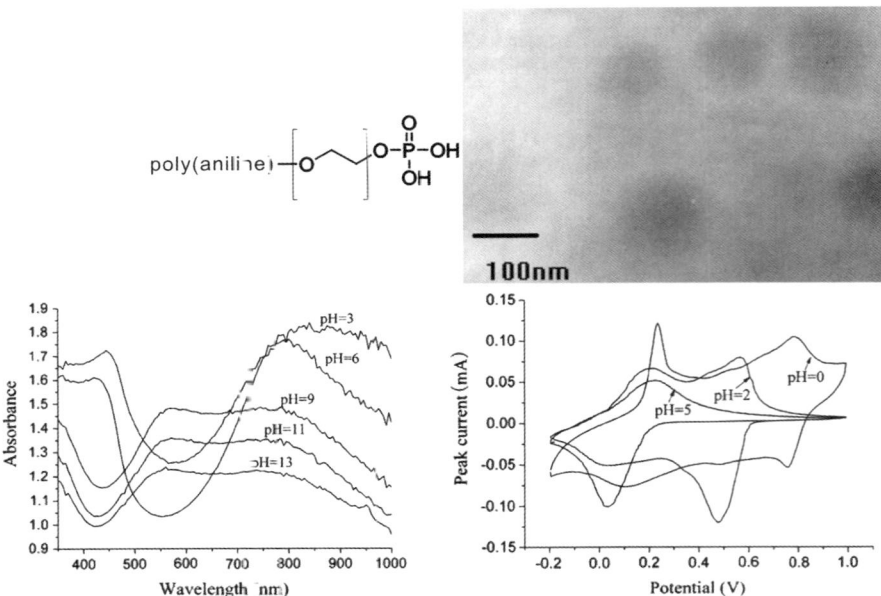

Figure 2.82 Strong near-IR bands and cyclovoltammograms indicate the survival of metallic polyaniline at pH values above the pK_a of aniline, which is extraordinary. The phosphate obviously wraps round the polarons and keeps the protons there, while the OEG chain keeps the polymer water soluble. © American Chemical Society.[134]

conductivity of 4–5 S cm^{-1} at acid pH and of 0.15 S cm^{-1} after 1 day at pH 8. The IR band and conductivity disappeared after 8 days at the high pH only.[132]

When the pH value was raised from 3 to 6, the broad band at 850 nm underwent a red shift accompanied by a colour change from green to turquoise. When the pH value further increased from 6 to an alkaline value, a broad shoulder at 570 nm appeared, and the peak at 440 nm disappeared. The band at 750–800 nm remained even at pH 14, and the relative intensity of the 570 nm compared to the 750–800 nm bands remained almost constant. Similar spectroscopic phenomena were observed in colloidal polyaniline–poly(phenol-p-tetrasulfonic acid), but these assemblies were neither conducting nor soluble.[133,134]

The conductivity or resistance of semiconducting materials should be always measured by the four-point probe technique (Figure 2.83), which was originally developed by geologists Four needle-like electrodes in a linear arrangement are injected into the material. The outer two electrodes measure the current produced by an applied potential, the inner two determine the electric potential distribution. The application of separate electrodes for current injection and potential measurement eliminates the contact resistance between the metal electrodes and the material and since the contact resistance can be large and difficult to interpret, data measured by the four-point probe technique are

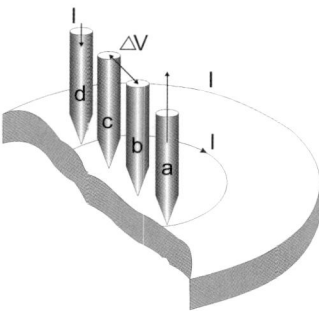

Figure 2.83 Four-electrode measurements of conductivity.[135]

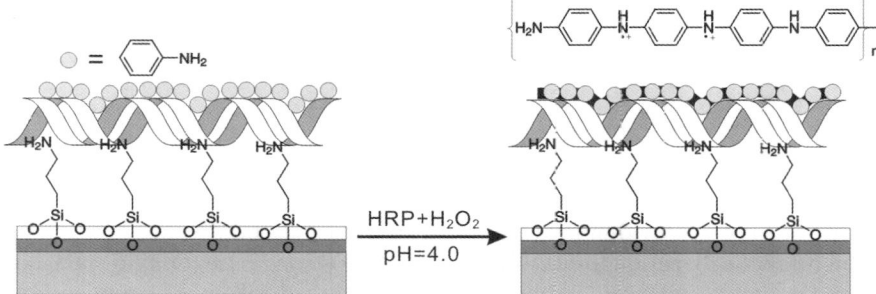

Figure 2.84 Fabrication of a polyaniline nanowire on a silicon surface with stretched double-stranded DNA as a guiding template, horseradish peroxidase as a catalyst and hydrogen peroxide as oxidant.[136]

usually much easier to understand than those gathered by two-point probe techniques.[135]

Enzymatic polymerization of aniline with peroxidases, *e.g.* horseradish peroxidase (HRP), requires only hydrogen peroxide as an oxidant, and reaction conditions are enzymatically mild. The polyaniline obtained was at first a disappointingly insulating mixture of *ortho*- and *para*-substituted C–C and C–N polymers with only few conducting head-to-tail components. The desired *para*-directed coupling was, however, greatly enhanced when linear, negatively charged λ-DNA was added as a template for the cationic aniline species. The DNA was first stretched on aminated silicate particles and pH 4 was acidic enough to protonate half of the aniline nitrogen atoms and leave negative charges on half of the phosphate diesters (Figure 2.84).[136]

A successful polymerization leading to poly(aniline) fibres on the gram scale occurred along the interface of silicon in immiscible solvents such as carbon terachloride, benzene, toluene or carbon disulfide with a solution of aniline, ammonium peroxydisulfate and camphorsulfonic acid in water. Dark green polyaniline formed in water with a total yield of 6–10%, while the organic layer contained un-wanted red–orange aniline oligomers.[137] Large (115 nm diameter) poly(aniline) nanoparticles were obtained with formic acid, and nanoparticle layers showed a minute conductivity.[138]

A simple aniline trimer with silyl ester terminals was deposited as a molecular layer on silica and had differentiating effects on growing neuronal cells similar to poly(pyrrole);[139] oxidase activity was transmitted by a poly(aniline) hydrogel.[140] The polymeric dye is thus compatible with isolated biological tissues and simple oligomers may be as versatile as the polymer. An electrochromic display exploiting redox-active poly(aniline)–Prussian blue $+$ vs. $-$ multilayers underwent a green to blue transition over the potential range from –0.2 to 0.6 V vs. a saturated calomel electrode.[141] A short review on poly(aniline) fibres is also available.[142] Aniline black, familiar to us in shoe polish as mentioned earlier, may become the major plaything of organic electronics because it looks less harmful, is more easily accessible and much less poisonous than all the competitors.

2.9 Dyes

Dyes should be electron conducting if they have to connect metallic or redox-active nanoparticles. Simple candidates are viologen and fulvalene. Viologen, a colourless 4,4'-biypyridinium dication, generates its eponymous colour only when it takes up one single electron either from a strong base (OH^- or NH_2^-) or from a reducing agent (H^-, $Mg(0)$). The dication then becomes a monocationic radical and its violet colour originates from charge transfer: upon irradiation with visible light the single electron of a neutral pyridine ring jumps to the pyridinium cation and absorbs the light quantum. The single electron of the viologen radical is also conductive; it hops from radical to radical, if a voltage is applied to bulk material. The fulvalene molecule, on the other hand, gives away one electron to oxidants to become conductive. The attached pyrrole ring stabilizes the radical.

Both chromophores were applied to connect two Au(0) nanocrystal electrodes. The viologen bridge became conducting at an applied potential of –0.4 V, the pyrrolo-tetrathiafulvalene bridge at $\sim +0.4$ V. Both types of single electrons managed to hop over two $(CH_2)_6$ bridges, but the jumps looked quite different due to the different thermal flexibilities of the chromophores. The conductance of the viologen increased as the potential became negative, but there was no peak around the V^{2+}/V^+ equilibrium potential. Only a broad rise occurred across the redox wave, which did not level off within the attainable negative potential limit of –0.8 V. The tetrathiafulvalene conductance rose from ~ 0.5 nS in its neutral state to a maximum of 2.5 nS, when the monocation radical concentration reached its maximum and fell again to the "off" conductance value of ~ 0.5 nS at positive overpotentials. The width of the conductance peak was ~ 0.2 V, as expected. The Au(0)–fulvalene single-molecule junction had a symmetrical peak close to the first oxidation potential for fulvalene, but the viologen radical became more and more conductive with rising

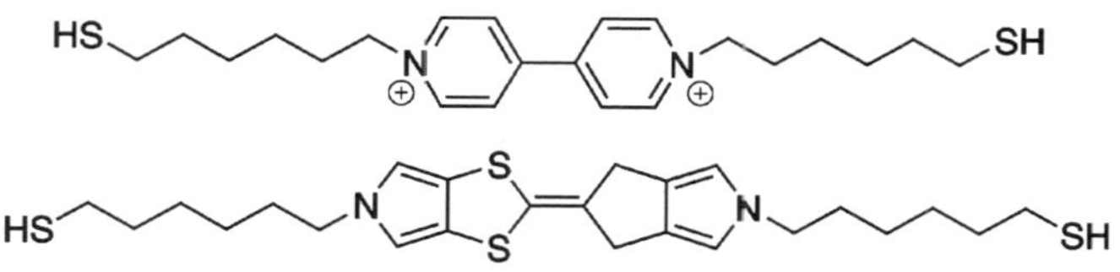

Figure 2.85 The viologen (*top*) becomes a radical upon reduction, the pyrrolofulvalene (*bottom*) upon oxidation. Both are typical redox-active building blocks of voltage-dependent switches between nanoparticles.[143]

negative potentials (Figure 2.85). Density function calculations showed that the pyrrolofulvalene does not change its planar conformation upon redox reactions, whereas the viologen dication tends towards a highly twisted C–C connection and the monocation radical shows an increased inter-ring C=C double bond. The conformation of the viologen changes every time the electron jumps back to the gold electrode at equilibrium potentials. Only at very low potentials does the planar conformation becomes stable and the tunnelling factor reach a maximum. The stereochemistry of flexible dyes or redox systems is reflected by voltage-dependent current flows: rigid dyes respond with sharp peaks close to the redox potentials.[143]

The application of oxidation–reduction potentials to redox pairs may be also translated into mechanical movements of molecules. An intertwined viologen–fulvalene macrocycle can function as part of a molecular machine, which is driven by changing electric potentials. One such redox-active macrocycle contained a thiafulvalene, a naphthohydroquinone and a dianiline as well separated donor units, and was placed into a *bis*-viologen acceptor cycle. Each of these molecules produced a radical at a different potential, and each of the charge transfers from the radical to the viologen produced absorption bands in the red, green or blue region of the visible spectrum. Red–green–blue colour switching was therefore obtained under three different applied voltages (Figure 2.86). The 2 nm nanoparticle is a possible pixel component in electronic displays.[144]

Another type of macrocycle is called a *calixarene* (from the Greek *calix* = vase, referring to its shape): four benzenes stand upright in a ring. They have often a hydrophobic base of alkyl groups and an almost cubic upper edge made of heavily substituted aromatic rings. Calixarenes are ideally suited (1) to entrap hydrophobic molecules and (2) to be fixated on functional nanoparticles. Not only can they entrap target molecules, they can also easily release them. For example, one calixarene produced a stoichiometric amount of linear hydrocarbons such as octane and sodium dodecylsulfate and both entrapped alkyl chains became helical, because of the limited height of the wall of the vase (Figure 2.87). NMR spectra showed that the octane became partly helical and coiled up in the centre of the cavity. For the dodecylsulfate, on the other hand, eight carbons formed a helix in the cavity, while the other four nearest to the sulfonate group remained in water.[145a,b]

In complex nanoparticle preparations, it is often mandatory that the chemical reactions involved must be quantitative, regioselective and run under

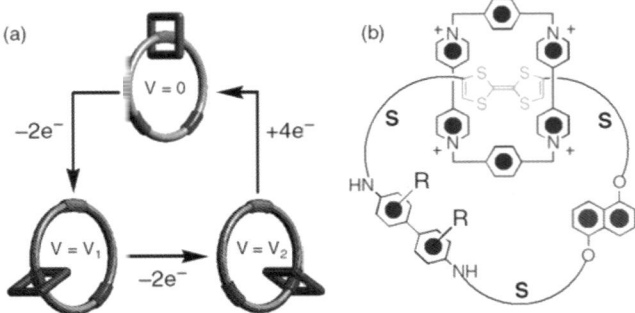

Figure 2.86 Three organic molecules, which can be oxidized to radicals that then donate the single electron to a viologen, are combined in one macrocyle: a dianiline (=benzidine), a diphenol (naphthohydroquinone) and a fulvalene. If any of the three molecules has lost an electron by an applied voltage, it rotates into the electron-accepting bis-viologen macrocycle and produces a colour. Different voltages thus colorize the nanoparticles either red, blue or green. © American Chemical Society.[144]

non-destructive conditions. Only very few reactions make C–C bonds in high yield, work under mild conditions, do not attack vulnerable functional groups and can therefore be applied again and again in sensitive systems. The 1,3-dipolar cycloaddition of azides to acetylenes giving 1,2,3-triazoles is today the most prominent representative of so-called click chemistry, which is quick, atom-efficient, and versatile (Figure 2.88). Both components, the azide and the acetylene, are compatible with enzymes under physiological conditions, and both are readily incorporated into organic halides by substitution. The yield of the cycloaddition is usually >90%, the reaction is water tolerant and its regioselectivity can be dramatically enhanced by localizing both components inside one host structure, e.g. on the surface of a copolymer or a nanoparticle. Furthermore, the triazole heterocycle is surprisingly stable; it is not degraded by exposure to dilute acid, base, or low-pressure hydrogenation conditions; and it acts as push-button after addition of multivalent cations, which are readily added. An example is the fourfold click on a calixarene periphery.[146]

Atomically smooth 100 nm silica nanoparticles with a monomolecular amino coating have been successfully applied in several ways:

- Flash photolysis experiments with covalently attached porphyrins (see p. 12,135) in water
- Enhancing the lifetime of the chemically reactive porphyrin triplet states by a factor of 10
- The study of 2D diffusion of fluorescence quenching molecules on a variety of surfaces and 1D diffusion in pores using standard spectrometers
- The analysis of reversible particle aggregation by UV/visual and fluorescence spectroscopy

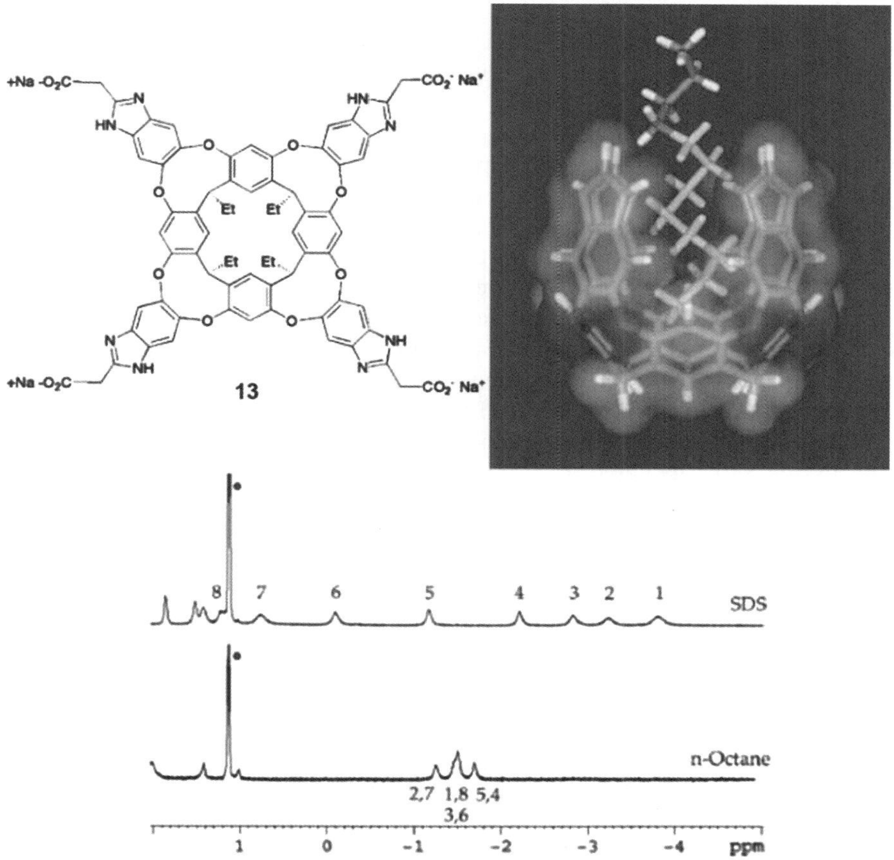

Organic Carbon 133

Figure 2.87 A calixarene with entrapped sodium dodecyl sulfate and its ^{1}H-NMR spectrum together with that of entrapped n-octane (bottom).[145a,b]

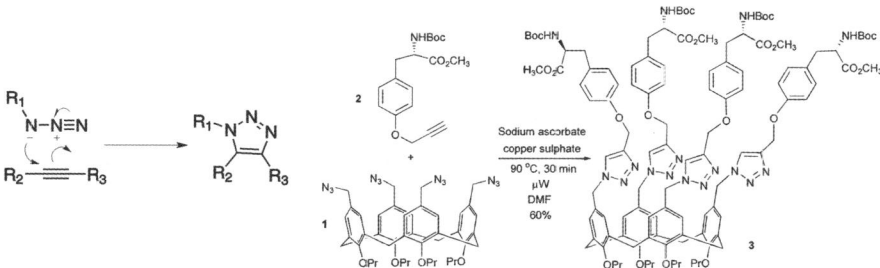

Figure 2.88 Click chemistry on a calixarene.[145,146]

- The establishment of nanosized containers that can be closed and opened by pH changes
- Most interestingly, the establishment of long-distance redox pairs in aqueous medium.

Their advantage over assembled polymer capsules, which serve similar purposes, is the constant size and separation of the 2 nm membrane gaps (yoctowells).

Porphyrins whose diameter is larger than that of the well should not reach the bottom at all. It proved, however, to be difficult to demonstrate this filter effect with the silicate particles. It did not work at all when the porphyrin acid chlorides were used in the first self-assembly step, which presumably formed porphyrin domains on the silicate surface rather than spots of monomeric porphyrins. More stable mixed anhydrides were much more reliable. Gaps based on the acid chloride porphyrins hardly differentiated between manganese porphyrins with diameters of 2.0 and 3.6 nm, whereas gaps produced with the corresponding anhydrides filtered out the large porphyrin with an efficiency of $>90\%$. It also turned out that charge interaction between bottom and quencher molecules was not necessary for efficient quenching; electroneutral partners interacted as well.

Diffusion of fitting porphyrins to the bottom of the wells was extremely slow. Time-constants between 1300 and 3700 s were found, corresponding to formal diffusion constants of $10^{-23}\,s^{-1}\,m^{-1}$. The slow kinetics of the migration was probably caused by adsorption of the large quencher molecules to the rigid, hydrophobic walls of the yoctowells and a very slow detachment. It allowed for quantitative "sorting" of three porphyrins in stacks in the orders A, B, C and A, C, B within the nanowells. Diffusion in wells with hydrophilic walls was ~ 1000 times faster (Figure 2.89).

1,2-*trans*-Cyclohexanediol, glucose, cellobiose, and tyrosine, together with some phenolic neurotransmitters, block hydrophobic yoctowells for the transport of small, water-soluble molecules or ions. The blocking effects of cellobiose, for example, remained unchanged for several weeks, when the yoctowells were in direct contact with bulk water volumes. The blocking molecules have three structural characteristics in common: (1) their backbone is a conformationally stable ring; (2) all hydrophilic substituents lie in an equatorial position of a cyclohexane-type ring or in a benzene plane; and (3) all molecules have a hydrophobic edge. Axial substituents, cyclopentane or furanose units, or open-chain compounds destroyed the blocking effect; galactose, for example, has no blocking effect and destroys that of cellobiose. Semi-quantitative radioactivity measurements of ^{14}C-labelled tyrosine on yoctowell-coated gold electrodes established a number of ~ 35 molecules per 8 yL well, which is comparable to the number of molecules in an 8 yL volume of a crystal. The formation of these "nanocrystals" took many hours, was totally inhibited by the addition of 1% of flexible molecules and was made reversible by brief heating to 72 °C, the melting point of the amide hydrogen bonds within the walls of the yoctowell [147a,b]

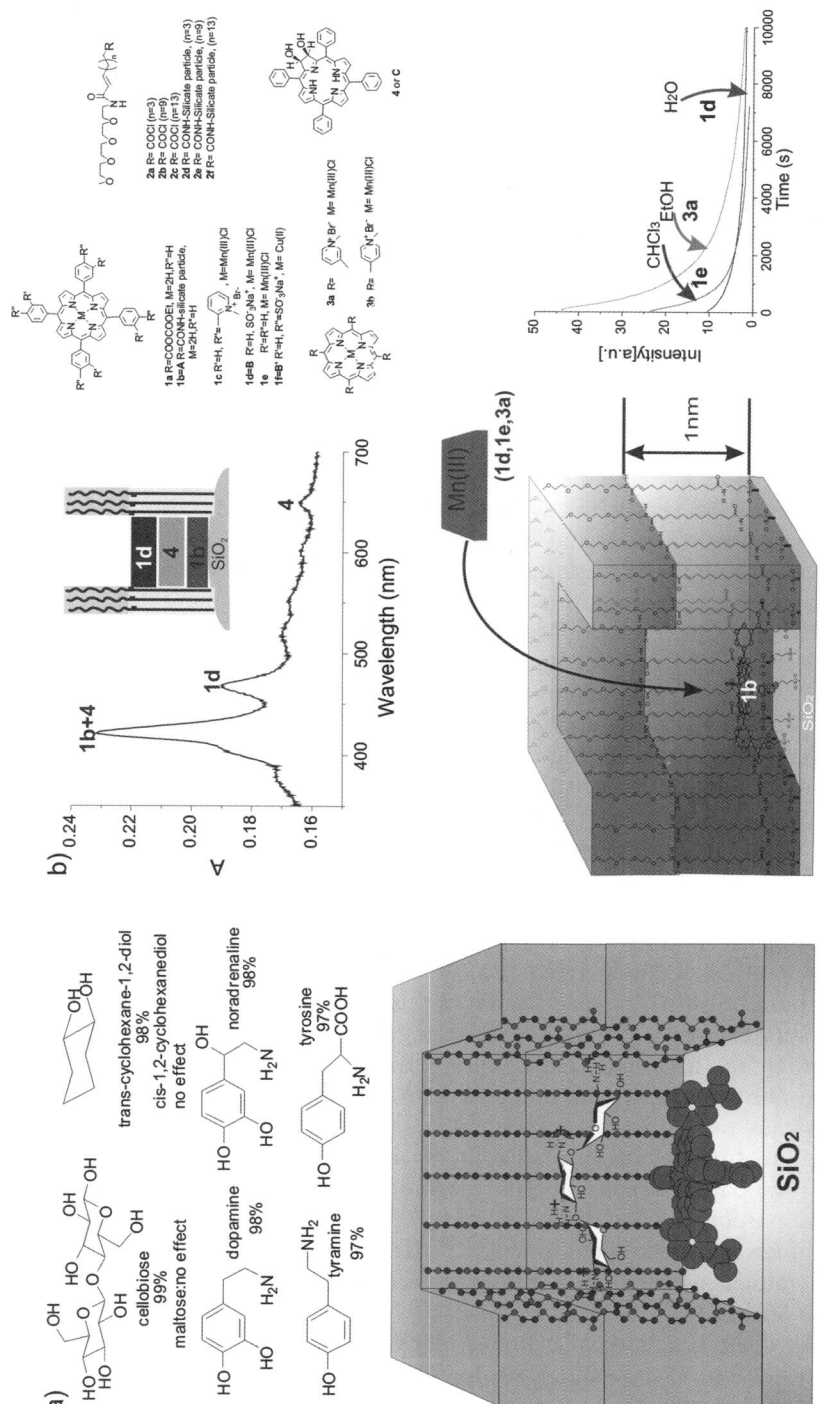

Figure 2.89 Hydrophobic yoctowells entrap small cyclic molecules with a hydrophobic edge as well as fitting molecules irreversibly in water. The letters indicate the sequence of addition.[113,147]

Another method used glucose pentaacetate (GPA, size ∼0.8 nm) or glucose pentabenzoate (GPB, size ∼1.3 nm) as pore-forming templates (Figure 2.90). These esters dissolved in the centre of vesicle bilayer membranes and left it upon hydrolysis with acetic or benzoic acid. Between the bilayers the membrane was loaded with a 1:1 mixture of 4-*tert*-butylstyrene and *p*-divinylbenzene, which also dissolved in the centre of the bilayer and could be copolymerized by UV-light. Removal of glucose then left a porous polymer membrane in the centre of the vesicle bilayer membrane. It was characterized by the usual entrapment and release of hole-fitting fluorescent dyes.[148]

2.10 Cellulose

Cellulose fibres are by far the most abundant natural products ($>10^{11}$ tonnes/ year) and have a porous nanofibril structure of 10–30 nm width, specific surface areas of 30–55 m^2 g^{-1} and a complex interior pore morphology. They provide plenty of reaction vessels for synthesizing nanoparticles and also function as particle stabilizers. The acetal and the hydroxyl oxygens anchor metal ions tightly on to cellulose fibres, stabilize metal nanoparticles similarly to crown ethers, and let aqueous solutions come close. 2–6 nm Ag(0) nanoparticles were thus formed in lint-free cellulose paper 50 μm thick with 30–70 nm nanopores from silver nitrate (AgNO$_3$) by reduction with sodium borohydride (NaBH$_4$) (Figure 2.91). The paper changed colour from white to yellow, and a broad band at 420 nm indicated a plasmon absorption.[149,150]

Dialysis of cellulose acetate dissolved in *N,N*-dimethylacetamide results in the formation of regular 100 nm porous nanoparticles (Figure 2.92).[151] Cellulose nanocrystals were obtained from paper by hydrolysis with 64% sulfuric acid at 45 °C, precipitation by dilution with water, filtration, washing and drying. 3–6 nm Au(0) and Ag(0) nanoparticles were precipitated on to these rodlike nanocrystals by reduction of metal salts with sodium borohydride and showed much more intense plasmon resonance bands for Ag(0) at 433 nm than for Au(0) at 525 nm. Ag(0)–, Au(0)– and Pt(0)–cellulose aerogels reduced with sodium borohydride also showed a relatively high transmittance, a large, porous surface area and good tensile strength (Figure 2.93).[152]

In order to flocculate clay, titania and other microparticles on the cellulose fibres of paper, the latter have to be stretched by shear and coated with patches of cationic polymers, allowing the anionic particles to attach smoothly to free spacings in between. Shear is produced by the 10–30 ms^{-1} speed of the paper suspension: all ordering processes of cellulose fibres, polymers and particles have to occur in a perfectly ordered and rapid manner. This shear-induced (= orthokinetic) polymer–particle interaction binds polystyrene and poly(ethleneglycol), which enfold the adsorbed particles and work into the cellulose fibres.[153]

Titania (TiO$_2$) replicas of cellulose fibres were obtained by adsorption of titanium butyrate (Ti(O*n*-Bu)$_4$) and hydrolysis of the adsorbed layer on paper or with polypyrrole titania composites. Titania/cellulose composites and

Figure 2.90 Preparation and fixation of a pore in a lipid bilayer membrane (see text).[148]

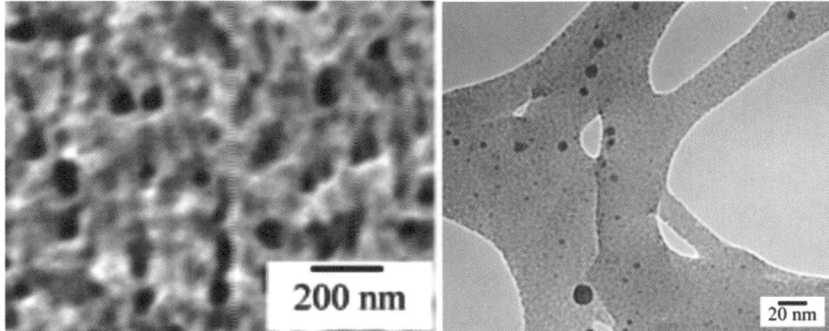

Figure 2.91 Scanning electron micrograph of a single cellulose fibre and transmission electron micrograph of Ag(0) nanoparticles within the pores. © American Chemical Society.[149,150]

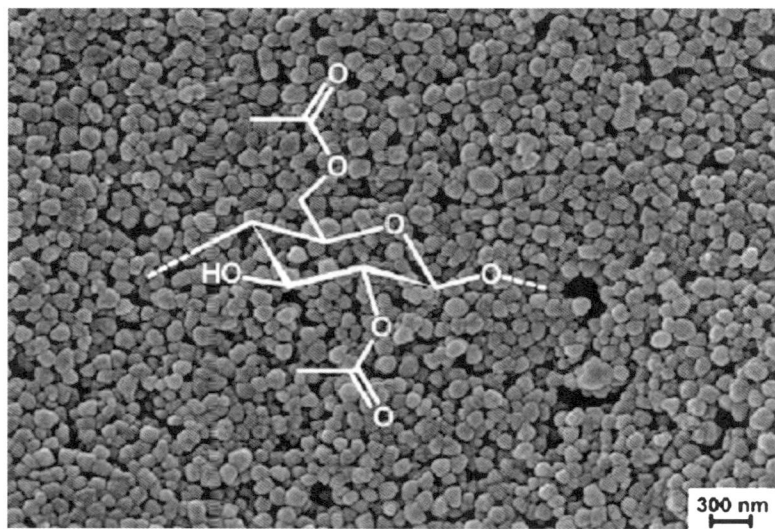

Figure 2.92 Structure of a cellulose acetate monomer and typical nanoparticles formed in N,N-dimethylacetamide and after precipitation with water. © American Chemical Society.[151]

replicas were prepared by depositing thin (30–55 nm) titania films on an ashless filter paper. The photocatalytic activity of the anatase titania/cellulose composite was verified by photocatalytic reduction of Ag(I) to Ag(0) by UV light. Ag(0) nanoparticles on titania nanotubes were thus formed on the cellulose matrix.[154]

Deposition of Fe_2O_3 nanoparticles in cellulose fibres followed by the removal of the cellulose matrix by calcination at 600 °C leads to micron-wide Fe_2O_3 fibres made up of nanoparticles (Figure 2.94).[155]

Organic Carbon 139

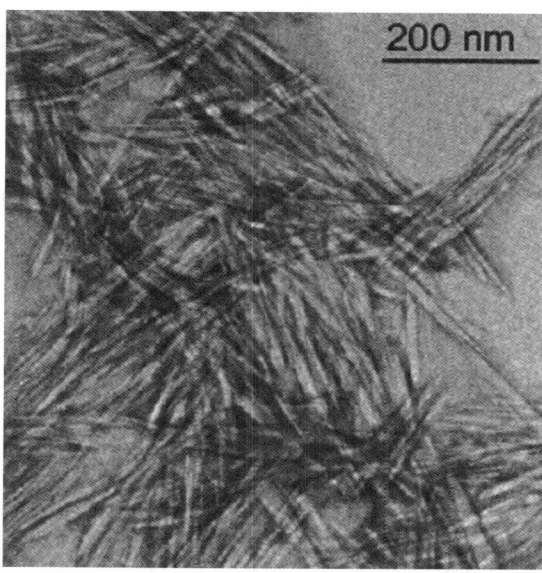

Figure 2.93 Transmission electron micrograph of cellulose nanocrystals as obtained by hydrolysis of cellulose fibres with sulfuric acid and precipitation. © American Chemical Society.[152]

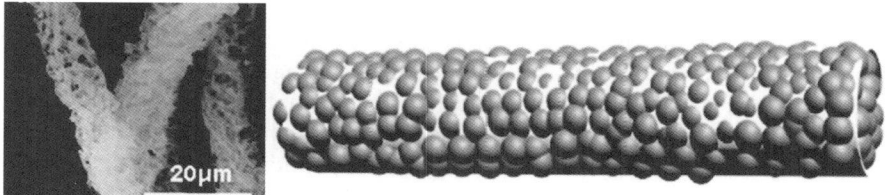

Figure 2.94 Scanning electron micrograph of the fibre-like Fe_2O_3 macroporous nanomaterial and a model of its formation on cellulose fibres.[155]

2.11 Non-covalent Twisted Ribbons and Proteins

The organic chemistry of chirality began in the 19th century with Pasteur's experiments on tartaric acid salts. In the 20th century, the concept of chirality became the dominating factor in the helices of cellulose, proteins and DNA. Now, in the 21st century, tartaric acid is still a most rewarding compound in nanochemistry. Monododecylamides of tartaric acid, for example, spontaneously form bundles of helical, 2 nm wide, fibres in water. Similar fibres in the form of twisted ribbons were obtained from gemini-hexadecyl diammonium tartrates in water.[156] Fibres sank when dispersed in D_2O (density 1.1), whereas they floated in H_2O and remained suspended in D_2O/H_2O mixtures. Fibre density was therefore estimated to be 1.05 ± 0.04, compatible with 4–6 water molecules in the asymmetric unit (Figure 2.95).[157] In both cases the pure

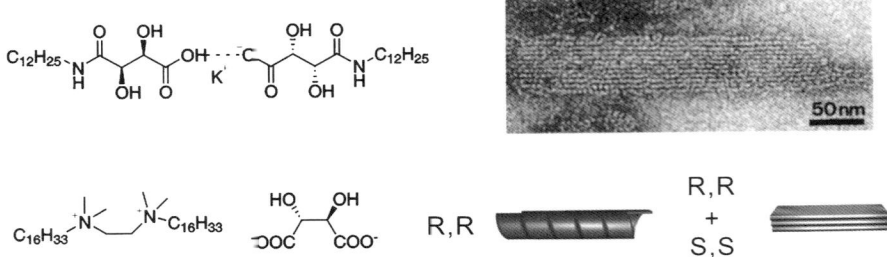

Figure 2.95 *Above*: Molecular structure of the potassium R,R-tartaric amide and electron micrograph of a helical fibre bundle.[156] *Below:* The gemini-diammonium R,R-tartrate, model of its twisted ribbon and of its racemic crystal.[157]

R,R- or S,S-tartrate produced bimolecular helical fibres in water, which precipitated as flat crystals without any curvature upon addition of the mirror image tartrate (chiral bilayer effect).[156]

Natural proteins are products of evolution and their structure is determined by intramolecular hydrophobic effects, hydrogen bonding, and charge interactions as well as by surface hydration. In terms of nanochemistry the most important aspect is the hydrophobic gaps or flat regions on the surface, which are stabilized and solubilized by hydrophilic domains surrounding them. These hydrophobic flat regions allow the proteins to adhere to nanoparticles, and small hydrophobic gaps are important parts of most reactive centres in enzymes. Furthermore they trigger the adsorption of water-insoluble gases, in particular oxygen and carbon monoxide, or benzene- or cyclohexane-derived pharmaceuticals.

Proteins are also structural elements in the animal body (bones, muscles, tendons, nerves) and functionalize cell membranes, where they must integrate into the lipid bilayer (p. 102 ff). An amphiphilic peptide with a palmityl amide ($=C_{16}$) end group, three cysteines for crosslinking, three glycines for flexibility, a phosphorylated serine an arginine and an aspartate for salt formation and water solubility was used as single building block for a protein block around apatite crystallites and did indeed form an artificial bone structure. Nothing in nature is more sticky than proteins and this is, at the same time, a very simple and very complex example: sticky inside and organizable with a lot of water outside.[158]

Coiled-coil double-helical peptides, *e.g.* the 21mers 1–4 in Figure 2.96, can be made to self-assemble into nanofibres 4 nm wide and a micron long without any "sticky ends". It is sufficient to apply blunt-ended dimers and to lower the pH of the solution to 3.5, where the glutamic acid units are protonated. The length and the lateral packing can be controlled through selection of amino acids that are not involved in the coiled-coil interface, using the known rules governing coiled-coil assembly. The four peptides shown in Figure 2.96 share three common primary structure features:

- Positions *a* and *d* are taken by the amino acids isoleucine (I) and leucine (L), respectively, to form the hydrophobic patch. These particular

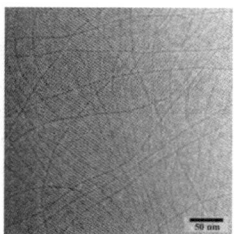

Figure 2.96 Four typical amino acid sequences that lead to nanorods and nanowires in water, and cryo transmission electron micrograph of a typical nanowire at pH 3.3. © American Chemical Society.[158,159]

hydrophobic amino acids were selected to favour the formation of a dimeric coiled coil.
- Glutamic acid (E) at positions *e* and *g* provides a uniformly acidic ionic region and the molecular switch of the pH neutralization. At low pH, ionic repulsion is eliminated and carboxylic acid side chains may hydrogen bond with one another.
- The key design features in engineering the morphology of nanofibres are the selection of amino acids at the *b*, *c* and *f* positions that control the length and the diameter of the fibres and the inclusion of lysine (K), glutamine (Q), serine (S) and tyrosine (Y), which provide attachment points on the outside.[159]

No self-assembly process is known in the chemical synthesis of covalent polymers connected with nanoparticles. A "wrong", unwanted attachment of molecules cannot be repaired by rearrangements, and causes permanent structural and functional defects. Only a living system can use enzymes to heal accidental crosslinks or reverse erroneous bond formation, or put destroyed or crosslinked structures out of action.[160] Chemists must protect, purify and deprotect the polymer after each synthetic step – and there are usually many steps. This tedious stepwise procedure has been automated in solid-phase DNA and protein syntheses, which function on the kilogram scale for medium-sized molecules. The protected monomers, nucleotides and amino acids, as well as the necessary apparatus, are available commercially. Synthesis is then in the hand of technicians, who just follow instructions. For the preparation of complex analytical nanoparticles (see below) it is mandatory to start with available proteins.

Single-molecule photobleaching of organic fluorophores makes it possible to count the number of proteins per nanoparticle at the single-particle level. For example, the protein R-bungarotoxin, already labelled with an organic fluorophore (Alexa488) was first bound to a lanthanide-ion doped, fluorecent oxide nanoparticle with a 1:1 stoichiometry and other labelled proteins were added step by step. Stepwise photobleaching at different wavelengths corresponding to the organic fluorophores then quantified the protein:nanoparticle ratio for each nanoparticle as well as its distribution (Figure 2.97).[161]

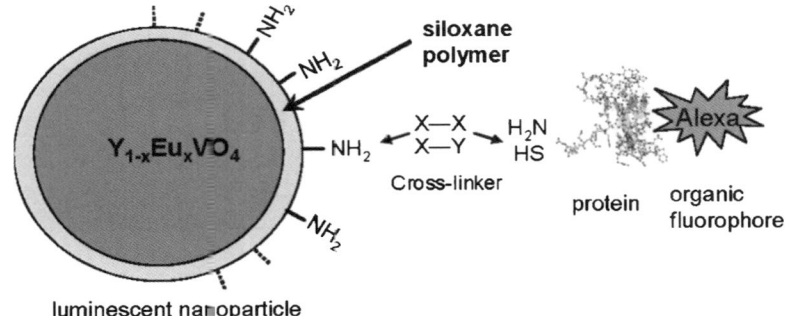

Figure 2.97 Model and first step of a luminescent nanoparticle loaded with different proteins connected to different fluorescent dye molecules. © American Chemical Society.[161]

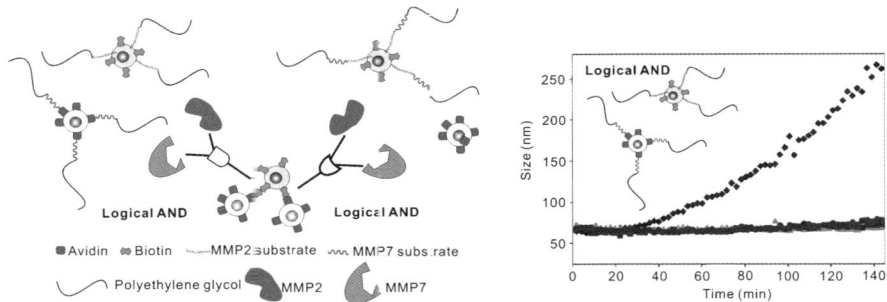

Figure 2.98 *Left:* model of logical nanoparticle sensors. *Right:* AND reaction: hydrodynamic radius in dynamic light scattering is increased only in the presence of both MMP2 and MMP7; either or none is insufficient to actuate assembly. © American Chemical Society.[162]

In medicine, the most important task for protein-loaded nanoparticles is the analysis and possible attachment of dangerous proteins in blood, in particular those connected with the spreading of cancer cells. Nanoparticles are suitable for such a delicate job because of their specificity. All kinds of substrates, ligands and dyes can be bound to the surface of one nanoparticle A, which should find and possibly neutralize protein A. At the same time, nanoparticle B with different receptors can be sent to deal with protein B. The surface agents on nanoparticle A and/or nanoparticle B may then react with solutes in bulk blood or may be steered magnetically to an area where the nanoparticle can be destroyed. Light microscopy or low-resolution AFM may then be sufficient for the detection of specific bimolecular reactions of isolated blood samples, and nanometre-smooth surfaces of the nanoparticles are not important. What makes this unique is that for the first time it is possible to gather several complementary binding and detection properties together on the same nanoparticle, and to analyse two different proteins in the blood simultaneously.

Two types of magnetic Fe_3O_4 nanoparticles (p. 276 ff) were used in this way to simultaneously monitor a multiple biological process associated with tumor growth. The enzyme matrix-metalloproteinase-2 (MMP2) is over-expressed in breast cancers, and is an indicator of cancer invasiveness, whereas matrix-metalloproteinase-7 (MMP7) is thought to facilitate early stages of mammary carcinoma progression. Their destructive cooperation implies a logical AND reaction: both must be present at the same time to start a disaster (Figure 2.98). This system can be detected and quantified with two different nanoparticles and can also be compared with cancers where only one of the two enzymes is present.[162]

2.12 DNA

DNA homopolymers were synthesized on an Au(0) surface by surface-initiated enzymatic polymerization using the terminal deoxynucleotidyl transferase, which repeatedly adds mononucleotides to the 3′-end. Monolayer stripes of 5′-SH(CH$_2$)$_6$-dTTP were thus first imprinted on Au(0) and then incubated at 37 °C for 2 h with dTTP and TdTase (Figure 2.99). The height of the stripes grew 20-fold, from 5.5 to 100 nm. The enzyme needed the 3′-OH group as an initiator; the 5′-end did not work.[163]

A molecular beacon is an oligonucleotide made up of of a loop of DNA: self-complementary regions of 5–6 nucleotides at opposite ends (the stem) and a fluorophore and a quencher at each end of the strand. In the absence of the loop's target molecule (usually the matching sequence of the oligonucleotide branch carrying the fluorescent dye on a surface), the fluorophore and the quencher are in close proximity and the fluorescence is quenched. If a target molecule attaches to the loop, it hybridizes, the beacon is separated and fluorescence becomes visible.[164] In biological systems false signals from interaction with proteins, membranes, intercalating agents in the cell cytoplasm and detachment of the fluorophore or quencher by enzymes are common.[165] On nanoparticles the beacon behaviour characterizes a surface quite sensitively (Figure 2.100).

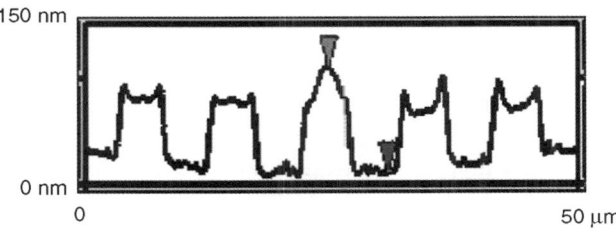

Figure 2.99 DNA synthesis with with dTTP and TdTase on an Au(0) surface containing imprinted 5′-SH(CH$_2$)$_6$-dTTP stripes. © American Chemical Society.[163]

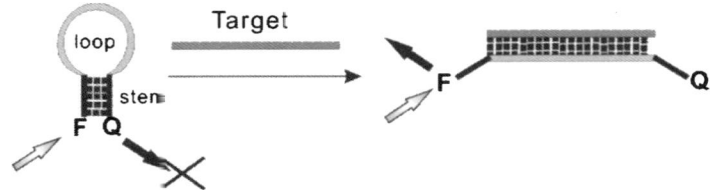

Figure 2.100 Model of a molecular DNA beacon formed on a surface.[164,165]

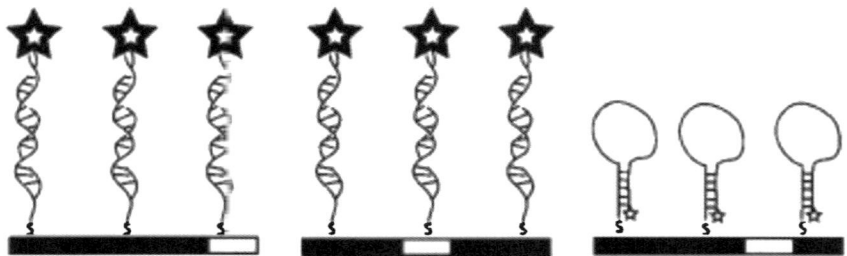

Figure 2.101 The three different barcode wires 000001, 00100, 00010 were covalently connected with the following thiols. TAMRA is the terminal fluorescing dye, the quenchers are Au and Ag of the wires, HCV thiol(CH2)6-*GCG AGC A*−*A GTG GTC T*G*C GGA ACC GGT GAC TCG C*-TAMRA, SARS thiol(CH2)6-*GCG AGA G*AT GCT GTG GGT ACT AAC CTA *CCT CTC GC*-TAMRA, HIV thiol(CH2)6-*GCG AG*T GTT AAA AGA GAC CAT CAA TGA *GCT CGC*-TAMRA. The wires with the given DNA beacons and loops were plunged into a suspension of a pathogenic virus. It opened the loop sequences typical for the viruses HCV and SARS, but not for HIV. © American Chemical Society.[166]

The quencher may also be the Ag(0) or Au(0) of a barcoded wire (p. 377 ff). The DNA beacon is attached to the metal wire at one end and the fluorophore on the other end is quenched by its close proximity to the metal surface, as long as the probe maintains its hairpin conformation. The striped nanowires were commercial 000001, 00100, 00010 Au(0) and Ag(0) patterns, 0.75 μm long, for each gold or silver segment (Figure 2.101),[166] and the DNA sequences were commercial disulfide dimers. The loops recognized 16 strains of a pathogenic virus in a single contact.

Carbon nanotubes have a similar quenching effect on the fluorescent dye, but here the flexible DNA loops wrap around the nanotube by π-stacking interaction and DNA hybridization again restores the fluorescence (Figure 2.102).[167]

Single-nucleotide point mutations of DNA constitute the most common type of genetic variation, and are often associated with susceptibility to diseases or difficulties with drug metabolism. Such minimally mutated DNA is chemically and structurally close to the non-mutated polymer and difficult to detect.

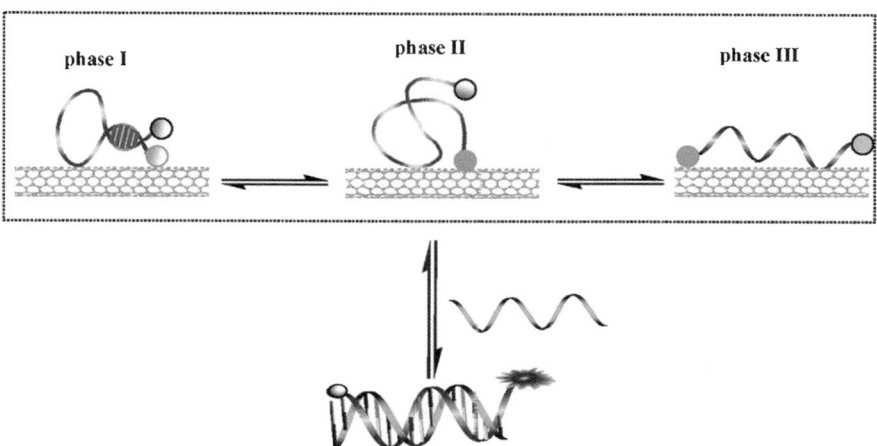

Figure 2.102 Model of an electrochemical fluorescence assay with a beacon DNA on a conducting carbon nanotube probe. © American Chemical Society.[167]

Apoferritin nanoparticles with massive amount of the matching base, *e.g.* guanine, adsorbed on the surface and Cd(II) phosphate inside the protein cage, were prepared in order to bind to the "open carbon" on the mismatch size and to release massive amounts of Cd^{2+}, which is easy to detect electrochemically, upon opening the apoferritin capsule at pH 4.5, where the Cd(II) phosphate is also dissolved.

Soluble Cd^{2+} salts were added first, then phosphate to precipitate them within the apoferritin nanoparticle. The guanine–apoferritin nanoparticle conjugate was prepared with guanosine 5′-monophosphate *via* phosphoramidite bonds with the free amino groups of the apoferritin protein shell. The mixture of intact and mutant single-stranded DNA was then combined with biotinylated complementary DNA to form the duplex, and avidin, a biotin-binding protein, was bound to a magnetic bead that was coupled to the DNA mixture. The combination of biotin and avidin then finally formed the magnetic nanoparticle with a cluster of DNA-occupied apoferritin nanoparticles. The guanine residues on the cadmium–ferritin nanoparticles detected the uncovered cytosine monomers on the DNA duplexes, bound their cadmium reservoirs to this site in the presence of DNA polymerase, were centrifuged, the ferritin capsules opened and the Cd^{2+} ions released by acidification to pH 4.2 and quantified by square wave voltammetry. The sensitivity was attomols (10^{-21} mol/L) of DNA. Nanoparticles with only "healthy" DNA did not bind any Cd^{2+}-containing nanoparticles[168] (Figure 2.103).

Au(0) nanoparticle–DNA conjugates were also prepared enzymatically *via* nanoparticles with complemetary single-stranded DNAs on the surface. The close dimers of hybridized Au(0) nanoparticles were then characterized by electron microscopy (Figure 2.104). The reaction should also be applicable to heterodimers of different metal and metal oxide nanoparticles.[169]

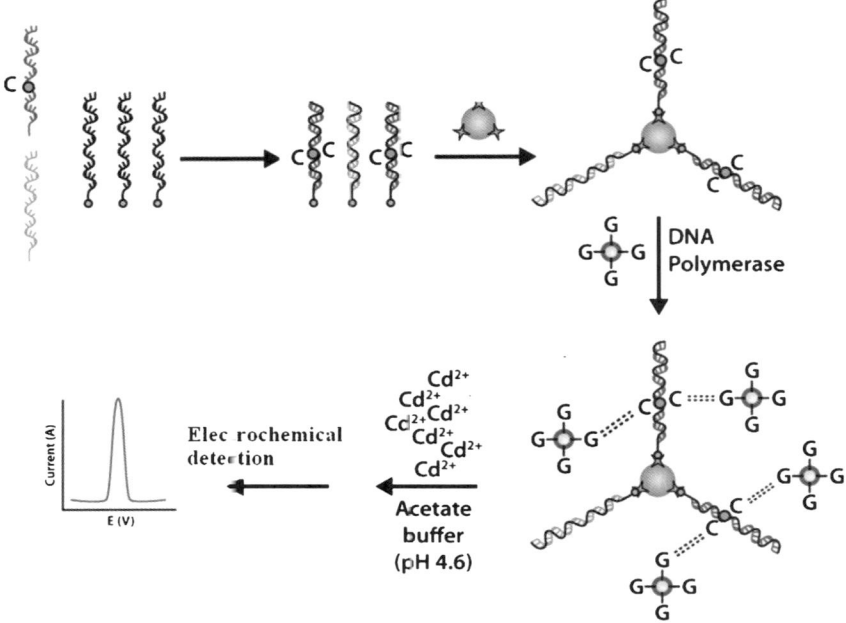

Figure 2.103 Model of the reaction sequence that coupled the release and detection of Cd^{2+} ions with the mismatched base cytosine (C) in a DNA double helix. © American Chemical Society.[168]

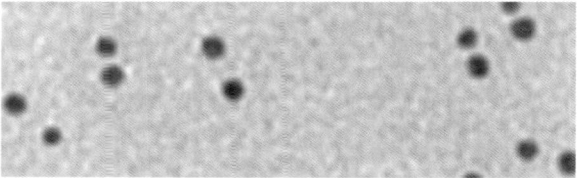

Figure 2.104 Transmission electron micrographs of Au(0) nanoparticle dimers connected by the double helix formed from complementary DNA single strands. © American Chemical Society.[169]

2.13 Fluoride

Fluorinated alcohols, in particular 1,1,1,3,3,3-hexafluoro-2-propanol ($CF_3CH(OH)CF_3$), tend to form hydrogen-bonded trimers that show a pronounced increase of the positive partial charge at the terminal free hydroxyl group and an enhanced hydrogen bond donor ability. These solvents activate oxidants, in particular hydrogen peroxide. Peroxidations of alkenes are accelerated by five orders of magnitude in $CF_3CH(OH)CF_3$ as compared to normal alcohols or ethers, so hexafluoropropanol is a very efficient catalyst.[170]

Covalent fluorination of amorphous CN film with $HOCH_2CF_2\text{-}(\text{-}OCF_2)_x\text{-}OCF_2CF_2)_y\text{-}OCF_2CH_2OH$ ($x/y = 2/3$ to 1) enhanced the surface wettability, adhesion, hardness, stress, elastic modulus, and density of the film and lowered the friction. Hardness and stress tests and density measurements of such $a\text{-}CFx$ films revealed that the diamond-like structure of slightly hydrogenated and aminated carbon films with very low fluorine content are transformed first into a graphite-like structure upon addition of the fluorinated ether and susequently to a polymer-like structure, when the fluorine content of the film is increased. Molecular oxygen oxidizes the amorphous $a\text{-}CH_x$ only very slowly. Physisorbed O_2 and H_2O are much more likely to desorb than to dissociate. Surface oxidation of $a\text{-}CH_x$ is helpful in the fixation of the fluorinated oligoethylene ether without any CN.[171]

The interaction of fluorinated hydrocarbons with normal alkyl surfactant layers, or of hydrocarbons with fluorinated detergents, leads to much more "solid" layers than observed for pure hydrocarbon or pure fluorocarbon systems. The mutual interfacial energy is much stronger in the $CF_2\text{-}CH_2$ mixture than in pure $CH_2\text{-}CH_2$ or $CF_2\text{-}CF_2$ mono- or bilayers. Rupture, coalescence and the flattening of droplets occurred under various conditions through strengthened interfacial adsorption which was elegantly quantified with a magnetic galvanometer.[172]

Fluorination of graphite gives black to gray to finally white materials depending upon the fluorine content. Reaction with alkyl lithium reagents yields alkylated graphene sheets; lithium dodecyl converted white fluorine graphite into black irregular graphite micron-sized nanoplatelets with a thickness of ~ 8 nm and no fluorine. The dodecylgraphite was soluble in apolar organic solvents.[173] Dodecyl radicals had replaced the fluorine groups. Electron transfer from lithium to the flucrinated graphite yielded first a black transient radical anion that decomposed rapidly to give a carbon-centred free radical and lithium fluoride. The alkylated graphite had one dodecyl group per 21 graphitic Cs, the white fluorine graphite probably 10 times more fluorine.

References

1. A. Krüger, *Neue Kohlenstoffmaterialien* (chapters on nanowires, nanotubes and nanodiamonds), Teubner, Wiesbaden, 2007.
2. F. Neugart, A. Zappe, F. Jelezko, C. Tietz, J. P. Boudou, A. Krueger and J. Wrachtrup, Dynamics of diamond nanoparticles in solution and cells, *Nano Lett.*, 2007, 3358–3561.
3. S. Schietinger, T. Schroder and O. Benson, One-by-one coupling of single defect centers in nanodiamonds to high-Q 2. Modes of an optical microresonator, *Nano Lett.*, 2008, **8**, 3911–3915.
4. S. Osswald, G. Yushin, V. Mochalin, S. O. Kucheyev and Y. Gogotsi, Control of sp^2/sp^3 carbon ratio and surface chemistry of nanodiamond powders by selective oxidation in air, *J. Am. Chem. Soc.*, 2006, **128**, 11635–11642.

5. J. Cheng, J. He, C. Li and Y. Yang, Facile approach to functionalize nanodiamond particles with V-shaped polymer brushes, *Chem. Mater.*, 2008, **20**, 4224–4230.
6. X. Wang, P. E. Colavita, K. M. Metz, J. E. Butler and R. J. Hamers, Direct photopatterning and SEM imaging of molecular monolayers on diamond surfaces: mechanistic insights into UV-initiated molecular grafting, *Langmuir*, 2007, **23**, 11623–11630.
7. A. Krueger, J. Stegk, Y. Liang, L. Lu and G. Jarre, Biotinylated nanodiamond: simple and efficient functionalization of detonation diamond, *Langmuir*, 2008, **24**, 4200–4204.
8. H. Huang, E. Pierstorff, E. Osawa and D. Ho, Protein-mediated assembly of nanodiamond hydrogels into a biocompatible and biofunctional multilayer nanofilm, *ACS Nano*, 2008, **2**, 203–212.
9. Coatings on enhancing the characteristics of ultrananocrystalline diamond films, *J. Phys. Chem. C*, 2008, **112**, 3759–3765.
10. Y. Dai, D. A. Proshlyakov, J. K. Zak and G. M. Swain, Optically transparent diamond electrode for use in IR transmission spectroelectrochemical measurements, *Anal. Chem.*, 2007, **79**, 7526–7533.
11. N. Yang, H. Uetsuka, E. Osawa and C. E. Nebel, Vertically aligned nanowires from boron-doped diamond, *Nano Lett.* 2008, **8**, 3572–3576.
12. J. C. Meyer, A. K. Geim, M. I. Katsnelson, K. S. Novoselov, T. J. Booth and S. Roth, The structure of suspended graphene sheets, *Nature*, 2007, **446**, 60–63.
13. A. A. Green and M. C. Hersam, Coloured semitransparent conductive coatings consisting of monodisperse metallic single-walled carbon nanotubes, *Nano Lett.* 2008, **8**, 1417–1422.
14. J. Crochet, M. Clemens and T. Hertel, Quantum yield heterogeneities of aqueous single-wall carbon nanotube suspensions, *J. Am. Chem. Soc.*, 2007, **129**, 8058–8059.
15. J. Zhou, C. Booker, R. Li, X. Zhou, T.-K. Sham, X. Sun and Z. Ding, An electrochemical avenue to blue luminescent nanocrystals from multiwalled carbon nanotubes, *J. Am. Chem. Soc.*, 2007, **129**, 744–745.
16. P. Poncharal, C. Berger, Y. Yi, Z. L. Wang and W. A. de Heer, Room temperature ballistic conduction in carbon nanotubes, *J. Phys. Chem. B*, 2002, **106**, 12104–12118.
17. N. Mingo and D. A. Broido, Thermal conductivity and the "problem of long waves", *Nano Lett.*, 2005, **5**, 1221–1225.
18. A. Striolo, The mechanism of water diffusion in narrow carbon nanotubes, *Nano Lett.*, 2006, **6**, 633–639.
19. M. Majumder, N. Chopra, R. Andrews and B. Hinds, Enhanced flow in carbon nanotubes, *Nature*, 2005, **483**, 44.
20. R. S. Voronov, D. V. Papavassiliou and L. L. Lee, Slip length and contact angle over hydrophobic surfaces, *Chem. Phys. Lett.*, 2007, **441**, 273–276.
21. F. Kuemmeth, S. Ilani, D. C. Ralph and P. L. McEuen, Coupling of spin and orbital motion of electrons in carbon nanotubes, *Nature*, 2008, **452**, 448.

22. A. Brataas, Spin surprise in carbon, *Nature*, 2008, **452**, 419.
23. L. Carpenter, W. J. Blau, J. J. Boland, J. P. Hamilton and J. N. Coleman, Towards solutions of single-walled carbon nanotubes in common solvents, *Adv. Mater.*, 2008, **20**, 1876–1881 and references therein.
24. Y. Sato, M. Ootsubo, G. Yamamoto, G. Van Lier, M. Terrones, S. Hashiguchi, H. Kimura, A. Okubo, K. Motomiya, B. Jeyadevan, T. Hashida and K. Tohji, Super-robust, lightweight, conducting carbon nanotube blocks cross-linked by de-fluorination, *ACS Nano*, 2008, **2**, 348–356 and references therein.
25. R. K. Wang, H. Park, W.-C. Chen, C. Silvera-Batista, D. Reeves, J. E. Butler and K. J. Ziegler, Improving the effectiveness of interfacial trapping in removing single-walled carbon nanotube bundles, *J. Am. Chem. Soc*, 2008, **130**, 14721–14728.
26. (a) H. Li, B. Zhou, Y. Lin, L. Gu, W. Wang, K. A. S. Fernando, S. Kumar, L. F. Allard and Y. P. Sun, Selective interactions of porphyrins with semiconducting single-walled carbon nanotubes, *J. Am. Chem. Soc.*, 2004, **126**, 1014–1015; (b) P. J. Boul, D. G. Cho, G. M. Aminur Rahman, M. Marquez, Z. Ou, K. M. Kadish, D. M. Guldi and J. L. Sessler, Sapphyrin-nanotube assemblies, *J. Am. Chem. Soc.*, 2007, **129**, 5683–5687.
27. C. Cumings, P. G. Collins and A. Zettl. Peeling, sharpening multiwall nanotubes., *Nature*, 2000, **406**, 586.
28. A. N. Khlobystov, D. A. Britz and G. A. D. Briggs, Molecules in carbon nanotubes, *Acc. Chem. Res.*, 2005, **38**, 901–909.
29. T. Okazaki, S. Okubo, T. Nakanishi, S. Joung, T. Saito, M. Otani, S. Okada, S. Bandow and S. Iijima, Optical band gap modification of single-walled carbon nanotubes by encapsulated fullerenes, *J. Am. Chem. Soc.*, 2008, **130**, 4122–4128.
30. A. Hirsch, Metallic and non metallic carbon tubes, *Angew. Chem., Int. Ed.*, 2002, **41**, 1853.
31. H. Dai, Carbon nanotubes: synthesis, integration, and properties, *Acc. Chem. Res.*, 2002, **35**, 1035–1044.
32. J. Cabana and R. Martel, Probing the reversibility of sidewall functionalization using carbon nanotube transistors, *J. Am. Chem. Soc.*, 2007, **129**, 2244–2245.
33. J. L. Bahr, J. P. Yang, D. V. Kosynkin, M. J. Bronikowski, R. E Smalley and J. M. Tour, Functionalization of carbon nanotubes by electrochemical reduction of aryl diazonium salts: a bucky paper electrode, *J. Am. Chem. Soc.*, 2001, **123**, 6536–6542.
34. H. Li, F. Cheng, A. M. Duft and A. Adronov, Functionalization of single-walled carbon nanotubes with well-defined polystyrene by "click" coupling, *J. Am. Chem. Soc.*, 2005, **127**, 14518–14524.
35. G.-H. Jeong, A. Yamazaki, S. Suzuki, H. Yoshimura, Y. Kobayashi and Y. Homma, Cobalt-filled apoferritin for suspended single-walled carbon nanotube growth with narrow diameter distribution, *J. Am. Chem. Soc.*, 2005, **127**, 8238–8239.

36. K. Wang, W. Zhang, R. Phelan, M. A. Morris and J. D. Holmes, Direct fabrication of well-aligned free-standing mesoporous carbon nanofiber arrays on silicon substrates, *J. Am. Chem. Soc.*, 2007, **129**, 13389–13395.
37. F. Marlow, B. Spliethoff, B. Tesche and D. Zhao, The internal architecture of mesoporous silica fibers, *Adv. Mater.*, 2000, **12**, 961–965.
38. P. Schmidt-Winkel, P. Yang, D. I. Margolese, B. F. Chmelka and G. D. Stucky, Fluoride-induced hierarchical ordering of mesoporous silica in aqueous acid-syntheses, *Adv. Mater.*, 1999, **11**, 303–307.
39. R. Gasparac, P. Kohli, M. O. Mota, L. Trofin and C. R. Martin, Template synthesis of nano test tubes, *Nano Lett.*, 2004, **4**, 513–516.
40. H. Hillebrenner, F. Buyukserin, M. Kang, M. O. Mota, J. D. Stewart and C. R. Martin, Corking nano test tubes by chemical self-assembly, *J. Am. Chem. Soc.*, 2006, **128**, 4236–4237.
41. A. Popp and J. Schneider, Ein nanostrukturierter chemischer Reaktor in Chipgroesse, *Angew. Chem.*, 2008, **120**, 1–5.
42. A. Barreiro, R. Rurali, E. R. Hernandez, J. Moser, T. Pichler, L. Forro, A. Bachtold, Subnanometer motion of cargoes driven by thermal gradients along carbon nanotubes, *Science*, 2008, **320**, 775–777.
43. G. Zubay, *Origins of life on the earth and in the cosmos*, Academic Press, 2000.
44. I. G. Brown, Cathodic arc deposition of films, *Annu. Rev. Mater. Sci.*, 1998, **28**, 243–269.
45. M. A. Petersen S J. Jenkins and D. A. King, Theory of methane dehydrogenation on Pt110. Part II: Microscopic reaction pathways for CH$x \rightarrow$ CHx-1 (x)= 1–3) *J. Phys. Chem. B*, 2004, **108**, 5920–5929.
46. Y. Yun, X. Ma, J. Gui, E. Broitman and A. J. Gellman, Oxidation kinetics of hydrogenated amorphous carbon (a-CHx) overcoats for magnetic data storage media, *Langmuir*, 2007, **23**, 5485–5490.
47. J. W. Y. Lam and B. Z. Tang, Functional polyacetylenes, *Acc. Chem. Res.*, 2005, **38**, 745–754.
48. H. Cao, T. Fang, S. Li and J. Ma, Interfacial charge transfer and transport in polyacetylene-based heteroionic junctions: quantum chemistry calculations and molecular dynamics simulation, *Macromolecules*, 2007, **40**, 4363–4369.
49. B. Frode Lutnaes, G. Kildahl-Andersen, J. Krane and S. Liaaen-Jensen, Delocalized carotenoid cations in relation to the soliton model, *J. Am. Chem. Soc.*, 2004, **126**, 8981–8990.
50. L. M. Tolbert, Solitons in a box: the organic chemistry of electrically conducting polyenes, *Acc. Chem. Res.*, 1992, **25**, 561–568.
51. C. H. W. Cheng, S. W. Boettcher and D. H. Johnston, Unidirectional current in a polyacetylene hetero-ionic junction, *J. Am. Chem. Soc.*, 2004, **126**, 8666–8667.
52. C. H. W. Cheng and M. C. Lonergan, A conjugated polymer pn junction, *J. Am. Chem. Soc.*, 2004, **126**, 10536–10537.

53. H. Mustroph, M. Stollenwerk and V. Bressau, Aktuelle Entwicklungen in der optischen Datenspeicherung mit organischen Farbstoffen, *Angew. Chem.*, 2006, **118**, 2068–2087.
54. D. Bratko, C. D. Daub, K. Leung and A. Luzar, Effect of field direction on electrowetting in a nanopore, *J. Am. Chem. Soc.*, 2007, **129**, 2504–2510.
55. K. C.-F. Leung, P. M. Mendes, S. N. Magonov, B. H. Northrop, S. Kim, K. Patel, A. H. Flood, H.-R. Tseng and J. F. Stoddart, Supramolecular self-assembly of dendronized polymers, reversible control of the polymer architectures through acid-base reactions, *J. Am. Chem. Soc.*, 2006, **128**, 10707–10715.
56. M. lvaro, J. F. Cabeza, A. Corma, H. Garcia and E. Peris, Electrochemiluminescence of zeolite-encapsulated poly(p-phenylenevinylene), *J. Am. Chem. Soc.*, 2007, **129**, 8074–8075.
57. N. Fujita, Y. Sakamoto, M. Shirakawa, M. Ojima, A. Fujii, M. Ozaki and S. Shinkai, Polydiacetylene nanofibers created in low-molecular-weight gels by post modification: control of blue and red phases by the odd-even effect in alkyl chains, *J. Am. Chem. Soc.*, 2007, **129**, 4134–4135.
58. S. Y. Noskov, G. Lamoureux and B. Roux, Molecular dynamics study of hydration in ethanol-water mixtures using a polarizable force field, *J. Phys. Chem. B*, 2005, **109**, 6705–6713.
59. M. Sol, M. Mejıa, J. F. Espinal, A. Restrepo and F. Mondragon, Molecular interaction of (ethanol)2-water heterotrimers, *J. Phys. Chem. A*, 2007, **111**, 8250–8256.
60. S. Park, K.-S. Lee, G. Bozoklu, W. Cai, S. B. T. Nguyen and R. S. Ruoff, Graphene oxide papers modified by divalent ions-enhancing mechanical properties via chemical cross-linking, *acsnano.org*, 2008, **2**, 572–578.
61. J. R. Lomeda, C. D. Doyle, D. V. Kosynkin, W. Hwang and J. M. Tour, Diazonium functionalization of surfactant-wrapped chemically converted graphene sheets, *J. Am. Chem. Soc.*, 2008, **130**, 16201–16206.
62. Y. Si and E. T. Samulski, Exfoliated graphene separated by platinum nanoparticles, *Chem. Mater.*, 2008, **20**, 6792–6797.
63. Y. Wang, Y Huang, Y. Song, X. Zhang, Y. Ma, J. Liang and Y. Chen, Room-temperature ferromagnetism of graphene, *Nano Lett.* 2009, **9**, 220–224.
64. J. Campos-Delgado, J. M. Romo-Herrera, X. Jia, D. A. Cullen, H. Muramatsu, Y. A. Kim, T. Hayashi, Z. Ren, D. J. Smith, Y. Okuno, T. Ohba, H. Kanoh, K. Kaneko, M. Endo, H. Terrones, M. S. Dresselhaus and M. Terrones, Bulk production of a new form of sp^2 carbon: crystalline graphene nanoribbons, *Nano Lett.* 2008, **8**, 2773–2778.
65. M. Kahlweit and R. Strey, Phasenverhalten ternarer Systeme des Typs H_2O – nichtionisches Amphiphil (Mikroemulsionen), *Angew. Chem.*, 1985, **97**, 655–669.
66. G. B. Sigal, M. Mrksich and G. M. Whitesides, Effect of surface wettability on the adsorption of proteins and detergents, *J. Am. Chem. Soc.*, 1998, **120**, 3464.

67. J.-H. Fuhrhop and T. Wang, Bolaamphiphiles, *Chem. Rev.*, 2004, **104**, 2901–2937.
68. J. T. Hirvi and T. A. Pakkanen, Wetting of nanogrooved polymer surfaces, *Langmuir* 2007, **23**, 7724–7729.
69. (a) T. M. Clausen, P. K. Vinson, J. R. Minter, H. T Davis, J. Talmon and W. G Miller, Viscoelastic micellar solutions: microscopy and rheology, *J. Phys. Chem.*, 1992, **96**, 474–484; (b) R. Oda, I. Huc, D. Danino and Y. Talmon, Aggregation properties and mixing behavior of hydrocarbon, fluorocarbon, and hybrid hydrocarbon–fluorocarbon cationic dimeric surfactants, *Langmuir*, 2000, **16**, 9759–9769.
70. J. S. Beck, J. C. V. Uli, W. J. Roth, M. E. Leonowicz, C. T. Kresge, K. D. Schmitt, C. T.-W. Chu, D. H. Olson, E. W. Sheppard, S. B. McCullen, J. B. Higgins and J. L. Schlenkert, A new family of mesoporous molecular sieves prepared with liquid crystal templates, *J. Am. Chem. Soc.*, 1992, **114**, 10834–10843.
71. M. R. Radowski A. Shukla, H. von Berlepsch, C. Boettcher, G. Pickaert, H. Rehage and R. Haag, Supramolecular aggregates of dendritic multishell architectures as universal nanocarriers, *Angew. Chem., Int. Ed.*, 2007, **46**, 1265–1269.
72. J. Keilitz, M. R. Radowski, J.-D. Marty, Rainer Haag, F. Gauffre and C. Mingotaud, Dendritic polymers with a core-multishell architecture: a versatile tool for the stabilization of nanoparticles, *Chem. Mater.*, 2008, **20**, 2423–2425.
73. O. Träger, S. Sowade, C. Böttcher and J.-H. Fuhrhop, Fluid and chiral ephedrinium myristate micellar fibers, *J. Am. Chem. Soc.*, 1997, **119**, 9120–9124.
74. G. Li, S. V. Bhosale, T. Wang, S. Hackbarth, B. Roeder, U. Siggel and J.-H. Fuhrhop, Nanowells on silica particles in water containing long-distance porphyrin heterodimers, *J. Am. Chem. Soc.*, 2003, **125**, 10693–10702.
75. H. Sellers, A. Ulman, Y. Shnidman and J. E. Eilers, Structure and binding of alkanethiolates on gold and silver surfaces: implications for self-assembled monolayers, *J. Am. Chem. Soc.*, 1993, **115**, 9389.
76. G. Li, M. Lauer A. Schulz, C. Boettcher, F. Li and J.-H. Fuhrhop, Spherical and planar gold(0) nanoparticles with a rigid gold(I)-anion or a fluid gold(0)-acetone surface, *Langmuir*, 2003, **19**, 6483–6491.
77. Y. Yokota, K. Fukui, T. Enoki and M. Hara, Strong intermolecular electronic coupling within a tetrathiafulvalene island embedded in self-assembled monolayers, *J. Am. Chem. Soc.*, 2007, **129**, 6571–657.
78. N. Murthy, Y. X. Thng, S. Schuck, M. C. Xu and J. M. J. Fréchet, A novel strategy for encapsulation and release of proteins: hydrogels and microgels with acid-labile acetal cross-linkers, *J. Am. Chem. Soc.*, 2002, **124**, 12398–12399.
79. O. Soubias and K. Gawrisch, Dependence of the correlation time of fast motions on the docosahexaenoyl, $(CH_2)_{22}$, chain carbon position. Docosahexaenoyl chains isomerize on the sub-nanosecond time scale, *J. Am. Chem. Soc.* 2007, **129**, 6678–6679.

80. P. A. Brugger, P. P. Infelta, A. M. Braun and M. Gratzel, Photoredox reactions in functional micellar assemblies, use of amphiphilic redox relays to achieve light energy conversion and charge separation, *J. Am. Chem. Soc.*, 1981, **103**, 320–326.
81. (a) F. M. Menger, The structure of micelles, *Acc. Chem. Res.*, 1979, **12**, 111–119; (b) K. A. Dill and P. J. Flory, Molecular organization in micelles and vesicles, *Proc. Natl. Acad. Sci. USA*, 1981, **78**, 676–680.
82. J.-H. Fuhrhop and J. Köning, *Molecular Assemblies and Membranes*, ed. J. F. Stoddart, Royal Society of Chemistry, London, 1994.
83. U. Zoller, *Handbook of Detergents*, Parts A–C, Taylor & Francis, Boca Raton FL, 2004.
84. Y. J. Na, H. S. Kim and J. Park, Morphology-controlled lead selenide nanocrystals and their *in situ* growth on carbon nanotubes, *J. Phys. Chem. C*, 2008, **112**, 11218–11226.
85. W. Lin, K. Fritz, G. Guerin, G. R. Bardajee, S. Hinds, V. Sukhovatkin, E. H. Sargent, G. D. Scholes and M. A. Winnik, Highly luminescent lead sulfide nanocrystals in organic solvents and water through ligand exchange with poly(acrylic acid), *Langmuir*, 2008, **24**, 8215–8219.
86. L. S. McCarty, A. Winkleman and G. M. Whitesides, Ionic electrets: electrostatic charging of surfaces by transferring mobile ions upon contact, *J. Am. Chem. Soc.*, 2007, **129**, 4075–4088.
87. A. Klyszcz, M. Lauer, M. Kopaczynska, C. H. Böttcher, F. Gonzaga and J.-H. Fuhrhop, Irreversible or reversible self-assembly procedures yield robust zirconium (IV)-porpyrinphosphonate cones or μm-long fibres of monomolecular thickness, *Chem. Comm.*, 2004, 2358–2359.
88. J.-H. Fuhrhop and D. Fritsch, Bolaamphiphiles form ultrathin, porous, and unsymmetric monolayer lipid membranes, *Acc. Chem. Res.*, 1986, **19**, 130–137.
89. J.-H. Fuhrhop and W. Helfrich, Fluid and solid fibers made of lipid molecular layers, *Chem. Rev.*, 1993, **93**, 1582–1585.
90. T. M. Fyles, C.-W. Hu and H. Luong, Solid-phase synthesis of oligoester ion channels, *J. Org. Chem.*, 2006, **71**, 8545–8551.
91. T. Mueller and G. Ceder, A Density functional theory study of hydrogen adsorption in MOF-5, *J. Phys. Chem. B*, 2005, **109**, 17974–17983.
92. D. Sun, S. Ma, Y. Ke, D. J. Collins and H.-C. Zhou, An interweaving MOF with high hydrogen uptake, *J. Am. Chem. Soc.*, 2006, **128**, 3896–3897.
93. S. S. Kaye, A. Dailly, O. M. Yaghi and J. R. Long, Impact of preparation and handling on the hydrogen storage properties of $Zn_4O(1,4$-benzenedicarboxylate$)_3$ (MOF-5), *J. Am. Chem. Soc.*, 2007, **129**, 14176–14177.
94. J. A. Greathouse and M. D. Allendorf, The interaction of water with MOF-5 simulated by molecular dynamics, *J. Am. Chem. Soc.*, 2006, **128**, 10678–10679.
95. S. Hermes, T. Witte, T. Hikov, D. Zacher, S. Bahnmueller, G. Langstein, K. Huber and R. A. Fischer, Trapping metal-organic framework nanocrystals: an in-situ time-resolved light scattering study on the crystal growth of MOF-5 in solution, *J. Am. Chem. Soc.*, 2007, **129**, 5324–5325.

96. J. Hafizovic, M. Bjørgen, U. Olsbye, P. D. C. Dietzel, S. Bordiga, C. Prestipino, C. Lamberti and K. P. Lillerud, The inconsistency in adsorption properties and powder XRD data of MOF-5 is rationalized by framework interpenetration and the presence of organic and inorganic species in the nanocavities, *J. Am. Chem. Soc.*, 2007, **129**, 3612–3620.
97. K. Matyjaszewski and J. Xia, Atom transfer radical polymerization, *Chem. Rev.*, 2001, **101**, 2921–2990.
98. B. Y. Lin, M. L. Coote, A. Gennaro and K. Matyjaszewski, Ab initio evaluation of the thermodynamic and electrochemical properties of alkyl halides and radicals and their mechanistic implications for atom transfer radical polymerization, *J. Am. Chem. Soc.*, 2008, **130**, 12762–12774.
99. H. Dong, M. Zhu, J. Ae Yoon, H. Gao, R. Jin and K. Matyjaszewski, One-pot synthesis of robust core/shell gold nanoparticles, *J. Am. Chem. Soc.*, 2008, **130**, 12852–12853.
100. (a) P. G. de Gennes, Interactions between polymers and surfactants, *J. Phys. Chem.*, 1990, **94**, 8407–8413; (b) H. H. S. E. Rankin, L. S. Penn, R. P. Quirk and H. T. Cheong, Transition from mushroom to brush during formation of a tethered layer, *Langmuir*, 2004, **20**, 5770–5775; (c) I. D. Hosein and C. M. Liddell, Convectively Assembled nonspherical mushroom cap-based colloidal crystals, *Langmuir*, 2007, **23**, 8810–8814.
101. (a) G. Sun, J. Xu, A. Hagooly, R. Rossin, Z. Li, D. A. Moore, C. J. Hawker, M. J. Welch and K. L. Wooley, Strategies for optimized radiolabeling of nanoparticles for *in vivo* PET imaging, *Adv. Mater.*, 2007, **19**, 3157–3162; (b) J. Xu, G. Sun, R. Rossin, A. Hagooly, Z. Li, K. Fukukawa, B. W. Messmore, D. A. Moore, M. J. Welch, C. J. Hawker and K. L. Wooley, Labeling of polymer nanostructures for medical imaging: importance of cross-linking extent, spacer length, and charge density, *Macromolecules*, 2007, **40**, 2971–2973; (c) Y. Li, W. Du, G. Sun and K. L. Wooley, pH-responsive shell cross-linked nanoparticles with hydrolytically labile cross-links, *Macromolecules*, 2008, **41**, 6605–6607.
102. G. Marque, S. Neyertz, J. Verdu, V. Prunier and D. Brown, Molecular dynamics simulation study of water in amorphous kapton, *Macromolecules*, 2008, **41**, 3349–3362.
103. M. J. Percy, Review of black surfaces for space-born IR systems, *Rev. Sci. Instr.*, 1999, **70**, 2193–2217.
104. L. S. McCarty, A. Winkleman and G. M. Whitesides, Ionic electrets: electrostatic charging of surfaces by transferring mobile ions upon contact, *J. Am. Chem. Soc.*, 2007, **129**, 4075–4088.
105. N. Duxin, F. Liu, H. Vali and A. Eisenberg, Cadmium sulfide quantum dots in morphologically tunable triblock copolymer aggregates, *J. Am. Chem. Soc.*, 2005, **127**, 10063–10069.
106. B. Wang, B. Li, B. Zhao and C. Y. Li, Amphiphilic janus gold nanoparticles via combining "solid-state grafting-to" and "grafting-from" methods, *J. Am. Chem. Soc.*, 2008, **130**, 11594–11595.

107. G. G. du Sart, R. Rachmawati, V. Voet, G. A. van Ekenstein, E. Polushkin, G. ten Brinke and K. Loos, Poly(*tert*-butyl methacrylate-*b*-styrene-*b*-4-vinylpyridine) triblock copolymers: synthesis, interactions, and self-assembly, *Macromolecules*, 2008, **41**, 6393–6399.
108. L. S. Ott, B. J. Hornstein and R. G. Finke, A test of the transition-metal nanocluster formation and stabilization ability of the most common polymeric stabilizer, poly(vinylpyrrolidone), as well as four other polymeric protectants, *Langmuir*, 2006, **22**, 9357–9367.
109. V. Boyko, S. Richter, W. Burchard and K.-F. Arndt, Chain dynamics in microgels: poly(*N*-vinylcaprolactam-*co*-*N*-vinylpyrrolidone) microgels as examples, *Langmuir*, 2007, **23**, 776–784.
110. Z. Feng, Z. Wang, C. Gao and J. Shen, Template polymerization to fabricate hydrogen-bonded poly(acrylic acid)/poly(vinylpyrrolidone) hollow microcapsules with a pH-mediated swelling-deswelling property, *Chem. Mater.*, 2007, **19**, 4648–4657.
111. A. Chen, K. Kamata, M. Nakagawa, T. Iyoda, H. Wang and X. Li, Formation process of silver-polypyrrole coaxial nanocables synthesized by redox reaction between $AgNO_3$ and pyrrole in the presence of poly(vinylpyrrolidone), *J. Phys. Chem. B*, 2005, **109**, 18283–18288.
112. Y. Sun, Y. Yin, B. T. Mayers, T. Herricks and Y. Xia, Uniform silver nanowires synthesis by reducing $AgNO_3$ with ethylene glycol in the presence of seeds and poly(vinyl pyrrolidone), *Chem. Mater.*, 2002, **14**, 4736–4745.
113. Y. Gao, P. Jiang, D. F. Liu, H. J. Yuan, X. Q. Yan, Z. P. Zhou, J. X. Wang, L. Song, L. F. Liu, W. Y. Zhou, G. Wang, C. Y. Wang, S. S. Xie, J. M. Zhang and D. Y. Shen, Evidence for the monolayer assembly of poly(vinylpyrrolidone) on the surfaces of silver nanowires, *J. Phys. Chem. B*, 2004, **108**, 12877–12881.
114. S. J. Lee, J. M. Baik and M. Moskovits, Polarization-dependent surface-enhanced raman scattering from a silver-nanoparticle-decorated single silver nanowire, *Nano Lett.*, 2008, **8**, 3244–3247.
115. Y. L. Hsin, K. C. Hwang and C.-T. Yeh, Poly(vinylpyrrolidone)-modified graphite carbon nanofibers as promising supports for PtRu catalysts in direct methanol fuel cells, *J. Am. Chem. Soc.*, 2007, **129**, 9999–10010.
116. A. N. Zelikin, G. K. Such, A. Postma and F. Caruso, Poly(vinylpyrrolidone) for bioconjugation and surface ligand immobilization, *Biomacromolecules*, 2007, **8**, 2950–2953.
117. W.-C. Chen, S.-W. Kuo, U. Jeng and F. C. Chang, Self-assembly through competitive interactions of miscible diblock copolymer/homopolymer blends: poly(vinylphenol*b*-methyl methacrylate)/poly(vinylpyrrolidone) blend, *Macromolecules*, 2008, **41**, 1401–1410.
118. S. M. Borisov, T. Mayr and I. Klimant, Poly(styrene-block-vinylpyrrolidone) beads as a versatile material for simple fabrication of optical nanosensors, *Anal. Chem.*, 2008, **80**, 573–582.
119. A. Pich, A. Tessier, V. Boyko, Y. Lu and H.-J. P. Adler, Synthesis and characterization of poly(vinylcaprolactam)-based microgels exhibiting

temperature and pH-sensitive properties, *Macromolecules*, 2006, **39**, 7701–7707.
120. (a) Y. Ono and T. Shikata, Hydration and dynamic behavior of poly(N-isopropylacrylamide)s in aqueous solution: a sharp phase transition at the lower critical solution temperature, *J. Am. Chem. Soc.*, 2006, **128**, 10030–10031; (b) Y. Ono and T. Shikata, Contrary hydration behavior of N-isopropylacrylamide to its polymer, P(NIPAm), with a lower critical solution temperature, *J. Phys. Chem. B*, 2007, **111**, 1511–1513.
121. S. Schachschal, A. Pich and H.-J. Adler, Aqueous microgels for the growth of hydroxyapatite nanocrystals, *Langmuir*, 2008, **24**, 5129–5134.
122. C. J. Hawker, "Living" free radical polymerization: a unique technique for the preparation of controlled macromolecular, *Acc. Chem. Res.*, 1997, **30**, 373–382.
123. Z. Li, Z. Chen, H. Cui, K. Hales, K. L. Wooley and D. J. Pochan, Controlled stacking of charged block copolymer micelles, *Langmuir*, 2007, **23**, 4689–4594.
124. C. J. Hawker, K. L. Wooley and J. M. Fréchet, Unimolecular micelles and globular amphiphiles: dendritic macromolecules as novel recyclable solubilization agents, *J. Chem. Soc., Perkin Trans., 1*, 1993, 1287–1297.
125. S. Hecht and J. M. Fréchet, An alternative synthetic approach toward dendritic macromolecules: novel benzene-core dendrimers via alkyne cyclotrimerization, *J. Am. Chem. Soc.*, 1999, **121**, 4084–4085.
126. (a) A. D. Schlüter and J. P. Rabe, Grenzflächenverhalten und Manipulation, *Angew. Chem.*, 2000, **112**, 860–880; (b) C. Münzenberg, A. Rossi, K. Feldman, R. Fiolka, A. Stemmer, K. Kita-Tokarczyk, W. Meier, J. Sakamoto, O. Lukin and A. D. Schlüter, Synthesis of compounds presenting three and four anthracene units as potential connectors to mediate infinite lateral growth at the air/water interface, *Chem. Eur. J.*, 2008, **14**, 10797–10807.
127. S. Nad, S. Roller, R. Haag and R. Breinbauer, Synthesis of N-substituted pyrroles electrolysis as an efficient key step in the homogeneous polymer-supported, *Org. Lett.*, 2006, **8**, 403–406.
128. J. L. Bredas and G. B. Street, Polarons, bipolarons, and solitons in conducting polymers, *Acc. Chem. Res.*, 1985, **18**, 309–315.
129. X. Wang, X. Gu, C. Yuan, S. Chen, P. Zhang, T. Zhang, J. Yao, F. Chen and G. J. Chen, Evaluation of biocompatibility of polypyrrole *in vitro* and *in vivo*, *Biomed. Mater. Res.*, 2004, **68A**, 411.
130. H. H. Fong, V. A. Pozdin, A. Amassian, G. G. Malliaras, D.-M. Smilgies, M. He, S. Gasper F. Zhang and M. Sorensen, Tetrathienoacene copolymers as high mobility, soluble organic semiconductors, *J. Am. Chem. Soc.*, 2008, **130**, 13202–13203.
131. D. Ofer, R. M. Crooks and M. S. Wrighton, Potential dependence of the conductivity of highly oxidized polythiophenes, polypyrroles, and polyaniline: finite windows of high conductivity, *J. Am. Chem. Soc.*, 1990, **112**, 7869–7879.

132. N. Kohut-Svelko, S. Reynaud, R. Dedryvère, H. Martinez, D. Gonbeau and J. Francois, Study of a nanocomposite based on a conducting polymer: polyaniline, *Langmuir*, 2005, **21**, 1575–1583.
133. J. Luo, H. Zhang, X. Wang, J. Li and F. Wang, Stable aqueous dispersion of conducting polyaniline with high electrical conductivity, *Macromolecules*, 2007, **40**, 8132–8135.
134. M. Trchova, J. Prokes, I. Krivka, E. Tobolkova, J. Stejskal and I. Sapurina, Solid-state protonation and electrical conductivity of polyaniline, *Macromolecules*, 1998, **31**, 2218–2222.
135. S. C. Moulzolf, D. J. Frankel and R. J. Lad, In situ four-point conductivity and hall effect apparatus for vacuum and controlled atmosphere measurements of thin film materials, *Rev. Sci. Instr.*, 2002, **73**, 2325–2330.
136. Y. Ma, J. Zhang, G. Zhang and H. He, Polyaniline nanowires on Si surfaces fabricated with DNA templates, *J. Am. Chem. Soc.*, 2004, **126**, 7097–7101.
137. J. Huang, Shabnam. Virji, B. H. Weiller and R. B. Kaner, Polyaniline nanofibers: facile synthesis and chemical sensors, *J. Am. Chem. Soc.*, 2003, **125**, 314–315.
138. G. Li, C. Martinez and S. Semancik, Controlled electrophoretic patterning of polyaniline from a colloidal suspension, *J. Am. Chem. Soc.*, 2005, **127**, 4903–4909.
139. Y. Guo, M. Li, A. Mylonakis, J. Han, A. G. MacDiarmid, X. Chen, P. I. Lelkes and Y. Wei, Electroactive oligoaniline containing self-assembled monolayers for tissue engineering applications, *Biomacromolecules*, 2007, **8**, 3025–3034.
140. N. Mano, J. E. Yoo, J. Tarver, Y.-L. Loo and A. Heller, An electron-conducting crosslinked polyaniline-based redox hydrogel, formed in one step at pH 7.2, wires glucose oxidase, *J. Am. Chem. Soc.*, 2007, **129**, 7006–7007.
141. D. M. DeLongchamp and P. T. Hammond, Multiple-color electrochromism from layer-by-layer-assembled polyaniline/Prussian blue nanocomposite thin films, *Chem. Mater.*, 2004, **16**, 4799–4805.
142. D. Li, J. Huang and R. B. Kaner, Polyaniline nanofibers: a unique polymer nanostructure for versatile applications, *Acct. Chem. Res.*, 2008, **42**, 135–145.
143. E. Leary, S. J. Higgins, H. van Zalinge, W. Haiss, R. J. Nichols, S. Nygaard, J. O. Jeppesen and J. Ulstrup, Structure-property relationships in redox-gated single molecule junctions – a comparison of pyrrolotetrathiafulvalene and viologen redox groups, *J. Am. Chem. Soc.*, 2008, **130**, 12204–12205.
144. W.-Q. Deng, A. H. Flood, J. Fraser Stoddart and W. A. Goddard III, An electrochemical colour-switchable RGB dye: tristable [2]catenane, *J. Am. Chem. Soc.*, 2005, **127**, 15994–15995.
145. (a) J. Rebek, Jr., Contortions of encapsulated alkyl groups, *Chem. Commun.*, 2007, 2777–2789; (b) R. J. Hooley, S. M. Biros and J. Rebek, Jr., Normal hydrocarbons tumble rapidly in a deep, water-soluble cavitand, *Chem. Commun.*, 2006, 509–510.

146. S. P. Bew, R. A. Brimage, N. L'Hermite and S. V. Sharma, Upper rim appended hybrid calixarenes via click chemistry, *Org. Lett.*, 2007, **9**, 3713–3716.
147. (a) S. Bhosale S. Bhosale, T. Wang, G. Li, U. Siggel and J. Fuhrhop, Slow motion, trapping, and sorting of water- and chloroform-soluble porphyrins in nanowell, *J. Am. Chem. Soc.*, 2004, **126**, 13111–13118; (b) S. Bhosale, Sh. Bhosale, T. Wang, M. Kopaczynska and J.-H. Fuhrhop, Hydrophobic and Hydrophilic Yoctowells, *Accts. Chem. Res.*, 2006, **39**, 498–508.
148. (a) L. T. Barner, D. C. Danila, K. Sharpe, M. Durkin, B. Clayton, B. Anderson, A. Richter and E. Pinkhassik, Controlled loading of building blocks into temporary self-assembled scaffolds for directed assembly of organic nanostructures, *Langmuir*, 2008, **24**, 11464–11473; (b) D. C. Danila, L. T. Banner, E. Karimova, L. Tsurkan, X. Wang and E. Pinkhassik, Directed assembly of sub-nanometer thin organic materials with programmed-size nanopores, *Angew. Chem. Int. Ed.*, 2008, **47**, 7036–7039.
149. (a) J. He, T. Kunitake and A. Nakao, Facile in situ synthesis of noble metal nanoparticles in porous cellulose fibers, *Chem. Mater.*, 2003, **15**, 4401–4406; (b) J. Huang and T. Kunitake, Nano precision replication of natural cellulosic substances by metal oxides, *J. Am. Chem. Soc.*, 2003, **125**, 11834–11835.
150. J. Huang, I. Ichinose and T. Kunitake, Nanocoating of natural cellulose fibers with conjugated polymer: hierarchical polypyrrole composite materials, *Chem. Commun.*, 2005, 1717–1719.
151. S. Hornig and T. Heinze, Efficient approach to design stable water-dispersible nanoparticles of hydrophobic cellulose esters, *Biomacromolecules*, 2008, **9**, 1487–1492.
152. Y. Shin, I.-T. Bae, B. W. Arey and G. J. Exarhos, Facile stabilization of gold-silver alloy nanoparticles on cellulose nanocrystal, *J. Phys. Chem. C*, 2008, **112**, 4844–4848.
153. T. G. M. de Ven, Orthokinetic heteroflocculation in papermaking, in *Highlights in Colloid Science*, ed. D. Platikanov and D. Exerowa, Wiley-VCH, Weinheim, 2009, 1–20.
154. M. Kemell, V. Pore, M. Ritala, M. Leskel and M. Lindn, Atomic layer deposition in nanometer-level replication of cellulosic substances and preparation of photocatalytic TiO_2/cellulose composites, *J. Am. Chem. Soc.*, 2005, **127**, 14178–14179.
155. S. Liu, L. Zhang, J. Zhou, J. Xiang, J. Sun and J. Guan, Fiberlike Fe_2O_3 macroporous nanomaterials fabricated by calcinating regenerate cellulose composite fibers, *Chem. Mater.*, 2008, **20**, 3623–3628.
156. J.-H. Fuhrhop, C. Demoulin, J. Rosenberg and C. Boettcher, Cloth-like aggregates of micellar fibers made of N-dodecyltartaric acid monoamides, *J. Am. Chem. Soc.*, 1990, **112**, 2827–2829.
157. R. Oda, F. Artzner, M. Laguerre and I. Huc, Molecular structure of self-assembled chiral nanoribbons and nanotubules revealed in the hydrated State, *J. Am. Chem. Soc.*, 2008, **130**, 14705–14712.
158. J. D. Hartgerink, E. Beniash and S. I. Stupp, Self-assembly and mineralization of peptide-amphiphile nanofibers, *Science*, 2001, **294**, 1684–1688.

159. H. Dong, S. E. Paramonov and J. D. Hartgerink, Self-assembly of R-helical coiled coil nanofibers, *J. Am. Chem. Soc.*, 2008, **130**, 13691–13695.
160. D. C. Danila, L. T. Banner, E. J. Karimova, L. Tsurkan, X. Wang and E. Pinkhassik, Directed assembly of sub-nanometer thin organic materials with programmed-size nanopores, *Angew. Chem.*, 2008, **120**, 7144–7148.
161. D. Casanova, D. Giaume, M. Moreau, J.-L. Martin, T. Gacoin, J.-P. Boilot and A. Alexandrou, Counting the number of proteins coupled to single nanoparticles, *J. Am. Chem. Soc.*, 2007, **129**, 12592–12593.
162. G. von Maltzahn, T. J. Harris, J.-H. Park, D.-H. Min, A. J. Schmidt, M. J. Sailor and S. N. Bhatia, Nanoparticle self-assembly gated by logical proteolytic triggers, *J. Am. Chem. Soc.*, 2007, **129**, 6064–6065.
163. D. C. Chow and A. Chilkoti, Surface-initiated enzymatic polymerization of DNA, *Langmuir*, 2007, **23**, 11712–11717.
164. S. Tyagi and F. R. Kramer, Molecular beacons: probes that fluoresce upon hybridization, *Nat. Biotechnol.*, 1996, **14**, 303–308.
165. A. A. Marti, S. Jockusch, N. Stevens, J. Ju and N. J. Turro, Fluorescent hybridization probes for sensitive and selective DNA and RNA detection, *Acc. Chem. Res.*, 2007, **40**, 402–409.
166. R. L. Stoermer, K. B. Cederquist, S. K McFarland, M. Y. Sha, S. G. Penn and C. D. Keating, Coupling molecular beacons to barcoded metal nanowires for multiplexed, sealed chamber DNA bioassays, *J. Am. Chem. Soc.*, 2006, **128**, 16892–16903.
167. R. L. Stoermer and C. D. Keating, Distance-dependent emission from dye-labeled oligonucleotides on striped Au/Ag nanowires: effect of secondary structure and hybridization efficiency, *J. Am. Chem. Soc.*, 2006, **128**, 13243–13254.
168. G. Liu and Y. Lin, Electrochemical quantification of single-nucleotide polymorphisms using nanoparticle probes, *J. Am. Chem. Soc.*, 2007, **129**, 10394–10401.
169. S. A. Claridge, A. J. Mastroianni, Y. B. Au, H. W. Liang, C. M. Micheel, J. M. J. Fréchet and A. P. Alivisatos, Enzymatic ligation creates discrete multinanoparticle building blocks for self-assembly, *J. Am. Chem. Soc.*, 2008, **130**, 9598–9605.
170. A. Berkessel and J. A. Adrio, Dramatic acceleration of olefin epoxidation in fluorinated alcohols: activation of hydrogen peroxide by multiple H-bond networks, *J. Am. Chem. Soc.*, 2006, **128**, 13412–13420.
171. N. Shukla and A. J. Gellman, Interaction of alcohols and ethers with a-CFx films, *Langmuir*, 2000, **16**, 6562–6568.
172. E. G. Shchukin, E. A. Amelina and A. M. Parfenova, *Stability of fluorinated systems*, in: *Highlights in Colloid Science*, ed. D. Platikanov and D. Exerowa, Wiley-VCH, Weinheim, 2009, 41–54.
173. S. Chakraborty, W. Guo, R. H. Hauge and W. E. Billups, Reductive alkylation of fluorinated graphite, *Chem. Mater.*, 2008, **20**, 3134–3136.

CHAPTER 3
Main Group Elements

3.1 Introduction

The organic nanochemistry of covalent carbon was dealt with in the previous chapter. This chapter focuses on other non-metals and metals without or with fully filled d-orbitals, which are compared with carbon and the elements in the centre and at top of the group. The accessibility of the elements is proportional to their price (Table 3.1). High prices mean that the elements are either rare (germanium) or difficult to prepare and store (alkali metals, except sodium).

All main group elements on the left-hand side of the periodic table and below carbon are electropositive with respect to carbon; those on the right-hand side are mostly electronegative. The most important exception is amino nitrogen in water, which takes up protons to form positively charged ammonium ions.

The oxides (borates, aluminates, silicates) often occur as polymers. The nitrides and phosphines of various main group elements form hard and occasionally luminescent materials. Most luminescent materials are made of III–V or IV–VI compounds that are covalent but polar.

Applications rely mainly on the following individual properties of the elements. Doped Si(0) or Ge(0) layers and many III–V and IV–VI compounds behave as semiconductors, which either take up or give away electrons upon application of an electric potential, but not both. Thermoelectric wires, with high electric conductivity and low thermal conductivity, cool down if an electric current flows through them. Such wires are dream materials for computers and cooling devices. Research on such materials currently concentrates on the heavy main group metals like caesium, bismuth, tellurium, and lead as well as on combinations with the non-metals sulfur and selenium. Many semiconducting micron- and nanosized elements of electron-steering chips integrate billions of npn-, pnp- or field effect transistors (FETs) within very small "gates"

Table 3.1 Cost of main group elements (pure powder or similar form) in €/100 g, in increasing order. Data from the 2008 Alfa Aesar catalogue.

Zn	Zinc	1.5
S	Sulfur	3
Pb	Lead	5
Al	Aluminium	6
C	Carbon	11
I	Iodine	13
Ca	Calcium	18
Sn	Tin	20
Na	Sodium	21
P	Phosphorus	24
Se	Selenium	24
Ce	Cerium	36
Sb	Antimony	37
Bi	Bismuth	37
Mg	Magnesium	50
Cd	Cadmium	51
Si	Silicon	56
Ba	Barium	105
Te	Tellurium	202
Li	Lithium	210
As	Arsenic	220
Ta	Tantalum	243
Sr	Strontium	272
Ga	Gallium	332
Be	Beryllium	508
B	Boron	626
Ge	Germanium	1080
K	Potassium	1100
Rb	Rubidium	1600
Cs	Caesium	2560

or light-measuring photodiodes and luminescence lamps. The band gap energies of the semiconductors decrease with atomic number, and range from a few millielectronvolts to 4.5 eV. Gallium arsenide and gallium selenide semiconductors provide extremely variable energy gaps here. In–SnO_2 (ITO) alloys form transparent, electroactive and stable thin films for all kinds of displays and sensor electrodes.

Heavy main group elements are often poisonous (*e.g.* As_2O_3, PbO_2, selenium, Tl(I), Hg(0)), because they interfere with respiration and/or ion transport in nerves and muscles.

More special material properties may be summarized as follows:

- All three of the light elements hydrogen, lithium and magnesium are possible converters of electric to chemical energy and *vice versa*. Hydrogen can be burnt to water and obtained from water; lithium and magnesium provide dense ion currents.

- Sodium, potassium, magnesium and calcium are abundant in aqueous environments and their interactions with phosphate dominate the biochemistry of all living things.
- Boron forms carbide nanoparticles and integrates into molecular carbon skeletons. Its isotopes, 18.8% ^{10}B and 81.2% ^{11}B, are easily distinguished by NMR. The ^{10}B isotope effectively captures neutrons.
- Aluminium is light and abundant. Its electrolytic oxidation ("anodization") punches regular patterns of $\sim 10\text{--}100$ nm wide holes covered with alumina (Al_2O_3) into its surface. Tantalum does the same, but the walls of the surface holes are more stable.
- Monovalent thallium (Tl^+) can displace K^+ in nerves and is therefore poisonous.
- Silicon is the most abundant semiconductor, n-doped, for example, by phosphorus and p-doped by germanium. Irregularities in the symmetry of silicate tetrahedrons in SiO_2 produce piezoelectric quartz crystals, which vibrate upon application of an alternating voltage. The dense coverage of silicate surfaces by OH groups allows a close to quantitative coating with acid or silyl chlorides. On SiO_2 nanoparticles with diameters ~ 100 nm the surface is so smooth that tight hydrogen bond chains in monolayers can be established. Zeolites contain long, regular tubules with a width < 1 nm, e.g. 0.75 nm in zeolite L.
- Germanium is an n- or p-semiconductor like silicon. $GeTe\text{--}Sb_2Te_3$ crystals and melts are used in rewritable laser disks.
- Tin telluride (SnTe) has a band gap of 0.17 eV, and ITO (see indium) is the most abundant photoelectrode material. Lead selenide (PbSe) nanocrystals provide the least expensive material for thermoelectric layers.
- Nitrides, such as BN, AlN, Si_3N_4, Sc_3N, GaN_3 and FeN do not occur in nature, but can be prepared in large quantities and in the form of hard, heat-stable nanoparticles. They are useful as polishing agents and luminescent materials.
- Phosphides are also only accessible by synthesis and may luminesce (indium phosphide), add hydrogen (boron phosphides) or be active as a catalyst in desulfurization of oil (MoP).
- Arsenide (As–) is a typical anion of III–V semiconductors. The most important one is gallium arsenide ($GaAs_3$) which is used in high electron mobility transistors, where it alloys well with metals and can be operated at high frequency without producing low thermal noise.
- Sulfur is the most aggressive element towards noble metals, e.g. mercury, copper, silver, platinum, palladium and gold. The soft sulfide anion obviously dissolves best in soft metals and a high oxidation potential is not necessary, because the noble metal does not lose its electrons, but forms covalent bonds with sulfur. The most important role of nanoparticles in the chemistry of sulfur is catalytic reduction of sulfides in gasoline. Lead sulfide (PbS) is a luminescent.
- Selenium or, nowadays, arsenic triselenide (As_2Se_3) is the basis of xerography in photocopiers. High light intensities carry electrons from the

valence band into the conductivity band of selenium, an electric field produces positive charges and graphite powder ("toner") sticks to the drum. Cadmium selenide (CaSe) and $ZnS_{1-x}Se_x$ are photoconductors Lead selenide (PbSe) has a narrow bulk gap band of 0.28 eV.
- Tellurium alloys with gold. The alloy bismuth telluride (Bi_2Te_3) is used in most electrical cooling devices, because it has the very low thermal conductivity of glass and a high electric conductivity.
- Fluoride/hydroxide exchange in the hydroxyapatite of bone takes place only at low pH. Treatment of titania by hydrogen fluoride improves its photocatalytic performance in water splitting.
- Zinc oxide (ZnO) is an important wide-band gap semiconductor with a band gap energy of 3.37 eV. $Zn_xCd_{1-x}S$ is a light emitter used in television screens, which covers the whole visible spectrum. Luminescent quantum dots are often made of cadmium selenide (core)–zinc sulfide (shell) assemblies.
- Mercury is the most poisonous of all elements. It is easily reduced in the body to the elemental fluid Hg(0), which dissolves as a metal in nerve tissue and induces electric currents there.

3.2 Hydrogen

Hydrogen is by far the most abundant element in the universe (92.6%), occurring mostly in the form of protons in the stars. Elementary hydrogen molecules (H_2) do not, however, occur naturally on Earth. The gravitational pull on this lightest of all molecules is too weak to maintain it in our atmosphere.

The simplest method of obtaining H_2 is electrolysis of a solution of potassium hydroxide or sodium hydroxide in water. The cathodic reduction produces OH^- and H_2 from water ($2H_2O + 2e \rightarrow 2OH^- + H_2$), the anodic oxidation gives oxygen and water exclusively from OH^- ($4OH^- - 4e \rightarrow 2H_2O_2 \rightarrow O_2 + 2H_2O$). Only water is used up, hydrogen and oxygen are the only products, and the pH does not change.

Today hydrogen is commercially produced primarily from water and methane by steam–methane reformation, which releases 2.5 times more oxides of carbon by mass than hydrogen. The electrochemical route is not feasible because of the price of electricity. Low-cost, renewable energy sources such as solar light need to be implemented (p. 310).

Hydrogen gas is a reducing agent: the 1 M H_2/H^+ potential determines the 0 V potential of electrochemistry. Hydrogen gas can be stored in solid metals or metal alloys, e.g. titanium or $LaNi_5$, which contain more hydrogen per volume unit than liquid hydrogen does (71 g/L). Simple ball-milling of graphite produces of dangling bonds, which appear as radicals on the edges of graphene sheets. Electron spin resonance detects them quantitatively, about 60 g of hydrogen are reversibly bound per kg of graphite.[1b] Hydrogen may well become the most important guest in pores of pressurized solids and nanoparticles. In order to load metals with hydrogen, they are warmed up together at high pressure. The hydrogen gas is reduced by the metal to a mixture of hydride ions, and free electrons and the metal is oxidized to a cation, e.g. Ti^{4+}. The release of molecular hydrogen and re-formation of the metal occurs at high temperatures. The catalytic combustion of hydrogen to water with molecular oxygen is used in fuel cells to produce electricity. So far, fuel cells are so heavy and expensive that they have been used only in submarines, where weight is a virtue.

Most hydrogen occurs as the isotope H, with a single proton as nucleus, but 0.015% is deuterium (D) with a neutron and a proton. In water the O–H bond is more easy to separate by electrolysis than the O–D bond and upon electrochemical reduction H_2 is exclusively formed and no DH until the remaining water is almost pure HDO. The proton has a nuclear spin of either $+1/2$ or $-1/2$ and NMR spectroscopy is the most powerful analytical method for carbon compounds with a stable C–H bond (p. 52 ff) in dissolved or solidified molecules as well as in the environment of paramagnetic nanoparticles in living tissues.[2]

3.3 Alkali Metals

The alkali metals lithium, sodium and potassium are the most important ions for ion currents in graphite (Li^+) or water within biological membranes or geological silicate pores (Na^-, K^+). Na^+/K^+ dominate the ion currents in nerves and muscles, as well as in rivers and oceans.

3.3.1 Lithium

Lithium metal is the lightest crystalline solid ($\sim 0.5 \text{ g/cm}^3$). An alloy of 14% lithium, 1% aluminium and 85% magnesium has a density of 1.35 g/cm^3, which is half that of pure aluminium. Lithium has the most negative standard reduction potential of all elements: $E° = -3.05 \text{ V}$. The light metal releases more energy upon oxidation to Li^+ than any other metal – more per mole, much more per gram.

The metal ions Li^+ and Mg^{2+} both form poorly soluble phosphates (Li_3PO_4 0.38 g/L; $Mg_3(PO_4)_2$ 0.0026 g/L). This property is the basis of the successful treatment of bipolar disorder with lithium chloride, which calms down nerve impulses. The concentration in the blood serum must, however, be strictly limited to 10^{-3} M, since higher concentrations cause heart failure.

Lithium precipitates with phosphate, carbonate and fluoride – anions that remain in solution with sodium and potassium as counterions. Lithium is also the only alkali metal to form a covalent nitride (Li_3N) and a normal oxide (Li_2O): sodium oxidizes to Na_2O_2, potassium to KO_2, both with O_2^- ions.

Lithium stearate ($C_{17}H_{35}COOLi$) is not suitable as a soap, because the Li^+ ion binds too tightly to COO– and lithium carbonate precipitates easily in hard water. Nevertheless, the stearate is the most important lithium compound, because it forms a stable mixture with oil that is used as axle grease for cars.

Main Group Elements

The mixture with oil is extremely stable at high temperature and does not freeze out in the cold, and the suspensions also remain uniform and stable in the presence of some water. Suspensions of pure lithium stearate in hydrocarbons are loose aggregates of the primary particles and spontaneously form gels. A concentration of 1.25 g/L is enough to form volume-filling end-to-end lithium–stearate chains ($-COO-Li^+-OOC-$). Electron microscopy showed lithium stearate rods with diameter of a few nanometres and lengths up to a micron (Figure 3.1). The consistency of the grease increased linearly with the length/width ratio of the elongated particles. Lithium–carboxylate and rod–rod interactions thus determine the rheological properties of greases containing lithium soap.[3]

The cathodes of lithium batteries are made of porous aluminum foils containing micron-sized graphite particles with entrapped Li(0) or lithium carbide, *e.g.* LiC_6. The particles hardly change their volume upon up- and downloading, the formation of Li^+ and Li(0). The anode is made of $LiCo(III)O_2$, and only the Co(III) reacts. The overall cell reaction is $2Li_{0.5}Co(II)O_2 + LiC_n \rightarrow 2LiCo(III)O_2 + C_n$, and the cycle runs at least 1000 times before the electrodes begin to degrade. The main reason for the final failure of graphite electrodes is not the suspected exfoliation of graphite particles, but their cracking and electrical isolation by organic surface films stemming from the decomposition

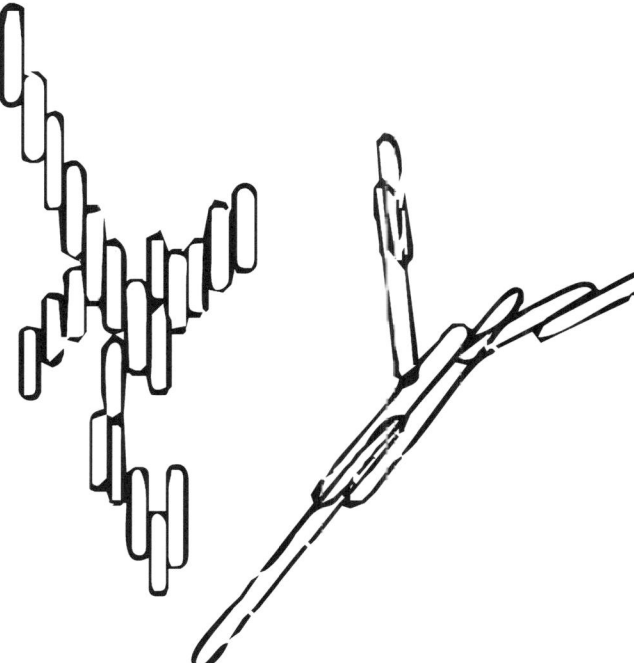

Figure 3.1 Schematic drawing of an electron micrograph of lithium stearate nanorods in oil. The longest particle is ~ 1 μm long. © American Chemical Society.[3]

of ethylene or propylene carbonate solvents for the Li$^+$ ions. Single-wall carbon nanotubes have also been characterized with respect to Li$^+$ loading.[4,5]

3.3.2 Sodium and Potassium

Sodium and potassium are almost equally abundant in the minerals of the Earth's surface and are washed out of sand and rocks to the same extent by rainwater. The chlorides of Na$^+$ (diameter 200 pm) and K$^+$ (266 pm) are not miscible within one crystal. Na$^+$ and K$^+$ phosphates are also both readily water-soluble (Na$_3$PO$_4$ 121 g/L; K$_3$PO$_4$ 923 g/L), so there is no obvious reason why they should precipitatate with silicates, algae or other living organisms. Potassium phosphate is six times more water soluble than sodium phosphate, so we might expect more potassium salts to end up in the sea. In fact, when rivers reach the sea almost all of the potassium has disappeared, sedimented to the bottom with minerals and microorganisms, whereas almost all of the sodium invariably does reach the sea. Sodium chloride is 10 times more abundant in sea water than potassium. The same is true in our bodies: sodium is again 10 times more abundant in blood and interstitial water ($\sim$1.5 wt% of NaCl), whereas potassium is 10 times more abundant than sodium inside nerve and muscle cells (Figure 3.2). These 10:1 ratios of sodium and potassium in the interior and exterior of cells are caused by the selective dehydration of the potassium tetrahydrate by polyelectrolytes within the cells. Sodium ions bind their 4–6 water molecules much more tightly and cannot pass biological membranes without the active help of proteins, nor are they immobilized by the carboxylate and phosphate groups of proteins or the silicate of sediments. Whereas sodium binds its water tightly, potassium loses it easily and prefers direct contact with polyanions. A chloride or carboxylate monoanion is not sufficient for differentiation, but the three oxygens of phosphate distinguish sharply between K$^+$ and Na$^+$.[6]

Early conductance measurements of sodium and potassium dihydrogen phosphates (NaH$_2$PO$_4$ and KH$_2$PO$_4$) in water indicated $\sim$30% more

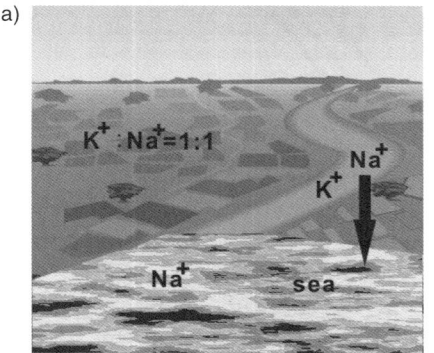
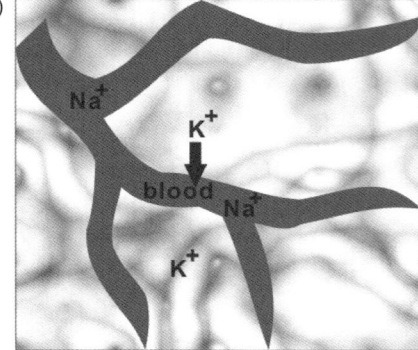

Figure 3.2 Distributions of Na$^+$ and K$^+$ in rivers and in the brain.[21b]

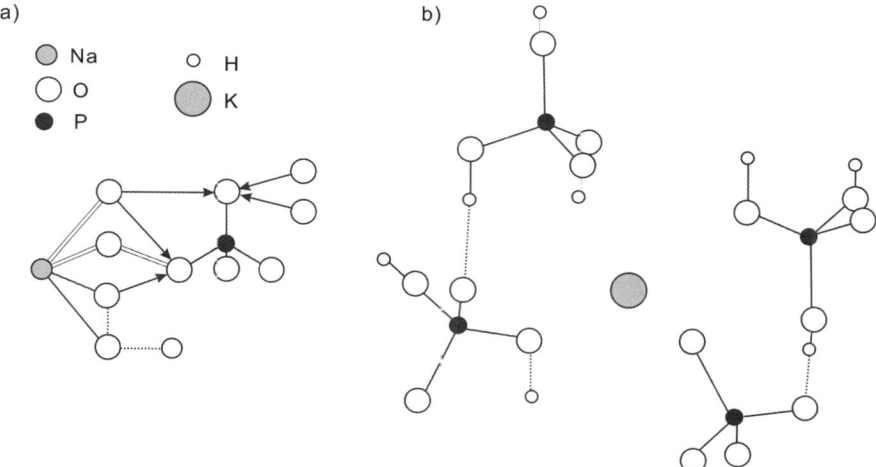

Figure 3.3 Schematic X-ray crystal structures of hydrated sodium phosphates and water-free potassium phosphates.[6]

conductance for the potassium salt. For sodium palmitate and potassium stearate the same difference of $\sim 30\%$ was measured. The difference is even greater in gaps in hydrophobic membranes and polymers. The crystal structure of a potassium phosphate ($KH_5(PO_4)_2$) shows no hydrate water, whereas the 3 Na^+ of a crystalline phosphate ($Na_3PO_4.12H_2O$) retain 12 hydration water molecules in the solid state[6,7] (Figure 3.3).

Sodium and potassium chlorides can be separated by flotation or electrical methods. In flotation, the salt mixture is dissolved with water to produce an oversaturated sol. This is mixed with a detergent and foam is then produced with an air stream. Crystals of potassium chloride are adsorbed selectively to the soap bubbles, float to the surface and are removed. The sodium chloride, on the other hand, crystallizes, sinks to the bottom and is dredged out.[8] Once again, the K^+ is selected by a polyelectrolyte surface.

With respect to anions in water, the decreasing relative preference for Na^+ and increasing preference for K^+ is

$$HCOO^- > OH^- > HSO_4^- > F^- > Cl^- > Br^- > H_2PO_4^- >$$
$$NO_3^- > SCN^- > ClO_4^- > I^-.[6]$$

Electrical separation takes advantage of the fact that solid potassium chloride becomes negatively charged by friction upon strong stirring with small amounts of salicylic acid whereas all other minerals, including sodium chloride, take up positive charges. When a stream of powder ripples down between two charged plates (voltage difference $\sim 100\,000$ V), the potassium chloride is separated out. Electron transfer between sodium chloride and potassium chloride crystallites is necessary for contact charging of potassium chloride and sodium chloride and depends on the adsorbed conditioner molecule, *e.g.* salicylic acid. The adsorbate generates characteristic electronic states with an effective band gap of 1–2 eV between the HOMOs and LUMOs of the aromatic ring, whereby the effective band gaps of the alkali halides are also reduced dramatically. The occupied molecular orbitals of the salicylate are conditioned by OH–OOC hydrogen bonding and line up with the upper edge of the valence bands of potassium chloride, which can then take up negative charges from the sodium chloride crystallites.

Migration of Na^+ and K^+ through and along the surface of lipid membranes is the basis of all electric currents in animal brains, nerves and muscles. Without a fast and efficient control of equilibration and fast active transport of these two ions we would have no vision, no thought, no feeling and no movements. The enzyme Na^+/K^+-ATPase, together with ATP, establishes the Post–Albers cycle by an active transport first of three Na^+ and then of two K^+. This cycle has been established exclusively by electrochemical measurements, but there is not the slightest structural hint of a selective binding site for Na^+. The phosphate ions of phosphorylated amino acids, the carboxylates of aspartate and glutamate side chains and the phenolic OH group of tyrosine are feasible candidates for the selective binding of Na^+, which is necessary for its active transport. None of them, however, is likely to prefer Na^+ over K^+.[9]

Not only water but also benzene derivatives differentiate strongly between Na^+ and K^+. K^+ binds directly to the oxygen and carbon atoms of hexakis(methoxymethyl)benzene, but Na^+ always keeps a connecting hydration water and the metal ion never comes near the π-system of benzene under the same conditions[10] (Figure 3.4).

For anions in aqueous solution the series with respect to a decreasing relative preference for Na^+ and increasing preference for K^+ is

$$HCOO^- > OH^- > HSO_4^- > F^- > Cl^- > Br^- > H_2PO_4^- > NO_3^- > SCN^- > ClO_4^- > I^-.$$

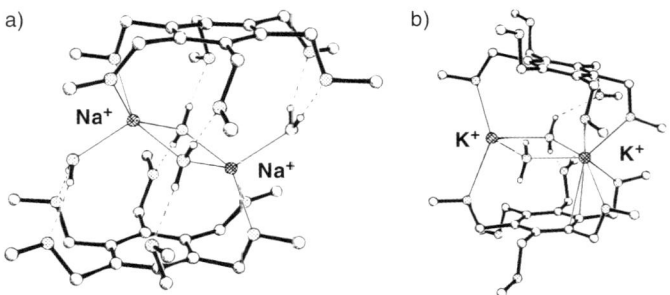

Figure 3.4 Even in symmetric $-CH_2-O-CH_3$ ether complexes the sodium ion uses water bridges, whereas potassium only uses it to separate charges. © American Chemical Society.[10]

In contrast to alkali halides, the mean activity coefficient of alkali hydroxide solutions of equal concentration decreases in the order CsOH > KOH > NaOH > LiOH. Hydroxides undergo local hydrolysis because the polarizing ability of the alkali cations is larger for the smaller ions. 200 ns force field simulations indicated that a larger number of Na^+ ions than K^+ ions binds to phospholipid heads in water.[11]

The gramicidin A channel is a well-known ion-selective transmembrane channel. It is a head-to-head dimer of a 15-mer polypeptide with alternating L and D amino acids in each monomer. Its primary sequence has been characterized as

$HCO - L - Val_1 - Gly_2 - L - Ala_3 - D - Leu_4 - L - Ala_5 - D - Val_6 - L - Val_7$

$- D - Val_8 - L - Trp_9 - D - Leu_{10} - L - Trp_{11} - D - Leu_{12} - L - Trp_{13}$

$- D - Leu_{14} - L - Trp_{15} - NHCH_2CH_2OH$.[12]

The membrane pore is a helical dimer. Hydrogen bonds formed between the backbone amide hydrogen and the carbonyl oxygen extend parallel to the pore axis and stabilize a transmembrane pore that is 2.5 nm long from Trp_9 to Trp_{11} oxygens in the dimer, with a pore diameter of 0.4 nm. This single-transmembrane structure is impermeable to anions and divalent cations. However, it exhibits selectivity among monovalent cations with permeability ratios $K^+(3.9) > Na^+$ (1.0). The pore is so narrow that only one cation or one water molecule can pass a given section. Thermodynamic and kinetic model calculations yielded the repulsive short-range forces of the gramicidin helices against the ion in their centre, as well as the most favourable path of the ion in the pore (Figure 3.5). This closely follows the helical structure, with slow movement in the wide and attracting carbonyl regions and fast movement in the space filled by side chains.[13] Potential of mean forces model calculations for gramicidin A showed that the binding sites for monovalent ions are near the channel entrances, and that Na^+ dehydration starts here and is then re-coordinated by two water molecules in the interior of the channel. The actual transport is a dynamic process dependent upon a concerted carbonyl bond pushing the Na^+ coordination from one energy well to the next (Figure 3.6). The flickering seen in the thermodynamic model is gone.

The water sphere also carries Na^+ ions into hydrophobic phospholipid bilayers, which contain polar ester domains in their centre. The distribution of Na^+ between the combined inner and outer water volumes of corresponding vesicles was monitored by sodium NMR using thulium(III)(1,4,7,10-tetraazacyclododecane-1,4,7,10-tetramethylenephosphonate $(Tm[DOPTP]^{5-})$ as a shift reagent. It shifted the ^{23}Na NMR signals of the ions in the outer and inner water volumes downfield by 30 ppm, whereas two weak signals for Na^+ remained unperturbed. Three observations then suggested that the signals were caused by "membrane Na^+", lying within the bilayer and not at the vesicle's surface: (1) the ion was not rapidly exchanging on the NMR time scale (in ms) with Na^+ in water; (2) it never came into close contact with the polar,

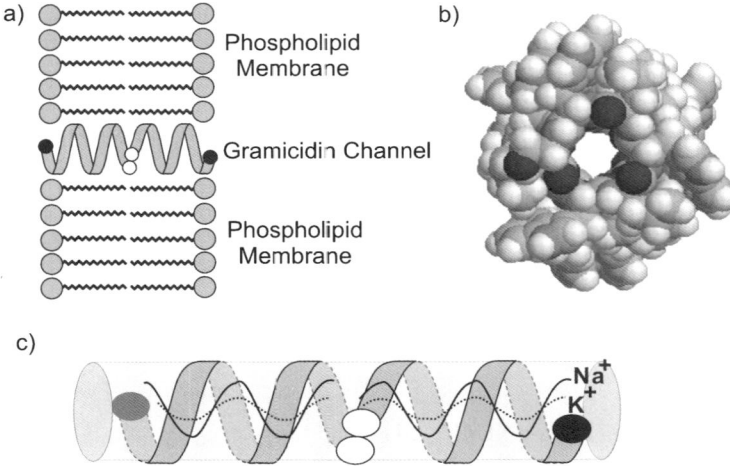

Figure 3.5 (a) Model of the gramicidin A dimer in bilayer membranes. (b) Molecular model of the gramicidin A helix. (c) Helical pathways of Na^+ and K^+ inside the gramicidin A pore as calculated by classical thermodynamic and kinetic models. © American Chemical Society.[12]

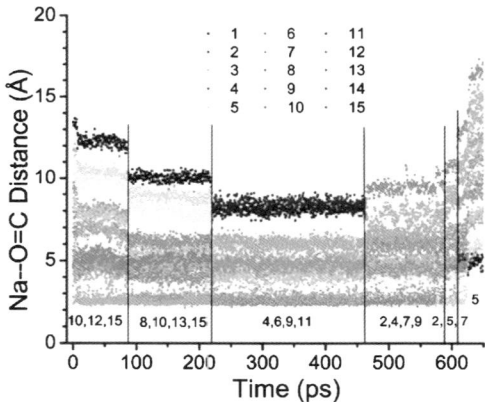

Figure 3.6 Calculated distances from a Na^+ ion to all carbonyl oxygens of the 15 residues (grey) in one monomer of gramicidin A recorded with 1 ps resolution. Because the ion moves in discontinuous steps, the distances exhibit distinct bands. The shortest distance band at 0.25 nm (black) remains constant along the pore. The Na^+ ion is always at equal distance to one coordinating carbonyl oxygen at all travel sites. © American Chemical Society.[13]

penta-anionic shift reagents in bulk water; and (3) the width and shift of its signal changed with time, which was best explained by slow rearrangements of the ester–lipid domains.[14]

Hydrated Na^+ may thus bind to CO groups of esters in a hydrophobic medium. It may be that hydrogen bonds together with distant charge

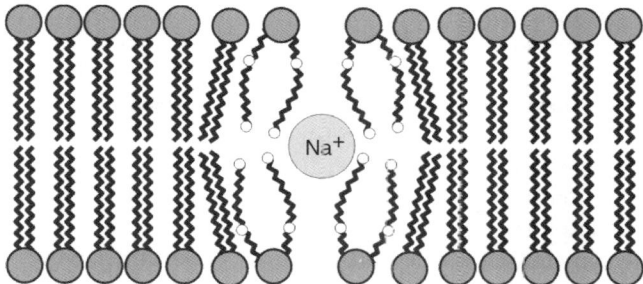

Figure 3.7 Schematic structure of a membrane-entrapped Na^+ ion close to an ester domain. The assumed water domain around Na^- is not shown.[14]

interactions provide an optimum for Na^+ binding and transport (Figure 3.7). This hint may help in solving the problem of desalination of seawater. Hydrophobic membranes can be mimicked by hydrophobized silicate pores; the long-chain ester groups may also be integrated there and hydrated Na^+ could in principle be bound by derivatized Sahara sand, evaporated in the heat of the Sahara sun and used again.

Biological K^+ channels are known in structural detail: one molecular structure has been resolved with a resolution of 0.35 nm. In this case the molecular origin of the experimental 10^4 margin of K^+ vs. Na^+ selectivity as well as the enormous throughput rate of 10^8 ions per second were qualitatively rationalized. The pore is made up exclusively of protein helices in form of a funnel. The pyramidal top is a negatively charged ring and entraps several proteins, which produce a hydrophobic 3 nm wide pore in the centre. A 1.2 nm wide entrance into this pore constitutes the potassium-selective filter. Although the amino acids here are hydrophobic, the $=C^+-H-O^-$ groups of the amide bonds form a row of ligands for the dehydrated K^+ ions and pull them into the pore. In crystals treated with 0.15 M KCl solution, the pore contained two K^+ ions separated by 0.75 nm, located near the opposite ends of the filter. The distance of 0.75 nm would be the average distance of K^+ ions in a 4 M KCl solution. The carbonyl groups in a hydrophobic environment have thus concentrated K^+ by a factor of 27 with respect to the bulk solution and the channel becomes positively charged. The outer water volume with the Cl^- counterions is, of course, very close. The second K^+ at the bottom is then attracted by the negative charge of the remote protein helices of the wall. Combined repulsion by an entering K^+ ion and the jump of the outside ion into the pore then allows ionic conduction to occur.

The conformation of the pore peptide is somewhat strange. All four neighbouring carbonyl oxygens point in the same direction, which does not occur in the linear *all-anti* conformers of proteins, where one carbonyl always points to the right and the next one to the left, because the $-(CO-NH-CHR)-$ unit contains an uneven number of chain links. It is also not present in the α-helix, or any other helix, where all four CO groups would point into different directions. A strange *gauche* conformer of the tetrapeptide sections at the four walls makes up the potassium pathway and it is favoured by attraction to the positive K^+ charges. This would never happen in water, but it happens in the narrow pore, where the hydrophilic K^+ repels the water, because there is not enough space and because the half-naked K^+ just binds more strongly to the CO chain than to the outside water. In a model yoctowell (see p. 12, 134 ff) for K^+ binding and release one would probably be best advised to start with β-peptides, where the *all-anti* conformer would have all carbonyl oxygens automatically on the same side and no *gauche* conformer would have to be created.

Of course the natural pore contains some water for the transport of K^+: one water molecule at the top even occurs in the crystal structure. But no negative charges neutralize the upper K^+ within the selective filter. Therefore two K^+ within the selectivity filter repel each other and thus overcome the carbonyl–K^+ binding. Ion transport is thus speeded up by K^+–K^+ interactions. Furthermore, the outer negative charge of the funnel's wall helices pulls the K^+ out of the carbonyl pore into the water volume inside the cell. In the wide part of the funnel this effect is negligibly small, but at the top of the water-filled tube, far away from the hydrophobic selectivity filter, it is very strong. If, however, the less polar carbonyl group of an ester was artificially introduced at the entrance of the selectivity filter instead of the natural amide, one hydrate water molecule disappeared, the ion conduction rate was reduced by a factor of 2 and became independent of an applied membrane voltage. The binding of two K^+, which occurs either at positions 1 and 3 or 2 and 4 in the wild-type amide pore, was also disturbed (Figure 3.8).

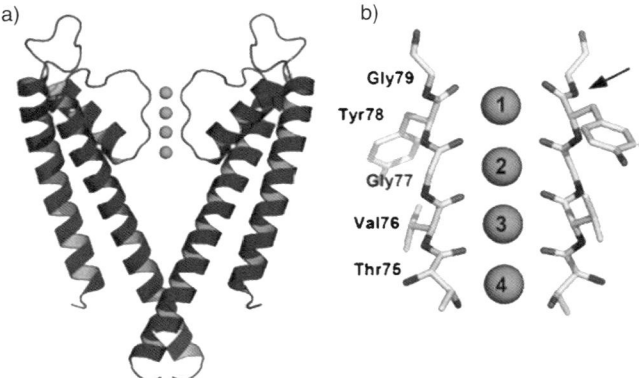

Figure 3.8 Schematic drawing of the selectivity filter of a K^+ channel with only two opposite subunits of the tetrameric protein shown. The four potassium binding sites are depicted with ions 1–4, but in X-ray structures only 2 and 4 or 1 and 3 were occupied. The close-up view in stick representation indicates the amino acids of the selectivity filter, the arrow indicates a peptide bond which was also artificially replaced by an ester bond (see text). The length of the pore is 1.4 nm. © American Chemical Society.[15]

With a 0.6 nm diameter and a height of 1.4 nm, the volume of the K^+ pore ($= \pi d^2 h$) is close to 1 yL ($= 1 \times 10^{-24}$ L $= 1$ nm^3), a typical yoctopore. Hydrated Na^+ ions are much too large to enter; naked Na^+ ions are too small to be bound effectively and are presumably never formed anyway, because the hydration water is bound too strongly.[15]

The Na^+ hydrate must also be carried selectively through the membrane pore and the driving force must be phosphate provided by ATP. This is an extremely difficult process, because phosphate dehydrates and then binds K^+ much more tightly than Na^+. In the Na^+ pore, which was identified by ion-selective electrodes, four helical transmembrane domains I–IV, each made up of segments 1–6, form a hydrophilic pore for Na^+ with the four segments 6, which also bind to channel-blocking molecules, *e.g.* cocaine. Segments 5 provide the membrane voltage *V* for Na^+ active transport, and segments 1–3 integrate the pore into the hydrophobic membrane. Voltage from segments 5 and phosphorylation of 6 must pull the ions against a concentration gradient (Figure 3.9). This process can only be selective for Na^+ if its hydration sphere is somehow recognized: phosphate charges and voltage must somehow act together in the 4 nm voyage.[16]

Upon membrane depolarization, a natural Na^- channel becomes "open", allowing a rapid influx of Na^+ into the cell. Within a few milliseconds, the channel undergoes a conformational state change involving isoleucine, phenylalanine, and methionine between domains III and IV that closes the channel to the "inactive" state. Repriming of the channel is another conformational state change that requires repolarization of the membrane. The channel then returns to the "resting" state where it is once again ready to open. Why K^+ does not enter, and how the Na^+ is actively transported back into the nerve cells, is not known. Several drugs such as local anesthetics, anticonvulsants and antidepressants bind to the open and/or inactivated state of the channel and block it. Cocaine is the classical example.

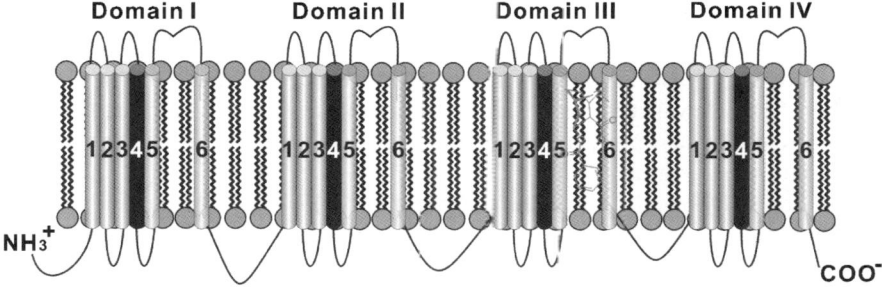

Figure 3.9 Scheme of the voltage-gated sodium channel in a rolled-up conformation. Tubes 5 and 6 make up the hydrated pore, tube 4 provides the voltages that drive the Na^+ against concentration grades and tubes 1–3 integrate the channel within the lipid bilayer membrane. The model has to be seen in relation to the K^+ pore shown above. Size selectivity plays no obvious role here and is replaced by voltage and a magic "double pore". © American Chemical Society.[16]

The major problem in the sodium–potassium world, however, remains unsolved: how do the Na^+/K^+-ATPases transport both ions, in particular Na^+, actively through membranes? What is the magic binding site for (presumably hydrated) Na^+ that permits the fast ion currents required to keep our hearts beating and our brain working?

3.4 Magnesium, Calcium and Barium

3.4.1 Magnesium

Mg^{2+} is the counterion of phosphate diesters in DNA and anhydrides in ATP. In the latter it stops the hydrolysis of ATP to form ADP, by forming a stable complex. Further hydrolysis to AMP requires strong acids. Within the chlorin ligand of chlorophyll, Mg^{2+} stabilizes the highly anionic dye and lowers its oxidation potential in the primary reaction cycle of photosynthesis.

The thermal conductivity of MgO is high. Magnesium dihydride (MgH_2) contains 7.6 wt% hydrogen and is attractive as a hydrogen storage material. Slow desorption kinetics and a high operating temperature of 603 K prevent applications, however. Magnesium powders were therefore transformed into nanowires and nanoparticles. The magnesium powder was evaporated at 1203 K, deposited from an argon gas stream as 30–50 nm magnesium wires and loaded with 2.93 wt% hydrogen at 373 K or 7.60 wt% at 573 K within 30 min. Hydrogen absorption required a hydrogen pressure of 0.4–2.0 MPa and desorption occurred at 0.02–0.6 MPa within 10 min at 573 K. After 10 cycles the nanowires collapsed to nanoparticles, but absorption and desorption cycles were not disturbed.[17]

3.4.2 Calcium

The bivalent Mg^{2+} and Ca^{2+} ions occur in concentrations of $1-4\,mol\,m^{-3}$ in normal tap water. They precipitate from water with carbonate, phosphate and fatty acids and must be solubilized when "hard" tap water is used in washing machines or dishwashers. This is done by soluble zeolite A and phosphonates, which are major components of washing powders and produce corresponding phosphate waste water.

The poly(ethyleneglycol) (sulfonatostyrene) diblock copolymer is a versatile blocking agent for the fast removal of insoluble metal salts from tap water and produces a wide range of micro- and nanoparticle structures depending on the conditions of salt formation, *e.g.* of calcium carbonate. A small amount of the copolymer added during aggregation–crystallization leads to clean single-crystalline products. With 10 times as much copolymer the crystalline nanoparticles aggregate to mesocrystals with rough, porous surfaces, entrapping nanoparticles. At even higher copolymer concentrations, coated polycrystallites are favoured (Figure 3.10). Such additive-controlled crystallization is quite common, and the result is always determined by relative solubility and time.

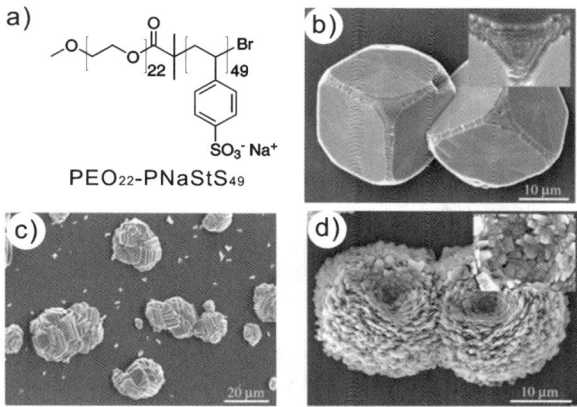

Figure 3.10 Crystallization of calcium carbonate in the presence of copolymer: (a) precipitated at Ca/polymer ratios of (b) 250 (single crystals), (c) 125 (mesocrystals) and (d) 12.5 (polycrystallites). © American Chemical Society.[18]

The distinction between a single crystal, a mesocrystal, and a polycrystal should always be based solely on crystallographic perfection, not on the mechanism by which they are presumably formed.[18]

A mixture of calcium hydroxide (lime, $Ca(OH)_2$) and sand forms mortar, which sticks to the surface of stones and bricks and solidifies there as insoluble carbonate after a reaction with the carbon dioxide of air. For luminescence, see p. 37.

Friction between inorganic nanogranules is a major origin of the strength of bones and was studied in "deorganified" adult bovine cortical bone mineral particles. In such a collagen-free bone material the strength depends on the density and fitting deformations of interparticle contacts. Local pressure in micron dimensions on collagen-free bone material did not cause any cracks, but a flattening of the 1 μm particles filled indented areas. Fast "stick–slip" equilibria and smooth sliding occurred between particle calcium phosphate layers.[19–21]

The organic matrix of bones is 90% made up of triple-helical molecules of collagen, with a large uniaxial dipole. The collagen is piezoelectric and pressure-induced electricity attracts platelets of hydroxyapatite ($Ca_5[CO_3/2, OH(PO_4)_3]$).[22] The application of mechanical stress at first results in the displacement of charges in bone tissue: pressure creates a negative polarity on the compressed side and a positive polarity on the other side of the collagen–calcium phosphate tubules. Mineralization of the space between the tubules then occurs predominantly over the compressed side of bone collagen and uses collagen-adsorbed apatite nanoparticles, as well as the water molecules and the glass state of the collagen triple helix, in order to stabilize the bones to external pressure, such as is applied in walking (Figure 3.11).[21,22]

Toughness and stiffness are thus supplied to bones by a matrix of collagen which takes up plate-shaped hydroxyapatite particles ~25–35 nm long and

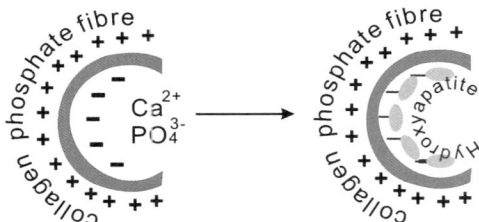

Figure 3.11 Schematic model of the biomineralization process of hydroxyapatite within collagen-Ca-phosphate fibres.[22]

wide and 2.5–3.5 nm thick. But is there really a strong contact between the protein and the calcium phosphate, since bones also contain proteoglycans, rich in acidic glycosaminoglycans? What is the role of these surface carbohydrates, if not adhesion to polar surfaces?

This question was answered by a comparison of bone and cartilage using a special solid-state NMR technique called rotational echo double resonance (REDOR). The embryonal skeleton consists mainly of cartilage, much of which later ossifies to become bone. Cartilage and bone minerals both contain ^{31}P nuclei, and REDOR re-establishes the $^{31}P-^{13}C$ through-space coupling, which is removed in high-speed magic angle spinning NMR. The spectra of bones and cartilage showed that the relative prominence of the sugar signals at 76, 175, 182 and 103 ppm (anomeric carbon) increased from bone, through mineralized cartilage, to hyaline cartilage (Figure 3.12). The farther away from bone, the closer to pure cartilage, the more sugar was seen. The biomolecules most intimately associated with the growing bone are sugars, and not proteins. The same was found to be true for teeth. The glycosaminoglycans are obviously better suited to propagate the mineralization process than the hydrophobic collagen, but the pressure experiments seem to characterize mainly short-term effects. Acidic glycosaminoglycans are also the most important connecting materials in teeth.[23,24]

Diffusion of Ca^{2+} and phosphate ions in a viscous, acidified gelatine gel leads to fractal structures, in particular to dumbbell-shaped particles made of branched rods. In order to avoid a fast precipitation of calcium phosphates, the buffered calcium and phosphate/fluoride solutes were diffused into the acidified gel from opposite sites. The viscous gel with carboxylate groups induces fibre growth and also aligns the fibres. Since the small Ca^{2+} ions migrate faster than the phosphate ions, the elongated hexagonal prismatic seeds are first formed near the phosphate phase. The fractal structure is presumably caused by the mobility of gelatine fibres, which first offer nucleation sites for the apatite mineralization and then facilitate and steer the sliding of the growing mineral particles with negative surface charges. As in dendritic polymers (p. 120ff), there is much denser packing in the outer regions than in the centre of the spheres. Needle-shaped prisms appeared at both ends of the seeds, leading slowly to dumbbell-shaped aggregates (Figure 3.13). The particles of successive generations are arranged in a way that (1) the angle between their long axes and

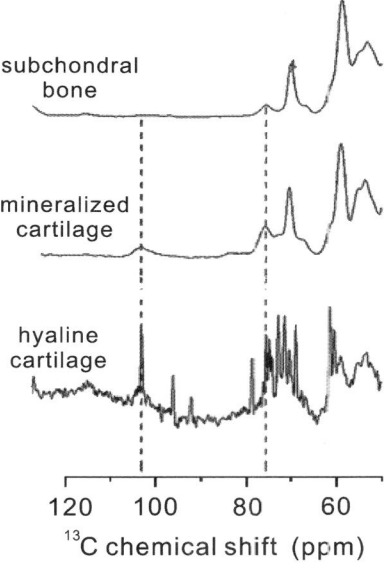

Figure 3.12 A comparison of the spectra of embryonal (subchondral) bone, young mineralized cartilage and hyaline cartilage. The vertical dotted lines connect signals ascribed to glycosaminoglycans; the most solidified bones contain the most glucose derivatives. The other signals are from collagen. © American Chemical Society.[23,24]

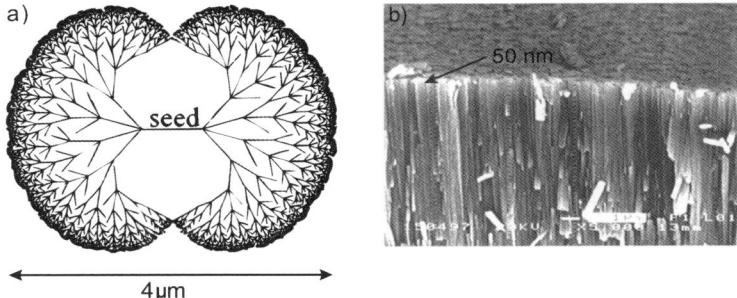

Figure 3.13 (a) Two-dimensional model (width 4 μm) for the crystal growth of the fluoroapatite–gelatine composites. (b) Electron micrograph of the outer shell apatite rods, 50 nm wide, which look similar to the enamel rods in teeth. © American Chemical Society.[25]

the prism axis of the mother individuum does not exceed 45° and (2) subsequent particle generations scale down by a factor of ∼0.7. Following rules (1) and (2), the spheres come together after the 10th generation, the diameter of the surface crystallites is limited to ∼50 nm and the diameter of the dumbell reaches 4 μm.[25]

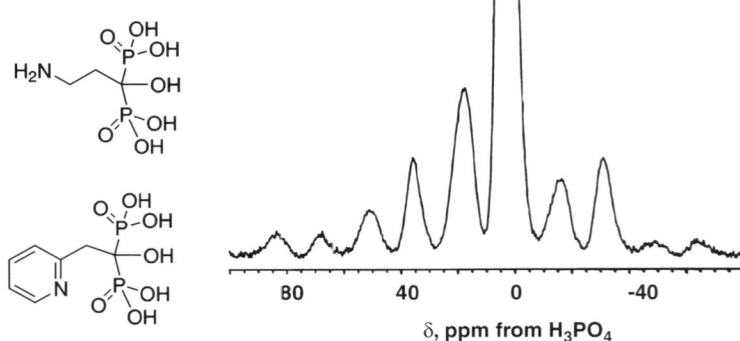

Figure 3.14 Structure of two aminophosphonates used in medicine and a ^{31}P MAS NMR spectrum of a bisphosphonate monolayer bound to bone. The overall span of the chemical shielding tensors is 120 ppm, similar to those observed in crystalline bisphosphonates. © American Chemical Society.[26]

The visible surface of human teeth also consists of ordered crystal bundles of hydroxylapatite (95 wt%), a pentacalcium triphosphate with one hydroxyl group, $Ca_5(OH)(PO_4)_3$. The hydroxyl ion is easily exchanged by fluoride, which hardens the surface. The viscous matrix described above was also sufficient to induce enamel formation from calcium chloride, phosphoric acid and fluoride ions (Figure 3.13b).

Aminobisphosphonate drugs are used to treat osteoporosis and to kill tumor cells. In model experiments it was shown that at pH 7 such bisphosphonates chemisorb to human bone surfaces and form molecular monolayers or bind proteins there (Figure 3.14).[26,27]

Teeth are also natural photonic crystals, like opals and butterfly wings. There is only a small difference in refractive index between the content of the apatite tubules and the surrounding mineral, and such index equalization strongly increases the translucency of materials. Light is transmitted exclusively *via* the hard tissue and not, as one might expect, through the holes. The latter have a core with varying low refractive indices and act as antiwaveguides shortening the pathway of transmitted light. Light is "collected" on the surface of teeth, and we like to show that when we smile.

3.4.3 Barium

The nuclei of barium are heavy, and aqueous solutions or suspensions absorb X-rays. Water-insoluble (3 mg/L) barium sulfate ($BaSO_4$) provides a good contrast for X-rays in internal organs and is harmless, although free barium ions are poisonous. Nanocrystals of barium titanate ($BaTiO_3$) integrated into molten poly(methacrylate) and processed to multilayers give high refractive indices for terahertz (10^{12} Hz = 300 µm) waves, which may be applied in medical imaging and wireless communication. Scattering of the long waves by 100 nm nanoparticles within the polymer layers was negligible.[28]

3.5 Boron, Aluminium, Gallium, Indium and Thallium

3.5.1 Boron

Boric acid (B(OH)$_3$) is a weak acid and polymerizes like silicic acid (Si(OH)$_4$) to polymers using oxygen bridges between the boron atoms. Boron exists in trigonal and tetrahedral coordination. Borates exhibit anionic arrangements that combine B–O triangles and tetrahedra linked through bridging oxygen atoms. For boron nitride see p. 202 ff.

Electron-deficient threefold coordinated boron substituents in organic, conjugated networks are strong p-electron acceptors, which are able to participate in charge delocalizations. Protection against nucleophilic attack is usually achieved with *ortho*-alkylated phenyl substituents, which provide steric protection of the boron atom and yield dyes with high fluorescence quantum yields. Boron within conjugated polymer skeletons yields luminescent materials of any colour, which also transport electrons or emit light. Boron *o*-bis(mesitylene) substituents of poly(thiophene) and or poly(acetylene) units are efficient spacers, which prevent any intermolecular coupling in condensed phases and thereby transport the high emission intensities of solutions to solid, light-emitting molecular layers.

The electroneutral phenalenyl biradical with a boron anion centre crystallized with a separation of 0.38 nm between the closest spin-bearing C–C pairs along the *y*-direction (Figure 3.15). Its reduction and oxidation potentials were –1.20 and –0.76 V and the conductivity of a crystalline powder was 1.5×10^3 S cm^{-1}. Flexible cycloheptyl groups kept the monomeric biradicals apart.[29]

The ionization potentials of boron are high (> 8.2 eV) and electron loss plays no part in boron chemistry. Boron is always trivalent and forms only covalent bonds. Chemically, boron behaves like a non-metal and resembles silicon more

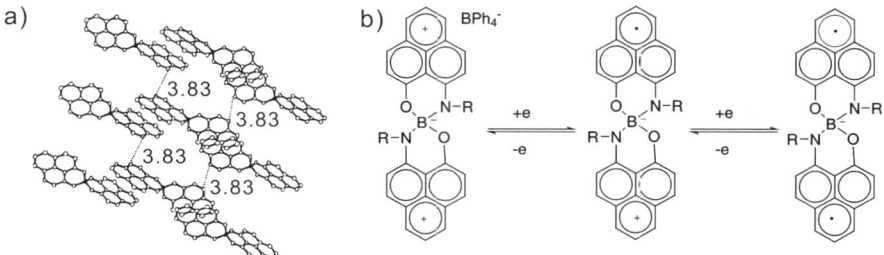

Figure 3.15 (a) Closest intermolecular C–C contacts (3.83 Å) among the spin-bearing carbon atoms of the phenalenyl boron complex (b). Cycloheptyl groups are omitted. The redox potentials of the biradical in the centre are –1.20 and –0.76 V$_{SCE}$.[29]

closely than aluminium: B(OH)$_3$ is an acid like Si(OH)$_4$, B$_2$O$_3$ and SiO$_2$ behave similarly, and the hydrides B$_2$H$_6$ and SiH$_4$ are both volatile.

Natural boron consists of 18.8% ^{10}B and 81.2% ^{11}B isotopes. The ^{10}B isotope effectively captures neutron radiation to reach the ^{11}B-state. Neutrons pass readily through biological tissue and do not produce ion pairs in air. They are extremely destructive to living organisms, however, because they split C–H bonds, thus releasing ion-producing protons. "Neutron cannons" are used in cancer therapy when the malignant cells survive X-ray radiotherapy. Boron neutron capture therapy of tumours relies on (1) methods that provide a significant ^{10}B contrast between malignant and normal tissue and (2) a high absolute ^{10}B concentration only in the malignant tissue. If both are achieved, the cells loaded with boron-containing nanoparticles may not only be killed, but the destroyed cells may also hinder the growth of neighbouring cells after the neutron irradiation.

For efficient neutron treatment the powerful X-radiation must be focused by massive boron concentrations at the cancerous site. Nanocrystals of ^{10}B give the highest possible concentration, because not only the surface molecules, but the whole body of the nanocrystal, is active. Boron carbide nanocrystals were therefore selected as a material. This is a hard, high-melting, electropositive p-type semiconductor with a wide gap, based on boron-rich icosahedra connected by carbon and boron atoms. It is used as an abrasive in thermoelectric devices and in control rods for nuclear power generation. Its formula is often close to B$_4$C, and it thus contains less carbon than Fe$_3$C (p. 248). The carbon content in non-stoichiometric nanocrystals ranges from $\sim 20\%$ (B$_4$C) to only 9% (B$_{10.5}$C). The (B$_{12}$C$_3$)$_8$ unit (only the centre of which is shown in Figure 3.16), whose stoichiometry is identical to that of Fe$_3$C, has a volume of ~ 1 yL or 1 nm^3. Each boron atom that is not connected to a carbon atom is electron deficient and can add nucleophiles, *e.g.* amino groups.[30]

Commercially available solid boron carbide with the approximate stoichiometry of B$_4$C was processed by ball milling in order to reduce and adjust the particle size. The B$_4$C nanocrystals were then treated with 1,10-diaminodecane as a coating material. The electron-deficient boron acquired an electron pair from the amine, forming a monolayer. The remaining amine group on the surface of the nanocrystal was labelled with a rhodamine fluorescence dye (lissamine) for detection and a fluorescein dye for targeting. Only the latter fluorescing dye was close to a peptide (TAT in Figure 3.16), which bound selectively to the malignant cells. The dye that was far away from the binding site of the nanocrystal acted as a reference: the relative change of fluorescence of the dye close to the cell was taken as a sign of binding to the cell. The viability and proliferation of the cells before and after neutron irradiation was then tested by measurements of the uptake of radioactive thymidine, as part of the DNA newly synthesized by living malignant cells.

The results showed that neutron irradiation not only killed the cell that was directly bound to the aminated boron nanocrystal, but also several unloaded neighbours. Only the bound nanocrystals, as quantified by the fluorescing markers, had an effect on the cancer cells.[31,32]

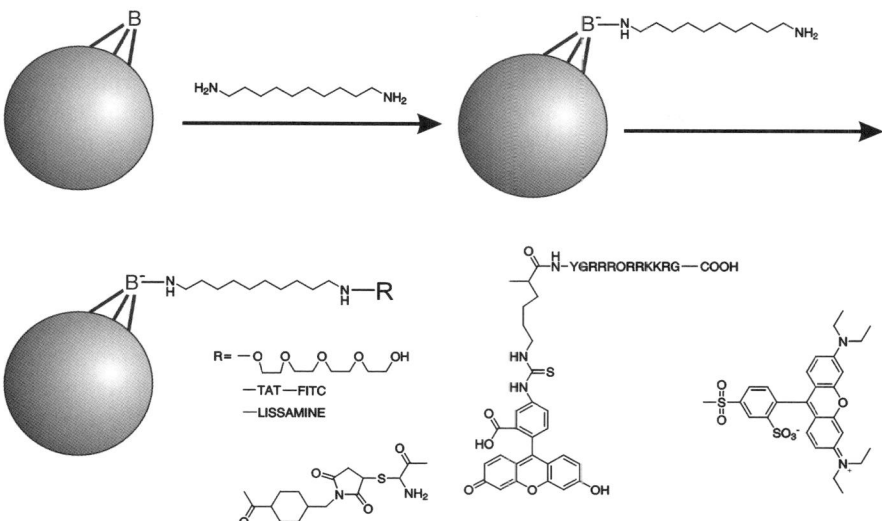

Figure 3.16 Sketch of a nanoparticle made of several boron carbide (BC$_4$) units and a diamine addition combined with further derivatizations for the attachment to malignant cells and neutron irradiation.[30]

3.5.2 Aluminium

Aluminium is a silver-white metal with face centred cubic (fcc) crystallization. It has a density of 2.7 g/cm^3, a melting point of 660.4 °C and a boiling point of 2467 °C, and self-ignites in air. Nevertheless it is less sensitive to oxygen than iron is, because it forms a surface layer of oxide (Al$_2$O$_3$) which is only a few molecular layers thick (5–10 nm after weeks) and remains unchanged even upon heating to 400 °C. At 550 °C the layer doubles in thickness. Aluminium is stable at pH 4.5–8.5, but zinc, tin, lead, silver and copper salts, sodium hydroxide, hydrochloric acid and phosphoric acid destroy the oxide layer and dissolve the metal.

Aluminium makes up 8.13% of the 16 km deep crust of the Earth, making it the most abundant element there after oxygen and silicon, and before iron. Common aluminium minerals are feldspars, *e.g.* orthoclase, K[AlSi$_3$O$_8$]; albite, Na[AlSi$_3$O$_8$]; anorthite, Ca[Al$_2$Si$_2$O$_8$] and mica, *e.g.* muskovite, KAl$_2$[AlSi$_3$O$_{10}$](OH,F)$_2$; margarite, CaAl2[Al$_2$Si$_2$O$_{10}$](OH)$_2$. Their weathering products are clays. Blue or green aluminium phosphate with some copper sulfate is turquoise. Corundum (Al$_2$O$_3$) is coloured red by Cr$_2$O$_3$ (ruby) and blue by V$_2$O$_5$ (sapphire); as a hard mass product it is called emery. Bauxite, the main ore of aluminium, is a mixture of aluminium hydroxides.

Aluminium salts in water prolong the lifetime of flowers, and the human body contains 50–150 mg aluminium. In food aluminium is harmless and mostly goes directly into the faeces: the liver contains 0.16 mg/100 g, the heart 0.056 mg/100 g. Aluminium may, however, play a role in Alzheimer's disease (dementia) by precipiting cells, neurons, proteins (β-protein, m.w. 4000) and phosphates (senile plaques).

Nanosized powders of alumina (Al_2O_3) have lower light absorption than crystals but increased dispersion. They are used as chemical mechanical planarization slurries for the polishing of computer chips; as coatings for light bulbs and fluorescent lamps, where they introduce uniform light emission; and as a component of vinyl floors, where they improve hardnes, flame stability and friction.

Industrially, an oxide layer 10–25 µm thick is usually established on aluminium metal by an electrochemical oxidation in acidic media called anodization. The layers are microporous and must be sealed by boiling water or hot vapour. Thin (~ 0.1 mm) aluminum sheets are commercially available and may be converted in the lab to porous anodic alumina membranes (PAAM) by electrolysis in oxalic, phosphoric or sulfuric acid. The aluminium foil is used as a cathode and the electrochemical oxidation produces highly ordered arrays of orthogonal pores. Pore diameter and interpore spacing increase almost linearly with the voltage of the electrochemical cell and decrease with increasing acid strength of the electrolyte.

An aluminum sheet was connected to two electrodes, anodized at 60 V for 1 h, and ultrasonicated in a phosphochromic acid solution (6% H_3PO_4 + 1.8% H_2CrO_4) at 60 °C for 30 min to remove the preformed alumina layer only, leaving the aluminum base intact. The aluminium sheet was oxidized for 4 h under the same conditions. In order to obtain a free-standing through-hole membrane, perchloric acid (72% w/w) and ethanol (v/v 1:1) were added at a voltage of 70 V for 3 s. The porous alumina film separated from the barrier layer immediately and was rinsed with water. 60 nm pores with a length of 100 µm were obtained.

The formation of the pores upon anodization of aluminium starts with or without mechanical nanoindentations. At first the applied voltage is concentrated at the tips of the growing tubes, leading to orthogonal orientation of the formed holes. Charge repulsion then leads to a distance between growing holes in hexagonal patterns. Conversion of aluminium to alumina by the acid (*e.g.* $2Al + 3PO_4^{3-} \rightarrow Al_2O_3 + 3PO_3^{3-}$) needs space, which means mechanical stress and again favours distance and regular patterns. In contrast to macroporous silicon (p. 189), porous alumina consists of two different materials: an aluminium metal substrate and a porous alumina layer. When aluminum is anodized to alumina, the self-ordering in hexagonal patterns requires a porosity of 10%, corresponding to a volume expansion of the alumina to aluminium sheets of ~ 1.2. Calculations showed that an expansion of < 1.2 would not cause pore formation and a factor > 1.3 would induce the formation of small porous domains to begin with, which would then combine. It is exactly the 1.2-fold volume expansion that provokes pore formation over the whole area.

The inside of the pores is always coated with an alumina layer. Mechanical stress at the metal–oxide interface causes repulsive electric forces between neighbouring pores during the anodization process and enforce a hexagonal pore pattern. The insulating character of the oxide surface and its acid stability under the given conditions, *e.g.* 0.3 M oxalic acid, 40 V and 1°C, prevents an infinite oxidation of the aluminium. The final treatment with strong perchloric acid, or mercuric chloride and phosphoric acid, is critical in destroying the bottom of the well.[33]

It is easy to etch wide capillaries (> 50 nm) in aluminium foils, which provide very stable templates at low cost. Nowadays a wide choice of porous foils is commercially available and only the very narrow pores (< 10 nm) are "home made". The pores usually consist of highly doped, polycrystalline alumina, and are rigid, well-defined, self-supporting, shelf-stable and optically transparent if annealed. The porous portion of the film consists of close-packed hexagonal cells, each with a central straight hole perpendicular to the surface of the aluminium substrate. The oxide along the hexagonal ridge separating the pores is far more compact and crystalline than the more amorphous oxide lining the walls of the pores. The pore diameters can be set in the range 5–200 nm. Uniformity and regularity also depend on the variation of the physical, chemical and electrochemical parameters. Electroosmotic flow experiments are easy to perform and are most helpful for pores with charged surfaces such as alumina (Figure 3.17). The applied electric field does not speed up the filling of the pore very much, but it tells the important details of the pore surface. Phosphoric acid, for example, loads the alumina with negative charges down to pH 3.5, but oxalate has no such effect.[34]

A 42 nm diameter iron nanowire grown in sulfuric-anodized pores was single-crystalline and had a coercivity of 1654 Oe. Increasing pore diameters resulted in improved crystallinity for the nanowires, but their coercivity was reduced because of the decreased aspect ratio. A wide variety of nanorods and nano-tubes has been formed within these pores, but their width was usually > 50 nm.[35] The increasing attraction of porous alumina as a template is mainly due to its low-cost processing – it has become a technological material on the nm–µm borderline. As a scientific nanometre tool it needs further development.

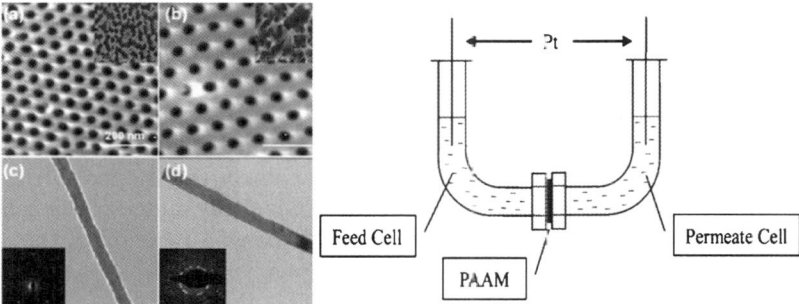

Figure 3.17 *Left:* (a, b) Scanning electron micrographs of a 60 nm porous anodic alumina membrane (thickness 80 µm) (b) the permeation cell for transport experiments and (c, d) a crystalline 42 nm iron nanowire. *Right:* Experimental setup for aluminium anodization. © American Chemical Society.[34,35]

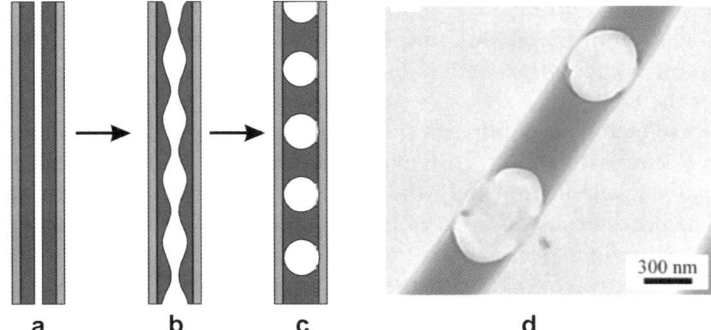

Figure 3.18 (a) Sketch of a polymer nanotube coating the alumina pore wall, which is (b) briefly heated to 250 °C surface to induce undulation of the tube, and (c) cooled down to form "holey" tubules with capillary bridges. The grey alumina wall is the removed with NaOH (not shown). (d) Electron micrograph of two holes in the polymer rod with walls of nanometre thickness. © American Chemical Society.[36,37]

A special technique for producing windows in nanotubes and holes in nanorods is based on thermal Rayleigh instabilities or undulations of the liquid films that coat capillary walls. These movements increase the surface area, and the surface tension of the liquid then favours the formation of droplets. The undulation occurs for each "string" with a well-defined wavelength and the drops are formed with a regular spacing between them. As the amplitude of undulations grows with time, crests merge to form bridges across the tube and produce a rod or tubule with periodic encapsulated holes, which may then be colourized or metallized.[36,37] Films of poly(methyl methacrylate) on the walls of alumina cylinders were heated to 150–200 °C, then shortly to 250 °C and rapidly cooled to room temperature. The alumina template was dissolved in 5 wt% NaOH(aq) and the final poly(methyl methacrylate) nanostructures were released and examined by scanning electron microscopy (Figure 3.18).

3.5.3 Gallium

The electronic energy-band structure of III–V compounds, in particular gallium arsenide (GaAs) and gallium selenide (GaSe), is clearly different from their major competitors in the semiconductor field, namely the (doped) group IV elements silicon and germanium. The major technical advantage of the gallium chalcogenides is the variability of energy band gaps as a function of the composition. This is most pronounced in the complex In–Ga–As–P system. The In–Ga–As alloy gap may be as low as 0.6 eV, compared to 1.8 eV for the Ga–As–P analogue of similar crystal structure.

Gallium arsenide and selenide consist of hexagonal layered sheets. The gallium atom is usually tetragonally coordinated to three arsenic or selenium atoms and to one gallium atom, and adjacent layers are held together by weak van der Waals interactions.

The loose layer structure of gallium selenide lets all elements sublime congruently. The vapour consists of predominantly of Ga_2Se and Se_2 as well as a small amount of Ga_2. Volatilization starts at ~800 °C and complete evaporation is reached at ~950 °C. Oxidized silicon ⟨100⟩ covered with 20 nm diameter gold particles was placed in a 950 °C gallium selenide stream carried by nitrogen gas and induced the formation of square-section gallium selenide nanowires in cooling zones ranging from 600 to 400 °C. The length of the gallium selenide nanowires reached tens of microns and the width was typically 70–80 nm (Figure 3.19).

The difference between strong intralayer covalent bonding and weak interlayer interaction leads to highly anisotropic electrical, optical, and mechanical properties which are particularly useful in solar cells and solid-state batteries, where ion transport may be important. The bonding anisotropy also implies that there are no surface dangling bonds, which may rupture. Photoexcited antibonding states, for example, will not destroy the layer structure and a 1D nanowire structure providing a large surface area for charge carrier separation will also be quite photostable. These thin nanowires may at first confine the quanta in one dimension and then transport them within an integrated nanoscale system.

Particularly successful vapour–liquid–solid growth procedures start with a solid metal particle as a seed, use solvents as medium for the starting materials and let a solid wire grow within this solvent. Gallium arsenide, for example, was made in "supercritical" hexane solution which was pressurized at 370 bar, above its critical point, and heated to 500 °C (Figure 3.20). (^{t}Bu)$_3$Ga and As(SiMe$_3$)$_3$ were selected as precursors, the coupling was catalysed by 7 nm dodecanethiol capped gold nanoparticles, and the isolated yield of pure 8–60 nm thick wires of micron length was 60%. The covalent organometallic

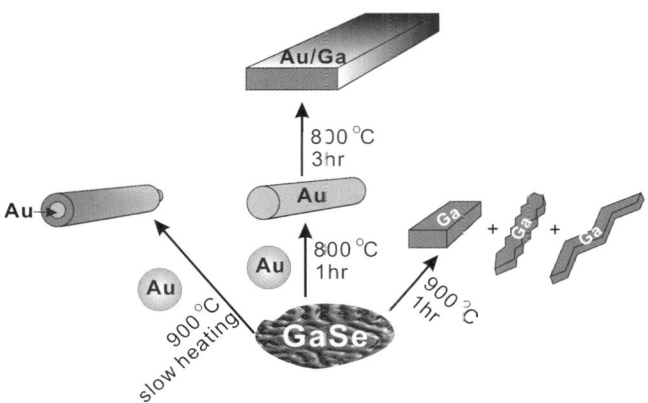

Figure 3.19 Different nanowires grow from gold–gallium nanoparticles at 800–900 °C: wires, bands of different forms and tubes. The form and composition of the gold nuclei change and so do the elongating gallium selenide products. Fresh gold nanoparticles lead to nanotubes again. © American Chemical Society.[38]

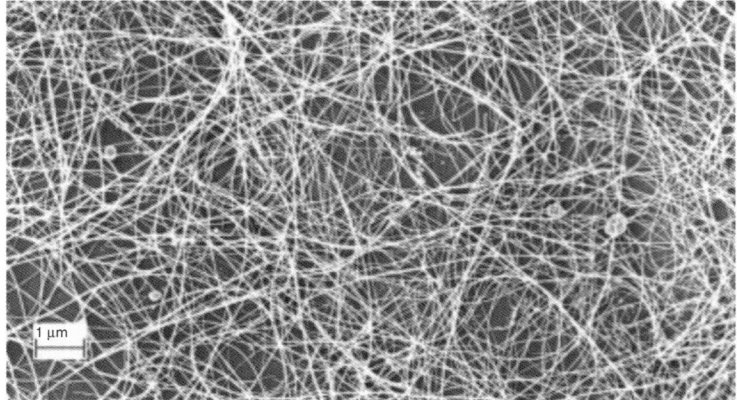

Figure 3.20 Scanning electron micrograph of gallium arsenide nanowires made by pyrolysis of (^{t}Bu)$_3$Ga and As(SiMe$_3$)$_3$ in hexane at 500 °C and 370 atm ("supercritical conditions").[40]

educts, gallium and arsenic silyl compounds, were practical, because they degraded slowly enough to allow the formation of long wires in the apolar medium. The variable width of the nanowires was caused by the unstable gold seed. When the (^{t}Bu)$_3$Ga was replaced by GaCl$_3$, the reaction speeded up and the wire yield dropped dramatically in favour of ill-defined nanoparticles. The resistivity of the crystalline wire was of the order of $2 \times 10^4 \, \Omega\,\text{cm}$, as compared to 10^7–$10^8 \, \Omega\,\text{cm}$ for bulk gallium arsenide.[39]

3.5.4 Indium

Indium metal is a superconducting metal, active in surface plasmon resonance spectroscopy and a useful component of low-melting solders and solid-state lubricants. The most common use of indium is in the indium tin oxide (ITO) electrode, which has a transparent, 200 nm thick surface layer of ~90% In$_2$O$_3$ and 10% SnO$_2$. ITO exhibits higher conductivities (2000–4000 S cm^{-1} for polycrystalline thin films) and carrier concentrations than pure In$_2$O$_3$ due to cationic (Sn^{4+})·In^{3+} species, which are neutralized by oxygen anions but structurally still ill-defined. In$_2$O$_3$ itself is usually non-stoichiometric (In$_2$O$_{2.99}$) because of oxygen vacancies, which contribute up to two electrons per vacancy to the conduction band.

Reduction of InCl$_3$ by borohydride is the simplest reaction leading to shape-controlled In(0) nanoparticles. They are magnetizeable at 3.5 K and showed a 380 nm UV absorption band at room temperature. InCl$_3$ and poly(vinylpyrrolidone) in isopropanol were reduced under argon with NaBH$_4$–tetraethylene glycol at room temperature, stirred and centrifuged. The resulting nanoparticles were washed several times with ethanol. Depending on the addition procedure of the borohydride, 50 nm wide cubes, octhedra or micron-long rods were obtained.[40]

Transparent ITO electrodes are used in liquid crystal screens, light-emitting diodes, touch screens and solar cells. In photovoltaic elements, which produce electric current from light, the current density may reach A/cm^2. The electrical activity of the ITO surface is not uniform, however, due to variable near-surface Sn/In ratios and hydroxylation of the oxide surface which lead to oxygen vacancies and tin doping. Adsorption of water vapour, carbon monoxide and/or carbon dioxide also disturb conductance and light conversion. The electroactivity can be increased, however, by brief strong-acid etching, which carries some more of the less soluble SnO_2 to surface sites. Conducting polymer thin films made, for example, of ferrocene dicarboxylic acid (p. 266 ff) also activate the ITO surface.[41,42] 6.6 nm In_2O_3 nanoparticles also show photoluminescence around 3.67 eV.[43]

The solubility limit for SnO_2 in In_2O_3 is 6 cation%, and the solubility of ZnO is even lower. When both metal oxides are co-substitued into In_2O_3, however, their solubilities increase dramatically. The indium content can be as low as 60 cation% and the conductivity and transparency is still comparable to ITO, which contains 30% more of the expensive indium. The band gap can be tuned by the ZnO and SnO_2 content. Crystals do not dissolve homogenously like solvents – each new solid "solute" changes the "solvation" in the environment of each component drastically.[44]

Organometallic chemical vapour deposition with hexane–acetone solutions of indium, tin and zinc diones in a stream of oxygen and pyrolysis at 500 °C gave 200 nm Zn–In–Sn–O films made of >100 nm nanoparticles. The conductivity of the ZnITO (ZITO) films with a nominal composition of $ZnIn_2Sn_{1.5}O_z$ was $2150\,S\,cm^{-1}$ and the optical transmittance of a 200 nm film was >80% over a wide spectral range.[44]

The protein-repulsive effect of poly(ethylene glycol) films (p. 85 ff) on ITO reduces the background level of non-specific binding of proteins and cells. The entropic cost caused by the binding of a biopolymer to flexible poly(glycol) chains, the electrostatic repulsion induced by the adsorption of hydroxide ions on to the polymer surface and the presence of a localized, high-viscosity water layer combine to attach an undisturbed protein to the ITO surface. Four different coupling chemistries were tried: (1) direct reaction of a poly(ethyleneglycol)-bis(amine) with the ITO surface; (2) Schiff base formation by silanization using 3-glycidoxypropyl trimethoxysilane (GPS), converting the epoxide to an aldehyde by diol formation and oxidative splitting with periodate; (3) adding a polyglycol methacrylate to the hydroxyl groups of ITO to form an ether bridge *via* oxo-Michael addition; and (4) direct coupling of trimethoxysilane–polyglycol to the hydroxyl groups of ITO (Figure 3.21). Procedure (4) produced the most homogeneous and effective films, followed by the Michael addition (3). The four-step aldehyde synthesis and imine formation with diols (1 and 2) did not work on the acidic surface. Surface coating should be a one-step reaction, relying only on fast, irreversible condensations such as –SiCl + ROH.[45]

Poly(aniline) is electroactive only in acidic conditions (electrolyte pH < 4), but electropolymerization of aniline in the presence of poly(acrylic acid) yields

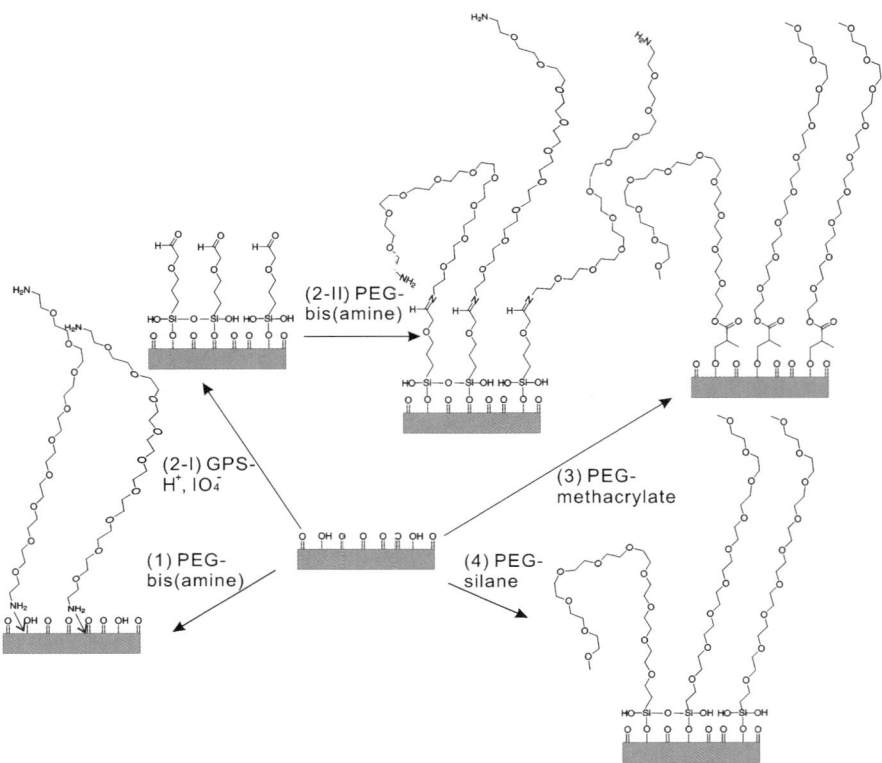

Figure 3.21 Four procedures to fixate poly(ethyleneglycol) on an indium tin oxide (ITO) electrode via amine, silylchloride, Schiff base or methacrylate end groups (see text).[45]

a composite film that is also active in aqueous solutions at pH 7.0. Such a film was deposited on ITO in a layer-by-layer deposition procedure, where the acidic In–OH surface renders the adsorbed poly(aniline) film electroactive even at neutral pH. Near-Nernstian potentiometric responses were observed at pH 3–9, so that the films could be used for quantitative optical pH sensing using strong absorption changes at 800 nm.[46,47]

3.5.5 Thallium

$Tl(I)^+$ resembles K^+ in its chemistry and biochemistry. It passes through cell membranes along the K^+ pathways (p. 166 ff), substitutes for K^+ in phosphorylating enzymes and damages all cells, particularly nerves and muscles. Diagnosis of Tl(I) poisoning is difficult. Precipitation with Prussian blue and administration of massive amounts of potassium salts help to remove it.

Since alkali metal chalcogenides have promising thermoelectric properties (p. 35 ff), the less electropositive and heavier thallium should have even less electrical resistivity and lower thermal conductivity and perform better.[48] Furthermore, unlike alkali metal ions Tl$^+$ has a lone pair of electrons which can be stereochemically active and can influence coordination geometries. Metal ions with the closed shell and pseudo-closed shell d^8, d^{10}, and s^2 electronic structures, such as Pt(II), Au(I) and Tl(I), are known to interact with one another to form weak metal–metal bonds. The formation of red-coloured extended –Pt(II)–Tl(I)–Pt(II)–Tl(I)– chains with pyridine and cyanide ligands is an example.[49]

3.6 Silicon, Germanium, Tin and Lead

3.6.1 Silicon

Lithographically prepared templates on silicon wafers have been extensively used to order micellar polymers in rows and 2D layers. Such elevated arrays are possible templates for pattern transfers or ordered quantum dot arrays.

The electronic properties of silicon wafers can be modified by polar monolayers, which introduce a net electrical dipole perpendicular to the surface/ interface, modify the work function and electron affinity, and change the band offset and band bending. At a specific coverage, negative molecular dipoles (Cl$_3$Si(CH$_2$)$_3$–CH$_3$; –2.80 D) decreased the work function of SiO$_x$/Si and positive molecular dipoles (Cl$_3$Si(CH$_2$)$_3$–Br; +2.05 D) increased it. These polar domains affect the semiconduction below them by electrostatic interactions with pinholes where there are no molecules, and enhance their ability to take up electrons. The extended space charge regions will then also increase the effective work function for neighbouring regions with nominally low work function.[50,51]

Nanowires with a crystalline 10–20 nm silicon core and a 5–10 nm oxide shell were grown by vapour transport catalysed by gold nanoparticles. After mechanical transfer on to a 200 nm thick silica layer on top of a heavily doped silicon wafer, the holes were introduced by boron ion beam doping and n-type wires followed from analogous phosphorus implantation. The observed large increase of the hysteresis gap for phosphorus- and boron-doped nanowires within a constructed field effect transistor showed no difference between phosphorus and boron implanted ions.[52] The n and p defects and traps created were thus localized within or close to the oxide shell. Strict stereochemical control is even possible with ion beams![53]

p$^+$-Doped silicon wafer crystals contain holes, *e.g.* boron or aluminium atoms, and upon anodization in aqueous hydrofluoric acid this random network of atomic pores is converted to nanoporous silicon with the pores running perpendicular to the surface as in nanoporous alumina (p. 182 ff). Beyond a critical pore density, silicon nanoparticles may become luminescent – the charge carriers are now "quantum confined". The surface of the silicon wafer may also be dented by lithography with nanowells, which focus the electric field of electrochemical oxidation.

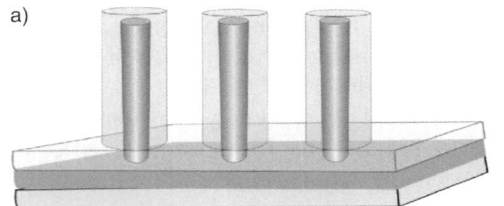 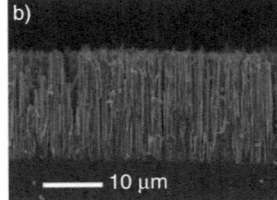

Figure 3.22 (a) Schematic cell design of a silicon nanowire solar cell. The n-Si nanowire core is given in stripe at the bottom, the metal contact in black and the polycrystalline p-Si shell in white. (b) Cross-sectional electron micrograph of a completed n-Si nanowire core. © American Chemical Society.[52–54]

Metallurgical grade silicon costs about $2/kg, solar grade silicon (sg-Si) $20–30/kg. A solar cell with a device geometry that allows the use of low-purity silicon should greatly reduce monetary and environmental costs. Vertically aligned silicon nanowire solar cells have indeed been shown theoretically to be much less sensitive to impurities than planar arrangements. The first such system to be was realized looks like the anodized alumina and silicate wire systems in Figure 3.22.[52–54]

The surface of elemental silicon oxidizes spontaneously. Its re-reduction with carbon at 1500 °C first produces black, glassy silicon monoxide (SiO), which is used to produce smooth surfaces and oxidizes at room temperature to silica.

Commercial silicon monoxide powder was irradiated under water with a pulsed Nd:YAG laser at room temperature to produce SiO gas, which immediately condensed to yellow nanocrystals. The dark-yellow suspension of SiO nanocrystals was washed with hydrogen fluoride to remove surface silica and filtered. Flat, circular nanocrystals varying from 10 to 400 nm in diameter and ∼0.2 nm thick were observed as products; these small particles quickly aggregated (Figure 3.22). The thin disks are presumably formed because silica adds to the (110) planes of SiO with the highest surface energy and blocks the addition of SiO gas there. The yellow nanocrystals fluoresced blue upon excitation with UV light (254 nm) and were then oxidized by air to silica.[55] (Figure 3.23)

Molecular oxygen (dry oxidation) and water (wet oxidation) diffuse through an existing silica layer, and the temperature-dependent overall growth of the silica layer leads to an expansion in volume of 125%. If the silicon surface was patterned before oxidation, both convex and concave curvatures show a retardation of the oxide growth with respect to planar oxidation: the larger oxide dianion is pushed outward as it forms at the silicon/silica interface, which is stretched in convex systems, but compressed in concave systems. Nanochannels are formed and at a width < 100 nm they are quickly filled with silica. Slow lateral growth, however, always leaves open triangular nanochannels with a uniform side length, *e.g.* 26 nm, close to the bottom (Figure 3.24). These pores are a perfect base for the formation of nanometre wires separated by a thin insulator layer from semiconducting silicon wafers.[55]

Figure 3.23 SiO–SiO$_2$ disks in SiO gas atmosphere. © American Chemical Society.[55]

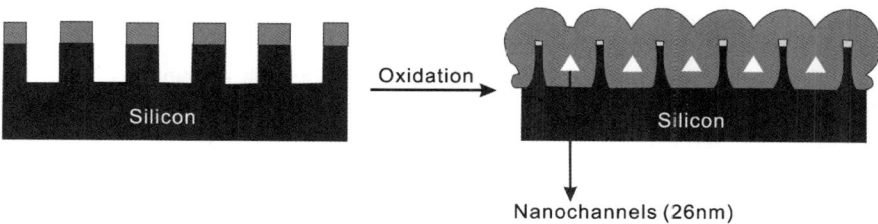

Figure 3.24 Formation of separate triangular channels, 26 nm wide, by the oxidation of surface channels filled up by large silica nanoparticles. © American Chemical Society.[55]

Colloidal silica particles are usually made by hydrolysis of tetraethyl orthosilicate and presumably form above a critical silicic acid concentration, where particle nucleation occurs. If all nucleation occurs within a short time and all particles grow at the same rate, monodisperse particles are the result. The particle size then depends inversely on the reaction temperature: at low temperature the hydrolysis of the ester becomes slower, fewer particles are nucleated and the final particle size must be large, if all tetraethyl orthosilicate reacts.

Colloidal particles are only stable if the interparticle repulsion is stronger than the van der Waals attraction forces Therefore charged colloidals are always stabilized by increasing particle size, because it lowers the interfacial free energy. Surprisingly, (aminopropyl)silica nanoparticles have their most stable dispersions at pH > 10, and precipitate at pH 3. Presumably the pK_a of the aminated spheres is so low that half protonation does not occur at pH 10, as for isolated amine groups, but seven orders of magnitude lower, because of intraparticle charge repulsion (p. 90 ff).[56,57]

α-Quartz is prepared by hydrothermal synthesis. Fine sand is mixed with water and some sodium hydroxide and then heated in an autoclave to to 400 °C until a pressure of 500 bar (50 MPa) is reached and the silica dissolves. In the upper part of the autoclave a small quartz crystal was previously fastened, and now the temperature is lowered slightly there. A large quartz crystal with a length of ~10 cm is formed. The same process is responsible for the formation of many large rock crystals during the cooling of the Earth's crust on a geological time scale, the "autoclaves" being the natural cavities of rocks.

Silicon usually appears on Earth in the form of silicate tetrahedron polymers, $(Si_2O_4)_n$. α-Quartz, the most important piezoelectric material, is made of SiO_4 tetrahedrons which are not completely regular and produce a dipole moment. These domain dipoles cancel within the whole crystal lattice, but upon pressure in a particular direction the cancellation is removed and electric charges appear on the crystal's surface. Conversely, it is possible to excite vibrations of the crystal by application of an alternating voltage. Oscillating quartzes are used in watches and computers.

Water-repellent surfaces are useful for many reasons, because water droplets do not wet them, but roll off and remove dirt and debris. They prevent the adhesion of snow to roofs and antennae, and repel oil and dyes on textiles and stone surfaces. Such surfaces are superhydrophobic and self-cleaning, because they are "rough" on the micro- and nanometre scale (the lotus leaf effect) and provide much air and little fibre surface for contact with the flat part of water droplets. The droplets remain spherical and the measurable contact angle is >150°. Such nanoscale rough surfaces appeared on silicon substrates after coating them with a 10 nm thick gold layer and heating to 1100 °C for 3 h. The silicon surface is dissolved by the hot gold particles and dendritic crystallization occurs upon thermal evaporation of the gold. The silicon became whitish-gray by intense light scattering and amorphous silicon oxide nanowires with diameters of 20–150 nm and lengths of 15–20 μm could be seem under the atomic force microscope.[58]

Natural laponite or clay consists of disk-shaped silicate nanoparticles ~30 nm in diameter and 0.96 nm thick. Its layer structure consists of six octahedral magnesium ions sandwiched between two layers of four tetrahedral silicon atoms. In aqueous solutions, water molecules are intercalated into the interlamellar space, magnesium may be replaced by lithium and sodium, and long-chain ammonium salts may separate the layers by a few nanometres. Molecular adsorption occurs through ion pair formation and van der Waals interactions between adjacent alkyl chains.[59] A host polymer was introduced to separate the clay stacks into individual layers and disperse them in water. Polymerizations on the clay surface with metal catalysts regularly failed, because the metals were inactivated on the oxide surface. On the other hand, organic radical polymerization as well as the activation/deactivation processes with nitroxide radicals coming from alkoxamines based on N-tert-butyl-N-[1-diethylphosphono-(2,2-dimethylpropyl)] and a terminal quaternary ammonium group for clay layer anchoring worked without problems (Figure 3.25). An almost complete exchange of the cationic sites was obtained and some physisorbed polymer chains could not be removed from the clay even after exhaustive dialysis.[60]

pH indicators cooperating with other organic molecules give approximate pK_a values of the surface of solid acids and the number of acid sites with different strengths can be determined. However, if the results are compared with catalytic activity, good correlations are not necessarily obtained because of

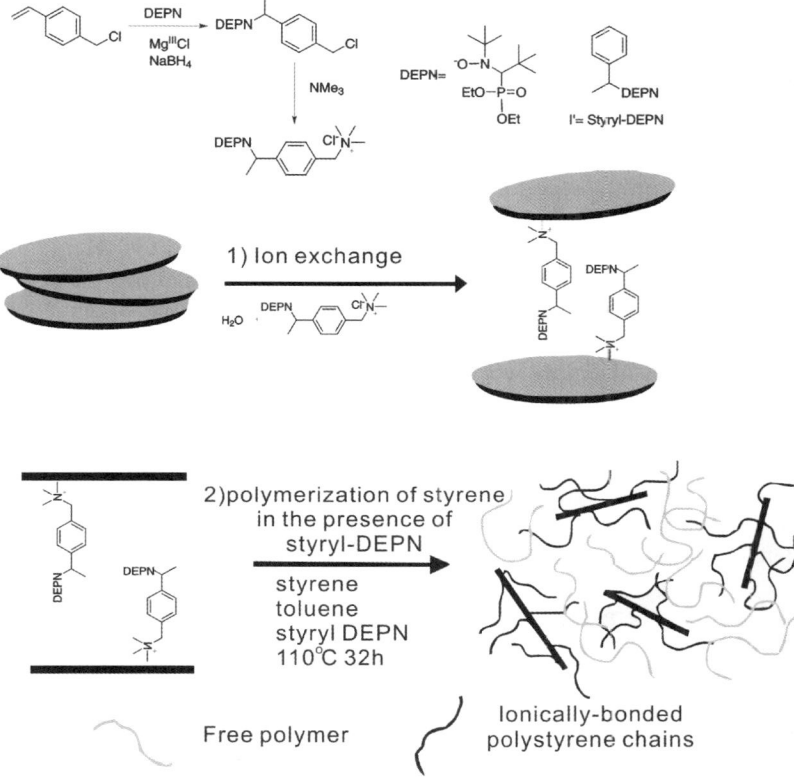

Figure 3.25 Living polymerization on clay nanoparticles with NO-terminated poly(styrenes) carrying ammonium groups in the *para*-positions.[60]

overtitration. Usually there are more basic molecules than catalytically active acid sites. Furthermore, only a small fraction of the total acid sites measured is usually active for a given reaction, and the low temperature at which acidity is measured favours indiscriminate adsorption of the basic molecules on all acidic sites.

Commercial silica sols with surface OH groups contain non-agglomerated, non-porous spheres of uniform diameters of 5–200 nm. These commercial nanoparticles are made by passing an alkaline sodium silicate solution through an ion exchange column in order to remove the Na^+ and lower the pH. At a particular pH the particles with a uniform size precipitate and coagulate to larger, porous particles. The surface of these silica particles is, however, never smooth enough to allow their coating with closed molecular monolayers. Colloidal silica particles made from tetraethoxysilane ($Si(OEt)_4$) by hydrolysis with sodium hydroxide in water are the carriers of choice as totally smooth (roughness <1 nm), slightly curved, reactive and photochemically inactive surfaces. They provide a perfect basis for the establishment of closed, reactive monolayers, *e.g.* an amine coating, addition of single, flat-lying dye molecules and rigid walls around them. Single molecular objects 1–3 nm high could be detected and measured (p. 16).

Such amine-coated smooth particles were prepared as follows. In a reaction vessel that had been dried for 3 h at 120 °C, 1.5 mL of tetraethoxysilane ($Si(OC_2H_5)_4$) and 3 mL of ammonia (28%) were dissolved in 50 mL of anhydrous ethanol, and the reaction mixture was slowly stirred at room temperature for 24 h in the dark. Water (Milli-Q, 400 μL) was added and stirred for a further 2 h. Then (3-aminopropyl)triethoxy silicate (($NH_2C_3H_6Si(OC_2H_5)_3$, 400 μL) was added, and the mixture was stirred overnight. The resulting silica sol was warmed to 80 °C and refluxed at this temperature for 10 h under an argon atmosphere. The amino-modified silica colloids with a diameter of 100 nm were then used for self-assembly work after cooling to room temperature.[57]

Polyethylene may disperse silica powders as fillers or be placed within the pores of silica by polymerization of ethylene. Blending of silica has been achieved by dissolving linear polyethylene in a boiling solution of tetraethoxysilane in xylene and emulsifying this hot solution in an alkaline water–alcohol dispersing solution (Figure 3.26). The base catalyses the polycondensation of the silane to silica within the emulsion droplets and entraps the polyethylene while the silica network forms. The final system resembles an interpenetrating network of an organic and an inorganic polymer, because the initial silica at the interface created a diffuse zone with both hydrophilic SiOH and hydrophobic SiOEt residues, the "cosolvent" needed in sol–gel processes. The silica network formed and the polyethylene chains could not separate, because they were entrapped in a porous inorganic matrix.[60]

Figure 3.26 Hydrophobic poly(ethylene) may appear within the pores of silica made by hydrolysis of tetraethoxysilane in water–toluene mixtures. © American Chemical Society.[60]

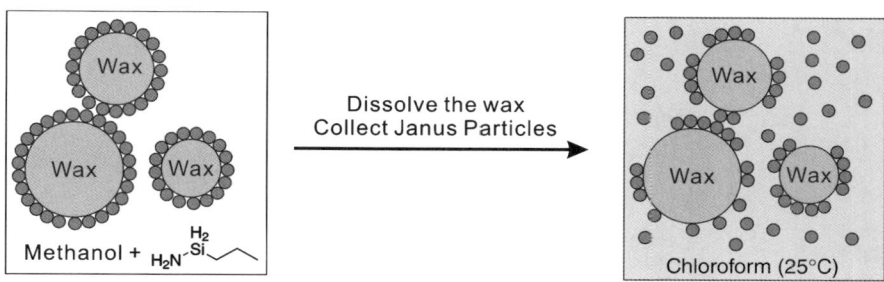

Figure 3.27 Wax may cover half of the silica nanoparticle surfaces and the other half may be coated with a cationic, anionic or neutral or hydrophobic mono- or multilayer. After dissolution of wax and nanoparticles, the latter may be coated with any other monolayer. The product is Janus nanoparticles, which may repel each other, form chains or precipitate. © American Chemical Society.[62]

Silica particles were dispersed in paraffin wax at 75 °C and then mixed with water by magnetic stirring at 1600 rpm for 1 h, more finely dispersed in methanol and reacted with aminopropylsilane. Only the outer half was silylated. The modified silica particles were separated from the wax by chloroform and reacted with octadecyltrichlorosilane (OTS) on the non-silylated hemisphere. The silica particles were now Janus particles, cationic on one side and hydrophobic on the other (Figure 3.27). The emulsification process that forms the basis of this synthesis strategy is achieved simply by mechanical mixing for a short time. The resulting emulsions are mechanically rigid and stable in moderately aggressive chemical environments. The ease of scale-up and low cost of production make this method attractive for products such as bipolar particles.[61]

The hydrophilic pore of a zeolite L crystal, with a diameter of 7.5 Å, is well suited to organize non-covalent rows of suitably sized molecules in an environment which is well protected against molecules outside the pore, in particular from the oxygen and water of air. One channel took up thousands of hydrated, electron-conducting methyl viologen molecules or energy-transporting oligophenyl derivatives, which were thin enough to enter the tunnel together with solvent molecules. Entrapped diphenyl hexatriene fitted only as a *trans* isomer and did not isomerize upon irradiation to the broader *cis* isomer within pores (Figure 3.28). Hydrophobic molecules were expelled from the hydrophilic pores upon addition of water and then settled on the mantle. Upon drying, the molecules migrated back into the pores. Because dye molecules did not pass each other in the pore, they could be ordered by stepwise addition. The number of molecules in each individual domain and intermolecular distances remained unknown. The pores in zeolites were also hydrophobized by the extraction of the alumina part with sodium hydroxide or by coating it with alkyl chains.[62] Aggregates in such zeolite subnanochannels were also prepared and the end-to-end exciton interaction was analyzed.[63]

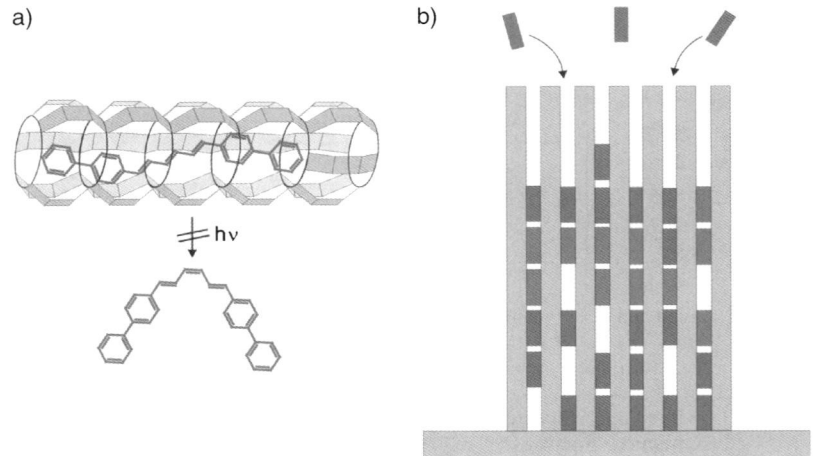

Figure 3.28 (a) Stabilization of fitting *trans* isomers in silicate nanotubes. (b) Ordering of molecules a, b and c by the time sequence of additions.[62]

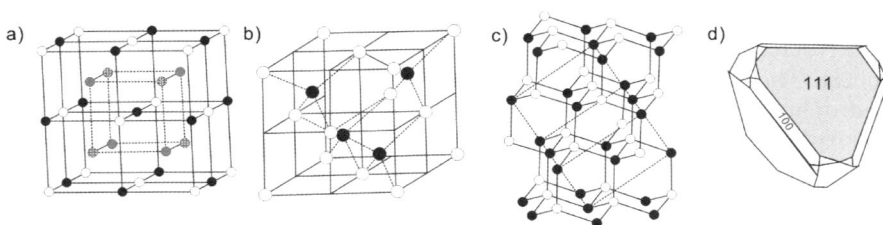

Figure 3.29 (a) The perfect fcc arrangement of germanium atoms. (b–d) Partial filling of the tetrahedral voids in the fcc germanium arrangement by six more germanium atoms produces a covalent semiconducting Ge(0) zinc blende (or diamond) type block.

3.6.2 Germanium

Crystals of Ge(0) occur as a fcc array in which only 6 of the 12 tetrahedral holes are filled with Ge(0) atoms in an alternating sequence (Figure 3.29). This crystal structure is called zinc blende or sphalerite, and is also found in diamond, gallium arsenide, zinc sulfide and zinc selenide. Diamond is usually drawn as an endless, three-dimensional network of carbon tetrahedra in the form of zigzag chains and six-membered rings and is the hardest of all natural materials. It is an orgy of covalent bonding. The other three compounds are soft, semi-conducting materials. The Ge–Ge crystal looks like a diamond and its Ge–Ge bonds are also purely covalent. Nevertheless the energy gap between valence and conducting band is relatively small (64 kJ/mol), Ge(0) is soft and it behaves as a semiconductor. The much higher energy of the C–C bond as compared to

Si–Si or Ge–Ge explains the huge chemical differences between carbon and the semiconductors silicon and germanium, which are so vulnerable to the polarization of their electron pairs.

Ge(IV)Cl$_4$ is reducible to Ge(0) metal by alkali naphthalides. The reaction rate with the different alkali metals was in the expected order K > Na > Li. The germanium nanoparticles rapidly achieved their maximum size based on the GeCl$_4$ concentration. A reaction time of 1 h at room temperature gave the same nanoparticles as after 10 min. Small (1–15 nm) germanium nanoparticles were also obtained by thermal decomposition of tetraethylgermane or tetrabutylgermane in solvents boiling close to 400 °C, such as trioctylamine, squalene and octacosane. Silica gel glasses doped with such germanium nanoparticles showed a strong room-temperature photoluminescence with peaks at 570, 605, 676 and 730 nm originating from <1–2 nm germanium clusters. Cubic, diamond-like germanium nanocrystals were made by hydrogen reduction of GeO$_2$ at 600 °C, but they did not luminesce.[64–66]

Ag$^+$ ions deposited on solid Ge(100) surfaces are immediately reduced to metallic Ag(0); the Ge(0) is oxidized to Ge(IV)O$_2$ by Ag$^+$ (Figure 3.30). If the Ag$^+$ salt was stirred in a toluene solution of a poly(styrene)–poly(vinylpyrrolidone) block copolymer, the silver was loaded on to the nitrogen units of the block copolymer. The Ag(0) nanoparticle arrays on Ge(100) then adhered strongly to the germanium surface after removal of the polymer matrix by sonication in a toluene bath. The hexagonal packing of the block copolymers as well as of the Ag(0) nanoparticles is the same on Si(0), Ge(0), Ga(III)As or In(III)P surfaces.[67]

Aluminium germanium hydroxide ("aluminogermanate") nanotubes 3.2 nm wide and 20 nm long were prepared from tetraethylorthogermanate (Ge(OEt)$_4$) and aluminium chloride in aqueous solution. At first amorphous 6 nm nanoparticles were formed, which then self-assembled to nanotubes.[68] In the presence of lysine, germania (GeO$_2$) crystals appeared.[69]

Opals are periodic arrays of nanoparticles of high refractive index wthin a transparent glass of lower refractive index. Inverse opals are periodic arrays of air pores within a transparent material of high refractive index (>2.9). Germanium has a refractive index of 4.12 At $\lambda = 2.0\,\mu m$. Germanium

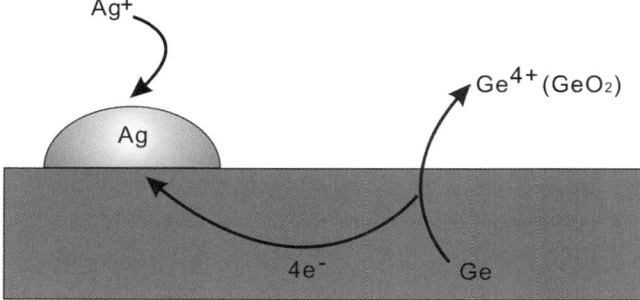

Figure 3.30 Reduction of Ag$^+$ ions to Ag(0) nanoparticles on the surface of Ge(0).[67]

nanoparticles infiltrated into polystyrene produced an opal made of 1.0 μm polystyrene sulfate microspheres. The germanium-in-polymer composite had a refractive index of 2.05. Commercial 1.0 μm polystyrene sulfate microspheres in water and the colloidal suspension of 25–100 nm germanium nanoparticles were placed in a polyethylene vial and maintained at 45 °C until the mixture evaporated to dryness after 22 h. The polystyrene particles "crystallized" with the infiltrated germanium nanoparticles on their surface, thus forming the opal, which was then sintered for 1 h at 80 °C. A drop of adhesive was then deposited on the sintered polystyrene opal to fill the empty spaces, making it translucent and opalescent.[70]

Germania self-assembles in basic aqueous solutions via GeO_2/OH^- into 1 nm nanoparticles corresponding to an octamer, a (GeO_2) sphere or a cube (Figure 3.31). The resulting germania nanoparticles are nearly four times smaller and more acidic (pK_a 7.1) than their silica counterparts. The nanoparticle size was independent of solution pH and the monovalent cation used to supply hydroxide to solution.[71]

The IV–VI compounds of the GeTeSb type usually have periodic cubic close packing and the stacking rules for the germanium, antimony, and tellurium layers are different from each other. They form a large group of "real" binary compounds existing in an NaCl-type structure and are of little practical use.

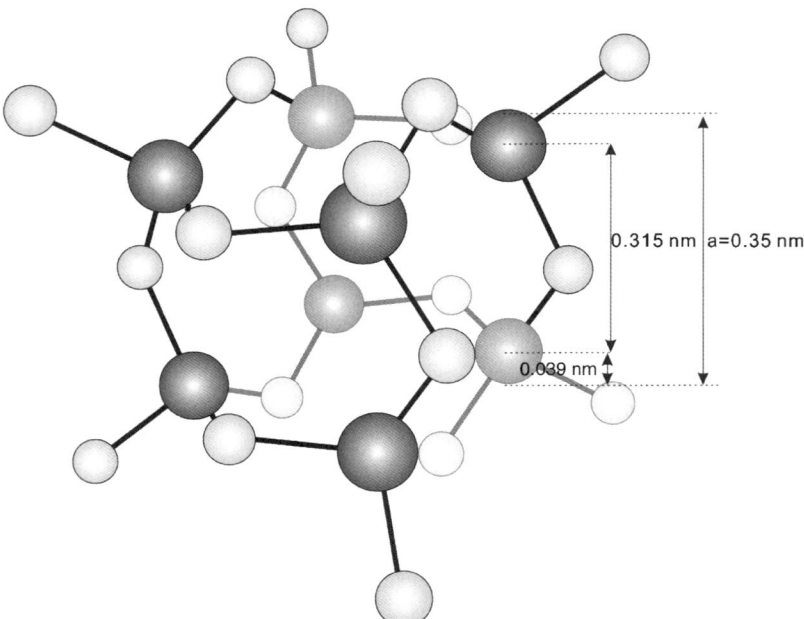

Figure 3.31 Diagram of the germania octamer with germanium (black) and oxygen (white) atoms (hydrogen atoms are omitted). The Ge–O bond length is 174 pm, the Ge–O–Ge bond angle is 130° and the Ge–O–Ge distance is 0.315 nm. © American Chemical Society.[71]

There is, however, one interesting variability. GeTe$_{(1-x)}$Sb$_2$Te$_{3x}$ crystals from $x = 0$ to $x = 2/3$ maintain their ordering when subjected to instantaneous heating by laser irradiation followed by equally rapid cooling. The atomic arrangements do not change, but the NaCl-type crystals become amorphous and remain so for long periods of time at room temperature. Surprisingly, if slightly heated they regain their previous crystallinity quickly and quantitatively. The compounds are said to be "pseudobinary", because the crystalline arrangement of atoms in them never really changes. Rewritable laser disks (DVD-RAM) use the changes in optical reflectivity and transmissivity caused by this "phase change" between crystalline and amorphous GeTe–Sb$_2$Te$_3$ to record information. The crystalline, metastable NaCl-type and the amorphous phase switch back and forth in less than a nanosecond under laser beam irradiation, and more slowly upon electric heating. In a blank disk the crystalline compound forms three cubic intermetallic compounds, which differ in the stacking of the germanium, antimony and tellurium layers. The components of GeTe-Sb$_2$Te$_3$ have altogether 38 valence electrons (Ge, s^{2}p^2 = 4; 2Sb, s^{2}p^3 = 10; 4Te, s^{2}p^4 = 24). N_{sp}, the average number of valence electrons per single atom, is 38:7 = 5.4 for the pure compound. In reality there is not as much GeTe, and N_{sp} is between 4.7 and 5.0 only. In the density functional theory calculation the cubic structure only needs an N_{sp} value greater than 4.25: below this value p–p bonds disappear and sp^3 bonding leads to hexagonal crystal lattices. The cubic GeTe–Sb$_2$Te$_3$ crystal remains stable over a composition range from 100% to 33 atom% GeTe. It allows for very fast crystallization, because the cubic structure is so simple, and also stabilizes amorphous states because the total energy per atom in the crystal lattice is particularly small. The "successful" pseudobinary materials used for rewritable laser disks are thus well understood.

The study of phase changes of nanoscale structures is just beginning, however, and deserves great efforts in material science. Antimony telluride (Sb$_2$Te$_3$) and germanium telluride (GeTe) are current of great interest, because the stable, slowly switching germanium telluride and the less stable, rapidly switching antimony telluride offer extremes and a special germanium antimony telluride (Ge$_2$Sb$_2$Te$_5$) is a compromise between them. Sb$_2$Te$_3$–GeTe core–shell nanowires offered both extremes in one wire. Core Sb$_2$Te$_3$ nanowires were first made by the vapour transport and metal catalysis in a horizontal tube furnace from bulk germanium telluride, antimony, and tellurium followed by vapour solidification.[72–75]

Electrical measurements on these Sb$_2$Te$_3$–GeTe core–shell nanowires were made between platinum electrodes which had been written by a focused ion beam directly on to the nanowires with a separation of 2 μm. The as-synthesized crystalline nanowire showed ohmic behaviour and low resistance (black squares in Figure 3.32). Upon the application of a 100 ns, 0.6 mA current pulse the I–V curve changed (white circles) to a highly resistive typical for the amorphous state. As the applied voltage reached a threshold value of 1.1 V, the current rose sharply owing to the recrystallization of the nanowire.[75]

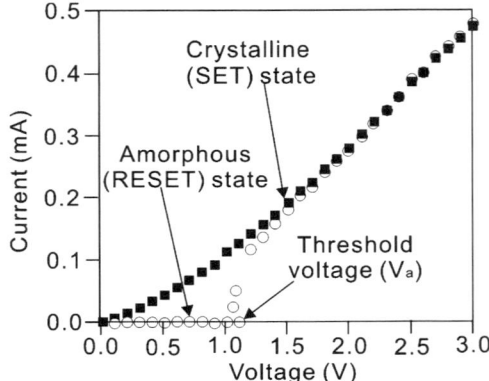

Figure 3.32 Resistance and current pulses of a $Ge_2Sb_2Te_5$ NW starting (1) from the crystalline state, which melts at 0.4 mA, and (2) from the amorphous state, which crystallizes there. © American Chemical Society.[75]

3.6.3 Tin

Tin is similar to titanium, which comes as a surprise, because one is a late main group metal and the other an early transition metal. The dioxides SnO_2 and TiO_2 (titania) are isomorphous; both oxides turn yellow upon heating; both tetrachlorides, $SnCl_4$ and $TiCl_4$, are fluid.

In dye-sensitized solar cells a porous electrode made of a wide-band gap semiconductor and long electron diffusion length (1) supports dye molecules and (2) transports photoinjected electrons. Binary oxides, including TiO_2, ZnO, SnO_2, Nb_2O_5, In_2O_3 and the ternary oxide Zn_2SnO_4, were used successfully. The latter is acid stable and was combined with a light stable, acidic Ru(II) dye. The Zn_2SnO_4 cell had an energy conversion efficiency of 3.8% with 5.6 μm film thickness, which was comparable with the highest efficiency reported for ZnO (4.1%) and much higher than that reported for SnO_2 (1.2%).[76]

Both Sn(0) and Te(0) are main group metals (IV and VI), their atomic diameter is similar (270–280 pm) and their electronegativity values are close: 1.7 for tin and 2.0 for tellurium. Bulk SnTe is a semiconductor with an extremely small band gap of 0.18 eV at room temperature. It absorbs mid-IR radiation and is used in thermoelectric heat converters.

A synthesis of SnTe nanocrystals was based on the reduction of a mixture of tin-silyl amide ($Sn[N(SiMe_3)_2]_2$) and trioctylphosphine telluride (($C_8H_{17})_3$PTe) with oleylamine in octadecene ($C_{18}H_{37}NH_2/C_{18}H_{36}$) at 150 °C. The almost instantaneous reduction by the amine and nucleation of SnTe elements led to a temperature drop to 120 °C. Dry oleic acid was then added to passivate the surface of the crystals, whose size could be varied from 4.5 up to ~15 nm by adjusting the injection and growth temperatures as well as the concentration of oleic acid.

SnTe nanocrystal films showed an extremely low electrical conductivity of 10^{-10} S cm^{-1} for mixed metals when an insulating oleic acid layer 1.5 nm thick

Main Group Elements

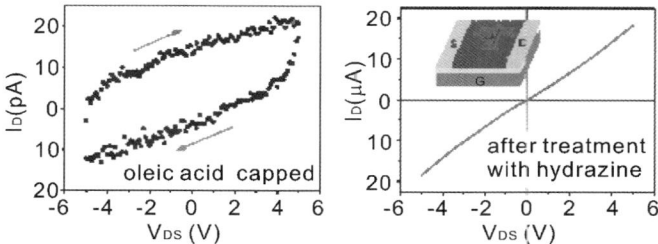

Figure 3.33 Oleic acid-capped 13 nm SnTe nanocrystals gave only 10^{-12} A currents, rising to 10^{-6} A after substitution of oleic acid by hydrazine. © American Chemical Society.[77]

separated the nanocrystals from each other. The conductivity increased by six orders of magnitude after treatment of the films with hydrazine in acetonitrile, which washed away a hydrazonium soap and also acted as an n-type transfer dopant on the surface, as a double-headed electron donor (Figure 3.33). The interparticle spacing was down to 0.4 nm.[77]

3.6.4 Lead

A level of $>100\,\mu g/kg$ of lead in the body is poisonous. Lead inhibits the insertion of iron into protoporphyrin and several other processes in the formation of haem in the red blood cells. Anaemia is a standard effect of lead poisoning, but brain damage is more dangerous. Therapy with $CaNa_2$ EDTA helps, because it replaces the Pb(II) by Ca(II) and the Pb(II) is then rapidly excreted.[78]

Lead vapour, like that of silver, is ideally suited for the formation of atomically flat layers on semiconductor surfaces. Such mirrors may be applicable for helium beam microscopy. A focused beam of helium atoms would be a unique tool for reflection or transmission microscopy, with a lateral resolution of 50 nm, and would be useful for easily oxidizable or heat-sensitive biological materials. A mirror with an atomically flat surface would be required to focus a beam of low-energy helium atoms into a small spot on the sample. Unusually strong quantum confinement of the electrons in ultrathin lead films deposited on silicon led to an electronic mode of growth, which enforced "magic island heights" or ultraperfect, atomically flat films of lead on Si(111) thin wafers. Four or six monolayers of Pb(0) were deposited on Si(111) and a very high helium atom reflectivity was observed.[79]

Tin and lead nanoparticles, as well as corresponding metal sponges, were prepared from $Me_2Si(NtBu)_2Sn$ or $Me_2Si(Nt\text{-}Bu)_2Pb$ with various metal hydrides, *e.g.* with $H_2AlOt\text{-}Bu$ in tetrahydrofuran. The reduction of the tin compound at $-115\,°C$ was accompanied by a colour change from red to dark brown. The crystallite size depended on the polarity of solvent and hydride used.[79] Porous alumina membranes of different pore diameters were also filled with tin and lead amides and the metal nanoparticles formed by hydride reduction. The filled alumina membranes were used as wavelength filters between 270 and 525 nm.[80]

Charge transport in an array of close-packed semiconductor nanocrystals separated by thin barriers can occur by hopping between quantum-confined orbitals with s and p symmetry. Due to the monodispersity and eightfold degeneracy of the 1s electrons of lead chalcogenide nanocrystals, they show the most promising charge transport properties. The hole states allow good energy-level alignment between adjacent particles, which leads to sharp peaks in the electronic density of states. Furthermore a very large number of electronic states is available for charge transport in the lead chalcogenides. In lead selenide, for example, the extraordinarily large static dielectric constant of ~ 250 also reduces the charging energy and prevents Coulomb blockade.[81]

Lead selenide nanocrystals of different sizes, usually 4–8 nm, and size distributions well below 10%, were synthesized from lead oleate and tri-*n*-octylphosphine selenide in squalane, a high-boiling hydrocarbon. Films of monodisperse nanocrystals were deposited by dropcasting hexane/octane (9:1 by volume) solutions on substrates with preformed parallel Cr/Au electrodes. In 50–200 nm films of 5 nm lead selenide nanocrystals the oleic acid was replaced by hydrazine to reduce the interparticle spacing from 1.1 nm to 0.4 nm and the carrier mobility in the superlattices rose to values above $1 \, \text{cm}^2/Vs$. Such assemblies of lead chalcogenide nanocrystals are successfully used in thermoelectric devices (p. 35).[81]

3.7 Nitrogen, Phosphorus, Arsenic, Antimony and Bismuth

Functional groups containing nitrogen and phosphorus are discussed in Chapter 2 (pp. 87 and 139). Only nitride and phosphide nanoparticles are discussed here.

3.7.1 Nitrogen

For carbon–nitrogen compounds, *e.g.* amines, proteins, dyes and DNA, see Chapter 2; in the following only a few nitrides of nanochemical interest are described.

Boron nitride (BN) has two representative crystal structures. One is the sheet-stacking-up hexagonal form (*h*-BN) with in-plane sp^2 bonding (graphite like), the other is the highly dense cubic form (*c*-BN) with tetrahedral sp^3 bonding (diamond like). *h*-BN is an insulating layered material with a band gap of $\sim 5.5 \, \text{eV}$ and is produced by arc discharge techniques, chemical vapour deposition, substitution reactions, and laser ablation. The *h*-BN structure looks like graphene and it is also made in the same manner. The preparation of porous BN "nanourchins" is given as a first example here.

Commercial mesoporous spherical silica (MCM 48) as a template was added to a solution of sucrose with trace amounts of sulfuric acid and some water. The mixture was heated and dried for 6 h at 100 °C under nitrogen and at 160 °C in a closed vessel for additional 6 h. The process was repeated, the silicate was dried and pyrolysed at 1000 °C under argon for 1 h and the products were stirred in 48 wt% aqueous hydrogen fluoride for 30 min in order to remove the silicate template. The resulting spherical carbon nanoparticles were placed in the centre of a 15 mm graphite disk which had a central depression with numerous perforations of 1.5 mm diameter around it. The disk was fitted into a cylindrical graphite crucible, which contained 400 mg of boron oxide (B_2O_3) powder at the bottom. The crucible was then slowly heated to ~ 1700 °C under nitrogen in a high-frequency induction furnace. The resulting material was subsequently annealed for 30 min in a flow of nitrogen, and cooled to room temperature. The boron in the nitrogen flow became porous boron nitride.

Scanning electron microscopy (SEM) images of the BN material showed spherical balls 100–400 nm in diameter with various protuberances (Figure 3.34). The centre of the spheres revealed interlinked BN layers forming cavities (= pores). Electron energy loss spectroscopy (EELS) or energy dispersive X-ray spectroscopy showed only B and N. Nitrogen adsorption measurements yielded a surface of $\sim 290 \, \mathrm{m\,g^{-1}}$. The urchin also emitted electrons with large current densities at low turn-on voltages, presumably originating from the BN ribbons protruding from the surface of the balls. Both BN ribbon zigzag edges provide B and N dangling bonds.[82]

A synthesis of single wire boron nitride nanotubes was based on the continuous heating of a h-BN target by a CO_2 continuous laser under nitrogen. The target was a commercial h-BN cylinder made of h-BN powder pressed with

Figure 3.34 Boron nitride sphere with various protuberances. © American Chemical Society.[82]

boron oxide as binding material, which was rotated under the laser beam and translated vertically at constant speed to ensure uniform vaporization. Liquid boron droplets were formed first, then reacted with the nitrogen gas to form boron nitride. Further incorporation of nitrogen atoms at the root of the boron particle allowed the boron nitride tube to grow.[83]

The edges of graphene sheets may be semiconducting zigzag ribbons with a spin-polarized ground state and opposite spins at each edge, which are coupled to the hexagonal carbon network. Such magnetism in low-dimensional systems involving s and p electrons, as in graphene and boron nitride, is still not really understood. Two-dimensional h-BN is a large band gap insulator and should remain so, and remain non-magnetic, when rolled up. Edge effects may, however, be as important as in graphene. Theory tells us that all spin configurations are thermally accessible at room temperature in boron nitride nanoribbons, and that they should be in the high spin state, $(++, ++)$, while in graphene they only occur in the low spin state $(++, --)$. The high spin state of boron nitride could be stabilized over the other magnetic configurations by an external magnetic field or by transition metal doping. The energy difference of the spin patterns vanishes at a diameter of 1.4 nm, where the interaction between both spin chains becomes negligible.[84]

If a strong enough electric field is applied in the transverse direction, zigzag graphene nanoribbons undergo a transition from semiconducting to semi-metallic due to charge reorganization in the edges. Density functional theory (p. 45 ff) says that zigzag boron nitride tubes should produce various magnetic patterns at the B and the N edges (Figure 3.35). The $(++, --)$ configuration, meaning all Bs are north poles and all Ns are south poles, should show a large band gap of 5.36 eV for the majority spin channel, and a small gap of 0.34 eV for the minority spin channel at zero field. When an electric field in the B←N direction is applied, in theory the electronic reorganization produces a rapid closing of the band gap for both spin channels, transforming the boron nitride tubes into metals. A field in the opposite direction (B→N) induces semi-metallic behaviour.[85]

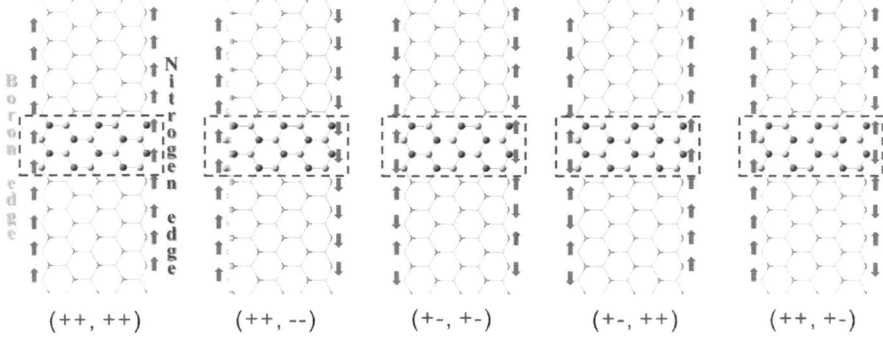

Figure 3.35 Schematic representation of the five simplest (domain wall) non-equivalent spin arrangements in zigzag boron nitride nanotubes. Above a width of 1.4 nm only $++, ++$ should occur.[85] Take notice of the boron and nitrogen edges indicated on the first model. © American Chemical Society.

The band gap of the armchair boron nitride nanoribbon and that of the zigzag boron nitride nanorod are determined by edge states and converge to values different from that of the bulk boron nitride sheet. The electronic and transport properties of boron nitride nanorods are therefore expected to be tunable by an external transverse electric field. Zigzag boron nitride nanorods should exhibit an asymmetric response. Boron nitride nanotubes and nanoribbons show gigantic Stark effects: their absorption peaks are shifted in response to a transverse electric field, from 4.5 eV to 1 eV in a field of 0.1 eV/Å.[85]

Elastic bending of boron nitride nanotubes between two AFM tips and under a transmission electron microscope converted the electrically insulating multiwalled tube into a semiconductor and induced piezoelectric behaviour. The bending of the nanotubes was induced either by a tip-attached piezoelectric crystal or mechanically and caused highly localized modifications of the electronic structure of the nanotubes (Figure 3.36). The effect on the electrical transport was much stronger in short nanotubes than in long ones. In the longer nanotubes, the local flattening effect near the contact points is obviously not effective over the whole length.[86] Introduction of boron nitride into graphene carbon nanotubes could produce exciting dye and electric current properties, provided one can learn to prepare and handle the 1.4 nm wide ribbons or tubules.

The high densities of surface atoms, as well as the cylindrical pore geometry of both carbon and boron nitrides, create high internal stress. This has an impact on the ordering and packing of adsorbed light particles. Very large potential gradients within small nanotubes quantize them in the radial direction, but they can move freely in the longitudinal direction. The combined effect of quantum delocalization and strong confinement differentiates hydrogen isotopes, for example, and can be used for their efficient separation: the quantized energy levels of the confined hydrogen isotopes are manipulated by the nanotube type, the external operating conditions, as well as the pore size. At cryogenic temperatures the molecules are localized closer to the surface and

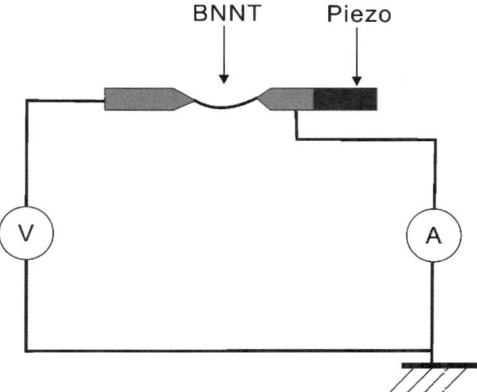

Figure 3.36 Mechanical bending of boron nitride nanotubes (BNNT) between two AFM tips made the nanotubes conductive. © American Chemical Society.[86]

the delocalization of the quantum particles means the heterogeneous external potential field generated by the boron and nitrogen surface atoms is large.[87] A path integral formalism combined with Monte Carlo simulations then separated the D_2/H_2 equilibrium selectivity in pore filling into three stages: (1) at first the enthalpy of H_2 and D_2 adsorption differentiates them; (2) the D_2/H_2 equilibrium selectivity drops more or less depending on the pore diameter; (3) the D_2/H_2 equilibrium selectivity is enhanced by the greater localization of hydrogen isotopes in the interior space of single-walled nanotubes. Boron nitride nanotubes are more selective and contain more D_2 than the equivalent carbon nanotubes, which favours them for industrial applications.

Regardless of the operating temperature, the boron nitride surface enhanced the equilibrium selectivity of D_2 over H_2 in comparison to the graphite surface (Figure 3.37). Regardless of the type of pore, deep minima of the D_2/H_2 equilibrium selectivity are observed near the pore width of 0.8 nm, where the solid–fluid potential minima generated from opposed carbon or boron nitride pore walls merge inside of the pore. The double-well solid–fluid potential is transformed at a distance of 0.8 nm into the single one. This means that adsorbed quantum paths find space for their delocalization, and are not affected by confinement. The 0.76 nm pores can thus be seen as virtual pores. Further extension of the pore size increases equilibrium selectivity due to the creation of a potential barrier in the middle of the pore.[87]

One-dimensional boron nitride nanotubes can be fabricated by rolling up the h-BN sheets. Because of the intertube van der Waals attraction they usually come in bundles. Theoretical simulations showed that fusion of adjacent boron nitride tubes can only happen if their polar B–N bonds are in an antiparallel arrangement (Figure 3.38). External pressure may overcome the deformation energy to form the intertube bonds, but only heteronuclear B–N bonds will form intertube connections, not the "frustrated" B–B and N–N bonds.[88]

In simulations a (5,5) boron nitride nanotube with a diameter of 0.69 nm and a length of 1.42 nm conducts water, but a corresponding (5,5) carbon nanotube does not, although a (6,6) carbon nanotube does. Van der Waals interactions

Figure 3.37 Equilibrium configuration of a mixture of hydrogen (dark spheres) and deuterium (light spheres) in a (12,12) single-walled carbon nanotube (*left*) and single-walled boron nitride nanotube (*right*) computed at 33 K and a total pressure of equimolar mixture of 0.1 Pa. Carbon tubes enrich H_2, BN tubes D_2. The equilibrium selectivity is higher for single-walled boron nitride nanotubes. © American Chemical Society.[87]

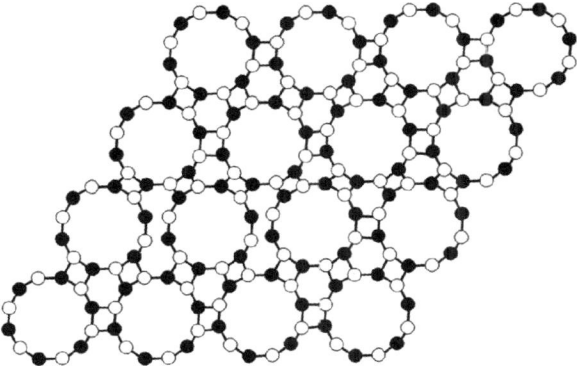

Figure 3.38 Since the B–N bond (B dark, N white) is polar, only antiparallel patterns aggregate. © American Chemical Society.[88]

between water molecules and nitride atoms are responsible for water conduction through the narrow boron nitride nanotube, making it a perfect candidate for a water channel in lipid membranes. Density functional theory computations predicted that, within uncharged (9,9) boron nitride nanotubes with a diameter of 12.18 Å, water molecules would form an ice-shell structure, analogous to the ice-like behaviour in a (9,9) carbon nanotube. In a (10,10) boron nitride nanotube an additional single-file water chain formed inside the surrounding ice-shell structure. Charges on nitrogen atoms attracted water molecules and stretched the ice-shell toward the tube wall. In uncharged boron nitride nanotubes there is no hydrogen bond to nitrogen, but the axial motion of water molecules is slowed down.[89]

Boron nitride nanotubes react in aqueous solution with anionic perylene-3,4,9,10-tetracarboxylic acid tetrapotassium salt to form a multiple μ-complex. Its water solubility is ∼0.3 g/L, and the bathochromic shift of the chromophore is 102 nm. Heating of the multiwalled nanotubular p-complex to 1180 °C and isothermal annealing at this temperature for 2 h gave B–C–N outer layers with a p-type semiconducting behaviour.[90]

Aluminum nitride (AlN) has the largest band gap of group III nitrides, which makes it useful as a short-wavelength light emitter. Chemical vapour deposition applied on Al/NH$_3$ or AlCl$_3$/NH$_3$ yielded nanowires, nanobelts, nanorods, and nanotubes and showed enhanced field emission properties in large-curvature geometries.[91] When the aluminium supply exceeded ammonia flow, excess aluminium was deposited on a silicon substrate and liquid aluminium etched it to form a solution of silicon in liquid aluminium. This process was not followed by any homogeneous alloy phase. When the aluminium was converted to aluminum nitride, most of the silicon precipitated as a separate phase; only trace amounts entered the aluminum nitride lattice and rendered the aluminum nitride nanowires semiconducting.[92]

Graphitic carbon nitride (g-C$_3$N$_4$) was prepared by heating cyanamide, NCNH$_2$, at 823 K for 4 h. The band gap energy of the material was estimated to be *ca.* 2.7 eV from the onset of the diffuse reflectance spectrum. It functioned as nonmetallic photocatalyst for H_2 or O_2 evolution from water under ultraviolet (UV) and visible light in the presence of a Pt (H_2) or RuO_2 (O_2) cocatalyst. Density functional theory calculations suggested an electron transition from the valence band populated by N2p orbitals to the conduction band formed by C2p orbitals. Nitrogen evolution from the nitride anions was also observed ($2N^{3-} + 6h^+ \rightarrow N_2$).[93]

Silicon nitrides are artificial materials: there is no Si–N covalent bond in nature. Their chemical compositions are diverse (*e.g.* Si_2N_3, Si_3N_4, SiN) but the most important compound is Si_3N_4. Silicon nitrides have a low density, high-temperature strength, superior thermal shock resistance, wear resistance and fracture toughness, and are chemically inert and biocompatible. They are applied in reciprocating engines, bearings, metal cutting, shaping tools, arc-welding nozzles, *etc.*

A highly ordered mesoporous silicon nitride framework with 32 wt% nitrogen was made from a mesoporous carbon as a hard template by pyrolysis of polycarbosilane (($CHCH_3Si$)$_n$; m.w. 1500, yield point 218–247 °C) as a ceramic precursor and *via* the nanocasting method in ammonia. This pyrolysis of polycarbosilane under ammonia gives a low ceramic yield (<60 wt%) and only ~30% of the mesopores were filled with silicon nitrides after the first impregnation pyrolysis cycle. a second cycle and high-temperature (1400 °C) treatment under nitrogen and removal of the carbon template in ammonia therefore followed. The use of hydrogen fluoride or sodium hydroxide solutions was avoided, because these not only remove silica but also dissolve silicon nitrides. The high-temperature treatment under nitrogen was a key factor for the formation of rigid silicon nitride frameworks, and the secondary impregnation–pyrolysis cycle improved the mesostructural regularity. The mesoporous silicon nitride material had a 3D bicontinuous cubic symmetry, a surface area of 384 m g^{-1}, a pore volume of 0.71 cm^3/g, and a narrow pore size distribution at a mean value of 5.7 nm.[94]

α-Germanium nitride (α-Ge_3N_4) was prepared from GeO_2 and NH_3 at 1173 K for >10 h. RuO_2-loaded Ge_3N_4 was photocatalytically active for the stoichiometric decomposition of pure water with UV light (>200 nm). Self-decomposition of the material and nitrogen release induced by photogenerated holes was suppressed when the crystallinity of the α-Ge_3N_4 increased, which was achieved with longer nitrification times of the GeO_2. The holes then reached the surface and oxidized surface water instead of nitride in the interior.[95]

Scandium nitride (Sc_3N) is thermally stable enough to survive the 1500 °C needed for fullerene syntheses and could make a high-temperature semiconductor. The yields of Sc_3N@C80 obtained are so far limited to a few tens of milligrams, but relative yields were promising (Figure 3.39).[96]

Gallium nitride (GaN_3) is not a molecule, but a dianion, and can therefore be isolated as a hydrobromide, $HBrGaN_3$. The parent acid H_2GaN_3 was prepared by reduction of Br_2GaN_3 with $LiGaH_4$, the bromide came from the Lewis

Main Group Elements

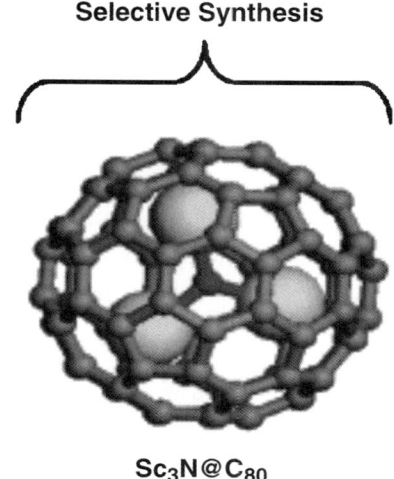

Selective Synthesis

Sc$_3$N@C$_{80}$

Figure 3.39 A scandium nitride carbon complex, where three scandium Lewis acids stabilize the carbon sphere, may be the basis of conducting layers and catalysts at 1000 °C. © American Chemical Society.[96]

acid-base adduct SiMe$_3$N$_3 \cdot$ GaBr$_3$, a stable crystalline solid:[97]

$$GaBr_3 + SiMe_3N_3 \rightarrow SiMe_3N_3 \cdot GaBr3$$
$$SiMe_3N_3 \cdot GaBr_3 \rightarrow SiMe_3Br + Br_2GaN_3$$
$$Br_2GaN_3 + LiGaH_4 \rightarrow HBrGaN_3 + LiBr + GaH_3$$
$$Br_2GaN_3 + 2LiGaH_4 \rightarrow H_2GaN_3 + 2LiBr + 2GaH_3$$

The Ga$_4$N$_4$ cyclic core of (HClGaN$_3$)$_4$ resembles cyclooctane.

The nitrides of 3d transition metals have been shown to exhibit antiferromagnetism (MnN, CrN, FeN), paramagnetism (CoN), and superconductivity (TiN, VN). An electron-doped, layered hafnium nitride (HfNCl) with a transition temperature (T_c) as high as 25.5 K, has stimulated interest in the superconductivity of metal nitrides. A corresponding theoretical study predicted a cubic MoN structure with $T_c = 29$ K. An experimental verification first developed the nitridation of Mo$_2$N and nickel with nitrogen in a high-pressure cell at 1300 K in order to obtain the desired α-MoN in a well-crystallized and stoichiometric state. It yielded superconductivity at $T_c = 14.0$ K, still far away from record values.[98]

3.7.2 Phosphorus

Carbon phosphates and phosphonates are discussed in Chapter 2 (p. 100) and sodium, potassium and calcium phosphates at the beginning of this chapter.

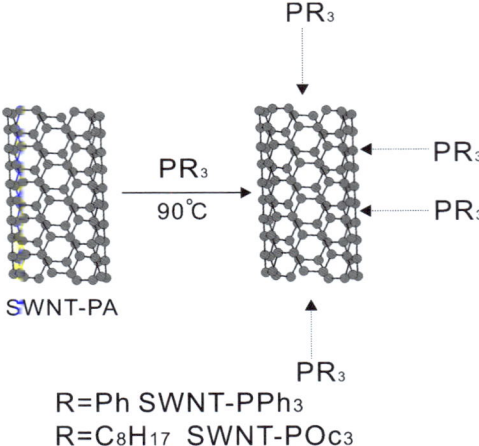

Figure 3.40 Phosphines add to carbon nanotubes like amines, but do not provide conducting electrons. © American Chemical Society.[100]

In this section we briefly describe some phosphides that may be, or have already been, applied in nanoparticle chemistry.

Boron phosphides activate molecular hydrogen at room temperature: an interaction with the Lewis acidic B centre is followed by intramolecular proton migration to phosphorus:[99]

$$BPh_3 + Pt - Bu_3 + H_2 \rightarrow [t - Bu_3 PH][BHPh_3]$$

Tertiary phosphines interact chemically with "electron-deficient" carbon nanotubes, presumably *via* the lone pair of electrons on the phosphorus, and dissolve them in dichloromethane, dimethylformamide, ethanol and 1,2-dichlorobenzene. Stable suspensions of ∼10 nm thick bundles are formed without any apparent change in the electronic structure of the nanotubes (Figure 3.40). Simple physisorption of the phosphine was ruled out, however, because the phosphines could not be removed by ultrasonication and washing with dichloromethane.[100] Dimethyl phosphine terminated alkanes had also the highest conductance between gold clusters. They bind and transfer electrons better than amines or sulfides.[101]

Metal phosphides are prepared either by hydrogenation of phosphates or by heating a metal acetylacetonate, *e.g.* Ni(II)(CH$_3$COCH–COCH$_3$)$_2$, in trioctylphosphine oxide to 300–360 °C. Nanoparticles of Ni$_2$P, PtP$_2$, Rh$_2$P, Au$_2$P$_3$, Pd$_5$P$_2$, and PdP$_2$ were prepared by the latter method, some of them polyphosphides with P–P bonds. Addition of dioctylether led to nanoparticles.

Nickel phosphide on silica was used as an efficient catalyst for the dechlorination of chlorobenzenes at 200 °C The Ni–P bond stabilizes the nickel 3d levels and a small Ni to P charge transfer allows a high activity for both the dissociation of the Ar–Cl bond and the addition of molecular hydrogen. The Ni–Cl bond is also weakened and the P sites provide only moderate bonding to the product HCl and the H adatoms necessary for hydrogenation.[102]

Colloidal indium phosphide quantum wires were grown by the "solution–liquid–solid" method, with indium myristate (In($C_{13}H_{27}COO$)$_3$) and trimethylsilylphosphine (P(SiMe$_3$)$_3$) as starting materials, solid bismuth nanoparticles as nucleation matrix and a polydecene melt as high-boiling solvent. Long-chain amines, phosphines and phosphinoxides helped to steer and to slow down the growth of wires and nanoparticles by selective "passivation" of surfaces. The optimal In/P precursor ratio was found to be 1.1–1.3, and 1-hexadecylamine and 1-octyl-phosphonic acid were essential to the production of indium phosphide wires. The amine stuck to the metal surface, favouring crystallinity and narrow-diameter distributions of the wires. The 1-octyl-phosphonic acid removed the amine from the surfaces, which grew to wires, and thereby quenched the growth of homogeneous nuclei. The optimal indium myristate/1-octyl-phosphonic acid ratio was found to be 1.5; more phosphonate disturbed the wire formation.

Indium phosphide quantum dots (QDs) were grown from bismuth nanoparticles in polydecene (boiling point >315 °C), again using P(SiMe$_3$)$_3$, In(myr)$_3$ and 1-hexadecylamine, but no octylphosphonic acid as a brake. Only the triple-chain phosphines tri-*n*-octylphosphine oxide and tri-*n*-octylphosphine were used to slow down growth. The bismuth nanoparticle/indium phosphide precursor ratio was also increased by 2–3 times, in order to limit the growth in length to rod size, and this also determined the shorter reaction time (Figure 3.41). Additional dioctadecylamine was needed to protect the bismuth nanoparticles from agglomeration at high temperatures.[103]

The 6 nm indium phosphide quantum wires were photochemically etched in the presence of hydrogen fluoride and air in order to show photoluminescence. Instead, however, the wires were photooxidized to yield amorphous indium phosphate wires with embedded dot- and rodlike domains of crystalline (residual) indium phosphide. Only these domains fluoresced. The wires were then redispersed in toluene witout hydrogen fluoride and allowed to stand under ambient light and atmospheric conditions. The photoluminescence of

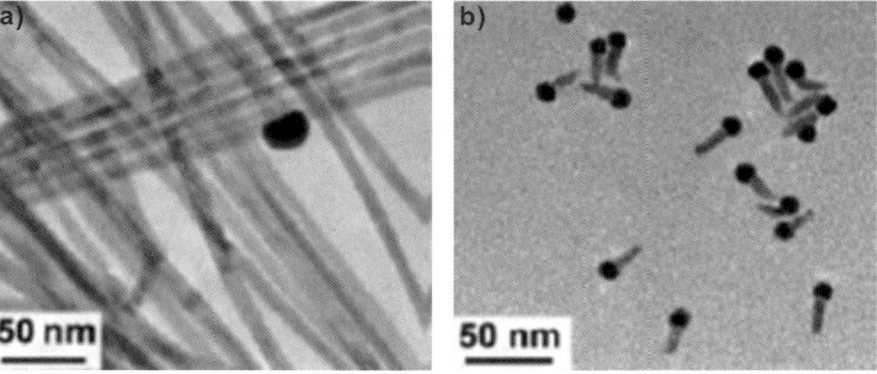

Figure 3.41 Synthesis and electron micrographs of indium phosphide wires and rods. © American Chemical Society.[104]

these isolated wires increased with time, but without additional blue-shifting. The sizes of the emitting indium phosphide domains no longer changed, but the interface between the amorphous matrix and the emitting domains changed and reduced the numbers of trap sites for photogenerated carriers.

Surface passivation routinely affords photoluminescence quantum yields of 60% in QDs, 15–30% in quantum rods, decreasing with rod length, and 0.1–2% in wires, where the third dimension of confinement vanishes completely and photogenerated electrons and holes diffuse along the wire and disappear in surface-trapping defects.[104]

Alumina-supported molybdenum phosphides (MoP) were found to be much more active catalysts for desulfuration and denitrogenation with molecular hydrogen than MoS_2 catalysts generally used. The 4.0–5.3 nm MoP/silica nanoparticles were stable over 100 h and displayed a high hydrogen adsorption capability. Their spherical shape was found to be especially stable in catalysis.[105]

3.7.3 Arsenic

Up to the 19th century compounds of arsenic were well established in medicine. They fell out of favour only after many fatal overdoses, accidental and otherwise. Arsenic oxide (As_2O_3) is tasteless and odourless, a perfect poison to administer in alcoholic drinks or coffee, but nowadays also very easy to detect by standard forensic tests. Arsine reacts with the SH groups of cysteine, but large doses (>100 mg) also block the respiration. Many deaths occurred, and it was soon replaced by penicillin. Nowadays arsenic has returned to favour as an anion in *in vivo* medicinal luminescent nanoparticles. Only cadmium is still avoided today as a possible fluorescence-enhancing component.

Gallium arsenide (GaAs) is one of the most commonly used materials for high electron mobility transistors (HEMTs), because it shows little lattice strain in alloys and can be operated at high frequency with much lower thermal noise than silicon-based devices. The water solubility of gallium arsenide prevents direct applications as sensors under biological conditions, but standard coating with organic thiol monolayers allowed the use of gallium arsenide electrodes and of indium arsenide QDs 10 nm away from water. The isolating character of the surface layer limits the measurements to surface potentials as indicated by capacitance changes. The ion current changes became measurable only after enzymatic degradation of the lipid monolayer by a phospholipase.[106]

Crystalline silicon wafers cannot be coated with III–V semiconductor layers by epitaxial growth. The massive mismatch between the lattice constants of the small silicon and the large chalcogenide atoms prevents this. Different thermal expansion and disparate crystal structures also disturb growth. With nanowires, however, such mismatches lose relevance because the matching surfaces are so small. Gallium arsenide nanowires could therefore be grown with high crystalline quality on silicon substrates.[108]

Indium arsenide (InAs) QDs emit in the near IR above 800 nm. Wires with diameters <2 nm containing zinc selenide (ZnSe) could be tuned to 750–920 nm, because ZnSe has only a small lattice mismatch of 6.44% with

InAs, but high bulk band offsets of 1.26 and 0.99 eV for the conduction and valence bands. The admixture of ZnSe thus increased the quantum yield of InAs by more than an order of magnitude. Longer emission wavelengths, particularly the biologically desirable 800–840 nm range, were achieved by increasing either the core size or the shell thickness. Both materials were water solubilized and exhibited stable emission in serum at 37 °C over 6 h. Quantum yields for (InAs)ZnSe (core)–shell QDs were 7–10% in hexane and 6–9% in water.

(InAs)ZnSe QDs were also water solubilized with dihydrolipoic acid and short poly(ethylene glycol) units. The coated QDs neither aggregated nor bound to proteins in rat serum. Their total diameter was only 8.7 nm, much smaller than commercial QDs used *in vivo*, which generally ranged from ~15 to 30 nm. Injected subcutaneously into the paw of a mouse or rat, the smaller QDs rapidly migrated to the lymph nodes and showed extravasation from the vasculature, which is the first step in tissue penetration. The coated (InAs)ZnSe QDs labelled and imaged up to five lymph nodes sequentially and were also visualized in the channels between the nodes. The exceptionally small size of the QDs allowed them to circulate for many minutes and migrate out of the blood vessels into the interstitial fluid.[108]

When indium arsenide nanowires were fixed on sharp tungsten support tips, a narrow electron beam with a small energy spread was emitted. This behaviour contrasts with the ususal unstable emission and large energy spread found for semiconductor emitters, and supports the concept of Fermi level pinning in indium arsenide nanowires. The indium arsenide nanowire presents a very special type of electron source, and quantum-confined states may occur when the emitters are externally cooled.[109]

3.7.4 Antimony

A direct antimonidization of gallium and indium started with molten droplets of the metals and antimony vapour. GaSb or InSb crystals grew on top of the droplets, which expanded to nanowires on appropriate surfaces. Vapor reactions between gallium and antimony, followed by wetting and and a tip-led growth, also gave GaSb nanowires with diameters from 30 to 700 nm and lengths up to hundreds of microns. The band gap of the GaSb nanowires was 0.72 eV.[110] Zinc antimonides Zn_4Sb_3 and ZnSb are among the best thermoelectric bulk materials between 150 and 400 °C, with a figure of merit of 1.3. These antimonides were prepared as nanocrystals in order to lower the thermal conductivity of thin films. The Zn_4Sb_3 nanocrystals were, however, much less thermostable than the bulk material and deteriorated to ZnSb at temperatures below 200 °C.[111]

3.7.5 Bismuth

Although bismuth is considered to be a metal and normally adopts positive oxidation states in oxides and halides, it undergoes very little electron transfer in intermetallic compounds and may also accept electron density from more electropositive elements to form polyanionic networks with extensive Bi–Bi bonding. In manganese bismuthates diverse manganese coordination geometries were found and Bi–Bi and Mn–Mn distances were often forced to be identical, because one set of distances was defined by matrix effects imposed by the other set.

Property measurements of rare earth manganese polybismuthates indicated metallic compounds which underwent magnetic ordering at modest temperatures.[112] Bismuth nanowires were predicted to exhibit a semimetal-to-semiconductor transition at a critical lateral width of 50 nm.[113]

3.8 Sulfur, Selenium and Tellurium

3.8.1 Sulfur

Sulfur is a remarkable element that is belched out by volcanoes and geysers as oxidized sulfur dioxide (SO_2) as well as reduced hydrogen sulfide (H_2S); neutral precipitates follow as S_8, among the volcanic ash. Hydrogen sulfide attacks noble metals such as copper, silver, mercury, gold, palladium and platinum and stays with them in air, because the metals react sluggishly and reversibly with molecular oxygen, a much stronger oxidant and even Cl_2.

The special behaviour of sulfur is obviously due to its softness: it forms highly covalent bonds with the noble metals and dissolves well in their metal lattices. The oxidation potential is not important, because the metals do not give away electrons – they just share them with sulfur. Nanolayers and nanocrystals of metal sulfides and selenides are interesting semiconducting materials and a wide variety of synthetic methods have been developed.[114]

Hydrodesulfurization of fluid, catalytically cracked gasoline converts thiols, sulfides, disulfides and thiophenes to hydrogen sulfide, and hydrocarbons ($RH:RS + H_2 \rightarrow RH + H_2S$). Raney Ni(0) or Co(0) usually cleaves the C–S bond, MoS_2 activates the H_2 molecule. The reaction takes place on coordinatively unsaturated sites of active metal catalysts such as CoMo/alumina or CoMo/silica. Three different *hydrogenation active sites* are differentiated here by technology, namely those for *n*-olefins, isoolefins and the sulfur compounds. The isoolefins and olefins may not only be hydrogenated, but may also be isomerized, which leads often to a "loss of octane": multibranched hydrocarbons rearrange to form less branched, more stable hydrocarbons. The CoMo/silica catalyst is, surprisingly, less acidic than the CoMo/alumina catalyst and therefore less active in the unwanted rearrangements. It was expected that the "basic" or electron-rich olefins would go to Lewis acid sites or acidic OH groups of the silicate, whereas the metal oxides would interact with MoS_2 *via* hydrogen bonds and be hydrogenated there. It was found, however, that the silicate catalyst had a lower hydrogenation activity than the CoMo/alumina catalyst. Minimizing octane loss (=isomerization), maximizing hydrodesulfurization (=sulfide reduction) and slower hydrogenation of olefins occurred together on the silicate support at 200 °C. At a slightly higher temperature of

240 °C, however, the hydrogen removed some of the of sulfur atoms, resulting in an increase in the number of coordinatively unsaturated sites. The CoMo/silica became more active, whereas the CoMo/alumina catalyst did not change. The alumina did not release sulfur. The commercial process operates with the aluminate catalyst at 300–425 °C.[115]

Lead sulfide (PbS) 6.0 nm nanoparticles coated with oleylamine exhibited sharp, symmetrical emission peaks at 1285 nm in toluene when excited at 831 nm and at 1209 nm in water where it moved after simple ligand exchange with poly(acrylic acid). The quantum yield was 82% in toluene in the presence of oleic acid and dropped to 24% in water.[116]

Cadmium sulfide (CdS) 6–8 nm nanocrystals photoluminesced at 510 nm and this hardly changed when the amine-coated nanocrystals were aligned along DNA. Dissolution in methylene dichloride yielded sevenfold enhanced electroluminescence due to the photogeneration of solvent radicals.[117]

3.8.2 Selenium

50 μm layers of selenium, Se(0), or, nowadays, arsenic triselenide (As_2Se_3) are the basis of xerography (Gr. *xeros*, dry, *graphein*, write) or photocopying. An electric field of 10^5 V cm^{-1} provides positive charges, and high light intensities carry electrons from the valence into the conductivity band of selenium. Graphite powder ("toner") sticks only to the positively charged zone and is transferred from there to the cellulose OH-groups of paper in the heat.

Cadmium selenide (CdSe) and zinc sulfide selenide ($ZnS_{1-x}Se_x$) are also photoconductive and are used in the photocells of warning signals or automatic door openers.

Zinc selenide (ZnSe) is a wide gap material operating in the blue-green colour region and the near IR in solar cells, light-emitting diodes and lasers. Sawlike, crystalline belts of zinc selenide with lengths of tens of microns and a thickness of 30–80 nm were prepared by thermal evaporation at 800–1000 °C.[118]

Cadmium selenide 3–6 nm nanocrystals were synthesized from dimethyl cadmium and *n*-octadecylphosphonic acid in tri-*n*-octylphosphine to form anhydrous Cd-*n*-octadecylphosphonate at 315 °C. ^{1}H NMR spectra of the nanocrystals in d_8-toluene showed broad resonances for methylene and methyl in the expected ratio of 17:1 representative of octadecyl chains. No sharp signals from "free" alkylphosphonates were detectable. Removal of the surface-bound ligands was achieved with bis(trimethysilyl)selenide and immediately yielded such sharp signals for the released ligands (Figure 3.42).[119] One-dimensional or "biaxial" Si–CdSe structures were obtained by a one-step, metal-catalysed thermal evaporation method and luminesced at 530 (Si) and 637 (CdSe) nm[120] without any mutual quenching.

Silver selenide (Ag_2Se) is, depending on its crystal structure, either a "super-ionic" (meaning conducting) solid-state electrolyte, where Ag^+ ions can diffuse freely, a mixed ionic conductor with a resistivity of 10^3 Ω cm, or an n-type semiconductor with a low resistivity of 10^{-3}–10^{-4} Ω cm. Transitions have been observed observed in crystalline silver selenide nanowires. The semiconducting

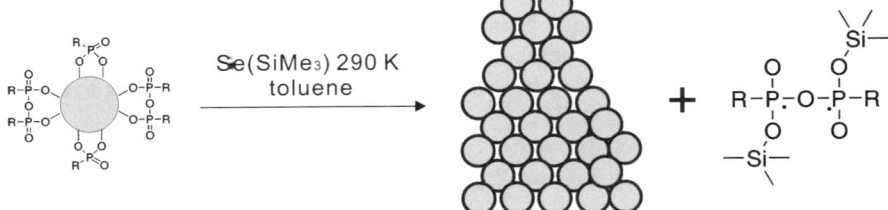

Figure 3.42 Model of the CdSe nanocrystals with surface-bound octadecylphosphonate (R = $C_{18}H_{37}$) ligands and their removal with bis(trimethysilyl)selenide leading to less toluene-soluble nanocrystals with Se(SiMe$_3$) surfaces.[119]

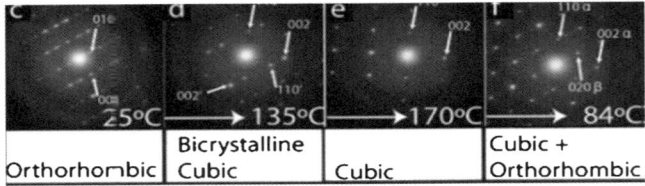

Figure 3.43 Electron microscopic images of Ag$_2$Se nanowires baked under resist at the indicated temperatures. © American Chemical Society.[121]

orthorhombic crystals of the nanowires rearranged to cubic superionic crystals at 135 °C (Figure 3.43). Upon cooling they remained stable at room temperature up to threshold potentials between 0.5 and 3 V, where recrystallization to semiconductors occurred. Only in nanostructures was the conducting cubic silver selenide structure stable at room temperature. The domain size in the wire was probably too small to induce a spontaneous rearrangement, and the surface energy of the nanowires may form too high a barrier.[121] Crystal structure analysis with an electron beam indicated that the silver ion migrated readily through the selenides, and many of the intermediate states crystallized immediately.

Group III monochalcogenides, *e.g.* gallium selenide (GaSe), have anisotropic electrical, optical, and mechanical properties due to strong intralayer covalent bonding. The interlayer van der Waals interactions are weak only, because there are no surface dangling bonds in chalcogenide nanolayers or nanowires except at the edges. Photoexcited antibonding states therefore do not weaken the metal–chalcogenide bonds and the crystals are soft, but photostable. Gallium selenide can therefore be used in solar cells and batteries. Layers of 1D nanowires with large surface areas for charge carrier separation and transport should be practical. Gallium selenide in particular crystallizes in hexagonal Se–Ga–Ga–Se sheets, each Ga(0) atom being tetragonally coordinated to three Se(0) atoms and one Ga(0) atom. The usual gold nanoparticles (p. 374) were used as matrices in the synthesis of gallium selenide nanowires and straight, zigzag, and saw-tooth morphologies were obtained in high yields under different conditions.[122]

Group IV monochalcogenides, in particular lead selenide (PbSe), have a narrow band gap of 0.28 eV in bulk and strong quantum confinement effects due to their large Bohr radius (46 nm for lead). Easy accessible lead selenide quantum dots (QDs) with photoluminescence from 1200 to 4000 nm are known and the quantum yields approach unity. This is especially useful in the telecom range of 1300–1550 nm. A carrier multiplication effect whereby a single, high-energy photon produces as many as seven electron–hole pairs will also be useful in solar cell technology. Lead selenide QDs are, however, sensitive to photooxidation by oxygen and need to be protected by lead sulfide and cadmium selenide surface layers. The QDs can be shaped by the addition of carbon nanotubes, temperature adjustments and long-chain amine and phosphinoxide concentrations. Their stereochemistry is extremely flexible and easy to handle. (p. 100): 15 different shapes were directly accessible.[123] Lead selenide photoluminesces in the IR.[124]

Silver selenide (Ag_2Se) is a mixed ionic conductor with a transition from a low-temperature orthorhombic phase to a high-temperature superionic conducting cubic phase at 135 °C. Silver selenide nanowires were produced by reacting selenium nanowires with silver nitrate: the Se(0) nanowires were simply grown from an amorphous Se(0) colloid in water. The crystalline nanowires varied in diameter from 40 to 200 nm, grew along the [001] direction, and showed high and low conducting phases corresponding again to the orthorhombic and cubic silver selenide crystal phases.[125]

3.8.3 Tellurium

Tellurium is one of the very few elements that combines with Au(0). Presumably for this reason, gold telluride (Au_2Te_3) is a major source of gold in the Earth's crust.

A photocopier drum with high red sensitivity and low residual potential after corona discharge may contain a thin Au(0) or Te(0) layer which injects holes into an upper, halogen-doped Se–Te alloy layer on an Al(0) substrate. Tellurium provides the long-wavelength sensitivity.[126]

Using cadmium telluride (CdTe) nanowires as an example, it was demonstrated that the electronic spectra of II–VI and III–V semiconductor nanowires can be calculated simply by a "particle in a cylinder" model, if the calculations are restricted to the unoccupied conduction-band energy levels. Valence-band electronic states are much more complicated because their electron states mix with each other, but the complex ground state can be neglected in analyses of excited nanowire states. These interactions simply lead to clusters of closely spaced transitions, which result only in broadening, not in shifts of absorption bands. Nevertheless, the observed absorption and emission bands are clearly grouped by energy spacings of the electron-accepting conduction band levels, and there is a close agreement between experimental spectra and those calculated solely from the conduction band levels (Figure 3.44). It became evident that the electrons in nanoparticles concentrate in dense clouds around the Fermi level. The "particle in a cylinder" model has become as simple and successful for semiconductor nanowires as the "electron in a box" for polyene

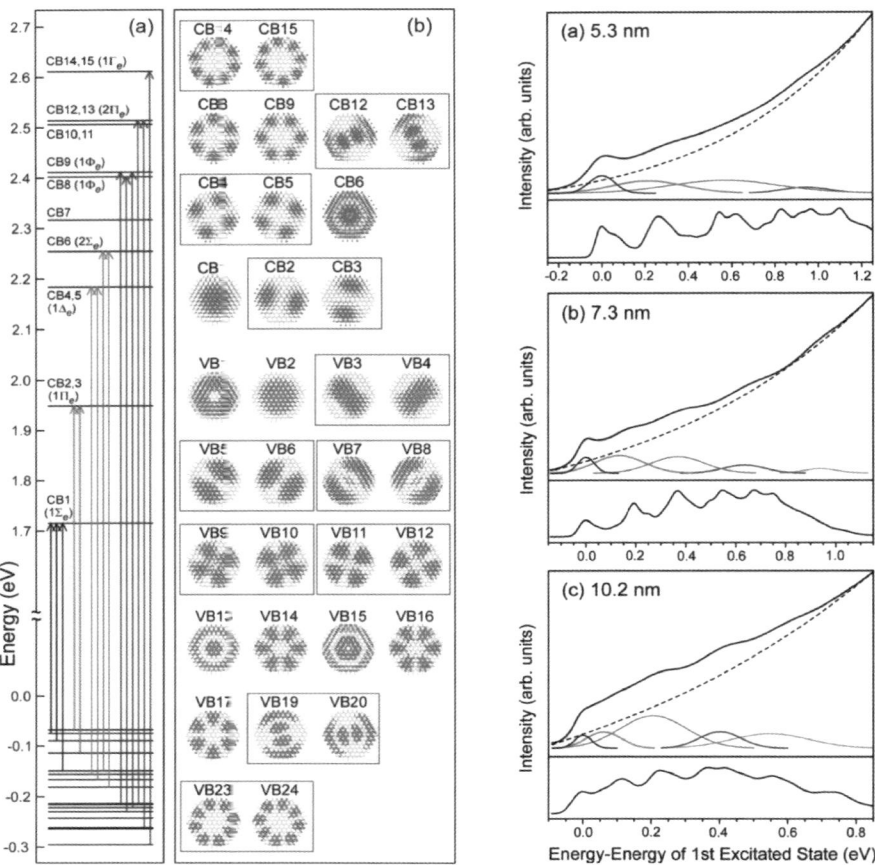

Figure 3.44 *Left:* Energy-level diagram for the 5.3 nm diameter wurtzite CdTe nanowire, with the transitions contributing to the broad absorption features identified by arrows matching the experimental spectra on the right for different sizes of nanowires and plots of the electron- and hole-density distributions for the conduction band (CB) states. Boxes indicate pairs of energetically degenerate levels. *Right:* States that do not make major contributions to the optical absorption are omitted. Experimental absorption spectra (upper panels) and calculated absorption spectra with average polarization (lower panels) of CdTe nanowires with diameters of (a) 5.3 nm, (b) 7.3 nm, and (c) 10.2 nm at 145 °C are shown. © American Chemical Society.[127]

and benzene-type molecules, although the semiconductor energy bands are, of course, not as well defined as molecular orbitals.[127,128]

Currently the alloy bismuth telluride (Bi_2Te_3) is used in most electrical cooling devices, because it has the thermal conductivity of a glass and an electric conductivity of 10^5 S m^{-1}. One alkali analogue, $CeBi_4Te_6$, was as good as bismuth telluride. Many others have been tried as well-characterized nanocrystals, but were not conductive enough. Among the various thermoelectric materials, bismuth telluride has a superior figure of merit of 1.14 for the p-type and 1.19 for the n-type material near room temperature.[129]

For lead telluride (PbTe) nanowire films, Seebeck coefficients of 628 μV K^{-1} were reported; bulk materials are lower by a factor of at least 3. No figures of merit are yet available for the nanowire films.

Ordered domain formation in thick films leading to enhancement of the Seebeck coefficient (which appears as a square), high electrical conductivity and low thermal conductivity are desperately needed in order to use electricity directly for cooling. This would enable us to get rid of the billions of electric pumps used in refrigerators, not to mention the problems of heat generation in computers.

Rewritable laser disks such as DVD-RAMs record information using changes of optical reflectivity, which is caused by the phase change of a material on the surface of a rotating disk. The most common commercial recording material are germanium telluride–antimony telluride (GeTe–Sb$_2$Te$_3$) films. Under nanosecond irradiation with a laser beam their crystalline NaCl-type phase switches to an amorphous phase, remains so after cooling and is converted back to the crystalline phase by heating to 145 °C. The NaCl-type crystal changes its phase so quickly because it is so simple and because it always contains an enormous number of vacancies. In IV–VI compounds with an NaCl-type structure, such as germanium telluride, it is expected that replacement of a group IV element by a group V element, *e.g.* antimony, stabilizes both metastable and crystalline phases over a wide composition range by its extra electron. The rewritable disk therefore remains in its crystalline or amorphous state for very long periods of time. Germanium telluride is the simplest prototype of such a crystalline/amorphous phase-change material. It has band gaps of 0.1 (0.8) eV and room temperature resistivities of 10^{-4} (10^3) Ω cm (Figure 3.45). The amorphous phase results from

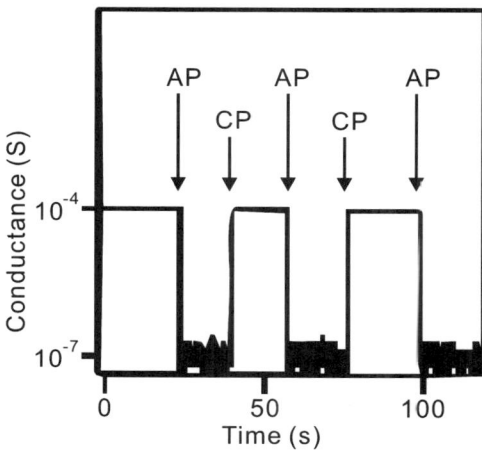

Figure 3.45 Conductance of amorphous and crystalline germanium telluride nanowires on application of voltage pulses (4 V and 500 ns and 1 V and 10 ms). © American Chemical Society.[131]

the rapid cooling of molten germanium telluride (10^{11} K s^{-1}), and can be converted back to the crystalline phase by heating to 145 °C[130,131]

A solution-phase synthesis of tin telluride (SnTe) nanocrystals started from bis[bis(trimethylsilyl)amino]tin(II) (Sn-[N(SiMe$_3$)$_2$]$_2$) and trioctylphosphine telluride in octadecene/oleylamine at 150 °C. Primary amines form strong complexes with Sn^{2+} ions, and small amounts of oleic acid were added to increase the nanocrystal nucleation rate. A larger number of nuclei was thus formed, yielding a smaller particle size, *e.g.* uniform diameters ~4.5 nm. Larger amounts of dry oleic acid were finally added to efficiently stabilize the nanocrystal surface with an oleic acid surface. The zero-dimensional nature of the quantum-confined electrons and holes in <10 nm nanocrystals leads to discrete energy levels. Defect-free nanocrystals also eliminate nonradiative exciton relaxation and the quantum yields of radiative recombinations approach unity. The optical band gaps of the tin telluride nanocrystals were close to 0.35 eV, which also makes them useful for solar cells, because coupling between such small QDs, arranged in layers, presumably allows efficient light harvesting.[132]

3.9 Fluorine

For organic fluorine compounds, see Chapter 2.13.

In nanobiochemistry, fluoride is important as a stabilizer of fibres in bones and teeth. Bone char is a mesoporous adsorbent with a specific surface area of 104 m g^{-1} and is mainly composed of hydroxyapatite (Ca$_5$OH(PO$_4$)$_3$). The concentration of the basic sites is higher than that of acidic sites and the adsorption capacity for F$^-$ increased drastically when the solution pH was lowered from 12 to 3 (Figure 3.46). The point of zero charge is pH 8.4; above

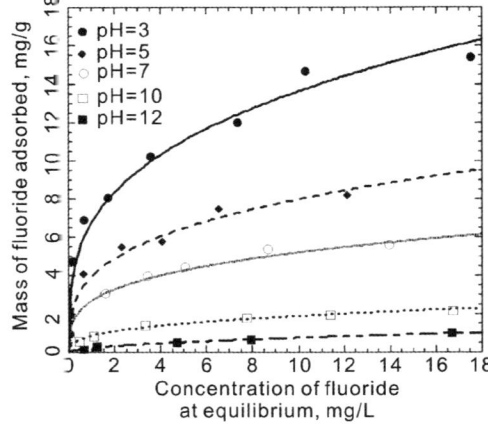

Figure 3.46 Fluoride/hydroxide exchange in the hydroxyapatite of bone char takes place only at low pH. © American Chemical Society.[133]

this pH the surface takes up less F^-. The adsorption capacity of bone char is independent of temperature since the adsorption occurs by ion exchange (F^- vs. OH^-) with a low heat of adsorption.[133]

Titania treated with hydrogen fluoride ($TiO_{(2-x)}F_x$) improves the photocatalytic performance of the bare oxide in the mineralization of various organic pollutants using UV and visible light. Electron paramagnetic resonance and theoretical calculations showed that insertion of fluoride into the titania matrix leads to Ti^{3+} reduced centres, which localize one electron in the t_{2g} orbitals of the metal without generating oxygen vacancies: fluoride anions rather substitute O^{2-} ions and yield bridging Ti–F–Ti bonds or Ti^{3+} species in an octahedral environment. On the titania surface the F^- ions at the solid surface substitute hydroxyl groups, yielding terminal Ti–F bonds without generating reduced centres. Anodic etching of titania in hydrogen fluoride produced a similar material, but also introduced pores on the surface. It was more efficient in the photoxidation of dyes.[134,135]

3.10 Zinc, Cadmium and Mercury

Zinc, cadmium and mercury are regarded as non-transition elements, because they have fully occupied d-orbitals and form no compounds in which the d shell is not full, they rarely exhibit multiple valence, they occur only in the +2 oxidation state, their tendency to form any d–π bonds to ligands is exceedingly low and no carbonyl or olefin compounds are known. In short, they behave like main group metals.

The crystal structure of zinc and cadmium metals shows strongly deformed hexagonal packings. The stable d electrons are tightly bound to the nucleus and metallic bonding is weak. Low density and melting points (Zn 420 °C, Cd 321 °C, Hg –39 °C) as well as low tensile strength follow from the weakness of their interactions with the outside world.

Zinc and cadmium react readily with oxygen to form 1:1 oxides ($E^0(M^{2+}/M^0)$ –0.76 and –0.40 V), whereas $HgO(E^0(M^{2+}/M^0) + 0.85$ V) is formed only above 300 °C. Cadmium and mercury also form monovalent salts. Mixed-valence M(I)/M(II) compounds are often yellow or green. Zn(II) ions are useful in organizing photoactive ligands without quenching their excited states. Zn(II) porphyrinates, for example, are as strongly fluorescent as Mg(II) porphyrinates.

Zinc, cadmium and mercury are used as dissolving electrodes in batteries. In zinc batteries the anodic zinc case envelops a graphite–manganese oxide cathode in a potassium hydroxide solution. Zinc is oxidized there to $Zn(OH)_2$, and $Mn(IV)O_2$ is reduced to Mn(III)OOH. OH^- is consumed at the anode and produced at the cathode. The battery potential remains constant until the oxidants and reductants are used up. In mercury cells the manganese oxide is replaced by mercuric oxide and the battery is much more compact. Nickel–cadmium (NiCd) batteries have the advantage of being rechargeable. Cadmium is oxidized to CdO, Ni(III)O(OH) is reduced to $Ni(OH)_2$. The insolubility of all three hydroxides is responsible for the successful back reaction and for charging memory effects in CdNi batteries: the reaction products stick to the surface of the electrodes, allowing for quick reverse reactions. Partially discharged batteries "remember" their oxide layers once formed and will never discharge any further. The remaining MnO_2 and $Cd(0)$ will always remain untouched. So these batteries have to be discharged completely every time.

3.10.1 Zinc

Zinc is the least poisonous of these three metals. It is an essential trace element for animals and also occurs in plants (beans, coffee, rice, oranges). More than 200 zinc enzymes are known, most of which depend on the rapid ligand exchange and easily distorted bond angles of the zinc tetrahedron. The flexible zinc sphere is usually involved in the condensation and splitting of esters and phosphoanhydrides.

Zinc oxide (ZnO, wurtzite) is an important wide band gap semiconductor, with a band gap energy of 3.37 eV. UV/blue lasing media depend on it. A standard preparation method is non-equilibrium processing by laser ablation in liquid medium. The applied surfactant concentration and the laser power must be adjusted in their competing capping and oxidation actions.

Zinc acetate $(Zn(OCOCH_3)_2)$ and sodium hydroxide in water were mixed at 65 °C to produce uniform zinc oxide nanoparticles. From the shifting edge of UV spectra toward lower energies after 3 min, it was clear that the growth of nanocrystals became strongly hindered. Increasing sodium hydroxide concentrations and acetate ions left on the nanocrystals attracted Na^+ ions and all three formed a "virtual" capping layer that hindered the transport of all reactants – the Zn^{2+}, acetate and hydroxyl ions – that formed the cap. Time-dependent UV spectra and X-ray scattering of the nanocrystals showed the expected fast growth only in the first 2–3 min. After that it slowed down by three orders of magnitude and crystal growth stopped at 6 nm; the nanoparticles were then "fully grown". Their saturated –Zn–O–Zn–O– network only fixated incoming OH^- ions and new nanocrystals built up from them and the Zn^{2+} acetate around then (Figure 3.47).[136]

A well defined Zn–ZnO core–shell nanoparticle was, for example, obtained from a zinc plate (99.99%) under an aqueous solution of 0.05 M sodium dodecyl sulfate with stirring and ablation at a distance of 4 mm with a 1064 nm Nd:YAG pulsed laser. After ablation the nanoparticles were centrifuged at 14000 rpm, rinsed several times with ethanol and dried. The zinc plasma was thus first produced at the solid–liquid interface, and cooling of the zinc plume region led immediately to zinc clusters. The final structure of the nanoparticles depended on the sodium dodecyl sulfate concentration (= protection) and the laser power (= oxidation).

If the concentration of the dodecyl sulfate was less than the critical micellar concentration (0.0008 m), the nanoparticles were mainly composed of wurtzite. When it was larger, metal zinc crystals grew and wurtzite decreased in X-ray

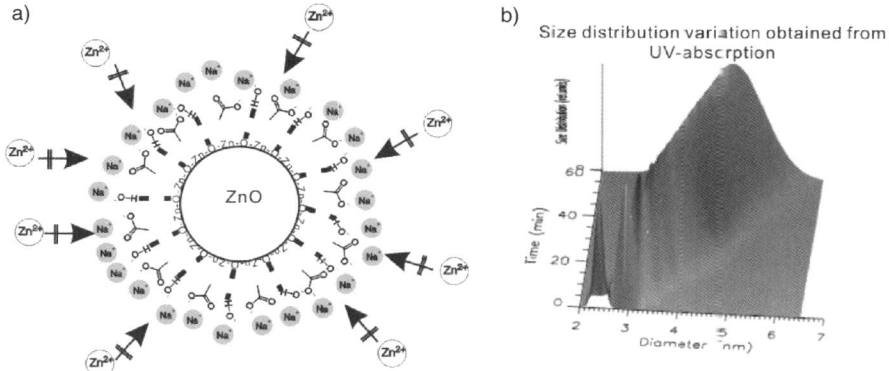

Figure 3.47 (a) Structure of the zinc oxide nanocrystal surface and the zinc acetate–hydroxide layer around it.[136] (b) Size distribution variation obtained from UV absorption.

diffraction patterns. Further, with a decrease of the applied laser power, the wurtzite shell thickness decreased to 2.5 nm, whereas the disorder degree increased. Above the critical micellar concentration of sodium dodecylsulfate the pure wurtzite nanocrystals changed to a Zn–ZnO core–shell structure. When the applied laser power was decreased the wurtzite shell thickness became as thin as 2.5 nm and its degree of disorder increased. The blue photoluminescence from the wurtzite nanoshells also showed an abnormal red shift first and then a blue shift when the temperature was raised from 8 to 300 K. These shifts were attributed to the extreme localization of single interstitial zinc atoms in the zinc oxide lattice of the nanoshells, an effect that was clearly due to the extreme heat in the laser ablation process[137] (Figure 3.48).

Arrays of 200 nm wurtzite nanorods in a porous polymethylmethacrylate mask on a thin gold film and a Kapton polyimide substrate were made by electron beam lithography at 70 °C (Figure 3.49). An applied potential at the gold film enhanced the nucleation process. No wurtzite seeds were required.[138]

Semiconductors in microprocessors process data at high speed and combined nanomagnets work as memories. Materials combining both capabilities would avoid the need for information transfer between semiconductors and magnets. One solution may be semiconductors containing magnetic impurities. Another, very surprising solution is capping semiconductors, *e.g.* wurtzite nanoparticles, with trioctylphosphine, octadecylamine or dodecylthiol. This may change their electronic structure in such a way as to induce ferromagnetic-like behaviour up to 300 K. Although the 4p level of isolated zinc atoms is empty, chemical bonding may lead to a charge transfer upon excitation with UV light, and electrons are pumped from the valence band to the conduction band and photoluminescence. The 2.3 eV emission (550 nm) is clearly observed for the trioctylphosphine sample, being weaker for the amine and absent in the thiol spectrum.[138] The magnetization curves of the samples exhibited the diamagnetic character of bulk wurtzite, but the amine- and thiol-capped nanoparticles

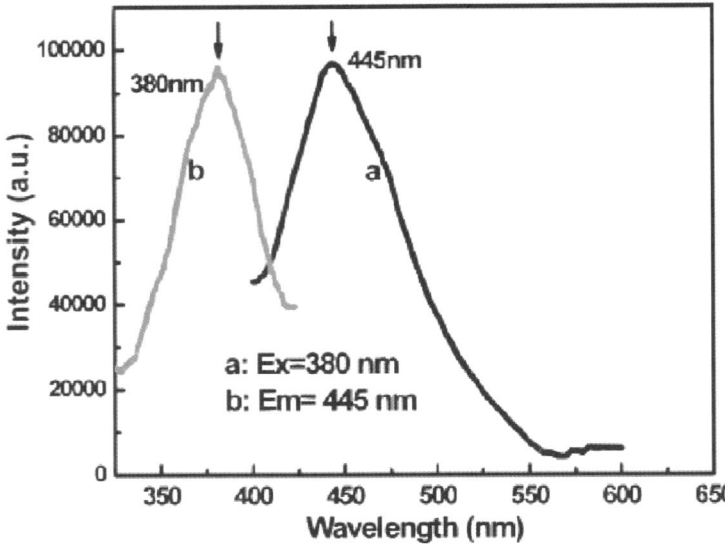

Figure 3.48 Photoluminescence excited at 380 nm and emitting at 445 nm of laser ablated ZnO/Zn nanocrystals. © American Chemical Society.[137]

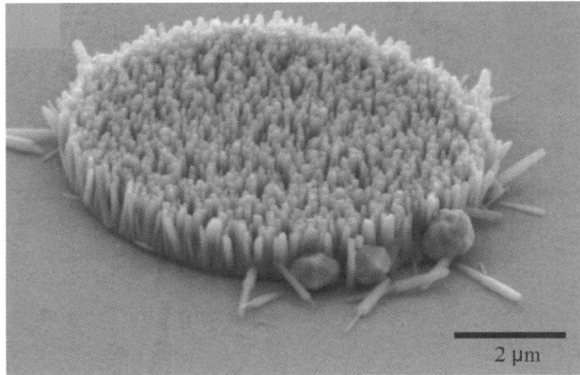

Figure 3.49 Scanning electron micrograph of a patterned zinc oxide nanorod array.[138]

showed a temperature-independent magnetism from 5 K to 300 K (Figure 3.50). The samples showed remanence, coercivity, and saturation behaviour similar to ferromagnetic materials. For an average nanoparticle size of 10 nm, the magnetic moment per surface atom was $2 \times 10^{-3}\,\mu B$ for the thiol and $0.5 \times 10^{-3}\,\mu B$ for the amine sample. Magnetic impurities were not responsible, because magnetic impurities dispersed in a non-magnetic matrix are paramagnetic and their magnetization decreases with temperature. The small magnetization of the

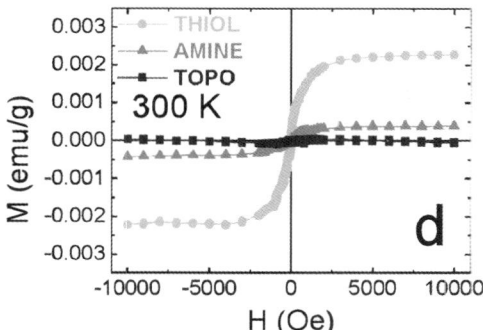

Figure 3.50 Magnetization curves for zinc oxide nanoparticles capped with different molecules (see text) after subtracting the diamagnetic/paramagnetic background. © American Chemical Society.[139]

amine probe was close to the detection limit of the superconducting quantum interference device (p. 326), but the sulfide coercivity was clearly measurable.[139]

3.10.2 Cadmium

Cd^{2+} is poisonous and has a retention time >10 years in the kidney. In kidney or liver amounts of 200 mg/kg are allowed: up to 3 mg in the liver and 50 mg in the kidney are normal values in a 50-year-old man. Kidney damage is the usual result of excess cadmium intake.[78] Like mecrury and lead, cadmium binds to the cysteine sulfide of proteins. It inhibits the uptake of iron in haem and calcium in bones.

Non-stoichiometric or defect semiconductors are salts with too many anions (p-conducting) or too many cations (n-conducting). The most common examples are II–VI compounds like zinc chalcogenides and the corresponding cadmium compounds. $Zn_xCd_{1-x}S$, for example, is a light emitter in television screens, which covers the whole visible spectrum: 430, 470, 490 520, 540, 580 and 610 nm and half-linewidths between 50 and 100 nm for $x = 0, 0.1, 0.2, 0.3, 0.4, 0.5$ and 0.6. The origin of this light emission are electron–hole pairs or excitons. Excess Zn^{2+} or Cd^{2+} ions act as the holes, which take up electrons from sulfur upon irradiation. The exciton formed is located in the conduction band, but moves freely in the crystal or nanoparticle. After nanoseconds the excited electron falls back into the valence band and light is emitted.[140,141] In ZnCdS nanoparticles, trapped electrons and holes are created on the crystal surface.

Reducing the diameter of QDs is also essential for *in vivo* applications, because nanoparticles >10 nm often do not pass through cell membranes, reach neural receptors or show renal clearance. The anionic micellar nanostructures in block copolymers, *e.g.* poly(sulfonated styrene-*block-tert*-butylstyrene), bind Cd^{2+} ions in clusters. Bubbling of H_2Se (from selenium and

NaBH$_4$) into the solution produced deep yellow cadmium selenide nanoparticles. Larger Cd^{2+}/SO$_3^-$ ratios gave larger aggregates in the block copolymer micelles and larger CdS/CdSe nanoparticles in the range of 1.5–4.5 nm upon reaction with S^{2-}/Se^{2-} anions (Figure 3.51). Cadmium sulfide nanoparticles were always larger than cadmium selenide nanoparticles, because the S^{2-} ions were made by a slow thermal decomposition of thioacetamide (NH$_2$COSH), giving time for the cadmium sulfide particles to grow, whereas the H$_2$Se gas employed for the synthesis of cadmium selenide reacted immediately.[142]

Solid cadmium nanoparticles, as obtained from dimethylcadmium (Cd(CH$_3$)$_2$) in trioctylphosphine oxide (OP(C$_8$H$_{17}$)$_3$) at 330 °C under nitrogen, were exposed to elemental sulfur gas. Time-dependent changes of the 350 nm nanoparticles at first produced a polycrystalline cadmium sulfide shell with a cadmium core, but Cd(0), Cd^{2+}, S and vacancies diffused quickly through the shell. Since cadmium was always faster than sulfur, it trapped all the sulfur and a hollow ball with a thick cadmium sulfide wall was the uniform result (Figure 3.52).[143,144]

Small, coated nanocrystals (3.7–4.1 nm) of cadmium sulfide were obtained from cadmium acetate (Cd(OAc)$_2$) and hydrogen sulfide in presence of poly(acrylic acid).[145] The photoluminescence occurred around 500 nm. The size distribution of the cadmium sulfide nanocrystals was fine-tuned by the length of the templating oligonucleotide sequences. Oligo(guanine)$_n$ with 5, 10, 15 and 20

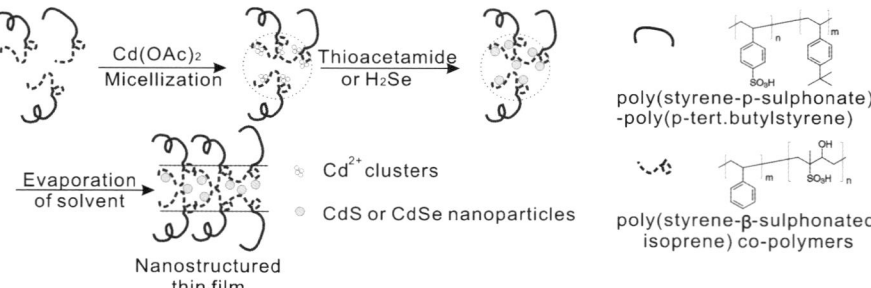

Figure 3.51 Synthesis of inverted CdS/CdSe micelles and formation of hydrated and isolated layers. © American Chemical Society.[142]

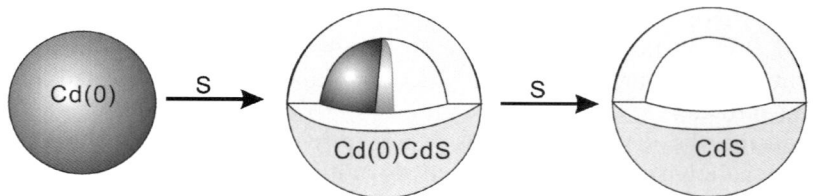

Figure 3.52 Coating of cadmium nanoparticles and removal of Cd(0) with sulfur vapour to yield cadmium sulfide only. © American Chemical Society.[144]

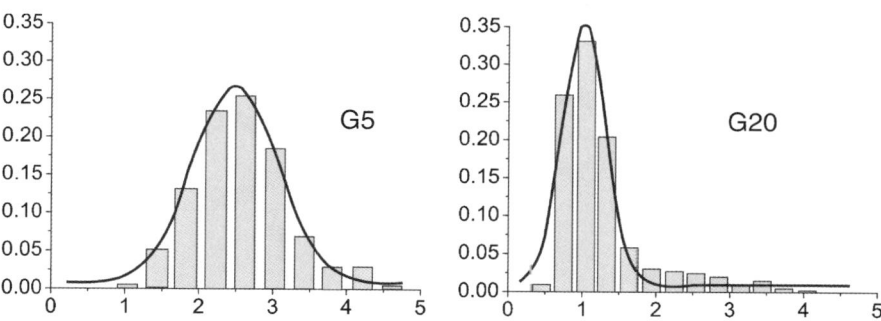

Figure 3.53 The guanine 20-mer leads to more uniform cadmium sulfide nanoparticles than the guanine 5-mer without making them larger. © American Chemical Society.[145]

nucleobases was used, because heavy metals, including Cd^{2+}, bind strongly to the NH-CO pairs of guanine. The surprising result was that the smallest oligomer gave rise to the largest nanoparticles, as seen under the atomic force microscope (Figure 3.53). Cadmium sulfide presumably aggregated quickly and statistically, but the large 20-mer fixated much of the Cd^{2+} immediately and blocked fast growth. Slow internal aggregation then gave small, uniform cadmium sulfide nanocrystals.[146]

Luminescent QDs are also attractive as donors in fluorescence resonance energy transfer (FRET). QD-bioconjugates are made by crosslinking with bifunctional linkers, *e.g.* click chemistry, direct thiolation or biotin–avidin interactions (see Chapter 2). Of all proteogenic amino acids, it is histidine that binds most tightly and selectively to CdSe(core)–ZnS(shell) QDs. Hexa (histidine), his$_6$, with a terminal SH-group, was, for example, connected with an SH-molecular beacon DNA to form a mixed disulfide that was subsequently reacted with a complementary DNA (Figure 3.54). A fluorescing rhodamine dye was also attached and it FRETted freely with a histidine-bound cyanine dye as well as with the QD before and after opening of the beacon DNA. A reduction of the FRET efficiencies from QD or cyanine dye to rhodamine from 70% to 35% was measured, but the overall interpretation remained unclear.[147] Carboxylates from dihydrolipoic acid, six histidine rings, DNA and complementary DNA, two dyes and a zinc sulfide surface are too complex to allow any predictions or conclusions.

Thiol-coated QDs that photoluminesce with good quantum yields are easily accessible by the same coating techniques, which were optimized on gold nanoparticles. Di sulfides are surprisingly effective directly on cadmium selenide surfaces; sulfur presumably substitutes selenium. The disulfide of the biradical 4-amino-2,2,6,6-tetramethylpiperidine oxide (TEMPO) is, for example, a much more efficient quencher of the QD fluorescence than TEMPO itself. In solution the biradical terminals interact twice and produce a five-line spectrum. On the cadmium selenide nanoparticles the S–S bond is cleaved and a clean triplet of NO˙ monomers appears. Cadmium selenide nanoparticles were also bound to SH groups of 16-mercaptohexadecanoic acid, which was

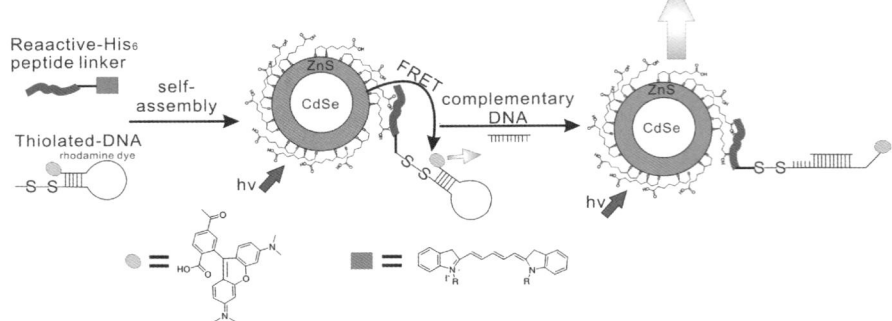

Figure 3.54 A His$_6$-peptide linker brings a dye-labelled S–S–DNA molecular beacon on to the QDs. With this hairpin DNA the acceptor dye is close to the QD and establishes acceptor FRET 1. DNA complementary to the molecular beacon detects the presence of its complementary sequence and unwinds it, changes the donor-acceptor distance to the dye and establishes FRETemission 2, this time from the QD. The measured effects were, however, small. © American Chemical Society.[147]

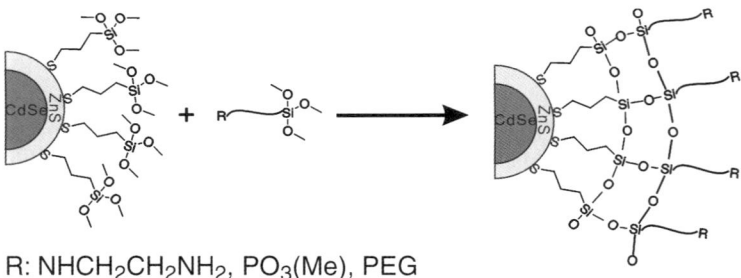

R: NHCH$_2$CH$_2$NH$_2$, PO$_3$(Me), PEG

Figure 3.55 Coating of <10 nm hydrophobic ZnS–CdSe nanoparticles with a thin silica multiplayer.[148]

adsorbed *via* the COOH group to nanocrystalline titania films. Here it was shown that it was not the S–S groups that reacted with cadmium selenide surfaces; first they had to be converted to SH. A CdSe–S bond has, however, not yet been characterized; presumably it is an –Se–S– or (–Se–S–S–Se–) bridge.

The hydrophobic ZnS–CdSe nanocrystals were also coated with a thin silica layer, in order to make them water-soluble and to stabilize and protect them. Trimethoxysilane or trihydroxysilane in toluene, low temperature and base were applied, which restricted the growth of thick silica shells. Silanization of nanoparticles <10 nm was most successful (Figure 3.55); nanoparticles >15 nm often precipitated. The water solubility of amine-, PEG-, and phosphonate-functionalized nanoparticles was 10–100 mg/mL, depending on their size, pH and ionic strength. Amine–phosphonate, amine–dimethylamine, and amine–PEG bifunctionalized nanoparticles were water-soluble and stable over a wide range of pH values.[148]

8 nm CdSe/CdS nanoparticles with organic coatings and NH_2, COOH and ethylene glycol surfaces had an overall hydrodynamic size of ~40 nm and perfused human skin within a few hours. The COOH-coated nanoparticles were deposited in large quantities on the inner surfaces of arteries.[149]

Cadmium telluride (CdTe) nanoparticles also show a strong, pH-dependent photoluminescence. It is strongest at pH 4.5 in the presence of thioglycolic acid ($HSCH_2COOH$) or mercaptoundecanoic acid ($HS(CH_2)_{17}COOH$), where sulfur presumably replaces tellurium at the surface and the fixated carboxyl groups provide water solubility (Figure 3.56). Such cadmium telluride nanocrystals may replace polynuclear aromatic dyes such as acridine derivatives or Methylene Blue as a photosensitizer for diphenyliodonium ($I^+(C_6H_5)_2$)-initiated cationic polymerizations.[150] This is triggered by the efficient photoinduced electron transfer between the nanocrystal and a diphenyliodonium in the same way as with dyes, but the nanocrystal did not bleach and was removed from the solution by centrifugation.

Tellurium powder was added to a suspension of sodium borohydride ($NaBH_4$) in water at 0 °C, warmed to room temperature and left standing for 8 h. The black tellurium powder had disappeared, white sodium tetraborate had precipitated and NaHTe was dissolved in the clear supernatant. Cadmium chloride ($CdCl_2$) was added at pH 9.0 in the presence of mercaptocarboxylic acids as a stabilizing agent. The molar ratio of Cd^{2+}/stabilizer/tellurium was 1:2.4:0.5. The resulting mixture was refluxed and 2.8–4.0 nm cadmium telluride nanoparticles were kept in aqueous stock solutions with the mercaptocarboxylic acids and with or without added poly(acrylic acid).

Cadmium telluride QDs induce cell death of human breast cancer MCF-7 cells *via* mechanisms involving both Cd^{2+} and reactive oxygen species, *e.g.* singlet oxygen, which are formed by cadmium ions and light. CdSe/ZnS QDs were

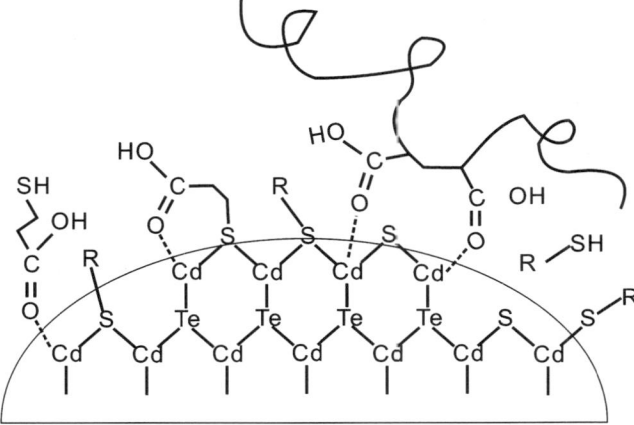

Figure 3.56 Cadmium telluride nanoparticles coated with mercaptocarboxylic acids, RSH, and polymeric hydrocarbons with a few carboxyl groups as anchors.[150]

non-toxic, but it seems that cadmium telluride is not stable in the biological environment. Biocompatible, fluorescent 6 nm CdSe–ZnCdS core–shell QDs with a DL-cysteine coat were injected intravenously into rats. The QDs accumulated in the bladder after 4 h and gel chromatography of the urine confirmed renal clearance. Photoluminescence was not detected in the animals. Similar CdSe–ZnS nanoparticles are the QDs with the longest lasting and most stable signals. They are therefore suitable for *in vivo* experiments and possibly for routine use in diagnosis and phototherapy. Characterized by their maximal emission wavelength, 565, 655 and 705 nm QDs can be distinguished: their excitation occurs slightly above 500, 600 or 615 nm. But how can one possibly reach a QD buried in a living organism with visible light, and how can one measure its emissions?[151–153] Some 705 nm QDs were intracortically injected in the brain of a living mouse and were detected immediately as well as at 1 h, 24 h, 72 h and 7 days after the injection (Figure 3.57). Astrocyte activation, a marker for inflammatory responses, could be detected in the living animals using bioluminescent imaging and a high-resolution CCD camera. *In vivo* light emission revealed the spatial and temporal expression of the selected gene promoters in the living mice.[151]

Colloidal cadmium telluride QDs are also useful because their large exciton Bohr radius (7.3 nm), the calculated distance between an electron in the conduction band and the hole it leaves behind in the valence band, allows for very small nanocrystals with continuous energy bands and semiconducting properties. A bulk band gap of only 1.475 eV ($\sim$12 000 cm^{-1} or 800 nm) renders it also optically active in the visible and near-IR region.

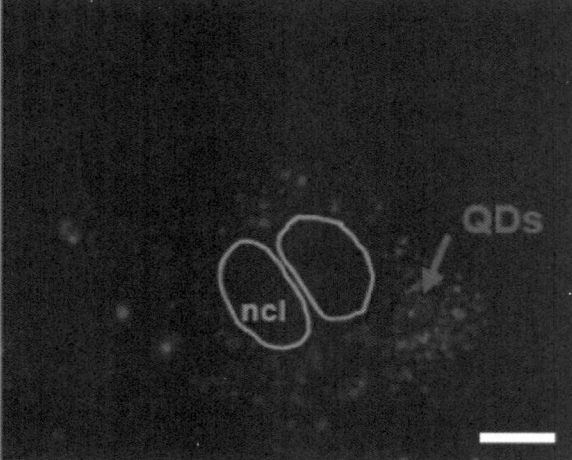

Figure 3.57 Representative pseudocolour fluorescence and bioluminescence images of PEGylated QD 705 injected in the right parietal cortex of the mouse, 1.5 mm below the surface of the skull. The luminescence indicates the locations of astrocytes, spider-like brain cells, which regulate the water supply of neurons. The scale bar is 10 μm (10 000 nm). The light signals are much larger than the QDs. © American Chemical Society.[151]

Cadmium telluride QD films are therefore used in photovoltaic cells and light-emitting diodes, as fluorescent labels for living cells and for the encoding of combinatorial libraries. The nanocrystals can be made, for example, from cadmium oxide (CdO), which was reduced at 310 °C by octadecene to Cd(0), and tellurium powder, which was solubilized by trioctylphosphine. The nanocrystals obtained were much smaller than the Bohr radius and produced emission bands between 2.31 and 1.98 eV as compared to 1.475 eV for bulk cadmium telluride. These nanocrystals absorbed in the visible region (2 eV or ~625 nm) (Figure 3.58).[154]

A protein from the *Listeria* bacterium contains a 4.5 nm biotemplate, which stores ~500 iron atoms (p. 272ff) as a ferric oxyhydroxide nanoparticle. The iron oxide was removed and the *Listeria* protein stored overnight in a zinc acetate–selenide or ammonium–cadmium acetate–thioacetic acid mixture in water. The expected 4 nm zinc selenide or cadmium sulfide nanoparticles were formed, showed photoluminescence at 600 nm and were, of course, biocompatible (Figure 3.59).[155,156]

3.10.3 Mercury

Mercury, once known as "quicksilver" (from the German *Quecksilber*) is the most dangerous of all metals and it is certainly not a desirable constituent of

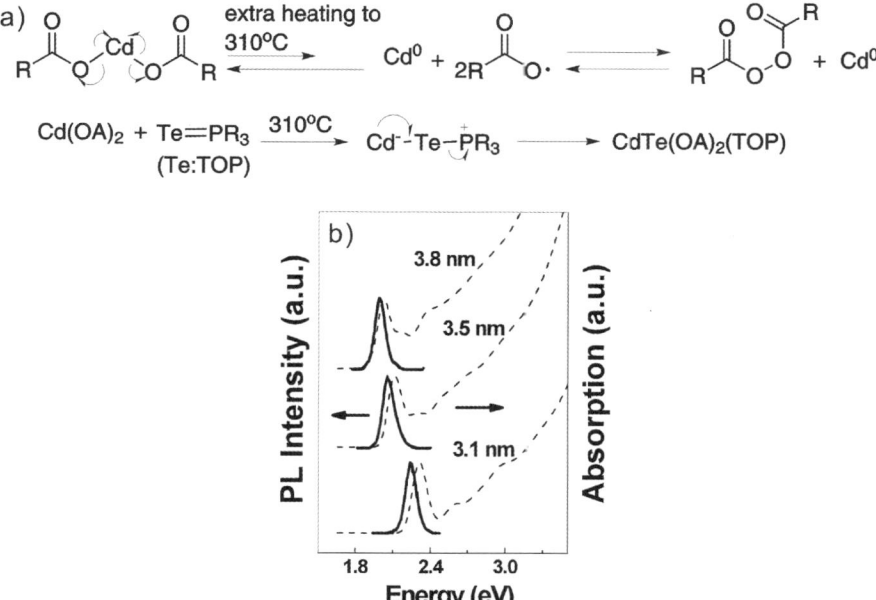

Figure 3.58 Absorbance (dashed) and photoluminescence spectra (solid) of cadmium telluride QDs with diameters between 3.1 and 3.8 nm. © American Chemical Society.[154]

Figure 3.59 Stepwise filling of a ferritin cavity: Zn^{2+}, Se^{2-}, and waiting.[155,156]

any nanoparticle intended for medical use. A few biomedical details of this unique metal are given here, because accidents with thermometers and batteries are common and because mercury is the only way to provide organisms with metallic sensitivity.

Hg(0) is hydrophobic and dissolves atomically in hydrocarbons (4×10^{-6} M or ~ 1 mg/L). It moves very rapidly into cells, diffusing like a gas over the hydrophobic blood–brain and placenta barriers and passing through cell membranes. Everywhere it goes, it adds to the cysteine SH groups of proteins and blocks them. Kidney, brain and testicles are the organs most damaged, and the mercury sulfides essentially never leave the body – the half-life in these organs is ~ 18 years. Mercury is also methylated by the body to $Hg(CH_3)_2$, which is even more poisonous than free Hg(0). A blast of highly concentrated Hg(0) is much more dangerous than small amounts over a long period, because the red blood cells oxidize Hg(0) to Hg(II) and excrete it into the urine.

Dimethylmercury ($Hg(CH_3)_2$) was an NMR standard until a few years ago, when a chemistry professor wearing latex gloves was killed by it in 1997. The organic Hg(0) passes through such gloves within 15 s.[157]

An even more frightening effect of mercury poisoning is electrosensibilization. People with mercury poisoning wake up in the night if the refrigerator in the kitchen switches on; computers and supermarket tills hurt them electrically, and they cannot stand to touch metals. Life is to terrifying that they are always at high risk of suicide. Hg(0) poisoning is the major origin of this electronic illness.[78,158,159]

Carboxylated peptide-functionalized Au(0) nanoparticles produce networks of 1D chains in the presence of Hg^{2+}. The colour of the aqueous suspension changes from red to purple to blue (670 nm). Even 10^{-6} M concentrations of Hg^{2+} are thus detectable in water by the naked eye.[160]

Model neurons were excited by action potentials coming from photocurrents generated by mercury telluride nanoparticle films. Light quanta were thus directly converted to nerve impulses in the neuronal cells and communicated to the outside system or to other neurons by nanoscale engineering.[161] A Hg(0) film may well contribute here.

References

1. (a) W. Lubitz and W. Tumas, Hydrogen: an overview, *Chem. Rev.*, 2007, **107**, 3900–3903 (Thereafter articles on: H_2 binding and reactivity on transition metals, Hydrogen production by molecular photocatalysis,

High-temperature solar hydrogen production, Membranes for hydrogen separation, Complex hydrides for hydrogen storage, Hydrogen storage in molecular clathrates, and other hydrogen themes); (b) C. I. Smith, H. Miyaoka, T. Ichikawa, M. O. Jones, J. Harmer, W. Ishida, P. P. Edwards, Y. Kojima and H. Fuji, Electron spin resonance investigation of hydrogen absorption in ball-milled graphite, *J. Phys. Chem. C*, 2009, **113**, 5409–5416.
2. (a) P. Vidossich, S. Piana, A. Miani and P. Carloni, Deuterium isotope effects in A:T and A:U base pairs: a computational NMR study, *J. Am. Chem. Soc.*, 2006, **128**, 7215–7221; (b) G. L. Olsen, D. C. Echodu, Z. Shajani, M. F. Bardaro, G. Varani and G. P. Drobny, Solid-state deuterium NMR Studies reveal μs-ns motions in the HIV-1 transactivation response RNA recognition site, *J. Am. Chem. Soc.*, 2008, **130**, 2896–2897.
3. M. Vold, Colloidal structure in Li stearate greases, *J. Am. Chem. Soc.*, 1956, **60**, 439–442.
4. H.-L. Zhang, F. Li, C. Liu and H.-M. Cheng, Poly(vinyl chloride) (PVC) coated idea revisited: influence of carbonization procedures on PVC-coated natural graphite as anode materials for lithium ion batteries, *J. Phys. Chem. C*, 2008, **112**, 7767–7772.
5. B. J. Landi, M. J. Ganter, C. M. Schauerman, C. D. Cress and R. P. Raffaelle, Lithium ion capacity of single wall carbon nanotube paper electrodes, *J. Phys. Chem. C*, 2008, **112**, 7509–7515.
6. (a) E. Tillmanns and W. H. Bauer, Crystal structures of phosphate and vanadate salts, *Acta Cryst. B*, 1971, **27**, 2124–2132; (b) E. Philippot and O. Lindquist, Crystal structure of $KH_5(PO_4)_2$, *Acta. Chem. Scand.*, 1971, **25**, 512–522.
7. B. Jagoda-Cwiklik, R. Vacha, M. Lund, M. Srebro and P. Jungwirth, Ion pairing as a possible clue for discriminating between sodium and potassium in biological and other complex environments, *J. Phys. Chem. Lett. C*, 2007, **111**, 14077–14079.
8. C. Tegenkamp and H. Pfnuer, Contact charging. An atomistic model, *Kali und Steinsalz*, 2004, **1**, 26–35.
9. M. Futai, Y. Wada and J. H. Kaplan, *Handbook of ATPases*, 2003, Wiley-VCH, Weinheim.
10. G. K. Fukin, S. V. Lindeman and J. K. Kochi, Molecular structures of cation (arene) interactions for alkali metals, *J. Am. Chem. Soc.*, 2002, **124**, 8329–8336.
11. A. Cordom, O. Edholm and J. J. Perez, Effect of force field parameters on sodium and potassium ion binding to dipalmitoyl phosphatidylcholine bilayers, *J. Chem. Theor. Comp.*, 2009, **1**, in print.
12. B. Roux, Computational studies of the gramicidin channel, *Acc. Chem. Res.*, 2002, **35**, 366–375 and references therein.
13. Z. Liu, Y. Xu and P. Tang, Steered molecular dynamics simulations of Na^+ Permeation across the gramicidin A channel, *J. Phys. Chem. B*, 2006, **110**, 12789–12795.

14. F. M. Menger, A. L. Galloway, M. E. Chlebowski and S. Wu, Sodium ion internalized within phospholipid membranes, *J. Am. Chem. Soc.*, 2006, **128**, 14034–14035.
15. F. I. Valiyaveetil, M. Sekedat, R. MacKinnon and T. W. Muir, Structural and functional consequences of an amide-to-ester substitution in the selectivity filter of a potassium channel, *J. Am. Chem. Soc.*, 2006, **128**, 11591–11599.
16. D. J. Kyle and V. I. Ilyin, Sodium channel blockers, *J. Med. Chem.*, 2007, **50**, 2583–2588.
17. W. Li, C. Li, H. Ma and J. Chen, Magnesium nanowires: enhanced kinetics for hydrogen absorption and desorption, *J. Am. Chem. Soc.*, 2007, **129**, 6710–6711.
18. A. N. Kulak, P. Iddon, Y. Li, S. P. Armes, H. Colfen, O. Paris, R. M. Wilson and F. C. Meldrum, Continuous structural evolution of calcium carbonate particles: a unifying model of copolymer-mediated crystallization, *J. Am. Chem. Soc.*, 2007, **129**, 3729–3736.
19. K. Tai, F.-J. Ulm and C. Ortiz, Nanogranular origins of the strength of bone, *Nano Lett.*, 2006, **6**, 2520–2525.
20. S. Weiner and H. D. Wagner, The material bone: structure-mechanical function relations, *Ann. Rev. Mater. Sci.*, 1998, **28**, 271–298.
21. G. E. Fantner, T. Hassenkam, J. H. Kindt, G. Weaver, H. Birkedal, L. Pechenik, J. A. Cutroni, A. Cidade, G. Stucky, D. E. Morse and H. G. Hansma, Sacrificial bonds and hidden length dissipate energy as mineralized fibrils separate during bone fracture, *Nat. Mater.*, 2005, **4**, 612–616.
22. K. Noris-Suarez, J. Lira-Olivares, A. M. Ferreira, J. L. Feijoo, N. Suarez, M. C. Hernandez and E. Barrios, In vitro deposition of hydroxyapatite on cortical bone collagen stimulated by deformation-induced piezoelectricity, *Biomacromolecules*, 2007, **8**, 941–948.
23. E. R. Wise, S. Maltsev, M. E. Davies, M. J. Duer, C. Jaeger, N. Loveridge, R. C. Murray and D. G. Reid, The organic-mineral interface in bone is predominantly polysaccharide, *Chem. Mater.*, 2007, **19**, 5055–5057.
24. D. G. Reid, M. J. Duer, R. C. Murray and E. R. Wise, The organic-mineral interface in teeth is like that in bone and dominated by polysaccharides: universal mediators of normal calcium phosphate biomineralization in vertebrates?, *Chem. Mater.*, 2008, **20**, 3549–3550.
25. S. Busch, U. Schwarz and R. Kniep, Morphogenesis and structure of human teeth in relation to biomimetically grown fluorapatite-gelatine composites, *Chem. Mater.*, 2001, **13**, 3260–3271.
26. S. Mukherjee, Y. Song and E. Oldfield, NMR Investigations of the static and dynamic structures of bisphosphonates on human bone: a molecular model, *J. Am. Chem. Soc.*, 2008, **130**, 1264–1273.
27. R. S. Ehrick, M. Capaccio, D. A. Puleo and L. G. Bachas, Ligand-modified aminobisphosphonate for linking proteins to hydroxyapatite and bone surface, *Bioconjugate Chem.*, 2008, **19**, 315–321.

28. J. Lott, C. Xia, L. Kosnosky, C. Weder and J. Shan, Terahertz photonic crystals based on barium titanate/polymer nanocomposites, *Adv. Mater.*, 2008, **20**, 3577–3586.
29. S. K. Pal, M. E. Itkis, F. S. Tham, R. W. Reed, R. T. Oakley, B. Donnadieu and R. C. Haddon, Phenalenyl-based neutral radical molecular conductors: substituent effects on solid-state structures and properties, *J. Am. Chem. Soc.*, 2007, **129**, 7163–7174.
30. M. Elbing and G. C. Bazan, Design organischer optoelektronischer Materialien durch laterale Borylsubstitution, *Angew. Chem.*, 2008, **120**, 846–850.
31. M. W. Mortensen, O. Bjorkdahl, P. G. Sørensen, T. Hansen, M. R. Jensen, H. J. G. Gundersen and T. Bjørnholm, Functionalization and cellular uptake of boron carbide nanoparticles. The first step toward T cell-guided boron neutron capture therapy, *Bioconjugate Chem.*, 2006, **17**, 284–290.
32. U. Anselmi-Tamburini, M. Ohyanagi and Z. A. Munir, Modeling studies of the effect of twins on the X-ray diffraction patterns of boron carbide, *Chem. Mater.*, 2004, **16**, 4347–4351.
33. K. Nielsch, J. Choi, K. Schwirn, R. B. Wehrspohn and U. Gosele, Self-ordering regimes of porous alumina: the 10% porosity rule, *Nano Lett.*, 2002, **2**, 677–681.
34. W. Chen, J. Yuan and X. Xia, Characterization and manipulation of the electroosmotic flow in porous anodic alumina membranes, *Anal. Chem.*, 2005, **77**, 8102–8108.
35. J. M. Baik, M. Schierhorn and M. Moskovits, Fe nanowires in nanoporous alumina: geometric effect versus influence of pore walls, *Phys. Chem. Lett. C*, 2008, **112**, 2252–2255.
36. J.-T. Chen, M. Zhang and T. P. Russell, Instabilities in nanoporous media between adjacent holes, *Nano Lett.*, 2007, **7**, 183–187.
37. Y. Qin, S.-M. Lee, A. Pan, U. Golsele and M. Knez, Rayleigh-instability-induced metal nanoparticle chains encapsulated in nanotubes produced by atomic layer deposition, *Nano Lett.*, 2008, **8**, 114–118.
38. H. Peng, S. Meister, C. K. Chan, X. F. Zhang and Y. Cui, Morphology control of layer-structured gallium selenide nanowires, *Nano Lett.*, 2007, **7**, 199–203.
39. F. M. Davidson, A. D. Schricker, R. J. Wiacek and B. A. Korgel, Supercritical fluid-liquid-solid synthesis of GaAs Nanowires seeded by alkanethiol-stabilized gold nanocrystals, *Adv. Mater.*, 2004, **16**, 646–649.
40. N. H. Chou, X. Ke, P. Schiffer and R. E. Schaak, Room-temperature chemical synthesis of shape-controlled indium nanoparticles, *J. Am. Chem. Soc.*, 2008, **130**, 8140–8141.
41. www.matsci.northwestern.edu/faculty/links/tom/gabyresearch.htm.
42. M. Brumbach, P. A. Veneman, F. S. Marrikar, T. Schulmeyer, A. Simmonds, W. Xia, P. Lee and N. R. Armstrong, Surface composition and electrical and electrochemical properties of freshly deposited and acid-etched indium tin oxide electrodes, *Langmuir*, 2007, **23**, 11089–11099.

43. A. Murali, A. Barve, V. J. Leppert, S. H. Risbud, I. M. Kennedy and H. W. H. Lee, Synthesis and characterization of indium oxide nanoparticles, *Nano Lett.*, 2001, **1**, 181–183.
44. J. Ni, H. Yan, A. Wang, Y. Yang, C. L. Stern, A. W. Metz, S. Jin, L. Wang, T. J. Marks, J. R. Ireland and C. R. Kannewurf, MOCVD-derived highly transparent, conductive zinc- and tin-doped indium oxide thin films: precursor synthesis, metastable phase film growth and characterization, and application as anodes in polymer light-emitting diodes, *J. Am. Chem. Soc.*, 2005, **127**, 5613–5624.
45. R. Schlapak, D. Armitage, N. Saucedo-Zeni, M. Hohage and S. Howorka, Dense passivating poly(ethylene glycol) films on indium tin oxide substrates, *Langmuir*, 2007, **23**, 10244–10253.
46. O. A. Raitman, E. Katz, A. F. Buckmann and I. Willner, Integration of polyaniline/poly(acrylic acid) films and redox enzymes on electrode supports: an in situ electrochemical/ surface plasmon resonance study of the bioelectrocatalyzed oxidation of glucose or lactate in the integrated bioelectrocatalytic systems, *J. Am. Chem. Soc.*, 2002, **124**, 6487–6496.
47. C. Ge, N. R. Armstrong and S. S. Saavedra, pH-Sensing properties of poly(aniline) ultrathin films self-assembled on indium-tin oxide, *Anal. Chem.*, 2007, **79**, 1401–1410.
48. M. A. McGuire, T. K. Reynolds and F. J. DiSalvo, Exploring thallium compounds as thermoelectric materials: seventeen new thallium chalcogenides, *Chem. Mater.*, 2005, **17**, 2875–2884.
49. J. R. Stork, M. M. Olmstead and A. L. Balch, Polymorphs with varying platinum(II)-thallium(I) interactions, *J. Am. Chem. Soc.*, 2005, **127**, 6512–6513.
50. S. Park, B. Kim, O. Yavuzcetin, M. T. Tuominen and T. P. Russell, Ordering of PS-*b*-P4VP on patterned silicon surfaces, *ACS Nano*, 2008, **2**, 1363–1370.
51. N. Gozlan and H. Haick, Coverage effect of self-assembled polar molecules on the surface energetics of silicon, *Phys. Chem. J. Lett. C*, 2008, missing.
52. (a) A. Lugstein, M. Steinmair, Y. J. Hyun, G. Hauer, P. Pongratz and E. Bertagnolli, Pressure-induced orientation control of the growth of epitaxial silicon nanowires, *Nano Lett.*, 2008, 2310–2314; (b) A. Lugstein, M. Steinmair, C. Henkel, E. Bertagnolli, Sculable approach for vertical device integration of epitaxial nanowires, *Nano Lett.*, 2009, **9**, 1830–1834.
53. A. Colli, A. Fasoli, C. Ronning, S. Pisana, S. Piscanec and A. C. Ferrari, Ion beam doping of silicon nanowires, *Nano Lett.*, 2008, 2188–2193.
54. E. C. Garnett and P. Yang, Silicon nanowire radial p-n junction solar cells, *J. Am. Chem. Soc.*, 2008, **130**, 9224–9225.
55. X. Chen, R. Ji, M. Steinhart, A. Milenin, K. Nielsch and U. Gosele, Aligned horizontal silica nanochannels by oxidative self-sealing of patterned silicon wafers, *Chem. Mater.*, 2007, **19**, 3–5.
56. R. D. Badley, W. T. Ford, F. J. McEnroe and R. A. Assinks, Surface modification of colloidal silica, *Langmuir*, 1990, **6**, 792–801.

57. G. Li, S. V. Bhosale, T. Wang, S. Hackbarth, B. Roeder, U. Siggel and J.-H. Fuhrhop, Nanowells on silica particles in water containing long-distance porphyrin heterodimers, *J. Am. Chem. Soc.*, 2003, **125**, 10693–10702.
58. J. Spadavecchia, P. Prete, N. Lovergine, L. Tapfer and R. Rella, Gold nanoparticles prepared by physical method on Si and sapphire substrates for biosensor applications, *J. Phys. Chem. B*, 2005, **109**, 17347–17349.
59. C. Konn, F. Morel, E. Beyou, P. Chaumont and E. Bourgeat-Lami, Nitroxide-mediated polymerization of styrene initiated from the surface of laponite clay platelets, *Macromolecules*, 2007, **40**, 7464–7472.
60. H. Sertchook, H. Elimelech, C. Makarov, R. Khalfin, Y. Cohen, M. Shuster, F. Babonneau and D. Avnir, Composite particles of polyethylene @ silica, *J. Am. Chem. Soc.*, 2007, **129**, 98–108.
61. L. Hong, Shan Jiang and S. Granick, Simple method to produce Janus colloidal particles in large quantity, *Langmuir*, 2006, **22**, 9495–9499.
62. C. Minkowski and G. Calzaferri, Forster-type energy transfer along a specified axis, *Angew. Chem.*, 2005, **117**, 5459–5463.
63. M. Busby, C. Blum, M. Tibben, S. Fibikar, G. Calzaferri, V. Subramanian and L. de Cola, Time, space, and spectrally resolved studies on j-aggregate interactions in Zeolite L Nanochannels, *J. Am. Chem. Soc.*, 2008, **130**, 10970–10976.
64. H. W. Chiu and S. M. Kauzlarich, Investigation of reaction conditions for optimal germanium nanoparticle production by a simple reduction route, *Chem. Mater.*, 2006, **18**, 1023–1028.
65. N. Zaitseva, Z. R. Dai, C. D. Grant, J. Harper and C. Saw, Germanium nanocrystals synthesized in high-boiling-point organic solvents, *Chem. Mater.*, 2007, **19**, 5174–5178.
66. H. Yang, X. Yao, X. Wang, S. Xie, Y. Fang, S. Liu and X. Gu, Sol-gel preparation and photoluminescence of size controlled germanium nanoparticles embedded in a silica matrix, *J. Phys. Chem. B*, 2003, **107**, 13319–13322.
67. M. Aizawa and J. M. Buriak, Block copolymer templated chemistry for the formation of metallic nanoparticle arrays on semiconductor surfaces, *Chem. Mater.*, 2007, **19**, 5174–5178.
68. S. Mukherjee, K. Kim and S. Nair, Short, Highly ordered, single-walled mixed-oxide nanotubes assemble from amorphous nanoparticles, *J. Am. Chem. Soc.*, 2007, **129**, 6820–6826.
69. T. M. Davis, M. A. Snyder and M. Tsapatsis, Germania nanoparticles and nanocrystals at room temperature in water and aqueous lysine sols, *Langmuir*, 2007, **23**, 12469–12472.
70. R. G. Shimmin, R. Vajtai, R. W. Siegel and P. V. Braun, Room-temperature assembly of germanium photonic crystals through colloidal crystal templating, *Chem. Mater.*, 2007, **19**, 2102–2107.

71. J. D. Rimer, D. D. Roth, D. G. Vlachos and R. F. Lobo, Self-assembly and phase behavior of germanium oxide nanoparticles in basic aqueous solutions, *Langmuir*, 2007, **23**, 2784–2791.
72. M. Luo and M. Wuttig, The dependence of crystal structure of Te-based phase change materials on the number of valence electrons, *Adv. Mater.*, 2004, **16**, 439.
73. T. Matsunaga, R. Kojima, N. Yamada, K. Kifune, Y. Kubota, Y. Tabata and M. Takata, Single structure widely distributed in a GeTeSb$_2$Te$_3$ pseudobinary system: a rock salt structure is retained by intrinsically containing an enormous number of vacancies within its crystal, *Inorg. Chem.*, 2006, **45**, 2235–2241.
74. Y. Jung, S.-H. Lee, D.-K. Ko and R. Agarwal, Synthesis and characterization of Ge$_2$Sb$_2$Te$_5$ nanowires with memory switching effect, *J. Am. Chem. Soc.*, 2006, **128**, 14026–14027.
75. J. S. Lee, S. Brittman, D. Yu and H. Park, Vapor–liquid–solid and vapor–solid growth of phase-change Sb$_2$Te$_3$ nanowires and Sb$_2$Te$_3$/GeTe nanowire heterostructures, *J. Am. Chem. Soc.*, 2008, **130**, 6252–6258.
76. B. Tan, E. Toman, Y. Li and Y. Wu, Zinc stannate (Zn$_2$SnO$_4$) dye-sensitized solar cells, *J. Am. Chem. Soc.*, 2007, **129**, 4162–4163.
77. M. V. Kovalenko, W. Heiss, E. V. Shevchenko, J.-S. Lee, H. Schwinghammer, A. P. Alivisatos and D. V. Talapin, SnTe nanocrystals: a new example of narrow-gap semiconductor quantum dots, *J. Am. Chem. Soc.*, 2007, **129**, 11354–11355.
78. M. Daunderer, *Metallvergiftungen*, ecomed, Landsberg, 1995.
79. D. Barredo, F. Calleja, P. Nieto, J. J. Hinarejos, G. Laurent, A. L. Vazquez de Parga, D. Farias and R. Miranda, A quantum-stabilized mirror for atoms, *Adv. Mater.*, 2008, **20**, 3492–3497.
80. M. Veith, J. Freres, P. Koenig, O. Schuett, V. Huch and J. Blin, Nanoscaled Sn and Pb particles aligned in Al$_2$O$_3$ tubes obtained from molecular precursors, *Eur. J. Inorg. Chem.*, 2005, **18**, 3699–3710.
81. R. Y. Wang, J. P. Feser, J. S. Lee, D. V. Talapin, R. Segalman and A. Majumdar, Enhanced thermopower in PbSe nanocrystal quantum dot superlattices, *Nano Lett.*, 2008, **8**, 2283–2288.
82. M. Terrones, J.-C. Charlier, A. Gloter, E. Cruz-Silva, E. Terrés, Y. B. Li, A. Vinu, Z. Zanolli, J. M. Dominguez, H. Terrones, Y. Bando and D. Golberg, Experimental and theoretical studies suggesting the possibility of metallic boron nitride edges in porous nanourchins, *Nano Lett.*, 2008, **8**, 1026–1032.
83. R. Arenal, O. Stephan, J.-L. Cochon and A. Loiseau, Root-growth mechanism for single-walled boron nitride nanotubes in laser vaporization technique, *J. Am. Chem. Soc.*, 2007, **129**, 16183–16189.
84. V. Barone and J. E. Peralta, Magnetic boron nitride nanoribbons with tunable electronic properties, *Nano Lett.*, 2008, **8**, 2210–2214.
85. C.-H. Park and S. G. Louie, Energy gaps and Stark effect in boron nitride nanoribbons, *Nano Lett.*, 2008, **8**, 2000–2003.

86. X. Bai, D. Golberg, Y. Bando, C. Zhi, C. Tang, M. Mitome and K. Kurashima, Deformation-driven electrical transport of individual boron nitride nanotubes, *Nano Lett.*, 2007, **7**, 632–637.
87. P. Kowalczyk, P. A. Gauden and A. P. Terzyk, Cryogenic separation of hydrogen isotopes in single-walled carbon and boron-nitride nanotubes: insight into the mechanism of equilibrium quantum sieving in quasi-one-dimensional pores, *J. Phys. Chem. B*, 2008, **112**, 8275–8284.
88. S. Hao, G. Zhou, W. Duan, J. Wu and B.-L. Gu, Transverse pressure induced phase transitions in boron nitride nanotube bundles and the lightest boron nitride crystal, *J. Am. Chem. Soc.*, 2008, **130**, 5257–5261.
89. (a) C. Y. Won and N. R. Aluru, Water permeation through a subnanometer boron nitride nanotube, *J. Am. Chem. Soc.*, 2007, **129**, 2748–2749; (b) C. Y. Won and N. R. Aluru, Structure and dynamics of water confined in a boron nitride nanotube, *J. Phys. Chem. C*, 2008, **112**, 1812–1818.
90. W. Wang, Y. Bando, C. Zhi, W. Fu, E. Wang and D. Golberg, Aqueous noncovalent functionalization and controlled near-surface carbon doping of multiwalled boron nitride nanotubes, *J. Am. Chem. Soc.*, 2008, **130**, 8144–8145.
91. Y. Taniyasu, M. Kasu and T. Makimoto, An aluminium nitride light-emitting diode with a wavelength of 210 nanometres, *Nature*, 2006, **441**, 325–328.
92. J. Zheng, Y. Yang, B. Yu, X. Song and X. Li, [0001] Oriented aluminum nitride one-dimensional nanostructures: synthesis, structure evolution, and electrical properties, *ACS Nano*, 2008, 2,134–142.
93. K. Maeda, X. Wang, Y. Nishihara, D. Lu, M. Antonietti and K. Domen, Photocatalytic activities of graphitic carbon nitride powder for water reduction and oxidation under visible light, *J. Phys. Chem. C.*, 2009, **113**, 4940–4947.
94. Y. Shi, Y. Wan, B. Tu, D. Zhao, Nanocasting synthesis of ordered mesoporous silicon nitrides with a high nitrogen content, *J. Phys. Chem. C.*, 2008, **112**, 112–116.
95. K. Maeda, N. Saito, Y. Inoul, K. Domen, Dependence of activity and stability of germanium nitride powder for photocatalytic water splitting on structural properties, *Chem. Mater.*, 2007, **19**, 4092–4097.
96. S. Stevenson, M. C. Thompson, H. L. Coumbe, M. A. Mackey, C. E. Coumbe and J. P. Phillips, Chemically adjusting plasma temperature, energy, and reactivity (CAPTEAR) method using NOx and combustion for selective synthesis of $Sc_3N@C80$ metallic nitride fullerenes, *J. Am. Chem. Soc.*, 2007, **129**, 16257–16262.
97. J. McMurran, D. Dai, K. Balasubramanian, C. Steffek and J. Kouvetakis, H_2GaN_3 and derivatives: a facile method to gallium nitride, *Inorg. Chem.*, 1998, **37**, 6638–6644.
98. K. Inumaru, T. Nishikawa, K. Nakamura and S. Yamanaka, High-pressure synthesis of superconducting molybdenum nitride δ-MoN by in situ nitridation, *Chem. Mater.*, 2008, **20**, 4756–4761.

99. G. C. Welch and D. W. Stephan, Facile heterolytic cleavage of dihydrogen by phosphines and boranes, *J. Am. Chem. Soc.*, 2007, **129**, 1880–1881.
100. H. Suri, K. Chakraborty and K. S. Coleman, A facile, solvent-free, noncovalent, and nondisruptive route to functionalize single-wall carbon nanotubes using tertiary phosphines, *Chem. Mater.*, 2008, **20**, 1705–1709.
101. Y. S. Park, A. C. Whalley, M. Kamenetska, M. L. Steigerwald, M. S. Hybertsen, C. Nuckolls and L. Venkataraman, Contact chemistry and single-molecule conductance: a comparison of phosphines, methyl sulfides, and amines, *J. Am. Chem. Soc.*, 2007, **129**, 15768–15769.
102. (a) X. Liu, J. Chen and J. Zhang, Hydrodechlorination of chlorobenzene over silica-supported nickel phosphide catalysts, *Ind. Eng. Chem. Res.*, 2008, **47**, 5362–5368; (b) A. E. Henkes, Y. Vasquez and R. E. Schaak, Converting metals into phosphides: a general strategy for the synthesis of metal phosphide nanocrystals, *J. Am. Chem. Soc.*, 2007, **129**, 1896–1897.
103. F. Wang, H. Yu, J. Li, O. Q. Hang, D. Zemlyanov, P. C. Gibbons, L. Wang, D. Wang, B. Janes and W. E. Buhro, Spectroscopic properties of colloidal indium phosphide quantum wires, *J. Am. Chem. Soc.*, 2007, **129**, 14327–14335.
104. F. Wang and W. E. Buhro, Determination of the rod-wire transition length in colloidal, indium phosphide quantum rods, *J. Am. Chem. Soc.*, 2007, **129**, 14381–14387.
105. A. Montesinos-Castellanos, T. A. Zepeda, B. Pawelec, J. L. G. Fierro and J. A. de los Reyes, Preparation, characterization, and performance of alumina-supported nanostructured Mo-phosphide systems, *Chem. Mater.*, 2007, **19**, 5627–5636.
106. D. Gassull, A. Ulman, M. Grunze and M. Tanaka, Electrochemical sensing of membrane potential and enzyme function using gallium arsenide electrodes functionalized with supported membranes, *J. Phys. Chem. B*, 2008, **112**, 5736–5741.
107. S.-G. Ihn, J.-I. Song, T.-W. Kim, D.-S. Leem, T. Lee, S.-G. Lee, E. K. Koh and K. Song, Morphology- and orientation-controlled gallium arsenide nanowires on silicon substrates, *Nano Lett.*, 2007, **7**, 39–44.
108. J. P. Zimmer, S. W. Kim, S. Ohnishi, E. Tanaka, J. V. Frangioni and M. G. Bawendi, Size series of small indium arsenide-zinc selenide core-shell nanocrystals and their application to in vivo imaging, *J. Am. Chem. Soc.*, 2006, **128**, 2526–2527.
109. E. C. Heeres, E. P. A. M. Bakkers, A. L. Roest, M. Kaiser, T. H. Oosterkamp and N. de Jonge, Electron Emission from Individual Indium Arsenide Semiconductor Nanowires, *Nano Lett.*, 2007, **7**, 536–540.
110. S. Vaddiraju, M. K. Sunkara, A. H. Chin, C. Z. Ning, G. R. Dholakia and M. Meyyappan, Synthesis of group III antimonide nanowires, *J. Phys. Chem. C*, 2007, **111**, 7339–7347.

111. S. Schlecht, C. Erk and M. Yosef, Nanoscale zinc antimonides: synthesis and phase stability, *Inorg. Chem.*, 2006, **45**, 1693–1697.
112. O. Y. Zelinska and A. Mar, Ternary rare-earth manganese bismuthides: structures and physical properties of $(La-Nd)_3 MnBi_5$ and $Sm_2Mn_3Bi_6$, *Inorg. Chem.*, 2008, **47**, 297–305.
113. A. Boukai, K. Xu and J. Heath, Size-Dependent transport and thermoelectric properties of individual polycrystalline bismuth nanowires, *Adv. Mater.*, 2006, **18**, 864–869.
114. J. J. Vittal and M. T. Ng, Chemistry of metal thio- and selenocarboxylates: precursors for metal sulfide/selenide materials, thin films, and nanocrystals, *Acc. Chem. Res.*, 2006, **39**, 869–877.
115. T. Mochizuki, H. Itou, M. Toba, Y. Miki and Y. Yoshimura, Effects of acidic properties on the catalytic performance of CoMo sulfide catalysts in selective hydrodesulfurization of gasoline fractions, *Energy & Fuels*, 2008, **22**, 1456–1462.
116. W. Lin, K. Fritz, G. Guerin, G. R. Bardajee, S. Hinds, V. Sukhovatkin, E. H. Sargent, G. D. Scholes and M. A. Winnik, Highly luminescent lead sulfide nanocrystals in organic solvents and water through ligand exchange with poly(acrylic acid), *Langmuir*, 2008, **24**, 8215–8219.
117. C. Ge, M. Xu, J. Fang, J. Lei and H. Ju, Luminescent cadmium sulfide nanochains templated on unfixed deoxyribonucleic acid and their fractal alignment by droplet dewetting, *J. Phys. Chem. C*, 2008, **112**, 10602–1060.
118. S. Y. Liu, W. C. H. Choy, L. Jin, Y. P. Leung, G. P. Zheng, J. Wang and A. K. Soh, Triple-crystal zinc selenide nanobelts, *J. Phys. Chem. C*, 2007, **111**, 9055–9059.
119. J. S. Owen, J. Park, P.-E. Trudeau and A. P. Alivisatos, Reaction chemistry and ligand exchange at cadmium-selenide nanocrystal surfaces, *J. Am. Chem. Soc.*, 2008, **130**, 12279–12231.
120. X. H. Sun, T. K. Sham, R. A. Rosenberg and G. K. Shenoy, One-dimensional silicon-cadmium selenide heterostructures, *J. Phys. Chem. C*, 2007, **111**, 8475–8482.
121. D. T. Schoen, C. Xie and Y. Cui, Electrical switching and phase transformation in silver selenide nanowires, *J. Am. Chem. Soc.*, 2007, **129**, 4116–4117.
122. H. Peng, S. Meister, C. K. Chan, X. F. Zhang and Y. Cui, Morphology Control of Layer-Structured Gallium Selenide Nanowires, *NanoLett.*, 2007, **7**, 199–203.
123. Y. J. Na, H. S. Kim and J. Park, Morphology-controlled lead selenide nanocrystals and their in situ growth on carbon nanotubes, *J. Phys. Chem. C*, 2008, **112**, 11218–11226.
124. J. M. Pietryga, D. J. Werder, D. J. Williams, D. L. Casson, R. D. Schaller, V. I. Klimov and J. A. Hollingsworth, Utilizing the lability of lead selenide to produce heterostructured nanocrystals with bright, stable infrared emission, *J. Am. Chem. Soc.*, 2008, **130**, 4379–4885.

125. D. T. Schoen, C. Xie and Y. Cui, Electrical switching and phase transformation in silver selenide nanowires, *J. Am. Chem. Soc.*, 2007, **129**, 4116–411.
126. H. Baumgartner, K. Ebner, P. Elsasser and P. Kersten, Brit. Pat. Appl. 1981.
127. J. Sun, W. E. Buhro, L.-W. Wang and J. Schrier, Electronic structure and spectroscopy of cadmium telluride quantum wires, *Nano Lett.*, 2008, **8**, 2913–2919.
128. G. Tai, B. Zhou and W. Guo, Structural characterization and thermoelectric transport properties of uniform single-crystalline lead telluride nanowires, *J. Phys. Chem. C*, 2008, **112**, 11314–11318.
129. S. Li, H. M. A. Soliman, J. Zhou, M. S. Toprak, M. Muhammed, D. Platzek, P. Ziolkowski and E. Müller, Effects of Annealing and Doping on Nanostructured Bismuth Telluride Thick Films, *Chem. Mater.*, 2008, **20**, 4403–4410.
130. T. Matsunaga, R. Kojima, N. Yamada, K. Kifune, Y. Kubota, Y. Tabata and M. Takata, Single structure widely distributed in a $GeTe-Sb_2Te_3$ pseudobinary system: a rock salt structure is retained by intrinsically containing an enormous number of vacancies within its crystal, *Inorg. Chem.*, 2006, **45**, 2235–2241.
131. D. Yu, J. Wu, Q. Gu and H. Park, Germanium telluride nanowires and nanohelices with memory-switching behavior, *J. Am. Chem. Soc.*, 2006, **128**, 8148–8149.
132. M. V. Kovalenko, W. Heiss, E. V. Shevchenko, J. S. Lee, H. Schwinghammer, A. P. Alivisatos and D. V. Talapin, SnTe nanocrystals: a new example of narrow-gap semiconductor quantum dots, *J. Am. Chem. Soc.*, 2007, **129**, 11354–11355.
133. N. A. Medellin-Castillo, R. Leyva-Ramos, R. Ocampo-Perez, R. F. Garcia de la Cruz, A. Aragon-Pin, J. M. Martinez-Rosales, R. M. Guerrero-Coronado and L. Fuentes-Rubio, Adsorption of fluoride from water solution on bone char, *Ind. Eng. Chem. Res.*, 2007, **46**, 9205–9212.
134. A. M. Czoska, S. Livraghi, M. Chiesa, E. Giamello, S. Agnoli, G. Granozzi, E. Finazzi, C. Di Valentin and G. Pacchioni, The nature of defects in fluorine-doped TiO_2, *J. Phys. Chem. C*, 2008, **112**, 8951–8956.
135. X. F. Cheng, W. H. Leng, D. P. Liu, Y. M. Xu, J. Q. Zhang and C. N. Cao, Electrochemical preparation and characterization of surface-fluorinated TiO_2 nanoporous film and its enhanced photoelectrochemical and photocatalytic properties, *J. Phys. Chem. C*, 2008, **112**, 8725–8734.
136. A. S. Ratkovich, R. L. Penn, Controlling nanosized ZnO growth kinetics using various Zn:OH concentration ratios, *J. Phys. Chem. C*, 2007, **111**, 14098–14104.
137. H. Zeng, Z. Li, W. Cai, B. Cao, P. Liu and S. Yang, Microstructure control of Zn/ZnO core/shell nanoparticles and their temperature-dependent blue emissions, *J. Phys. Chem. B*, 2007, **111**, 14311–14317.
138. B. Weintraub, Y. Deng and Z. L. Wang, Position-controlled seedless growth of ZnO nanorod arrays on a polymer substrate via wet chemical synthesis, *Phys. Chem. C Lett.*, 2007, **111**, 10162–10165.

139. M. A. Garcia, J. M. Merino, E. F. Pinel, A. Quesada, J. de la Venta, M. L. Ruz Gonzalez, G. R. Castro, P. Crespo, J. Llopis, J. M. Gonzlez-Calbet and A. Hernando, Magnetic properties of ZnO nanoparticles, *Nano Lett.*, 2007, **7**, 1489–1494.
140. M. Furis, J. A. Hollingsworth, V. I. Klimov and S. A. Crooker, Time- and polarization-resolved optical spectroscopy of colloidal CdSe nanocrystal quantum dots in high magnetic fields, *J. Phys. Chem. B*, 2005, **109**, 15332–15338.
141. W. Liu, H. S. Choi, J. P. Zimmer, E. Tanaka, J. V. Frangioni and M. Bawendi, Compact cysteine-coated CdSe(ZnCdS) quantum dots for in vivo applications, *J. Am. Chem. Soc.*, 2007, **129**, 14530–14531.
142. K. D. Gatsouli, S. Pispas and E. I. Kamitsos, Development and optical properties of cadmium sulfide and cadmium selenide nanoparticles in amphiphilic block copolymer micellar-like aggregates, *J. Phys. Chem. C*, 2007, **111**, 15201–15209.
143. K. D. Gatsouli, S. Pispas and E. I. Kamitsos, Development and optical properties of cadmium sulfide and cadmium selenide nanoparticles in amphiphilic block copolymer micellar-like aggregates, *J. Phys. Chem. C*, 2007, **111**, 15201–15209.
144. A. Cabot, R. K. Smith, Y. Yin, H. Zheng, B. M. Reinhard, H. Liu and A. P. Alivisatos, Sulfidation of cadmium at the nanoscale, *ACS Nano*, 2008, **2**, 1452–1457.
145. S. Celebi, A. K. Erdamar, A. Sennaroglu, A. Kurt and H. Y. Acar, Synthesis and characterization of poly(acrylic acid) stabilized cadmium sulfide quantum dots, *J. Phys. Chem. B*, 2007, **111**, 12668–12675.
146. L. Berti, A. Alessandrini, M. Bellesia and P. Facci, Fine-tuning nanoparticle size by oligo(guanine)n templated synthesis of CdS: an AFM study, *Langmuir*, 2007, **23**, 10891–10892.
147. I. L. Medintz, L. Berti, T. Pons, A. F. Grimes, D. S. English, A. Alessandrini, P. Facci and H. Mattoussi, A reactive peptidic linker for self-assembling hybrid quantum dot-DNA bioconjugates, *Nano Lett.*, 2007, **7**, 1741–1748.
148. P. S. Billone, L. Maretti, V. Maurel and J. C. Scaiano, Dynamics of the dissociation of a disulfide biradical on a CdSe nanoparticle surface, *J. Am. Chem. Soc.*, 2007, **129**, 14150–14151.
149. J. R. Mann and D. F. Watson, Adsorption of CdSe nanoparticles to thiolated TiO_2 surfaces: influence of intralayer disulfide formation on CdSe surface coverage, *Langmuir*, 2007, **23**, 10924–10928.
150. T. Nakashima, M. Sakashita, Y. Nonoguchi and T. Kawai, Sensitized photopolymerization of an ionic liquid-based monomer by using CdTe nanocrystals, *Macromolecules*, 2007, **40**, 6540–6544.
151. D. Maysinger, M. Behrendt, M. Lalancette-Hebert and J. Kriz, Real-time imaging of astrocyte response to quantum dots: in vivo screening model system for biocompatibility of nanoparticles, *Nano Lett.*, 2007, **7**, 2513–2520.

152. S. Ju Cho, D. Maysinger, M. Jain, B. Roder, S. Hackbarth and F. M. Winnik, Long-term exposure to CdTe quantum dots causes functional impairments in live cells, *Langmuir*, 2007, **23**, 1974–1980.
153. W. Liu, H. Soo Choi, J. P. Zimmer, E. Tanaka, J. V. Frangioni, M. Bawendi, Compact Cysteine-Coated CdSe(ZnCdS) Quantum Dots for in vivo Applications, *J. Am. Chem. Soc.* 2007, **129**, 14530–1453.
154. V. Kloper, R. Osovsky, J. Kolny-Olesiak, A. Sashchiuk and E. Lifshitz, The growth of colloidal cadmium telluride nanocrystal quantum dots in the presence of Cd(0) nanoparticles, *J. Phys. Chem. C*, 2007, **111**, 10336–10341.
155. K. Iwahori, T. Enomoto, H. Furusho, H. Miura, K. Nishio, Y. Mishima and I. Yamashita, Cadmium sulfide nanoparticle synthesis in Dps protein from Listeria innocua, *Chem. Mater.*, 2007, **19**, 3105–3111.
156. K. Iwahori, K. Yoshizawa, M. Muraoka and I. Yamashita, Fabrication of ZnSe nanoparticles in the apoferritin cavity by designing a slow chemical reaction system, *Inorg. Chem.*, 2005, **44**, 6393–6400.
157. M. Danderer, Handbuch der Amalgamvergiftung,4.Erg., Lief.3/95, Elektrosensibilitaet, ecomed 1995, 11–61.
158. G. Toelg, Quecksilber-ein Problemelement für den menschen, *Chem. uns. Zeit*, 1977, **11**, 150–166.
159. S. Si, A. Kotal and T. K. Mandal, One-dimensional assembly of peptide-functionalized gold nanoparticles: an approach toward mercury ion sensing, *J. Phys. Chem. C*, 2007, **111**, 1248–1255.
160. T. C. Pappas, W. M. Shan Wickramanyake, E. Jan, M. Motamedi, M. Brodwick and N. A. Kotov, Nanoscale engineering of a cellular interface with semiconductor nanoparticle films for photoelectric stimulation of neurons, *Nano Lett.*, 2007, **7**, 513–519.

CHAPTER 4
Iron

4.1 Introduction

Iron starts life in the stars, without an electron shell. There it constitutes the heaviest nucleus whose production by the fusion of smaller nuclei still releases energy. The binding energy per proton and neutron has its shallow maximum in iron: the formation of all heavier elements costs energy. These heavy elements then tend to decompose again within the stars to smaller nuclei. Iron thus became the most common metal in the universe.

Some of the iron nuclei reach the surface of the stars and are expelled into interstellar space, where they cool down and react with the electron clouds to form metal atoms. Subsequently most of these react with water to become oxides and end up as some metallic iron and a lot of oxide in interstellar dust, meteors and planets.

On Earth metallic iron does not occur naturally on the planet's surface, but in the Moon's crust $\sim 0.5\%$ of metallic iron was detected in several minerals. The core of the Earth presumably contains massive amounts of iron metal, but in the Earth's crust we find only the oxides haematite (Fe(III)$_2$O$_3$), magnetite (Fe$_3$O$_4$=Fe(III)$_2$O$_3$ Fe(II)O) and limonite (2Fe(III)$_2$O$_3$ 3H$_2$O), and the carbonate siderite (FeCO$_3$). Some 600 million tonnes of iron ore is used for steel-making each year, reduced by heating with carbon (in the form of coke):

$$2Fe_2O_3 + 3C \rightarrow 4Fe + 3CO_2$$

Some chalk or limestone (CaCO$_3$) is added to the process in order to convert silica (SiO$_2$) to calcium silicate (CaSiO$_3$, "slag") whose melt serves as a solvent for the iron oxides:

$$SiO_2 + CaCO_3 \rightarrow CaSiO_3 + CO_2$$

Iron is also the most common metal in the human body. Respiration depends on the Fe^{2+} ions of the blood and cell pigment haem, which carries and activates molecular oxygen within a highly reductive biological environment.

Human civilization also depends on metallic iron. About 4000 years ago it began its career as the cheapest and most versatile of all metals for tools and weapons. Iron made ploughing, digging, hunting and fighting more efficient, faster and tougher. In the last two centuries the peculiar hardness of nano- and microcrystals of iron carbide (Fe_3C) within the iron–carbon alloy familar to us as steel has led to the construction of cars, ships, airplanes, rockets, machines, and bridges. The ferromagnetism of iron produced the compass; in recent times the paramagnetism of thin iron oxide layers and nanoparticles has allowed the development of the hard disks and magnetic memories of computers. Human life depends on iron for every breath and every computer click.

For the nanochemists of today and tomorrow iron is a major field of research:

- Ferrocene is a cheap and unique basis for the synthesis of metal polymers.
- The controlled growth of iron carbide crystals at the border of steel domains is the basis of material engineering.
- Magnetite and ferritin provide unique magnetic and chemical properties for the attachment of proteins and drugs and their direction in the bloodstream by external magnetic fields.
- Memory devices based on thin magnetic iron oxide layers or nanoparticles are organized, fixated and stabilized to form computer disk reading/writing systems.
- Iron oxide nanoparticles on porous alumina and related carriers serve as catalysts in the synthesis of ammonia from nitrogen and gasoline synthesis from coal (coal liquefaction) as well as in the water gas shift reaction, the reduction of water to hydrogen with carbon monoxide.
- Coated metallic iron nanoparticles alone reduce chlorocarbons (C–Cl) to hydrocarbons (C–H) in soil.

4.2 Steel Domains and Magnetism

Pure iron is relatively soft and malleable at room temperature, with a bcc crystal structure (α-iron). Above 910 °C it becomes fcc (γ-iron), above 1390 °C, *i.e.* ~145 °C below its melting point, bcc again (δ-iron) (Figure 4.1). The relative positions of the iron atoms change, but the volume per iron atom is the same in bcc and fcc lattices. The overall volume of solid iron grows only upon increasing the temperature, not upon recrystallization.

The most prominent form of metallic iron, however, is the iron–carbon alloy we know as steel. Addition of 4.3% of carbon to iron lowers the melting point from 1535 °C to 1015 °C, but only 0.4% of carbon remains dissolved in the solid γ- and α-iron phases. So little carbon dissolves because the interstitial empty

Iron 247

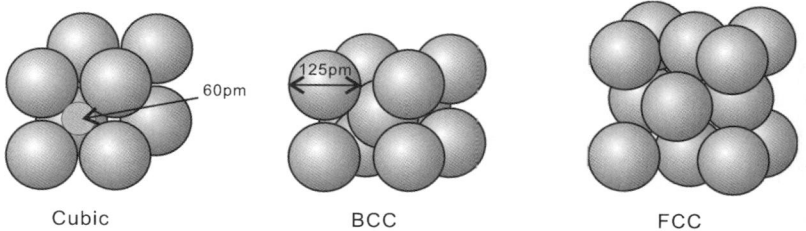

Figure 4.1 Schematic structures of cubic, body-centred-cubic and face-centred cubic metal lattices.

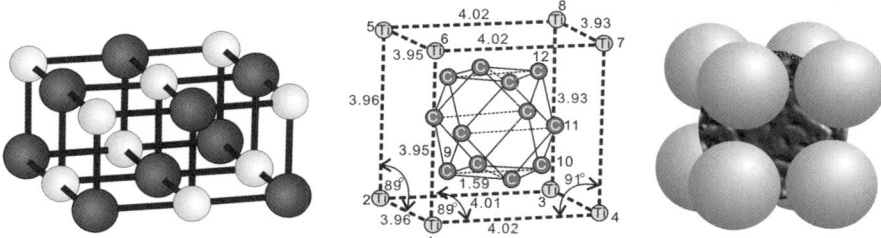

Figure 4.2 The Ti_9C_9 cluster shows its (100) surface. The metallocarbohedrene Ti_8C_{12} cage structure and its space-filling model indicate how titanium and carbonform networks more or less independently.

space between iron atoms with a diameter of 126 pm in the crystalline solids has only a diameter of 80 pm, whereas the diameter of a carbon atom is 154 pm. Each carbon atom in solid iron is therefore rejected by a strong local strain field, and has to search for disturbed regions. A carbon atom does, however, fit into octahedral interstices of metal lattices when the atom radius of the metal is >135 pm. A stoichiometry of 1:1 is then possible, *e.g.* titanium carbide (TiC) for titanium with a radius of 147 pm. The hardness and high melting point of titanium carbide made from the elements at high temperatures allowed the development of wear-resistant coatings for machine tools. Neither the carbon atom of titanium carbide nor that in other interstices reacts with acids. It behaves like neutral carbon, not like an electron-rich carbide.

Laser vaporization of titanium in hydrocarbon carrier gases yielded a completely different material, namely Ti_8C_{12} cations (also obtained by electron bombardment in the mass spectrometer) with a C_{12} centre, as a "supermagic" peak in the spectra. The quadrivalent titianium metal atoms cover an electron-rich C_{12} molecule[1,2] (Figure 4.2)

However, when carbon is added to molten iron it forms irregular crystallites of hard and insoluble iron carbide called *cementite*. The surprising stoichiometry of cementite is Fe_3C, and originates from the unit cell $Fe_{12}C_4$ shown in Figure 4.3. Each iron atom is trivalent, each carbon is surrounded by eight iron atoms. This crystal structure, as well as mathematical models of steel structures,

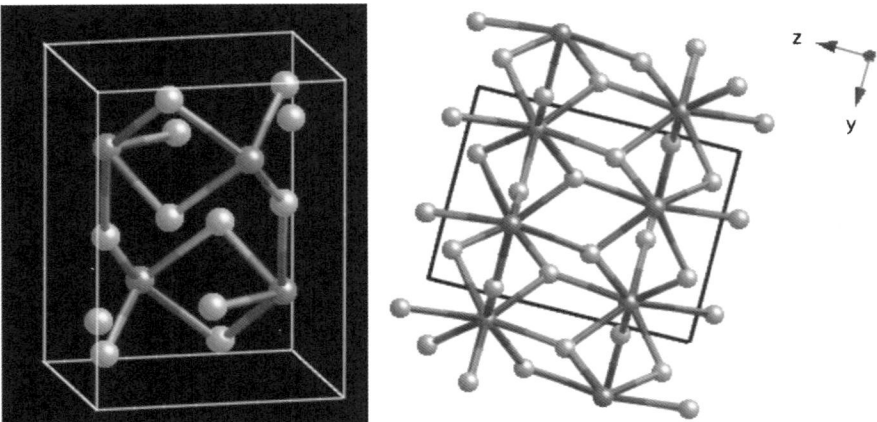

Figure 4.3 Cementite crystal structure.[3] Each carbon atom is surrounded by eight iron atoms, each iron atom is connected to three carbon atoms. These connections are not obvious from this figure, which only illustrates the unit cell $Fe_{12}C_4$. C, dark atoms; Fe, light atoms

are shown in the internet citation given.[3] Cementite is an interstitial carbide and is formed in molten iron from elemental carbon and iron at carbon concentrations >6.3%. Here the diameter of iron atoms is only 125 pm and the 77 pm carbon has some anionic salt character. HCl forms, correspondingly, CH-bonds with carbon and H_2 with iron.

The C–Fe distance of ~1.0 Å is about the same as the 1.1 Å of the covalent C–H bond. Nevertheless, Fe_3C exists at temperatures > 1000 °C, whereas C–H bonds fade away at 500 °C. This stability is as unexpected from chemical first principles as the Fe_3C stoichiometry. Both simply reflect the "chemically impossible" mixture of metallic and covalent bonding in Fe_3C. Metal–metal bonds break only above the melting point, whereas organometallic compounds usually pyrolyse quantitatively above 500 °C. The metallic iron and non-metallic carbon elements are thus neighbours, and only the stable crystal lattice prevents the thermolysis of the Fe–C "bonds". Furthermore, a strongly reducing electron cloud of Fe(0) surrounds the carbon atoms, keeps oxygen away from the isolated carbon atoms and prevents their oxidation to carbon dioxide.

As mentioned above, α-iron or ferrite has a bcc crystal structure. It usually contains at least 0.02 wt% of dissolved carbon at 723 °C, which originates from coal, the reducing agent for iron oxide minerals. At higher temperatures a solid state solution of Fe_3C in γ-iron (austenite) with <0.2% carbon is formed. With 0.8% of carbon in the γ-iron the Fe_3C precipitates in the form of micron-thick layers in metallic iron upon slow cooling and below 723 °C. This phase of iron, pearlite, is soft and technically useless because the metal slides easily along the carbide planes, leading to deformations. Fast cooling prevents the formation of pearlite and the separation of carbide: a hard and brittle martensite with carbide and carbon needle-shaped grains is produced instead. Renewed heating

Iron 249

and slow tempering, as well as adjusting the amount of carbon, then dissolves the carbon-containing nanoparticles, and the relative proportions of hard and malleable modifications determine the ductility, fracture and toughness of the material. Fe_3C is ferromagnetic like Fe(0), but the internal magnetic field of iron atoms neighbouring carbon is less than that in pure iron. All of these particles in steel are micron-sized and visible under a light microscope. Cementite-like phases in nanoparticles have not yet been realized, but FeC "steel nanoparticles" are known (see below).

Solid metallic iron and other metals contain not only these light-microscopic domains, but also nanocrystallites separated by high-angle planes.[4] This fine separation is a prerequisite for strength, ductility, fracture toughness and corrosion, resistance and high magnetic susceptibility. Crystalline grains with sizes between 5 and 100 nm are separated by disordered grain boundary regions, and their size grows linearly with time at high temperatures (Figure 4.4). The method of choice for studying steel domains is time-dependent synchrotron X-ray diffraction. The speed of grain growth accelerates abruptly after the transition from the nanocrystalline to the microcrystalline regime, since the diffusion of iron atoms from smaller to larger grains is relatively slow, whereas within large domains it occurs quickly. The theory of magnetism, which is currently being developed for isolated iron nanoparticles (see Chapter 1, ref. 29) should also be applicable to the nanodomains of bulk iron, because the atom exchange is slow.

Finally, the mechanical properties of bulk steel also depend on the form of the carbon-containing particles. Hard, round particles stabilize steel against

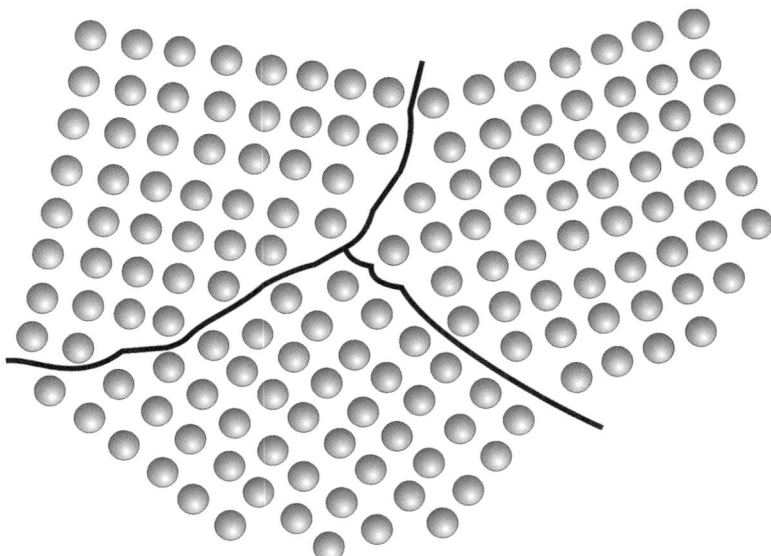

Figure 4.4 Model of a typical grain boundary in metals. Each point indicates a metal atom in one crystal plane; the lines are iron carbide (Fe_3C) or carbon nanoparticles.

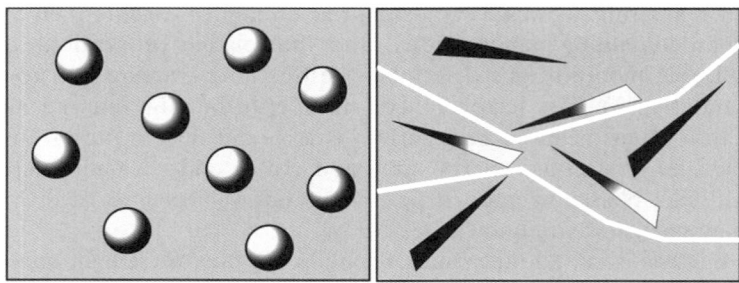

Figure 4.5 Spherical carbon or iron carbide nanoparticles (*left*) lead to soft, malleable steel; acicular nanoparticles (*right*) cause brittleness and favour rupture.

rupture; soft, thin needles, so-called acicular nanocrystals, favour ruptures (Figure 4.5).

Up to 768 °C (the Curie point), pure iron is ferromagnetic. The d-orbitals of iron atoms in bcc lattices do not point to each other. Single electrons that are not delocalized into the conducting band of the metal remain non-bonding and unpaired, with their spin moments all oriented parallel to each other, even if there are as many as five unpaired electrons per iron atom. The unpaired electrons in the upper three of the five d-orbitals of iron may, however, also participate in the formation of the conducting band (Fermi level) and such levels are primarily occupied by paired electrons. Ferromagnetism therefore does not occur with a strength of $1\mu_B$ per unpaired electron; it is usually weaker.

Above the Curie temperature the thermal energy abolishes the parallel orientation of single electrons, and their magnetic interactions as well as permanent magnetism disappear completely. Normal paramagnetism then replaces the ferromagnetism. Iron then only becomes magnetic in an applied magnetic field, *e.g.* of a DC current. This paramagnetic iron state has been called β-iron, although the bcc crystal lattice of α-iron has not changed.

The magnetic properties of iron are also strongly dependent on its purity and heat treatment. Cheap permanent magnets are usually cobaltic steels or cobalt–nickel alloys (p. 345 ff, 353); more powerful ones are neodyme boron ferrites, *e.g.* $Nd_2Fe_{14}B$. The addition of diamagnetic, quadrivalent silicon to iron disturbs the pairing of iron electrons (p. 251), and favours rapid changes of magnetization, which is useful for the soft magnets of rapidly moving electric motors and transformers.

Each single domain of iron is ferromagnetic. The magnetism of all domains may be oriented in one direction and this strong magnetism may remain in the absence of a magnetic field (permanent magnetism). "Hard" magnets are characterized and differentiated from "soft" magnets by a magnetization process in the form of a hysteresis loop (Figure 4.6). Hard magnetic materials do not change their magnetization under weak external fields. The magnetization stays constant at first and only when the external field becomes quite strong do the magnetic domains change the direction of their spin polarization.

Iron

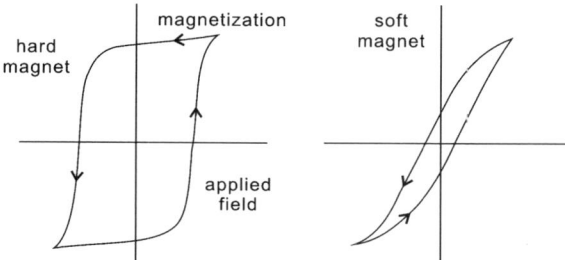

Figure 4.6 Magnetization curves of hard and soft ferromagnets. They are both different in the case of rising and decreasing external field, which leads to hysteresis loops. Hard magnets (*left*) are only influenced by strong electric fields. For soft magnets (*right*) the curves are steep, which indicates that electron pairing is strongly reduced even by weak external fields.

The overall magnetization of the treated material then rises strongly. When the external current is lowered, the hard magnets again does not react immediately, but only after the field has almost vanished. Soft magnets, on the other hand, react immediately to external fields. They usually start as weakly paramagnetic substances and become magnetized strongly only by external fields. Weak fields induce weak magnetism, strong fields strong magnetism. The effect is again non-linear and also takes the form of a hysteresis loop, but this is much narrower than in the case of hard magnets and approaches reversibility.

4.3 Layers on Steel Surfaces, Water and Corrosion

A steel blade was used to scratch a few lines on the surface of a quartz crystal. Steel microparticle traces were thus generated on the hard quartz surface and carbon vapour produced single-wall carbon nanotubes, which grew perpendicular to the scratches. Primary carbon nanoparticles within the steel microparticles probably functioned as crystallization nuclei. The first carbon shells were then formed on the steel surface, followed by a row of thousands of carbon nanoparticles on quartz. The fibres grew only there; nothing appeared in the air above the surface. Finally, the carbon nanoparticles arranged to form carbon nanotubes. The growth in just one direction indicates that neighbouring tubes reject each other, but tubes coming from parallel scratches seem to combine when they meet (Figure 4.7). End-on-addition, side-on-rejection may be a principle that allows the formation of uniform carbon nanotube populations.[5] This is a nice model for the development of prebiotic syntheses of the linear hydrocarbon chains of fatty acids.

The unsaturated fatty acids, which dominate the structure and flexibility of biological membranes around membrane proteins, were adsorbed on steel surfaces from *n*-hexadecane or 2,2,4,4,6,8,8-heptamethylnonane solutions. The time-dependent growth of surface layers was measured with a quartz balance, where a smooth quartz crystal was sputtered first with gold vapour, then with steel

Figure 4.7 Scanning electron micrograph of two horizontal steel blade scratches on quartz and several parallel carbon tubes connecting them. The steel served as first nucleation centres of carbon vapour, which condensed to form graphene sheets rolling up to growing tubes whose repulsive surface charges and interactions with the crystalline quartz surface caused the observed parallel alignment. © American Chemical Society.[5]

nanoparticles. The steel ground was chosen here because of its hardness and its frictionless quality. The components of the steel surface were also characterized so that all the variables of the system were known. X-ray photoelectron spectroscopy (XPS) indicated 25.9% of iron, 56.9% of oxygen (surprisingly high), 13% carbon and 1–2% each of chromium, nickel and silicon. The linear and branched solvents were selected because it was suspected that long-chain alkanes may organize parallel to the fatty acids at solid surfaces and trigger the formation of uniformly stretched conformers, whereas branched hydrocarbon solvents may favour statistical mixtures of conformers. Different solvents caused no significant change of the adsorption process. Oleic acid was adsorbed from both solutions as a monolayer with the alkyl chains parallel to the steel surface. Linoleic acid and linolenic acid did not reach the plateau of a single monolayer like oleic acid, but rather produced parallel-lying multilayers from the beginning, each layer being adsorbed with the same force. Oleic acid thus first covered the whole quartz surface before a second layer was started, whereas linoleic and linolenic acids at first produced multioccupied spots and empty quartz regions. With all three fatty acids, the presence of water did not make the acid layers stand upright. The flat-lying aggregates just smeared out, and the chemisorption of linolenic acid was even totally inhibited by water. Its flexibility removed it from the steel surface and transported it into micelles of the bulk solution. Without any water, the physisorbed amount in flat multilayers increased with increasing unsaturation of the fatty acids.[6] One may predict that these fatty acids and their esters behave similarly in the walls of red blood cells, the most prevalent and most mobile microparticles in blood vessels. Their membranes would then be fluidized by mobile linolenic ester domains, whereas the oleic esters would be evenly distributed and have a less fluidizing effect, because the most frequent saturated fatty acids would force them into a linear conformation.

Aerosol-assisted chemical vapour deposition of molybdenum diethylrhodanide ($Mo(S_2CNEt_2)_4$) has been used to deposit the standard solid lubricant molybdenum sulfide (MoS_2) on steel (Figure 4.8). The films formed consisted of a crystalline layer of iron sulfide (FeS) covered by molybdenum sulfide needles of micron length and nanometre height. Their more or less planar edges were predominately exposed perpendicular to the steel surface. The low coefficients of friction arose from the combined performance of the iron sulfide layer and burnishing of the molybdenum sulfide nanoparticle films.[7]

Tribology is the science of interacting surfaces in relative motion. A commercial nanotribometer measures friction and visualizes the rubbing effects with a 2 mm steel ball of 1.4 ± 0.2 nm roughness sliding over oxidized silicon wafers coated with octadecyltrichlorosilane. Such silane monolayers, self-assembled and covalently fixated on a solid surface, are contenders for the lubrication of engines and small electromechanical devices. The nanotribometer indicated that increased humidity led to a large reduction of the coefficient of friction on the hydrocarbon layers (Figure 4.9). This effect was most pronounced on surfaces containing nanosized defects, which were deliberately introduced either by heat treatment or by frequent repetition of sliding experiments. The disturbed surface itself remained hydrophobic, and the rubbing of the steel ball hardly changed. When there were holes in the

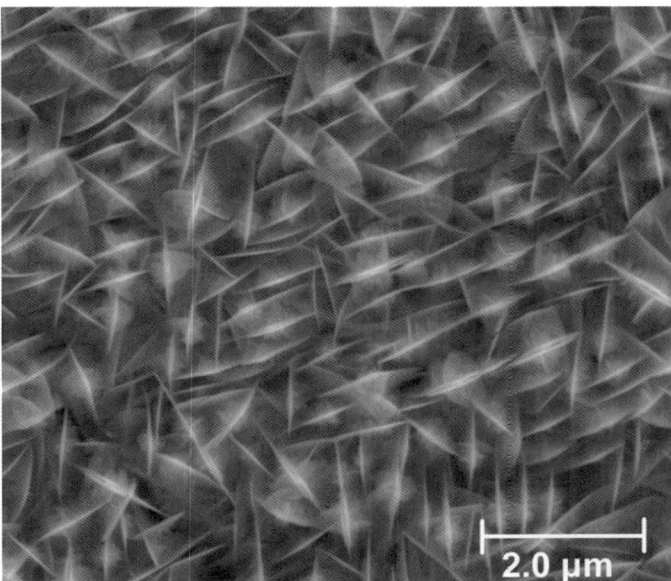

Figure 4.8 Scanning electron micrograph of an MoS_2 film grown on steel. © American Chemical Society.[7]

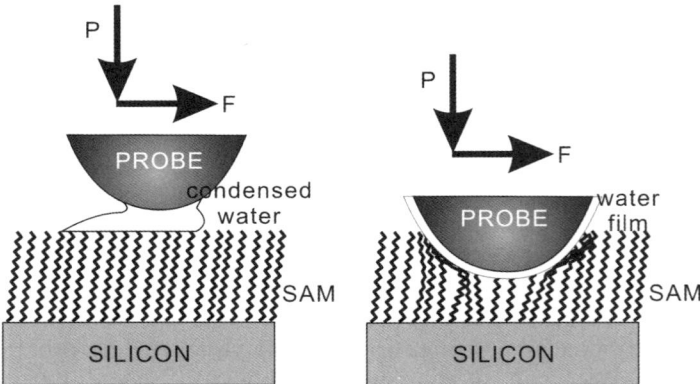

Figure 4.9 Configurations of an undisturbed (*left*) and a defect (*right*) octadecyl self-assembled monolayer (SAM), when a steel ball slides with a pressure P and a friction force F along the probe in the presence of water vapour. A sliding water film on steel is only formed in the fluid defect layer. © American Chemical Society.[8]

hydrophobic surface layer, however, a water film condensed on the steel ball, which was rubbed away by the intact layer, and the more distant oxidized silicone surface presumably also took up some water. The whole system was moist but there was only one closed water film. A more rapid relaxation of molecular deformations within the disturbed and moist film and the water film on the steel ball were thus thought to be responsible for the decrease in the friction coefficient from 0.04 in the pristine monolayer to <0.02 in the disturbed one.[8]

A plasma in the radiofrequency energy range generally enhances chemical vapour deposition on steel. Hydrophobic, highly adherent nanometre-thick crosslinked polymer films were, for example, formed there from hexafluoropropene ($CF_2=CFCF_3$). These films were thermally stable and had low surface energy, low coefficients of friction and low dielectric constants. The water contact angle was $115\pm6°$, and the plasma-deposited films showed continued excellent adhesion and stability even after treatment with ultrasound or hot water. Only a strong hot-water jet was able to remove the deposited nanometre-thick layers quickly by mechanical abrasion. It was not only the crosslinking of the fluorinated fragments, but presumably also covalent bonding to the iron and carbon of steel, which was responsible for the extreme adhesion of the plasma deposited-films.[9] In comparison, spin-coated nanometre-thick films of ready-made Teflon polymers were totally exfoliated from the steel surface even by gentle washing.

Fe(III)-polyphosphates provide another interesting type of nanometre-thick protective films on steel. At anodic sites of steel the tripolyphosphate anion ($P_3O_{10}^{5-}$) produces a hard, adhesive layer of ferric tripolyphosphate which covers the surface with negative charges in an analogous manner to biological bones (p. 100, 176). A zinc tripolyphosphate surface film was produced from $Zn_2NaP_3O_{10}\cdot 2H_2O$, which was obtained from zinc nitrate and sodium tripolyphosphate by precipitation at pH 4, and long-term storage at pH 5.1. Such tripolyphosphates are unstable on heating, since the anion undergoes self-hydrolysis. Its half-life is more than a year at 25 °C, but drops to 5 h at 80 °C.[10]

A 100 μm thick protective ferric phosphate film formed after application of zinc tripolyphosphate at pH 7 immediately inhibited all iron corrosion, but the painted panels developed blisters when exposed in a humidity chamber. The zinc tripolyphosphate was then applied in water-borne epoxy paints and the steel surface remained stable during 2400 h of exposure to a salt spray test, because it displaced the corrosion potential to much more positive values. The elastic epoxy layer did not blister and protected the steel against oxygen, water and salt. The charge transfer resistance of the polyanionic zinc tripolyphosphate still protected the iron from oxidation even when the epoxy layers showed cracks.[10] Nanometre layers have not yet been tested in this way. This thinning out could perhaps achieved by using phosphonate groups instead of phosphates, which in general provide good anticorrosion performance simply because the metal phosphonate films are formed *via* coordination to the steel surface and do not hydrolyse.

One attempt to form a protective layer for steel used insoluble strontium and barium phosphonates, which provided 1D, 2D and 3D supramolecular networks as steel coatings and were easy to prepare in quantity.[11] The synthetic principle of the applied metal phosphonate preparation was simply electroneutrality: at the synthesis pH, where the phosphonate ligand and the metal were mixed, both partners had the same number of charges but of opposite sign. The resulting frameworks were neutral; there was no phosphate ester or anhydride in the coating. An acid-stable C–P bond connected the anion to the carbon skeleton. At pH 2–2.7 insoluble Sr^{2+} or Ba^{2+} phosphonates appppeared as small $\{M^{2+}[R,S\text{-}(-O_2CCH(OH)-PO_3H)(H_2O)_x]$ crystals, which agglomerated at pH 7 to amorphous, highly hydrated polymers. At neutral pH the strontium or barium phosphonate hybrids then act as 100% effective anticorrosion films on the carbon steel surface for at least 7 years in humid air. The phosphate and carboxylate charges presumably fixate any cation on the surface and do not allow the migration of cations into the iron core.

Porous stainless steel filters are commercially prepared by precompressing and heating uniformly sized steel particles of less than micron size to millimetre-thick platelets. The size of the pores between the microparticles can be made uniformly as small as 100 nm (Mott Metallurgical, Farmington) and coverage of the steel filter platelets with silica sols made of 100 nm particles brought the pore size down to 11.2 nm (Figure 4.10). Smaller silica nanoparticles did not adhere strongly enough to steel; larger particles left too large interstitial voids. These steel–silica membranes were then used for gas separations. Hydrogen gas diffused 3.4–3.7 times faster than nitrogen.[12] The steel–silica combination guarantees longevity of the gas filter system.

In an acidic environment, neutral iron atoms attract protons to form the FeH^+ molecule in a quintet (d^5) electronic state. The experimental proton affinity of iron is as high as 750 kJ/mol. FeH^+ then readily reacts with water, producing $HFeOH_2^+$, and decomposes to $FeOH^+ + H_2$ with a barrier of only

256   Chapter 4

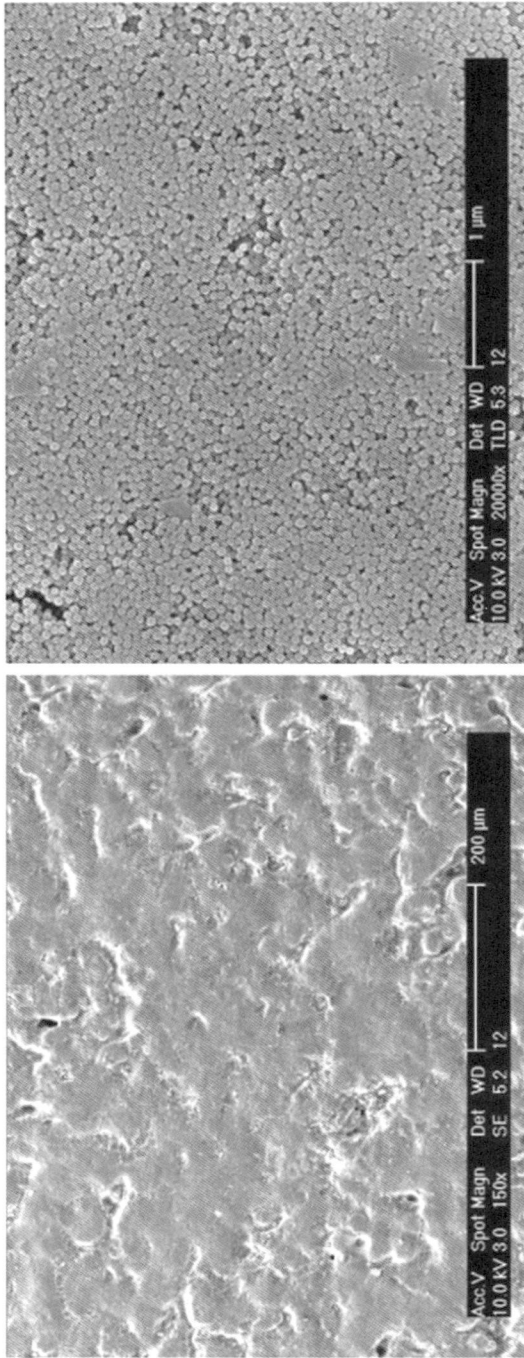

Figure 4.10 Light and electron micrographs of 100 nm silica nanoparticles on steel. © American Chemical Society.[12]

45.7 kcal/mol. Hydrogen evolution from acidic water is thus expected to be fast. Decomposition of the HFeOH, which is subsequently formed in neutral water, to FeO and H_2 is hindered by a high energy barrier and is not likely to occur even at high temperatures. Oxygen must now take over the HFeOH in order to produce water-insoluble Fe(III) oxides.

The three main steps of corrosion are therefore: (1) The iron surface takes up protons from carbonic acid and water and becomes an Fe^{2+} anode with a standard electromotive force of 1.67 V. (2) The Fe^{2+} is further oxidized by oxygen and ferric oxid (Fe_2O_3) hydrates, called rust, are formed. (3) The electric circuit is completed by the migration of electrons and ions; correspondingly, rusting occurs most rapidly in presence of salt water.[13,14]

$$4Fe(s) \rightarrow 4Fe^{2+}(aq) + 8e^-$$
$$4Fe^{2+}(aq) + O_2(g) + 2H_2O \rightarrow 2Fe_2O_3 \; x \; H_2O(s) + 8H^+(aq)$$

In dry air, *e.g.* in desert conditions, iron is therefore quite long-lived. A steel industry in the Sahara, based on solar energy, would be advantageous for both Africa and Europe.

The rather flaky and loose layer of rust is quickly differentiated from other surface layers by contact with glossy gelatine paper soaked with sodium dithionite and then dipped into ferrocyanide solution. Characteristic patterns of Prussian blue appear on the paper and identify the ferric oxide of rust.[15]

4.4 Green Rust

Wet soils close to rivers have organic top layers tens of centimetres thick consisting of black polyphenols (humic acids). Below that is an iron-containing mineral layer, usually several metres deep, which gets progressively more yellow-brown with scattered orange dots from top to bottom. Beneath this mineral layer is the *gley*, a soil whose properties are dominated by interactions with the groundwater. The typical gley horizon is at a depth of ~ 40 cm. After that follows a moist humus horizon, followed by the groundwater horizon and a blue-grey, homogeneous, ductile, pale and fully reduced Fe(II) horizon.

The most prominent mineral that incorporates Fe(II) cations belongs to the brucite-like double-layered hydroxide family $[FeII_{(1-x)}FeIII_x(OH)_2]^{x+}$ with additional hydrated anionic interlayers balancing the cation layer charge. The Fe(III)$_x$ ratio of this mineral always exceeds 0.3 and changes continuously up to 0.7 following the water redox front. The deeper one gets into the soil, the less biology there is and the less Fe(II). Reaction of the double-layered green rust (GR) hydroxide with carbonate anions from the soil gives the dianion $(GR(CO_3))^{2-}$, which then continuously transforms into the green rust oxyhydroxy–carbonate, $Fe(II)_{6(1-x)} Fe(III)_{6x} H_2O_{12(7-3x)}CO_3$. This reduces the nitrate of fertilizers effectively and becomes goethite (Fe(III)OOH):

$$Fe(II)_4Fe(III)_2(OH)_{12}CO_3 + (3/8)NO_3^- + (7/4)H^+$$
$$\rightarrow 5Fe(III)OOH + HCO_3^- + Fe^{2+} + (25/8)H_2O + (3/8)NH_4^+$$

Algae with nitrate concentrations of 30–40 m g^{-1} sometimes proliferate in river estuaries. This could be avoided if the existing soil were to be replaced by a mixture of iron ore and organic compost to enhance green rust carbonate. The catalytic effect of additional iron helps bacterial activity in reducing nitrate to the basic ammonium ion, which is rapidly oxidized to nitrogen gas in the soil (Figure 4.11).[16]

The green rust was also identified as Fe(II)$_4$Fe(III)$_2$OH$_{12}$Cl (H$_2$O)$_n$, a family of mixed-valent iron phases formed by a number of abiotic and biotic processes under alkaline conditions in the soil. All Fe(II) ions are in a 3d^6 high-spin configuration (four unpaired electrons), and the Fe(III) ions are in a 3d^5 high-spin configuration (five unpaired electrons). Such a high Fe^{2+} content is abnormal for geological iron salts, which are usually oxidized by molecular oxygen to Fe^{3+}. The essential Fe(II) component of green rust, which is formed by reduction of Fe^{3+} with carbon compounds of biological origin, is an Fe(OH)$_2$ sheet of Fe(II) octahedrons. Calculations modelling the Fe(II/III) oxide sheets with 4 × 4 slabs of iron ions showed that a hopping mobility of

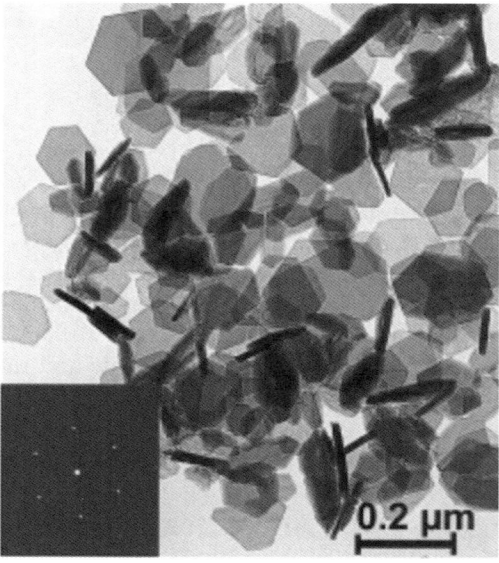

Figure 4.11 Transmission electron micrograph and X-ray diffraction pattern of green rust nanoparticles, Fe(II/III) (oxy)hydroxycarbonate, as obtained by co-precipitation of Fe(II) and Fe(III) salts. Such nanoparticles reduce the nitrate of fertilizers in the soil. © American Chemical Society.[16]

Iron

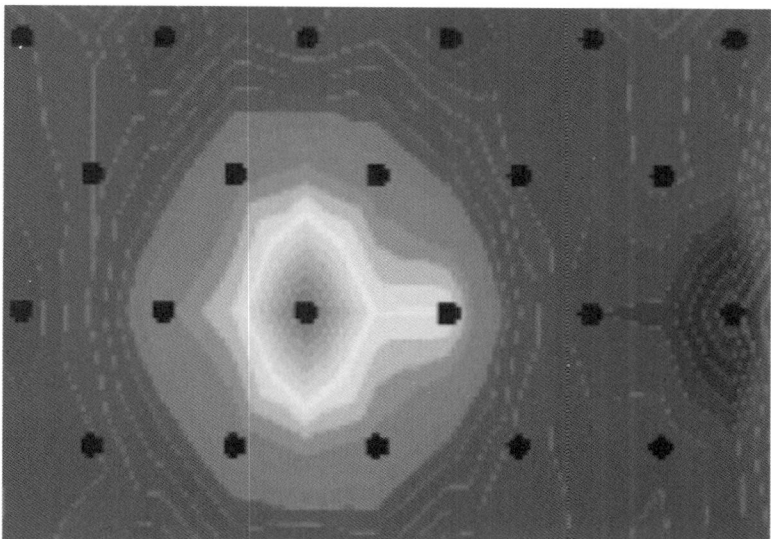

Figure 4.12 Ground-state crystal calculations produced a map of a Fe^{3+}-hole polaron in terms of Fe^{3+} ion displacements in the x and y directions. In the picture the left ring of seven nearest neighbour iron atoms shows an expanded electron cloud, while the ring of a nearest neighbour assembly is slightly compressed. Such maps are the basis for calculations of the Fe(III) hopping rates producing green rust. The distance between the peripheries of the undisturbed iron rings is about 0.6 nm. © American Chemical Society.[17]

Fe^{3+} hole polarons dominates the Fe^{2+}–Fe^{3+} valence interchange reactions (Figure 4.12). A hop to the next nearest neighbour sheet at an average distance of 0.56 nm takes only $\sim 10^{-10}$ s. This self-diffusion of holes along sheets regenerates Fe^{3+} sites and may be responsible for the generation of insoluble green rust in biological environments.[17]

The reduced iron of green rust may also be used for the production of hydrogen gas from water. A bismuth-doped anodic layer of titania deposited on titanium metal first converts solar light to an electric potential (p. 310), which may then be used to oxidize phenolic compounds of the green rust to carbon dioxide at a semiconductor anode. OH· and Cl· radicals from water and NaCl help in the partial oxidation of the phenols. Oxalic, maleic, and formic acids together with protons are primary oxidation products. The current efficiency obtained for the complete (and complicated) oxidation of phenolic compounds ranged from 3% to 17%. A coupled stainless steel cathode then reduces the protons of water to hydrogen gas and the cathodic current efficiency is, as usual, much better than for the more complex oxidation process, ranging from 68% to 95%. The energy efficiency for hydrogen gas generation was 30–70% (Figure 4.13). Hydrogen gas also has the useful property of suffocating bacteria without having the typical unhealthy side

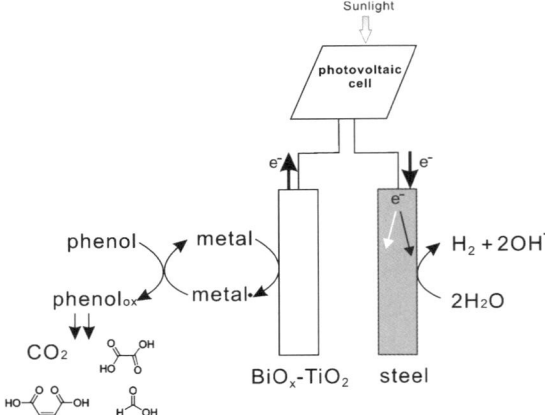

Figure 4.13 Schematic diagram of an electrochemical hybrid system with photovoltaic cell and titanium dioxide anode–steel cathode for oxidation of phenols and hydrogen production *via* water splitting.[18]

effects of oxidants on the organics.[18] Water photolysis should therefore replace the conventional UV, chlorination and ozone treatments for the disinfection of wastewater.

The surfaces of biological organisms, such as bacterial cells, carry a net negative charge. Quaternary ammonium groups therefore have a high binding affinity for bacterial cells and are rapidly adsorbed. They diffuse through the cell wall, bind to the cytoplasmic membranes, cause their disruption, affect cell metabolism and finally kill the cells. The release of protons and eventual precipitation of the cell contents are observed.

Surface layers that kill bacteria on contact were therefore formed by treating stainless steel with oxygen gas and hexamethyldisiloxane plasma, followed by ethylenediamine, hexyl or dodecyl bromide and subsequently methyl iodide, using cold plasma techniques throughout. A high concentration of ammonium groups was necessary to achieve good bactericidal potency (>99%) and the efficacy depended specifically on the length of the alkyl chain. A hydrophobic surface layer made up of hexyl chains, for example, was most potent against *Streptococcus aureus*, whereas dodecyl surface layers on steel exhibited a maximum efficacy against *Klebsiella pneumoniae*. Advanced medical devices and implants in the human body may thus be kept free of microrganisms if their steel is coated with such hydrophobic molecular monolayers terminating with ammonium groups.[19]

In soils further away from fresh water the Fe(II) disappears and only Fe(III) ions are found. They occur in two typical pools: either as mononuclear Fe(III) complexes of natural organic matter, mainly phenolic polymers, which accumulate in soils of low pH (4.5–6), or as precipitated inorganic Fe(III)(hydr)-oxides in soils of higher pH.[20]

4.5 Fe(0) Nanoparticles

Porous metallic Fe(0) nanoparticles are produced commercially from Fe_2O_3 nanoparticles by reduction with hydrogen gas at high temperature. The Fe(0) nanoparticles corrode anaerobically in water or moist soils by an autocatalytic reaction with water. In water $1.9\,mol\,H_2\,kg^{-1}\,d^{-1}$ is produced first, within 2 days ($3Fe(0) + 4H_2O \rightarrow Fe_3O_4 + 4H_2\uparrow$), and the release of hydrogen gas helps to mix the water and iron nanoparticles. The rate then drops sharply over the next 20 days and enters a period of uniformly decreasing rates.[21,22] In quartz sand the diffusion-controlled hydrogen production exhibits a double maximum over the first 20 days. The corrosion resistance of the Fe(0) nanoparticles in moist sand is four times greater than in pure water due to Fe(0) adsorption to silica. The highest rate attained was $<0.5\,mol\,kg^{-1}\,d^{-1}$, and no gas bubbles stirred up any dispersion. By 30 days, the rate of hydrogen production from iron in quartz sand exhibits the same uniform decrease as in the iron/water system.[23]

These Fe(0) nanoparticles are thus a long-term source of hydrogen gas for groundwater remediation, and are sold as such. However, Fe(0) nanoparticles obtained by the sodium borohydride reduction of oxides, which is the usual laboratory method, are not porous and are completely inactive in the production of hydrogen from water at pH 7.

Less reactive Fe(0) nanoparticles than the porous hydrogenated ones mentioned above were prepared from Fe(II) sulfate ($FeSO_4 \cdot H_2O$). The salt was mixed with poly(acrylic acid) in deionized water, brought to pH 6.2–7.0 with sodium hydroxide and quickly reduced with an excess of solid sodium borohydride in the presence of air. When visible hydrogen evolution had ceased, solid iron particles had formed and were centrifuged. The sulfuric acid formed was neutralized by the sodium hydroxide from the borohydride. The Fe(0) nanoparticles coated with poly(acrylic acid) were filtered, then washed with water, ethanol, and acetone and dried. Other metals, *e.g.* nickel, could be added as sulfates to the original Fe(II) sulfate and then also appeared as metals in the nanoparticles.[23]

Commercial dry cleaning and the manufacture of polymers and insecticides contaminate the soil worldwide with chlorinated hydrocarbons at or below the ground surface. The C–Cl bond is neither reduced nor digested by microorganisms. The hydrophobic chlorohydrocarbons are retained by capillary forces and by adsorption to organic or inorganic colloidal materials in the soil, and migrate slowly towards the groundwater level. Consequently, sites can concentrate these poisonous organic contaminants in soil layers and channel them into the groundwater of big cities. Reliable remediation technologies are needed.[24]

The metallic Fe(0) nanoparticles have many strongly reducing iron atoms on their surface and are therefore applicable only in reductive geological or biological environments. Naked Fe(0) nanoparticles are much too reactive for medical use in the bloodstream. They are useful for anoxic soils and in practice they need massive protection by anionic organic coatings, which keep them away from adsorbing sand surfaces but dissolve the chlorinated hydrocarbons. The hydrophobic chlorocarbons should diffuse into these coatings and be reduced by the iron core to carbon and hydrocarbons. The chloride is converted to Fe(II) chloride, which is then integrated into the silicates of the soil.[25]

A comparison showed that naked Fe(0) nanoparticles penetrated <1 cm into a glass chromatography column filled with sand. The nanoparticles then agglomerated and stopped moving. The transport of anionic, polymer-coated nanoparticles in soils of different textures showed, on the other hand, that the poly(acrylic acid) coating did not hinder the reduction with iron; however, this was always slow. Palladium was therefore added in the laboratory experiments, to speed up the reduction of the trichloroethylene substrate. The appearance and reactivity of the metallic iron–palladium nanoparticle remained unchanged during and after the passage of the suspension through the whole sand column (Figure 4.14). Furthermore, it was found that soil grains containing clay platelets had a effective high negative charge, which prevented blocking and crowding of the polyacrylate coating layer when the nanoparticles collected the carbon halides. The palladium-catalysed reduction by iron nanoparticles became more efficient and the subsurface deposits of hydrocarbons were easily excavated. In low-clay soils, on the other hand, the transport of the same nanoparticles was relatively poor. The coating of the the particles has thus to be adjusted to the character of the soil. In general, particles 10–30 nm in diameter were slow in comparison to 30–100 nm particles and 400–500 nm particles were

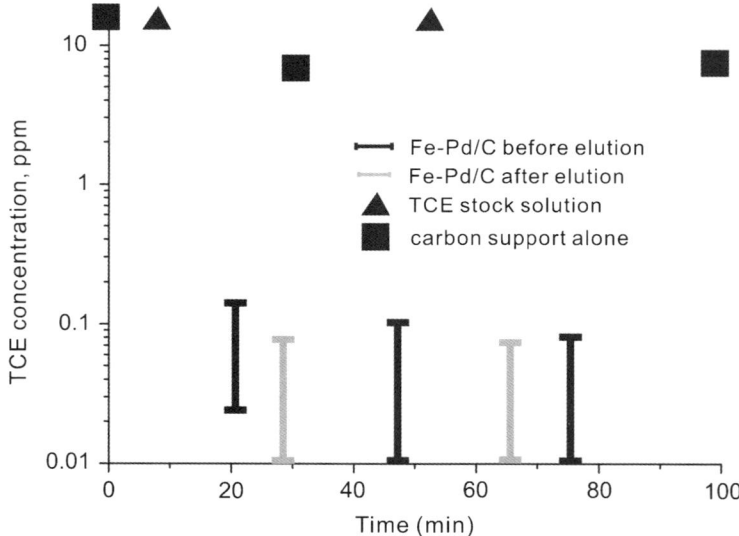

Figure 4.14 Dehalogenation of 8 mL of trichlorethylene (CCl_2CHCl), using Fe–Pd/C (50:1) before and after elution through an Ottawa sand column. The trichlorethylene was rapidly removed to 20–30 ppb levels. The carbon support itself adsorbed about 40% of the trichlorethylene. © American Chemical Society.[25]

still faster. The larger particles were much less reactive, however, because of their low surface area per unit mass, and 30–100 nm seems to be the optimal compromise.

Efficient C–Cl reduction is not the only problem for soil-cleaning nanoparticles. To be effective for groundwater remediation – the most important reason for using a cleaning procedure – the iron nanoparticles must not get lost in the terrain. They must be aggressive and stable at the same time. At first the reactive $Fe(0)/Fe_3O_4$ nanoparticles must be dispersible in water at varying conditions of pH and ionic strength. The solubilizing coatings of the nanoparticles must guarantee electrostatic repulsion in water and be hydrophobic at the interface between water and the non-aqueous phase. This was achieved by a poly(methacrylic acid)-*block*-poly(methyl methacrylate)-*block*-poly(styrenesulfonate) triblock copolymer. The poly(methacrylic acid) anchored to the magnetite shell; the hydrophobic attraction to the non-aqueous phase was provided by the poly(methyl methacrylate) ester block; and the final, strong polyanion poly-(styrenesulfonate) block gave the desired interparticle repulsion leading to the nanoparticles' solubility in water (Figure 4.15). The iron of the nanoparticle itself contained the electrons required for the reduction of -CCl to -CH, and the iron oxide coating provided the OH groups for anchoring the polymer and also protected the iron from oxygen.

Uniform goethite (Fe(III)OOH) acicular nanocrystals 300 nm long and 20 nm wide were prepared by air oxidation of an Fe(II) salt in the presence of carbonates and Al(III) salts. The acicular nanoparticles concentrated the Al(III) in the outer layers, because aluminium carbonate was less stable than Fe(III) carbonate. The aluminium enrichment became even greater after reduction to the metals, because the two metals do not mix well. The presence of aluminium was found to be essential for the acicular morphology of

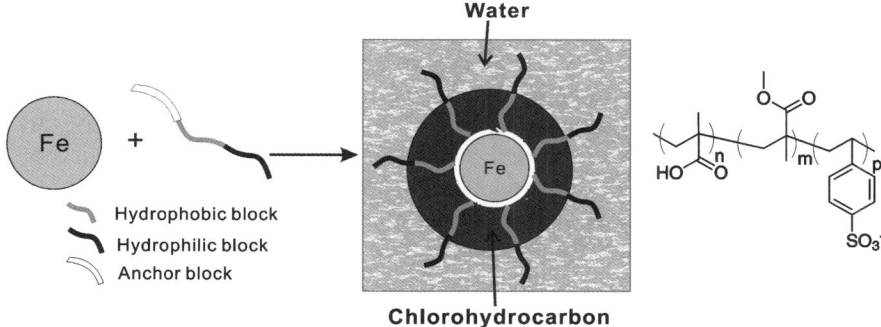

Figure 4.15 The number n of magnetite binding methacrylate units was small, m and p were large. The hydrophobic methyl ester section kept the particle near the chlorohydrocarbons, the sulfonate section took care of water solubility. The co-polymer-coated Fe(0) nanoparticle with a magnetite surface was thus held on the interface between water and a chlorohydrocarbon droplet. © American Chemical Society.[26]

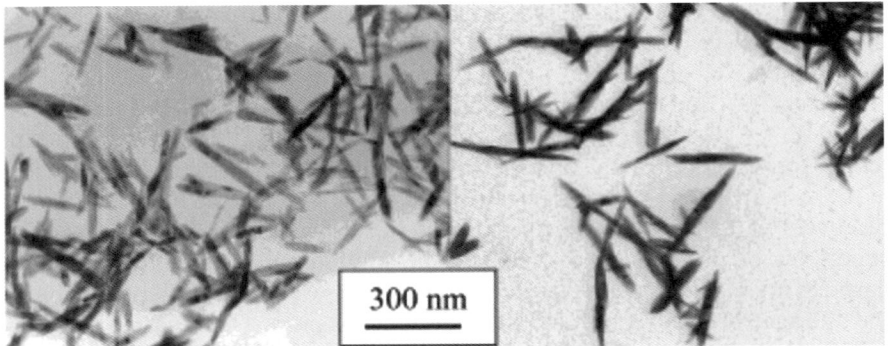

Figure 4.16 Acicular Fe(0) nanoparticles synthesized from goethite (FeOOH) particles in the absence and presence of aluminium. © American Chemical Society.[27]

the metallic particles and for an iron core with good magnetic properties (Figure 4.16).[27]

A preparation of uniform 23 nm iron nanospheres and cubes started from Fe(II) stearate dissolved in oleic acid. This mixture was heated slowly to 380 °C and aged at that temperature for 2 h. The Fe(II) chains of the $(C_{17}H_{35}COO-)_2Fe(II)$ double layers were thermolysed to carbon monoxide, hyrdogen and hydrocarbons, which then reduced the Fe(II) chains to Fe(0). The metallic Fe(0) chains collapsed to spherical Fe(0) nanoparticles with a slightly faceted surface. Faster heating produced preferably cubes, steered by the multiplication of the crystalline (100) faces. The oleate moiety in sodium oleate is not responsible for the formation of the hollow structure because the same moiety is also present in oleic acid, and the nanoframes can be obtained using sodium stearate. Molten salt corrosion (NaCl, NaOH) was assumed: the standard corrosion process that is catalysed by sodium chloride charge transport. Electron microscopy showed that the (110) facet underwent prominent etching, which explains that some parts, *e.g.* the (111) facet of the outer walls, survived, while the core with a smaller number of atoms dissolved completely.

This specific chemical sensitivity of the (110) facet is presumably caused by underpotential deposition of Na^+ species, which improve the kinetics of OH^- adsorption by Fe(0) (110).[28] Underpotential deposition means the reaction of a metal cation with a metal electrode at potentials below the Nernst potential, which is necessary to combine a bulk material electrode with these metals by reduction. Adsorbate–substrate interactions, *e.g.* Fe(0)–Na^+, must be stronger than the adsorbate–adsorbate ones, *e.g.* Na–OH. Formation of a layer of Na^+ on Fe(0) may involve the movement of the hydrated Na^+ ion from water to the Fe(0) surface, expulsion of the hydration sphere close to the iron electrons, and partial electron transfer from the Fe(0) to Na^+ leading to the entrapment of Na^+. Such underpotential deposition of metals is dictated by crystal imperfections: Fe(0) as well as Na^+ must lose the adsorbed solvent dipoles before underpotential deposition of Na^+ can lead the OH^- counterion to the disturbed reaction site of the Fe(0) nanocrystal (Figure 4.17). The final hollow-frame

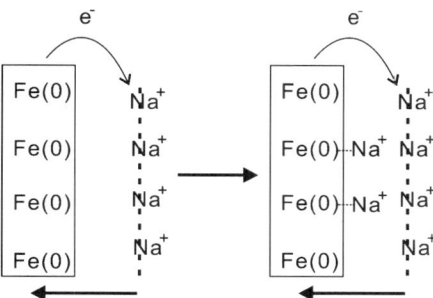

Figure 4.17 Model of the Fe(0) activation for OH⁻ addition by Na⁺ underpotential deposition.[28]

Fe(0) nanoparticles, after dissolution of the (110) facet, contain the highest possible proportion of reactive surface atoms, practically 100%.

6.9 nm iron–platinum nanocubes containing 50% of each metal were also synthesized by mixing oleic acid and iron pentacarbonyl ($Fe(CO)_5$) with a benzyl ether/octadecene solution of platinum acetylacetonate ($Pt(acac)_2$) and heating the mixture to 120 °C. Oleylamine was then added, the mixture was heated to 205 °C for 2 h, and addition of some more oleic acid followed. The faster reduction of platinum dominated the nanoparticle formation. Carboxylate (COOH) does not bind strongly to platinum, but amines form a stable Pt–NH_2 bond. Furthermore, the surface energy of the planes of a face-centred-cubic Pt(0) crystal also follow the trend of (111)<(100). The iron-rich Pt–NH_2R species thus prefers to deposit on the (100) plane, leading to the formation of a cube if the speed of growth is slow (Figure 4.18). Since the long and mobile C_{17} chain of oleylamine hinders the crystallization process, spherical iron–platinum nanoparticles, similar to micelles, are primarily formed. Uniform iron–platinum nanocubes or spherical iron–platinum nanoparticles are thus derived from atomic diffusion between a platinum-rich core and an iron-rich shell in a process where the platinum core is formed first and determines the outcome. High temperatures for longer time periods in general favour the crystallization process and often lead to cubes instead of spheres.[29]

Iron nanorods were prepared from spherical nanoparticles in trioctylphosphineoxide solution by adding iron pentacarbonyl ($Fe(CO)_5$) in trioctylphosphine was and heating to 320 °C for 30 min (Figure 4.19). Rods are preferentially magnetized along their axis, disks parallel to the surface.

Iron–platinum nanorods were also obtained by decomposition of $Fe(CO)_5$ and reduction of platinum acetylacetonate in a liquid-crystal mesophase made up of oleylamine and oleic acid at 300 °C in a sealed autoclave. Cylindrical micelles of the solvent worked as matrices. Gradient magnetometry of the annealed nanorodsm, which had a uniform width of 2.1 nm and a length of 11.3 nm (Figure 4.19), showed them to be ferromagnetic with a coercivity of 5 kOe at room temperature and the nanorods aligned in a magnetic field.[30]

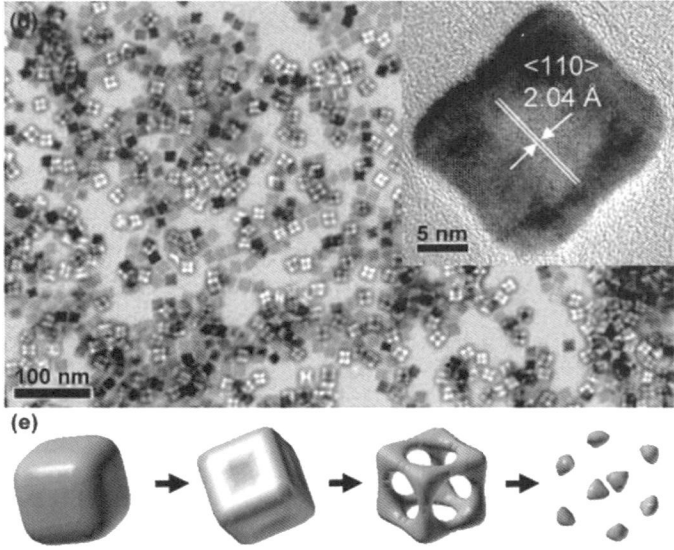

Figure 4.18 Iron(0) cubes made by Pt(0)-catalysed reduction of Fe(II) stearate dissolved in oleic acid. © American Chemical Society.[29]

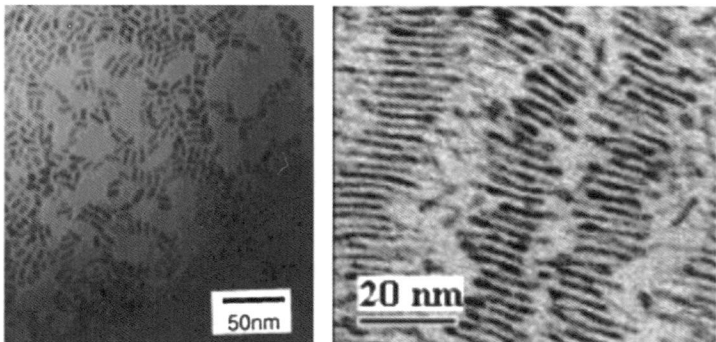

Figure 4.19 Transmission electron micrographs of rod-shaped iron nanoparticles (*left*) and iron–platinum nanoparticles (*right*). © American Chemical Society.[30]

4.5.1 Ferrocene

Ferrocene (Fe(cp)$_2$) is one of the most stable organometallic compounds: it is air stable at room temperature, water stable and thermally stable up to 500 °C. Other metallocenes and most compounds with σ- or π-metal carbon bonds decompose rapidly in air and water, but ferrocene is the steel of polymer chemistry. The cyclopentadienyl radical, $C_5H_5{}^\bullet$ or cp$^\bullet$, is a five-electron donor; elemental iron, Fe(0), has eight electrons. Fe(cp)$_2$ thus contains the noble gas

Iron

18-electron configuration. One may also count 3 electron pairs in each C_5H_5 group, which are connected to a Fe(II) ion and fill 12 electrons into 6 d^2sp^3-hybrid orbitals. The d^6 electrons of Fe(II) then occupy the remaining 3d-orbital, spin-paired. The 18-electron rule is again fulfilled and the observed diamagnetism is explained.

Dimethylferrocenyl silane polymerizes quantitatively at 200–235 °C by ring-opening polymerization with butyl lithium and addition of dimethylsilanes to a linear polymer *via* biradicals. Ferrocene and silica form the backbone, providing the reversible Fe(II)/Fe(III) couple needed for electron transfer, whereas the silica functions as a high-yield pyrolytic precursor to ceramics. This material then contains iron nanoparticles in a carbon–silicon carbide matrix. The polymer with a block ferrocene:silicon ratio of 1:12 coils up to form solvent-filled nanotubes in liquid hydrocarbons. The ferrocene part crystallizes in the apolar solvent to form a shell around a cavity in the middle of the tube, while the soluble poly(dimethoxysilane) blocks forms the surface. It is an inversed micellar tubule, which entraps large amounts of solvents.[31]

Extended metallic Fe(0) layers, wires and tubules are therefore mostly prepared from polymers of substituted ferrocenes, *e.g.* Fe(0)(cyclopentadienyl-R)$_2$. In such polymers an iron chain is already arranged, but several dimethylsilane and cyclopentadienyl groups separate the chains from each other, connected through covalent C–C bonds, thus obtaining the polymerization of the substituted cyclopentadienyl radical units. If one substituent is a silicone, the iron polymers may also become part of organized ceramics.[32]

Iron-containing films, 50 nm thick, of the two copolymer samples described above were made by solvent casting from tetrahydrofuran solutions followed by microtoming at –170 °C and collection on carbon-coated TEM grids. The diblock copolymer thin section showed black ferrocene and grey silane stripes. Freshly made ruthenium tetroxide (RuO$_4$) was then added for heavy metal staining (Figure 4.20). Although this is a vigorous oxidant and can react with

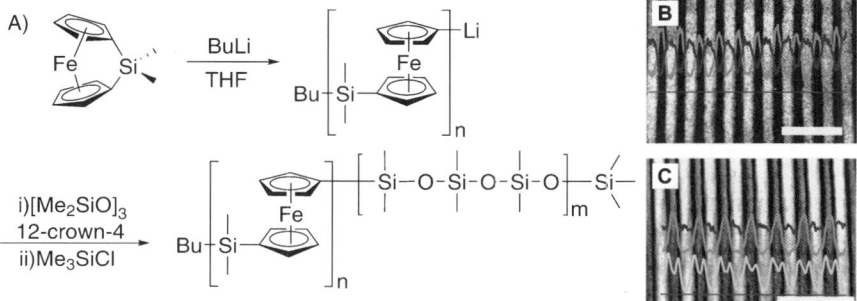

Figure 4.20 Ring-opening synthesis of a poly(ferrocenedimethoxysilane)$_{105}$–poly(dimethoxysilane)$_{300}$ film and the bright field electron density image of a thin section showing iron and silicon wires after RuO$_4$ staining.[33] Scale bar: 100 nm. © American Chemical Society.

both aromatic and unsaturated polymers it did not, as expected, stain the easily oxidizable cyclopentadiene microdomains but went exclusively into the interfaces between the ferrocene and the silicon domains. Ruthenium tetroxide is probably more soluble in the silane than in the ferrocene phase and does not react with the silane. It therefore diffuses to the silica–ferrocene interface, where ruthenium metal is precipitated by reduction through the first cyclopentadiene units that are reached. This regioselective oxidation of the ferrocene polymer at its ends may also occur in (aqueous) solutions of the copolymer, but is more difficult to characterize in bulk products and may be stained *e.g.* with ruthenium tetroxide.[33]

Water-soluble polyferrocenyl silanes have been used to produce multilayers of cationic and anionic Fe(0) ferrocenes. Electroneutral, crystalline polyferrocenyl silane was, on the other hand established as a core of stiff metallic rods with mobile, solubilizing polyisoprene coronas. Their self-assembly was carried out in a polyisoprene selective solvent such as hexane or toluene and led to long ferrocene cylinders (Figure 4.21). The peripheral vinyl groups of the polyisoprene corona were also crosslinked by a Pt(0)-catalysed hydrosilylation with 1,1,3,3-tetramethyl disiloxane and grew in length. The soluble polymer may be considered as an iron wire with an insulating coating.[34]

80 nm wide metallic iron rods are made routinely by the standard electrochemical deposition in poly(carbonate) membranes. These rods comprise long sections of ferromagnetic alloy (CoNi) separated by short sections (20–100 nm) of a diamagnetic metal (Au(0)). When the rods were magnetized, high magnetic field gradients formed at the boundaries between these sections and it is these regions that attract and trap other magnetic nanoparticles.[35,36]

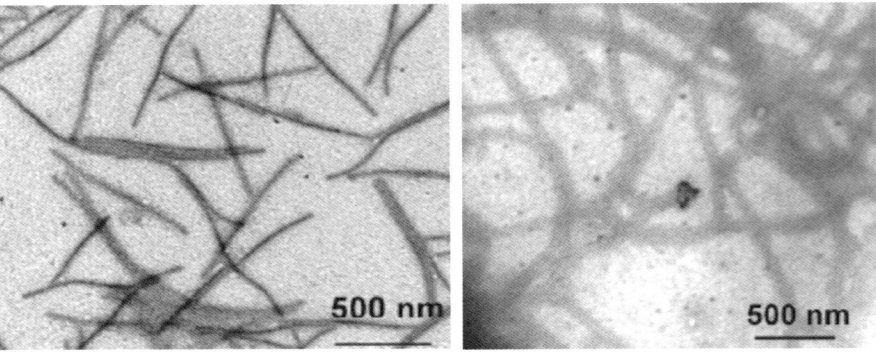

Figure 4.21 Electron micrograph of cylindrical micelle self-assembled from polyisoprene$_{250}$–polyferrocenylsilane$_{50}$ in hexane/tetrahydrofurane (200:1) and after cross-linking of the polyisoprene corona by a Pt(0)-catalysed hydrosilylation with 1,1,3,3-tetramethyl disiloxane. The much longer iron "wires" with an enlarged hydrophobic corona were more soluble in tetrahydrofurane and swelled. © American Chemical Society.[34]

4.6 Iron Oxide Nanoparticles

Fe(III) and Al(III) ions have the same charge and similar diameters (106 and 90 pm), both form MCl_4^- anions and acidic hydrates, but only Al(III) adds four OH groups to become soluble in strong alkaline solution. Fe(III) precipitates there as a trihydroxide, $Fe(OH)_3$, which rapidly dehydrates to Fe_2O_3 at lower pH values. Solid Fe(III) oxides are magnetic, because the Fe(III) contains unpaired electrons in half-occupied d-orbitals.

"Soft" and "hard" ferrites are commercially important. Soft $M(II)Fe_2O_4$ ferrites (M = manganese, magnesium, nickel, cobalt) are rapidly magnetized by electric currents and lose their magnetism if the current is switched off. They are used in the record and erase heads of computers. Hard ferrites, $M(II)Fe_{12}O_{19}$ (M = barium or strontium), retain their magnetism and are used in motors and other electrical equipment. The coatings on computer hard drives and floppy disks, and on audio and videotapes, are all composed of Fe(III) oxides combined with other metals. The soft ferromagnets are characterized by a narrow hysteresis loop in the magnetization process induced by external currents. Their magnetism is altered strongly and quickly by changing magnetic fields (p. 251).

The interaction of unpaired electrons in the Fe(III) orbitals must be strong enough to orient the spins, but not so strong as to enforce pairing. In the soft ferrites with their smaller ions, interactions that are too strong (or too weak) lead to rapidly fluctuating interactions with the environment. In hard ferrites, with their larger ions, the perfect spin–spin interactions keep them all in a given orientation.

Videotapes and computer disks consist essentially of acicular Fe(III) nanoparticles containing some Co(II). The needle shape of the particles, acting like a simple compass needle, strongly favours parallel orientation of electron spins. In the manufacturing process high temperatures first dehydrate the acicular iron oxyhydroxide precursors and then support their hydrogenation to metallic iron. The harsh thermal treatment promotes interparticle sintering, leading to a loss of the acicular shape that determines the high coercivity. Aluminium, boron, silicon, phosphorus or tin, either doped or adsorbed, help to preserve the acicular morphology. Doping is usually preferred, because it avoids expensive coating procedures.

The coprecipitation of a stoichiometric 2:1 (Fe^{3+}/Fe^{2+}) mixture of ferrous and ferric salts with sodium hydroxide in aqueous medium at pH 8–14 gives a precipitate of magnetite. This is, however, quickly transformed to maghemite (Fe_2O_3) in the presence of oxygen, which differs from magnetite not only by the oxidation state, but also in having unavoidable cationic vacancies within the octahedral sites. In nanocrystals with diameters <5 nm the ordering of these vacancy (= V) sites is purely statistical within the octahedral sites:

$$Fe^{2+} + 2Fe^{3+} + 8OH^- \rightarrow Fe_3O_4 + 4H_2O; Fe_3O_4: [Fe^{3+}]_{tetrahedral}[Fe^{3+}Fe^{2+}]_{octahedral}O_4$$

$$Fe_3O_4 + 2H^+ \rightarrow Fe_2O_3 + Fe^{2+}$$
$$+ H_2O; \gamma-Fe_2O_3: [Fe_{0.75}{}^{3+}]_{tetrahedral}[Fe_{5/3}{}^{3+}V_{1/3}]_{octahedral}O_4$$

The coprecipitation process produces large quantities of nanoparticles, but a short burst of nucleation of Fe(II)/Fe(III) mixtures leads to mixed sizes and the slow growth by diffusion of the solutes does not change this. Size control of monodispersed particles should always occur during the very short nucleation period, because the final particle number is determined here. The size and shape of the iron oxide nanoparticles is better tailored by adjusting pH, ionic strength, temperature, nature of the salts (perchlorates, chlorides, sulfates, and nitrates), and Fe(II)/Fe(III) concentration ratios. Organic counterions and ligands can either prevent nucleation, leading to larger nanoparticles, or inhibit the growth of the crystal nuclei, leading to small nanoparticles.

The diameter of magnetic nanocrystals for medical applications is usually 4–18 nm. Each nanocrystal is thus much smaller than the size of a magnetic domain in steel and is therefore a fully magnetized nanomagnet made of a single domain. Nanocrystals of metallic iron, Fe(0), are five times stronger magnets than Fe_3O_4, but chemically very unstable. In air they are oxidized and in biological environments they react with cysteine sulfide and many other reagents.[37]

The conversion of iron into iron oxide nanoparticles with water yields interesting mixed nanoparticles. At first a thin, low-density region oxide shell is rapidly formed at room temperature, which leaves many vacancies open. Transmission electron micrographs show a darker inner region corresponding to the iron core and a lighter, porous oxide shell. The iron atoms then diffuse outwards, the vacancies ultimately coalesce into a central void and the oxide shell grows thicker. Complete oxidation to Fe_2O_3 capsules would, however, take 600 years at room temperature. Only an increase in temperature accelerates the rates of iron diffusion and oxidation. Similar 20–80 nm $Fe_xO_y@Fe$ nanoparticles were prepared by sodium borohydride reduction of Fe(III) chloride in aqueous solution. They adsorbed added Ni^{2+} ions, which became partly reduced there and caused the nanoparticles to form chains as a result of their magnetism (Figure 4.22).[38]

Amorphous Fe(0) nanoparticles were made by ultrasonic decomposition of $Fe(CO)_5$ in the presence of carbon nanoparticles. The primary Fe(0) shell was

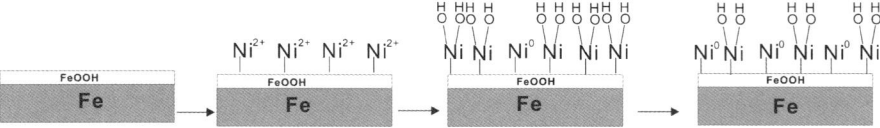

Figure 4.22 Fe(0) nanoparticles with a goethite (FeOOH) coating adsorb the Ni^{2+} ions with their negative surface charge and fixate them by reduction to Ni(0).[38]

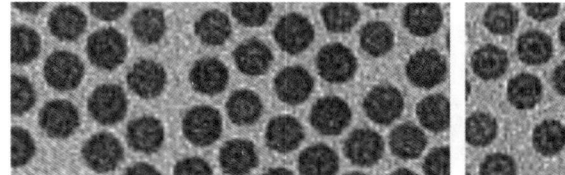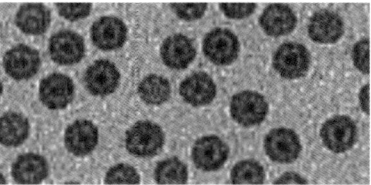

Figure 4.23 TEM micrographs of iron/iron oxide nanoparticles exposed to dry air (a) for 1 h at 80 °C and (b) for 12 h at 80 °C. © American Chemical Society.[39]

rapidly transformed to hollow haematite (Fe_2O_3) nanoparticles upon exposure to oxygen. The heat released from the rapid oxidation of the high-surface-area iron shells ignited the carbon nanoparticles inside, which generated enough additional heat to melt and crystallize the iron oxide shells, converting them to haematite with remnant hollow cores. Neither iron carbonate, which is heat labile, nor the cementite (Fe_3C) of steel was formed. In nanochemistry, it is always an iron oxide that remains in the presence of oxygen. Figure 4.23

Above 150 °C, an additional supply of electrons is provided by thermo-emission of electrons from the iron into the oxide conduction band. The oxide layer now grows to 10 nm thickness. The outward diffusion of iron cations then becomes faster than the inward flow of oxygen anions. Oxide growth slows down markedly if the shells are than >2.0 nm thick and the temperature <250 °C. Below this temperature, only Fe(0) nanoparticles <8 nm could be completely converted into hollow oxide particles. The oxidation of larger particles occurred only when there were cracks on the oxide shells. At 350 °C such cracks finally led to the formation of hollow iron oxide nanoparticles with diameters as large as 20 nm and shells 4–5 nm thick.[39]

Metallic Fe(0) as such does not occur in nature, but the positive charge of Fe(II) in haem, the carrier molecule of molecular oxygen in blood, is very close to it at $Fe^{+0.4}$. In cytochrome P450, which splits molecular oxygen into two oxygen atoms, the positive charge is completely removed by an extra S^- ligand. The iron atom carries a lot of electron density although it is formally an Fe(III) ion and cannot become "metallic", because it occurs as a single, isolated atom. The interplay of O_2^-, N_4^{2-} and RS^- on a single iron ion leading to atomic oxygen is certainly one of nature's most spectacular atomic feats (Figure 4.24). In technology, the applications of magnetic materials is equally marvellous. How remarkable it is that our life and civilization are so dominated by the magical electronic behaviour of that very familiar element, iron!

The porphyrin ligand and the proteins of haemoglobin and cytochrome P450 are vulnerable to rapid destruction by oxygen in the human body, with half-lives of the order of 10 days. The degradation products (bile pigments) are excreted in urine and faeces. Haem and blood proteins are, however, continuously produced in the bone marrow, which manufactures haemoglobin along with the red blood cells that contain it. The production and transport of red blood cells in the bone marrow is quantitatively the most important growth process in adults. This is the major reason why artificial chemical crosslinking

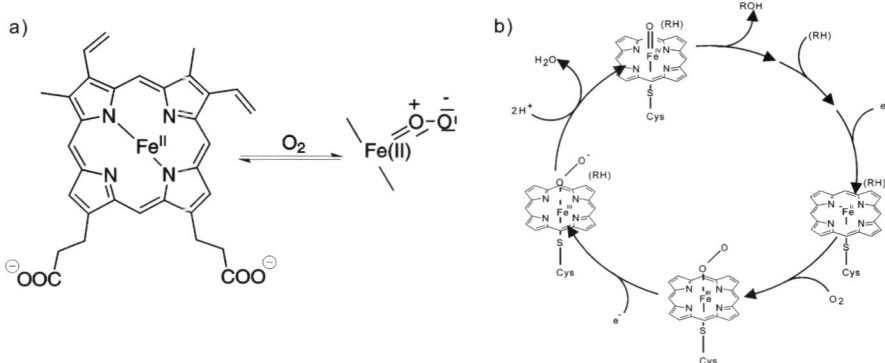

Figure 4.24 *Left:* in oxyhaemoglobin a porphyrin ligand neutralizes Fe(II), which then transfers an electron to the oxygen biradical and takes one in a back bonding. The oxygen molecule is bound by a double bond, has a negative charge, no unpaired electrons and no oxidative character. It can be transported in blood, which has an oxidation potential close to 0 V like the hydrogen/proton redox pair. *Right:* in cytochrome P450 the haem-bound oxygen molecule is reduced to water and an oxygen atom by hydride transfer from dihydronicotinamide. This oxygen atom oxidizes RH groups to ROH in a "respiration cycle" (see Chapter 1, ref. 1b).

of proteins and DNA in growing tumours by chemotherapy is allowed only for short time periods in the living human body. The brain and the body cannot live if the haemoglobin and P450, with half-lives ~10 days, are not replaced continuously.

The iron salts required for the continuous synthesis of haemoglobin and cytochrome P450 are collected and stored in the cage-like molecules of a protein called ferritin. Ferritin without its load of iron is called apoferritin. Ferritin is a non-haem iron protein made up of of iron–sulfur proteins of low molecular weight (6000–12 000). This water-soluble complex forms a spherical container that contains massive amounts of insoluble Fe(III) hydroxyphosphate. Ferritin is the most important nanoparticle of the human body. Its spherical shell is made of 24 helical polypeptides, which entrap a crystalline, superparamagnetic Fe(III) core with a diameter of 6.5 nm. It contains ~2000 Fe(III) and 100 phosphate ions. This is surprising, since the Fe(II)/Fe(III) oxidation potential is 770 mV and there should be no Fe(III) in biological tissues, whose oxidation potential is close to 0 V.

In ferritin and cytochrome P450, it is the sulfide counterion that lowers the oxidation potential of Fe(III). Cytochrome P450 serves in the biological electron transport chain and contains one, two, four or eight "Fe(III)" ions coordinated to four sulfur atoms in a tetrahedral arrangement. The reduction potential (0.05–0.49 V) and magnetic moments are low, which indicates strong electronic and spin–spin interactions to sulfur radicals. The Fe(III) has indeed only a formal oxidation number of 2.5, indicating Fe(III)/Fe(II) electronically delocalized clusters in which all iron ions are equivalent.[40]

Iron 273

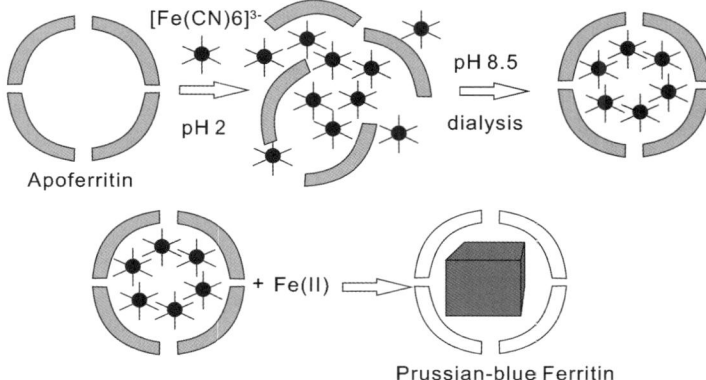

Figure 4.25 Schematic drawing of ferritin depleted of iron oxides and phosphates, its filling with ferricyanide and reaction with Fe(II) to give entrapped Prussian blue nanocrystals. © American Chemical Society.[42]

The trapping of other molecules within the apoferritin cavity is generally performed by dissociation of apoferritin into its 24 subunits at pH 2.0, addition of the new guest and self-reconstruction of the cavity with the new guest at pH 7.0–9.0. Dialysis at pH 7 then removes all the guest molecules and only the entrapped ones, *e.g.* [FeIII(CN)$_6$]$^{3-}$, remain. Addition of Fe(II) then produced the characteristic Prussian blue colour and 90 hexacyanoferrate ions were entrapped within the ferritin cage (Figure 4.25).[41,42]

Ferritin nanocrystals have been used as negative-differential resistors, where a current decreases with rising voltage. The resistor functions as additional power source at a given potential and the "differential" is defined here by the steepest slope in the $I(V)$ curve (in $-\mu\Omega\,\text{cm}^2$). In the case of a ferritin cage connected to electrodes *via* carbon nanotubes (p. 57 ff) the current only flows if the Fe(III) takes up electrons from the carbon radicals at a given potential and then becomes reduced to Fe(II) at a higher potentials. After that has happened, it cannot take up any more electrons and the current decreases to zero. In the experimental setup the chemical carbon nanotube–Fe(II)/Fe(III) connection crossed a gap of 20 nm between two metal electrodes perfectly at potentials where the ferritin shuffled electrons, but it did not work at potentials where all of the iron was reduced to Fe(II) or oxidized to Fe(III) (Figure 4.26). Technically, all that counts is the durability of the system. The ferritin–carbon nanotube nanogap device exhibited a reproducible negative-differential resistance behaviour with a peak positioned at 4.90 V upon countless switching events and no degradation within 1 month. The peak current density of 4.0×10^6 A cm^{-2} was the highest current density reported so far, and the differential resistance was $\sim -1.2\,\mu\Omega\,\text{cm}^2$. The reversible Fe(III)/Fe(II) redox reaction during the *I–V* scan is not destructive: pure Fe(II) or Fe(III) capsules blocked the current independent of its strength on both sides of the 4.9 V peak.[43]

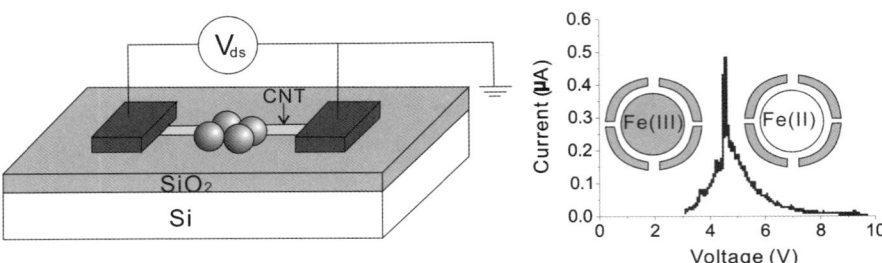

Figure 4.26 *Left:* model of a negative differential resistor between two metal electrodes based on carbon nanotubes and ferritin. *Right:* a current flows only if both Fe(II) and Fe(III) exist together and allow for reversible electron transfers. © American Chemical Society.[43]

Iron-doped carbon aerogels containing Fe(II) or Fe(III) ions as well as elemental iron between carboxylated polymers were made by the sol–gel polymerization of the potassium salt of 2,4-dihydroxybenzoic acid with formaldehyde, yielding K^+-doped gels with a kind of bakelite polymer. The K^+ was then exchanged with Fe^{2+} or Fe^{3+}, the doped gels were dried with supercritical carbon dioxide, and finally pyrolysed under an inert atmosphere. Metallic Fe(0) nanoparticles containing some iron carbide but no graphite were thus formed from the reducing organics and the iron oxides. Placed in a thermal chemical vapour deposition reactor and exposed to a mixture of methane, hydrogen and ethane at 600–800 °C for 10 min the carbon compounds pyrolysed and combined to form multiwalled carbon nanotubes 25 nm wide and up to 4 μm long, on an aerogel of Fe(0) nanoparticles (Figure 4.27).[44]

In magnetite (Fe_3O_4) the d^5-Fe(III)ions in both sublattices, tetrahedral and octahedral, are magnetically parallel coupled and yield a strong ferromagnetism. The spins of each sublattice are, however, statistically oriented and the large number of domains usually leads to a compensation of all magnetic units. The compound then becomes antiferromagnetic. Ferromagnetic materials with parallel spins only are used in high-frequency transformers and for data storage in computers, because they do not produce eddy currents. Black magnetite is not only strongly ferromagnetic, but also highly conductive, because electrons travel freely between Fe(II) and Fe(III), as shown in Figure 4.26. The corresponding magnetite nanoparticles are most useful for high-density information storage and medical imaging, which both rely on their ferromagnetism.

The monodisperse magnetite nanoparticles are usually prepared by high-temperature decomposition of iron compounds with oxygen-containing ligands, such as acetylacetonates, acetates, or oleates, in surfactant-containing solutions. The narrow particle size distribution is due to the separation of nucleation and growth processes on the temperature scale. Fe(II)oleate, for example, is first heated to 70 °C, which leads to removal of the crystal hydrate water, dissociation of oleic acid dimers, and finally the bis-oleate salt, RCOO–Fe(II)–OOCR, whose final decomposition occurs at ∼ 380 °C.

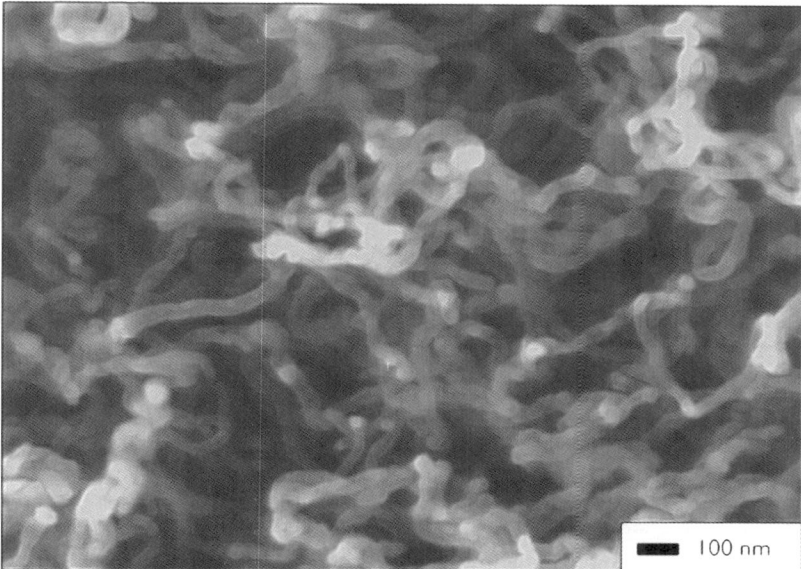

Figure 4.27 Scanning electron micrograph of an iron-doped carbon aerogel after 700 °C pyrolysis and 800 °C chemical vapour deposition. © American Chemical Society.[44]

In a typical preparation of 20 nm Fe_3O_4 nanoparticles, Fe(II) oleate was thermally homogenized to the bis-oleate at 70 °C in a vacuum oven for 24 h. Some more oleic acid and docosane ($C_{22}H_{46}$) were then added under vigorous stirring and the mixture heated to 370 °C at a rate of 3.3 °C/min, again under stirring, and finally refluxed for 3 min at 380 °C. The initial reddish-brown colour of the solution turned brownish-black. After cooling acetone was added at room temperature to precipitate the nanoparticles, which were then centrifuged or collected by a magnetic field.

These magnetite nanoparticles are favoured in biomedicine as a core for 5–10 nm nanoparticles, because they are readily degraded *in vivo* after the coating has been digested. In computers they work as reliable magnetic memory materials, because they are stable under air, strongly magnetic and easily accessible as thin, uncoated layers.

Fe(III)(acetylacetonate) also gave monodisperse, magnetic Fe_3O_4 nanoparticles with good crystallinity and magnetic properties. In a high-boiling mixture of 1,2-tetradecanediol, oleic acid, oleylamine and benzyl ether, heating to 100 °C at a heating rate of 5 °C/min gave 11 nm Fe_2O_3 nanoparticles, which were further oxidized at 200 °C for 3 h, then heated to 300 °C under nitrogen for 1 h. A black material precipitated upon cooling the solution and was centrifuged. This black precipitate was soluble in hexane, because ammonium oleate was adsorbed on the nanoparticle surface.

The advantage of procedures based on the decomposition of iron oleate is that iron oxide nanoparticles can be prepared in a wide range of sizes (6–30 nm) merely by varying the reaction time and detailed conditions.[45] The narrow particle size distribution is due to the separation of nucleation and growth processes on the temperature scale. For example, if the Fe(II) oleate is heated to 70 °C, the crystal hydrate water is removed first. Then follows the dissociation of hydrogen-bonded oleic acid dimers, and finally, at ∼380 °C, that of the bisoleate salt, RCOO–Fe(II)–OOCR. Up to that point no particle growth occurred, but the hydrophobic coating prevented further functionalization. It was therefore replaced *via* ligand exchange reactions with various capping agents bearing reactive hydroxyl moieties, which were either exploited to initiate ring-opening polymerization of polylactic acid esters or were acylated with acetic anhydride to permit the addition of alkyl halides, which prepared the nanoparticles for atom-transferradical polymerization (ATRP, p. 105 ff) (Figure 4.28).[46]

Large Janus nanoparticles, with hydrophobic and hydrophilic surface halves comparable to the corresponding Janus dendrimers (p. 119), confer exotic self-assembly characteristics and peculiar phase transitions on nanoparticle architectures. Typical assemblies in solution consist of 2-, 7- and 10-mers (Figure 4.29).[47]

The asymmetric functionalization of nanoparticles is possible by (1) masking or protecting a part of the particle surface, or (2) trapping the particle at the interface between two phases, so that only one half of the particle is exposed to chemical modification. Both methods are, however, limited to particles of micron size: nanoparticles are too mobile at interfaces and must be fixated.

Water-soluble 10 nm Janus nanoparticles have been made starting from 10 nm poly(acrylate)-coated magnetite species attached to the surface of aminated 700 nm silica beads (p. 90, 191). Subsequent addition of large,

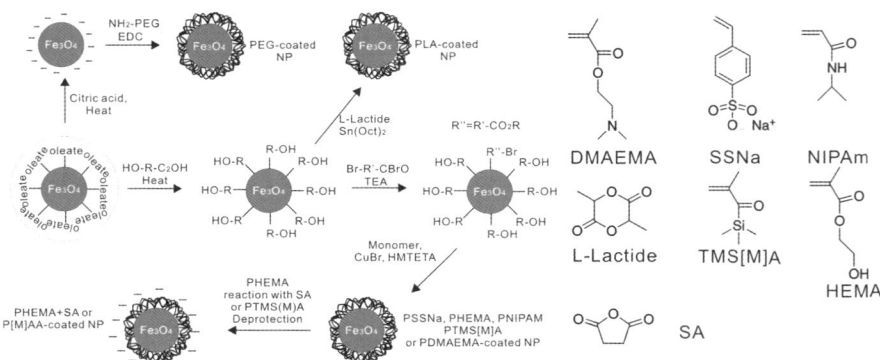

Figure 4.28 Typical functionalization of a magnetite (Fe$_3$O$_4$) nanoparticle with oleic acid, substituted amines, lactide and bromides.[46] The obscure series of abbreviations is typical for nanochemistry publications, the accompanying structures are exceptional, but most helpful. DMAEMA, SSNa, NIPAm TMS[M]A, HEMA and SA mean here, dimethyl aminoethyl methacrylate, p-styrene sodium sulfonate, N-isopropylcaprylamide, trimethyl silyl methacrylate, hydroxyethylmethacrylate and succinic anhydride. © American Chemical Society.

Iron 277

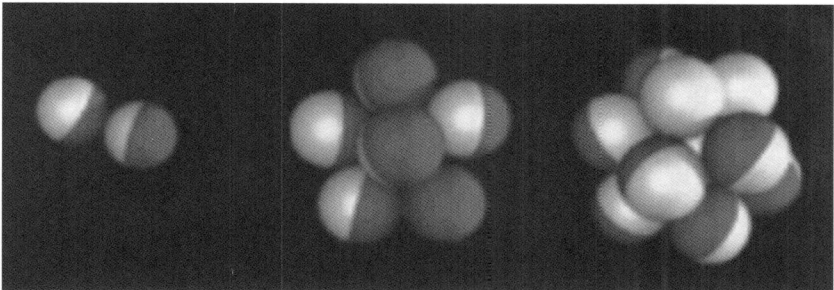

Figure 4.29 Schematic models of 2-, 7- and 10-mers of Janus nanoparticles. © American Chemical Society.[47]

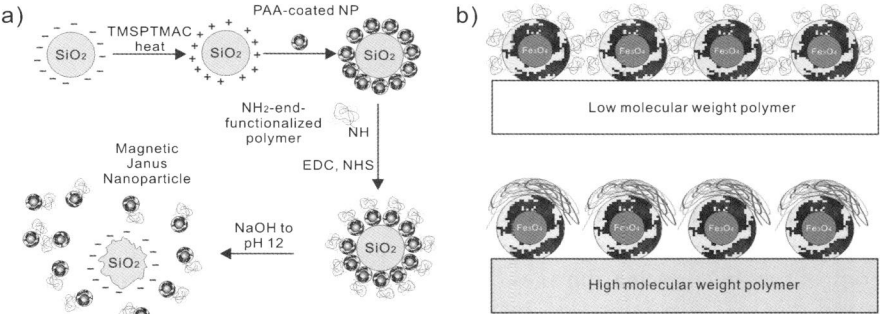

Figure 4.30 (a) Formation of Fe_3O_4 nanoparticles on SiO_2 micrometre spheres. (b) 80% coverage of a fixated Fe_3O_4 nanoparticle with a small polymer and 50% with a large one. © American Chemical Society.[48]

hydrophobic amino-ended polymers to the acrylate occurred only to a small extent; the attached silica particles prevented occupation of >50% of the surface area (Figure 4.30). After the pH of the solution was raised to 12, the acrylate half of the nanoparticles separated from the tetralkylammonium-coated silica particles and remained in solution, although their other half was hydrophobic.[48]

Magnetite nanocrystals are efficient negative magnetic resonance $T2$ contrast agents because of their small size and strong magnetism. Medically useful coating materials are dextran, albumin, silicones, and poly(ethylene glycol). The reported hydrodynamic mean diameter in NMR experiments ranged from 10 to 3500 nm. A small particle size (5–10 nm) usually excluded any use as $T1$ contrast agent. The size of the Fe_3O_4 nanocrystals should be as large as feasible, while keeping the organic coating as thin as possible.[49]

Magnetite (Fe_3O_4) and maghemite (γ-Fe_2O_3) are commonly used as ^{1}H-NMR contrast agents for the imaging of cancer or other abnormal tissues. Commercial nanoparticles are typically coated with poly(vinyl alcohol) or dextran, and the *in vivo* half-life is usually short. A few minutes after intravenous delivery the particles are selectively captured by the reticuloendothelial system

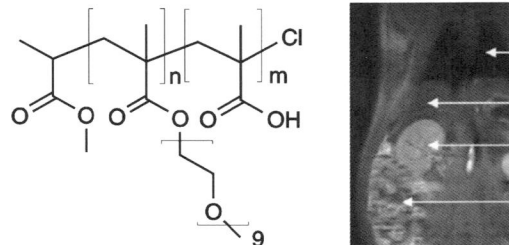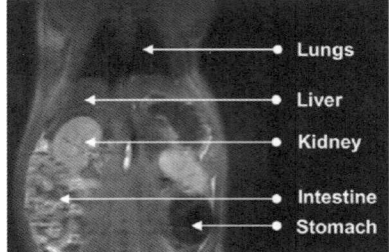

Figure 4.31 Magnetic resonance tomography of a rat coronal section measured 70 min after injection of 0.5 mL of a solution containing magnetite (Fe_3O_4) nanoparticles with the polymer coating on the left and an average diameter of 10.1 nm.[49] The picture shows no darkening of the liver image. Such images are never impressive; only a careful comparison of treated and non-treated sections and ill and healthy subjects provides information for the physician. © American Chemical Society.

and efficiently removed from the bloodstream. Only spleen and liver diagnostics are then feasible. Intravenous injection of nanoparticles coated with poly(-oligo(ethylene glycol) methacrylate-co-methacrylic acid), however, showed no detectable darkening of the liver within the first 2 h (Figure 4.31).[49]

The histidine hexamer (His_6) tags at the C- or N-terminal of a recombinant protein bind strongly to divalent metal chelates such as Ni(II) nitrilotriacetate. Four of the six coordination sites of octahedral Ni(II) are then occupied by the ligand, the two remaining sites by two of the six imidazole moieties in the His_6 tag. The His_6 unit is small, binds weakly to other amino acids and generally does not interfere with the native structure and function of the tagged protein.

Ni(II)-bis-and mono-nitrilotriacetic acid (NTA) chelates anchored with a catechol to a superparamagnetic iron oxide nanoparticle (SPION) bound to His_6-tagged proteins are most useful for the transport of proteins in the bloodstream. The NTA chelate density on the nanoparticles was so great that two neighbouring ligands worked together to bind to proteins in their native, folded conformations *via* their poly(histidine) tag (Figure 4.32). Simple addition of excess imidazole or histidine was then sufficient to release the tagged proteins quantitatively from the magnetically steered nanoparticles and let them search out a biological target.[50]

In diagnostic medicine *in vitro* nanoparticle systems are designed to sense single molecular targets, but *in vivo* sensors would benefit from the ability to simultaneously monitor two or three molecules or enzymes that cooperate in a disease (see Chapter 2, section 2.11 for an example). (Chapter 2, ref. 162).

Iron oxide nanoparticles with a chiral mannose surface were formed simply by *in situ* coating, namely precipitation of Fe(II) and Fe(III) salts with ammonium hydroxide in the presence of a D-mannose solution. The alternative postsynthesis coating of magnetite nanoparticles treated with sodium hypochlorite followed by addition of D-mannose solution gave similar results.[51] The first method gave tiny 2 nm particles, the second one detectable 6 nm nanoparticles. D-Mannose molecules are bound *via* the axial hydroxy group at C2,

Iron 279

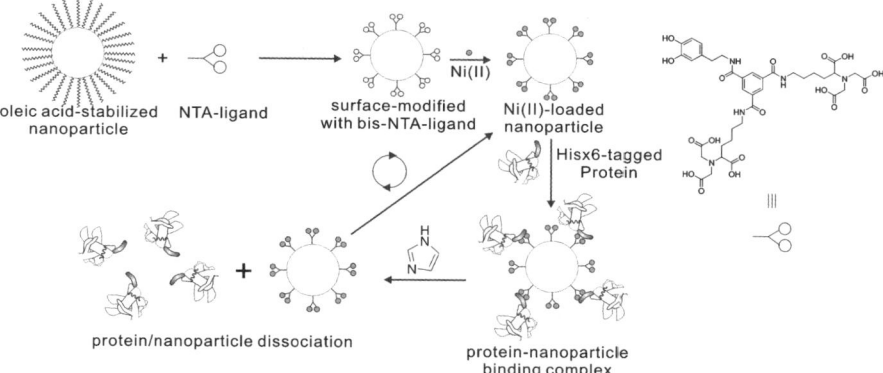

Figure 4.32 Coating of an Fe$_3$O$_4$ nanoparticle with oleic acid, a ligand (NTA) for Ni(II) attachment, Ni(II) and a his$_6$-tagged protein molecules. The latter was removed by addition of imidazole or histidine. The released protein then produced sharp ^{1}H-NMR signals. © American Chemical Society.[50]

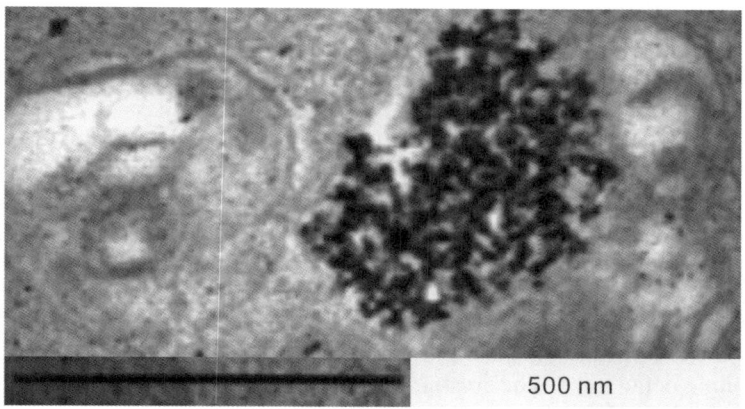

Figure 4.33 Transmission electron micrograph of bone marrow stromal cells labelled with D-mannose modified magnetite nanoparticles. © American Chemical Society.[51,52]

presumably *via* an Fe–O–C linkage. The D-mannose coating suppressed nonspecific sorption of serum proteins and allowed the magnetic Fe$_3$O$_4$ nanoparticles to pass through the membranes of rat bone cells (Figure 4.33). Minimal particle size was important for the ease of cellular internalization, but it reduced the magnetic susceptibility of the magnetite nanoparticles. The magnetization of protein-bound Fe$_3$O$_4$ nanoparticles could also be visualized with a light microscope. Applying a magnetic field in a horizontal direction resulted in a reversible linear alignment of dark nanoparticle assemblies.[52]

Cubic Fe$_3$O$_4$ nanoparticles of 9–23 nm side length were obtained in the presence of sodium oleate as stabilizer (Figure 4.34). The cubic shape of the

Figure 4.34 Electron micrographs of magnetite nanocubes © American Chemical Society..[53]

nanocrystals was a result of a slower growth rate of the 100 facets as compared to all other facets. This anisotropy was exclusively attributed to the adhesion of sodium oleate to this surface, since its addition converted the original spheres, which developed with oleic acid as well as its potassium or ammonium salts, into cubes. Presumably it is again the hydrate water on Na^+ (p. 14) that erodes the spheres. The presence of a carboxylate with an affinity to basic metal oxide surfaces together with a mobile non-polar chain generally prevents the rapid crystal growth of nanocrystals and stabilizes them in organic solvents such as high-boiling octadecene. The crystal structure of spherical and cubic nanocrystals was always that of an inverse spinel, and the diameter was always ~ 20 nm with size deviations $< 10\%$. Both types of nanocrystals are paramagnetic but interesting differentiations occur at 5 K. At this low temperature the magnetization saturates at 0.2 T for 13 nm cubes, but needs well over 1 T for 16 nm spheres with the same volume. The ordered crystal planes thus favor the parallel orientation of the electron spins, so information may be stored first in the cubes, and then at higher magnetization in the spheres also, provided both are kept cold enough.[53]

Single crystalline Fe_3O_4 nanowires were first grown on magnesium oxide nanowires on silicon/silica substrates. An Fe_3O_4 layer was deposited on to the magnesium oxide using the pulsed laser deposition technique (p. 35, 57). The magnesium oxide inner core was then removed with a $(NH_4)_2SO_4$ solution at pH 6.0 and 80 °C. This took 90 min because diffusion, surface dissolution and salt transport were slow in the nanotubes.[54,55] The main reason that the combined diffusion times were so much longer than the transport times of quencher molecules in hydrophobic yoctowells is, of course, the enormous difference between the

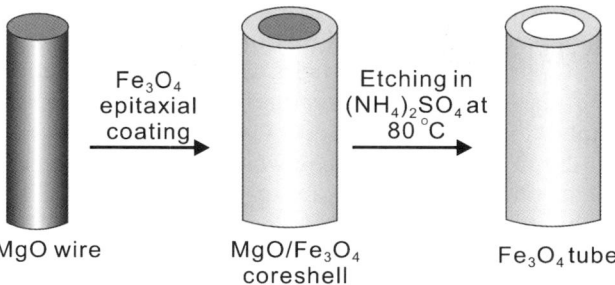

Figure 4.35 Fabrication of magnetite nanotubes. © American Chemical Society.[54,55]

distance in micron-long Fe_3O_4 capillaries as compared to the 2 nm deep yocto-wells (see Chapter 1, ref. 13). Electron microscopic images of the Fe_3O_4 nanotubes show an inside hollow region of ∼25 nm and the uniform sidewall had a thickness of 7 nm (Figure 4.35). Electron diffraction indicated a single-crystal spinel. The room-temperature resistivity was $4 \times 10^{-2}\,\Omega\,cm^{-1}$ for a single Fe_3O_4 nanotube, which is in good agreement with the value found for epitaxial Fe_3O_4 nanolayers.[56]

One-dimensional ferromagnetic nanowires and nanotubes may be applicable in high-density data storage and can be of various lengths, diameters and thicknesses, hopefully leading to an improved remnant magnetization. $Fe_2(O_tBu)_6$ was thermally decomposed to haematite (Fe_2O_3) and deposited in a porous anodic alumina membrane. Hydrogenation (5% H_2/95% Ar) at 400 °C converted the Fe_2O_3 to Fe_3O_4, accompanied by a colour change from golden or coppery brown to black, indicating a decrease of the band gap. The reduced tubes were air-sensitive and were kept under a macroscopic layer of polystyrene, which allowed for handling in air over a period of days to weeks.

Very thin Fe_3O_4 nanotubes showed a soft ferromagnetism with a vanishingly small remnant magnetization and coercive field. With increasing thickness of the nanotubes a monotonic improvement of their magnetic properties up to an optimum near a diameter of 13 nm was observed. Further increases in thickness were first accompanied by receding remanescence and coercivity, but above 20 nm the magnetism improved again. This behaviour is not inherent to single tubes, but was ascribed to antiferromagnetic coupling between neighbouring tubes through their stray fields.[56]

Without any catalyst, hydrocarbons and carbohydrates decompose to charcoal or graphitic carbon at temperatures between 500 and 1000 °C. Iron–carbon compounds, *e.g.* iron cyclooctatetraene, $Fe(COT)_2$, separate into iron oxide nanoparticles and graphite in the heat, but may also decompose to form a black, reflective graphite containing long tubes with a diameter of 4 nm in addition to magnetite (Fe_3O_4) nanoparticles. The condition for the latter process was to reflux $Fe(COT)_2$ in dimethoxyethane ($CH_3OCH_2CH_2OCH_3$) and dimethylsulfoxide (($CH_3)_2SO$) for 5 days. The process was slow – no graphite appeared after just 12 h at reflux, and at 300 °C in heptadecane the graphite black appeared only after 1 h. The iron oxide particles were removed with hydrochloric acid and graphite sheets and carbon nanotubes were isolated. Addition of Fe_2O_3 or Fe_3O_4 nanoparticles catalysed the formation and growth of carbon nanotubes at higher temperatures (400–1000 °C). The formation of crystalline graphite occurred even at 300 °C, an all-time low.[57]

Oxidation of Fe_3O_4 gives haematite in which the Fe(III) ions are statistically ordered in octahedral and tetrahedral gaps. α-Haematite is the thermodynamically most stable iron oxide. Bulk α-haematite is red and has a corundum (= aluminium oxide, Al_2O_3) structure. It can be described as a hexagonal close-packed array of oxygen atoms in which two-thirds of the octahedral holes are occupied by the Fe(III) ions; alternatively it can be viewed as an FeO_6 octahedron linked together by shared edges, faces, or vertices. Bulk γ-haematite or maghaemite is brown-black and has a spinel structure. Which form dominates in a nanocrystal population can at first be judged by the colour, secondly by the X-ray diffraction pattern. Hollow haematite nanoparticles were obtained by simple sonochemical decomposition of $Fe(CO)_5$. Iron shells that were first formed around carbon nanoparticles and burned upon exposure to oxygen. The heat released by the rapid oxidation of the high-surface-area iron shells ignited the carbon particles and their combustion generated even enough heat to form partly crystalline thin Fe_2O_3 shells (Fig 4.36).[58]

Haematite nanoparticles adsorb zinc chloride, $ZnCl_2$, in the form of four and six-co-ordinated hydroxides at pH 5.5[59] and were also tried as magnetic transport vehicles for dopamine. It was thought that their oxidative character

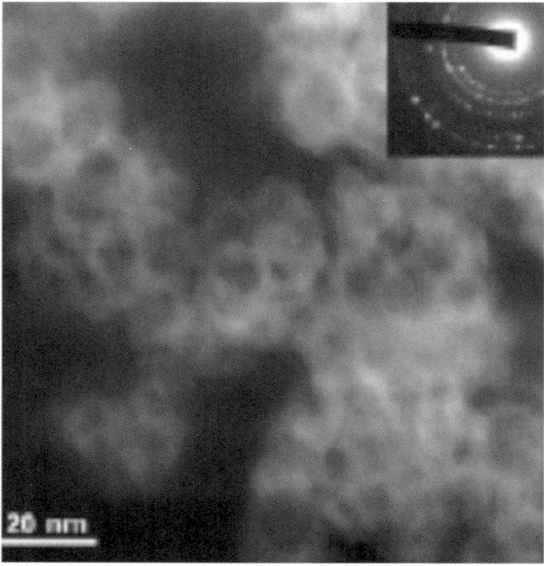

Figure 4.36 Dark-field electron micrograph and electron scattering pattern of hollow haematite nanocrystals. © American Chemical Society.[58]

Iron

would be neutralized by zinc(II) hydrates as well as by reducing biological fluids. In 2004 dopamine was even introduced into nanocrystal particle synthesis as a "stable anchor to present functional molecules on the surface of iron oxide nanostructures... due to the inherent biocompatibility of iron oxide". Dopamine was chosen because "the bi-dentate enediol ligand converted the under-coordinated iron surface sites back to a bulk-like lattice structure with an octahedron of oxygens around iron". Neither the oxidation of dopamine to a quinone on Fe(III) nanoparticles, nor a rapid decomposition of the nanocrystals, which should follow, was considered. This came only three years later. Attempts to use the bis-phenolic dopamine as a "robust" anchor on commercial γ-Fe_2O_3 nanoparticles led to a clear, blue/purple ferrofluid. After remaining in solution for 1 day, however, green particles made of black-magnetite/maghemite and Fe(II) hydroxide appeared, and turned slowly red/brown, typical of Fe(III) oxyhydroxide, Fe(III)OOH. The morphology of the particles was lost (Figure 4.37). This happened in neutral aqueous solutions under air. In the reductive medium of cell water or blood, which were not investigated, the Fe(III) oxide particles should be even more reactive. One should not combine hundreds of reactive Fe(III) ions in one particle with phenols and hope for inertness. Local oxidation potentials around Fe(III) nanoparticles will be high and an uncontrolled destruction of phenols, or more generally of organic tissue, is likely. In the dopamine–magnetite nanoparticle case, transmission electron microscopy showed spherical particles and concentric rings as well as a cloudy region around the particles, which was attributed to ferric catechol bound to the dopamine layer instead of two "robust" Fe–O bonds. UV-vis spectra revealed the quinone absorbance around 420 nm.[60]

Magnetic γ-Fe_2O_3 nanoparticles were also added to the surface of the temperature-dependent "smart" poly(*N*-isopropylacrylamide) microgels (p. 116 ff) in order to produce a magnetically and thermally responsive material. The acrylamide was redox-inactive and no problems occurred at first upon adding γ-Fe_2O_3. But then 38% (w/w) of γ-Fe_2O_3 blocked the desired thermal

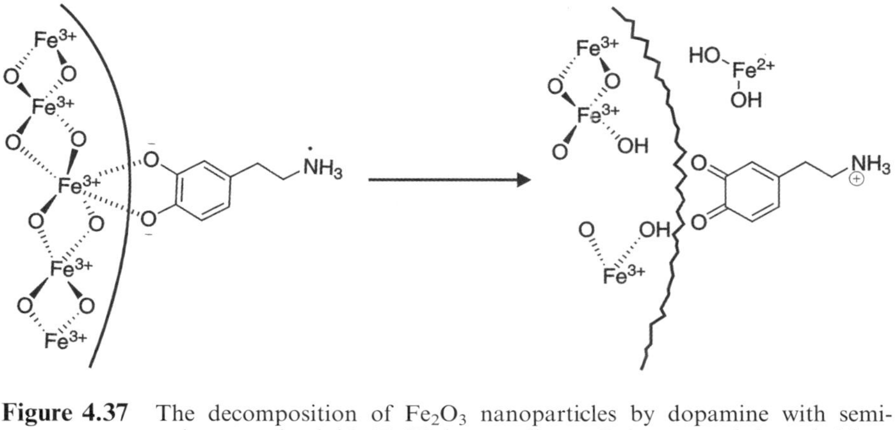

Figure 4.37 The decomposition of Fe_2O_3 nanoparticles by dopamine with semi-quinone and soluble Fe(II) intermediates. Quinones and degraded iron oxide nanoparticles are the products.[60]

isomerization of the polymer. Too many amide bonds were immobilized by the polar iron oxide nanoparticles. 18% of Fe_2O_3 (w/w) was just the right amount to raise the transition temperature of the polymer from 36 °C to the desired 40 °C, the temperature of inflamed tissue and growing cancer cells. External magnetic fields then located polymer-dissolved drugs, and the high temperature at sites of inflammation in the body released the drugs as planned.[61]

γ-Fe_2O_3 nanotubes 200 nm wide have been made in poly(carbonate) matrices. High magnetic anisotropy in parallel and perpendicular magnetic fields was achieved, together with high coercivity values. If every ferromagnetic nanotube within parallel arrays acted as a recording unit, very high magnetic recording density would also be attainable.[62]

4.6.1 Ferrites

Ferrites are mixed metalloxides, $M(II)Fe_2(III)O_4$, usually with a spinel or inverse spinel structure. In the spinel structure the oxygen anions O^{2-} form a cubic lattice (p. 247), the dications are in 1/8 of the tetrahedral gaps and the trications occupy one half of the octahedral gaps. An inverse spinel replaces the dications with one half of the trications: two of the three cations, namely M(II) and half of M(III), are in octahedral gaps. The energy splitting of the d-orbitals is smaller in tetrahedral gaps: only 4/9 of the splitting in the octahedral gaps. The paramagnetic Fe(III) ions (d^5) have a ligand field stabilization of zero, all five electrons are equal, and they tend to occupy tetrahedral gaps. Fe(II) d^6 prefers octahedral gaps(inverse spinel); Mn(III), on the other hand, is diamagnetic (d^4) and is preferably located in octahedral gaps; Mn(II) (d^5) correspondingly goes tetrahedral (spinel).

The most versatile iron oxides for chemical tuning of composition, structure and properties are the ABO_3 perovskites with iron as B, having a large variety in the geometry of the d electrons and an A site where size-related coordination preferences determine the properties of mobile oxide anions in ion-conducting systems. In the trilayered magnet structure $Ca_2Ba_2Nd_2Fe_6O_{15.6}$ three different Fe(III) ions are present: Fe(1) is square-based pyramidal, Fe(2) octahedral, and Fe(3) tetrahedral. Only the octahedral Fe^{3+}(2) (75%) and the tetrahedral Fe^{3+} (3) (25%) ions were found to be magnetic. The oxygen defects were characterized by diffuse electron density features in electron and synchrotron X-ray diffraction measurements of single layers. Point-like variations of the contrast were locally observed. The individual magnetism of the Fe(III) centres was deduced from magnetic susceptibility measurements, Mössbauer spectra and neutron powder diffraction patterns at different temperatures (Figure 4.38).[63] The distorted iron ions (1) and (3) did not contribute, since a parallel alignment of spins was not possible. Small amounts of aluminium oxide on iron(0) nanocrystals form a $Fe_2AlFe_6O_{12}$ layer, which prevents the oxidation of iron by oxygen.[64]

Ferrites may produce high values of magnetization at room temperature, which is needed when the operating frequencies increase. Molecular field

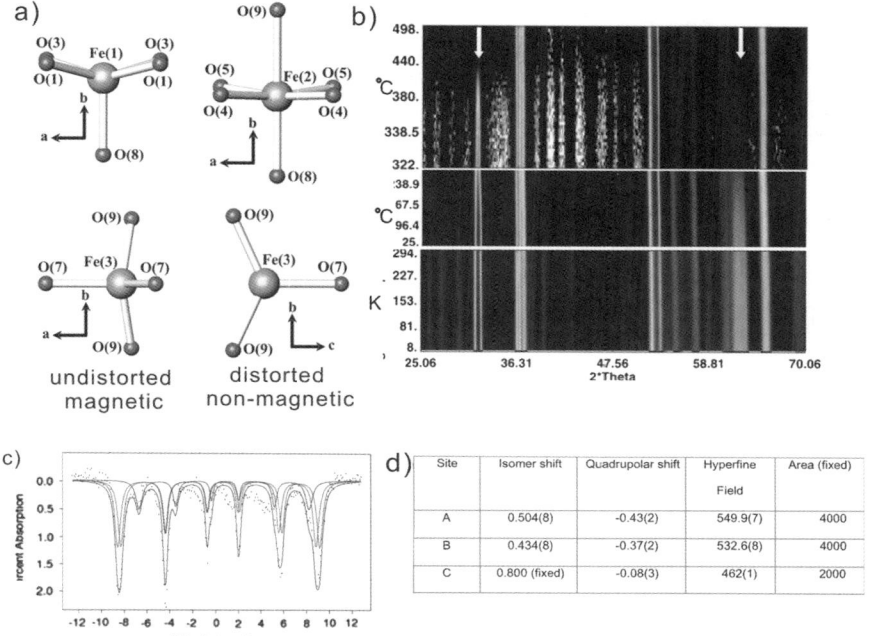

Figure 4.38 (a) Coordination of four iron centres in the ferromagnet $Ca_2Ba_2Nd_2Fe_6O_{15.6}$. (b) Evolution of the neutron powder diffraction patterns vs. temperature showing the disappearance of the magnetic Bragg peaks above 700 K. (c) Typical Mössbauer spectra at 77 K indicating 30–40% Fe^{4+} is present in the cubic materials produced in the initial firing in air. 150 K spectra show the onset of magnetic order and are consistent with a charge state of $Fe^{+3.33}$. (d) ^{57}Fe Mössbauer parameters at 77 K. © American Chemical Society.[63,64]

calculations showed a maximum magnetization of 8000 Gauss for the ideal, hypothetical case, where every magnetic cation is Fe^{3+} in a spinel unit $Fe^{3+}{}_{(1-x)}Q_x[Fe^{3+}]O_4$, x is close to 0.1 and the Q site is a diamagnetic metal, e.g. Zn^{2+} or Ba^{2+}. The hard magnetic material barium hexaferrite ($BaFe_{12}O_{19}$), which is accessible from iron nitrate and barium carbonate in ethylene glycol, comes close to this value. Its dispersion was distributed over a silicon surface, annealed at 450–950 °C and scanned with a magnetic AFM tip. Strongly magnetic micron-length bars of 5 nm height were thus formed.[65,66]

Lithium ferrite ($Li_{0.5}Fe_{2.5}O_4$) has a strong dielectric tensor, which results in a magneto-optical effect in the visible region. Transparent lithium ferrite nanoparticles in ethyl(hydroxy-ethyl)cellulose films showed a specific Faraday rotation of the plane of a polarized 600–1000 nm light beam under magnetic fields of 2.5, 5.0 and 10 kOe (Figure 4.39).[67]

In the case of cobalt ferrite ($CoFe_2O_4$) the rotation at 633 nm becomes enhanced by almost a factor of 10 in the presence of an silver nanoparticle close to the ferrite nanoparticle.[68] The magnetic field induces chirality in the ferrite

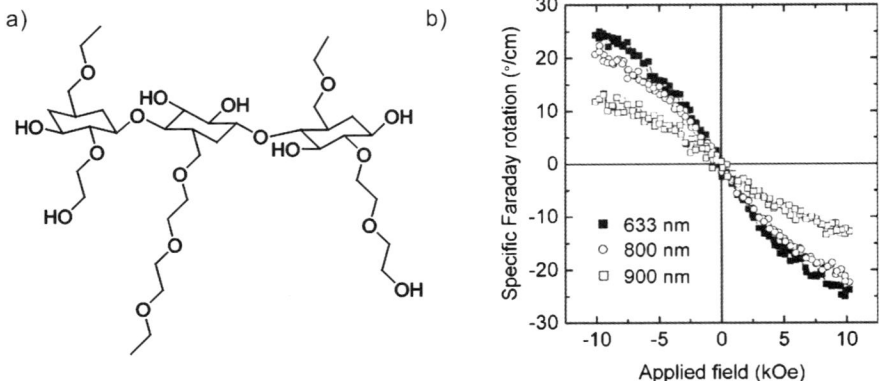

Figure 4.39 Structure of the cellulose polymer unit carrying the transparent lithium ferrite($Li_{0.5}Fe_{2.5}O_4$) nanoparticles and the rotation of the plane of polarized light. © American Chemical Society.[67]

nanoparticles, and dielectric contributions of the photon energy of a touching, single metal nanoparticle enhances it.

Mackinawite, tetrahedral FeS, is composed of FeS layers with iron sandwiched between sulfur sheets, which are held together by weak van der Waals interactions. Other heavy metal atoms are captured in vacancy sites between the layers. The absence of any magnetic moment and the close Fe–Fe distance (2.65 Å) within each layer indicate extensive iron d-orbital delocalization and metallic character.[69]

Fe_7S_8 and Fe_3S_4 (= greigite) nanosheets were synthesized by pyrolysing Fe(II) and Fe(III) complexes in oleylamine. The ferrous complex yielded Fe_7S_8, the ferric complex Fe_3S_4 nanosheets, up to 500×500 nm in size and ~ 50 nm thick. Both had a hexagonal shape as in graphene sheets.[70]

Magnetotactic bacteria produce greigite (Fe_3S_4) and pyrite (FeS_2) 20 nm nanoparticles instead of ferrites for iron storage.[71] All iron sulfide nanoparticles are readily, but often not quantitatively, oxidized to iron oxide nanoparticles in air.

4.7 Photosynthesis and Catalysis on Iron Oxide Nanoparticles

The first step in the photoelectrolysis of water by sunlight (a photosynthesis model) is the harvesting of the visible part of the solar spectrum by a two (or more) photon system (p. 23) and the reduction of protons on a light-driven cathode (p. 304, 310–311). The most difficult part, however, is the light-driven oxidation of water to molecular oxygen.

The classical titanium dioxide (titania) system only works with UV light, but the red α-haematite (α-Fe$_2$O$_3$) absorbs in the visible range, due to its indirect band gap (~2.1 eV). The absorption coefficient is, however, small and the hole diffusion length short. Furthermore, the high oxidation potential does not allow any direct use in proton reduction, although this may be overcome by an external bias voltage as supplied by a dye-sensitized solar cell in a tandem-cell configuration (p. 310).

The remaining problem, therefore, is an efficient haematite photoanode for water oxidation. Pure α-haematite is an inefficient photoelectrode, but translucent silicon-doped thin films made of haematite nanoleaflets provided photocurrents under illumination with sunlight. The electrical conductivity is up to four orders of magnitude higher along the (001) basal plane of silicon-doped haematite than perpendicular to it, which may be the main reason for the high water-splitting efficiencies obtained. The silicon acts here as an electron donor due to substitution of Fe^{3+} by Si^{4+} in the haematite lattice. It also reduces the grain size to a level that is commensurate with the hole diffusion length of only a few nanometres in haematite. Each hole can now reach the surface.

The specific surface area of the photoanode may be greatly increased by using a haematite microstructure with nanometre-sized branches (Figure 4.40). Hole capture by a surface site thus becomes competitive with recombination, even without an electrical field. A dendritic haematite microstructure, ~170 nm high with branches 20–30 nm thick, achieved photocurrent up to 1.45 mA/cm^2 at 1.23 V vs. the hydrogen electrode, which is sufficient for the oxidation of water.[72,73]

When sunlight strikes the cell, the top electrode will absorb the photons (hv) having a wavelength below 600 nm and produce electron–hole pairs (h$^+$e$^-$) in the γ-haematite film:

$$hv + Fe_2O_3 \rightarrow Fe_2O_3 + (h^+e^-)$$

Figure 4.40 Scanning electron microscope image of a silicon-doped dendritic haematite film. © American Chemical Society.[72]

The holes then migrate in the valence band to the surface where they react with water producing oxygen and protons:

$$4h^+ + 2H_2O \rightarrow 4H^+ + O_2.$$

The photogenerated electrons e^- then travel from the conduction band of the haematite to the conducting glass support, which now serves as a collector of electrical current and injects them into the cathode of a second cell, *e.g.* a dye-sensitized solar cell.

The ferromagnetism of iron is not restricted to the metal and its oxides, but is also present in some paramagnetic Fe(II)/Fe(III) salts. Prussian blue Langmuir–Blodgett films, for example, have been made by spreading a positively charged dioctadecyldimethylammonium ion monolayer on a very diluted colloidal dispersion of "soluble" Prussian blue with or without a mechanically stable clay mineral in between. The hybrid film was magnetized anisotropically by visible light at 5 K (Figure 4.41).

If Fe(III) ions are added to aqueous solutions of $K_4(FeII(CN)_6)$ an intensely blue precipitate of Prussian blue (KFe(II)[Fe(III)(CN)$_6$]) is formed. The Fe(III) ions were partially reduced by adding potassium metal and at K^+ ions, which converted the soluble Prussian blue reversibly to an insoluble modification. The polarizable K^+ ions contracted the crystal lattice and enhanced the magnitude of the magnetic field mostly at a Fe^{3+}/Fe^{2+} ratio close to 0.40. A thin layer of KFe(II)[Fe(III)(CN)$_6$] on a vibrating quartz resonator in air then increased the motional resistance of the crystal if an external magnetic field was applied. The coercive field associated with electromagnetic materials thus changed the the elasticity of the quartz crystal's surface.[74,75]

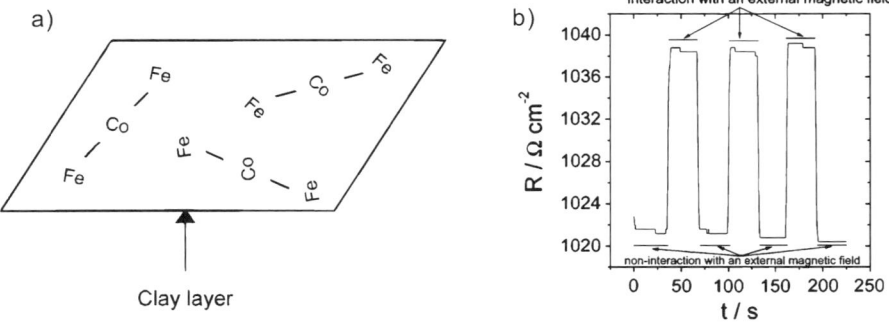

Figure 4.41 (a) A thin cobalt–ferricyanide (cyanide omitted for clarity) layer on clay becomes magnetic at 5 K, if irradiated with visible light. (b) A quartz crystal with a Prussian blue surface layer changes its motional resistance in a magnetic field. © American Chemical Society.[74]

4.7.1 Catalytic Iron Nanocrystals in Industrial Processes

For almost 100 years iron oxides have served as the catalytic basis for the synthesis of ammonia from nitrogen, water and carbon monoxide by the Haber–Bosch process. The conversion of coal to hydrocarbon (gasoline) using water (the Fischer–Tropsch process) came only 30 years later. Both of these technological breakthroughs had at first the unfortunate consequence of prolonging wars and killing millions of people – the typical disaster of progress with iron. Today the Haber–Bosch process requires $\sim 1\%$ of the world's total power generation and yields fertilizers and polyamides. The Fischer–Tropsch process may become an important source for the chemical industry in future.

In the Haber–Bosch process, high pressure (1500 bar) and temperature (400 °C) ensure that the surface of the iron oxide catalyst is covered with nitrogen (N_2), as well as converting its triple bond into a double bond, splitting into N atoms and hydrogenation to NH groups. Bond-breaking upon chemisorption is the most important task of iron catalysts in ammonia synthesis. In industrial catalysts the specific surface area of the iron oxide on porous alumina is $20\,m^2\,g^{-1}$. This surface area is exclusively obtained by active iron oxide nanoparticles on a heat-stable framework made of Al_2O_3–CaO, which stabilizes the active nanoparticles against sintering. At high temperatures and pressures, hydrogen is first formed by the reduction of water with carbon monoxide and the iron oxide nanoparticles are reduced to metallic iron. This is covered by a monolayer of K (+O), which acts as an "electronic" promoter.

The chemisorption of molecular nitrogen on the Fe(0) single-crystal surfaces causes the formation of N atoms. The probability for this process (= sticking coefficient) is typically of the order of 10^{-6}, and this step is rate limiting. The most densely packed (110) surface is least active. The open (111) plane exhibits the highest sticking coefficient and is indeed responsible for the overall activity of the industrial catalyst (Figure 4.42). This activity is further enhanced by the presence of the electronic promoter potassium, which stabilizes the intermediate $N_{2,ad}$ complex and increases its dissociation rate by pulling electrons into the system. Pressurized Fe–N_2–K chemisorption complexes offer reaction pathways, whose energy differences can be surmounted by the available thermal energy.[76]

With respect to the all-important nitrogen activation, three-coordinate iron nitrogen complexes with Fe(II) or Fe(I) are possible intermediates, where extensive π back-bonding lengthens the N–N triple bond from $1.10\,Å$ to 1.18–$1.19\,Å$. This would be consistent with an easily reducible N–N double bond,[77] but their importance in the Haber–Bosch process has not yet been demonstrated.

The Fischer–Tropsch process transforms synthesis gas (CO + H_2) at $\sim 250\,°C$ to an alkane mixture $(CH_2)_n$. The feedstock is simply pumped through a slurry reactor containing transition metal catalysts such as iron, cobalt, and ruthenium. Iron-based catalysts allow for a low H_2 content of the feed gas because iron generates H_2 from H_2O.-containing by-products. However, iron suffers from surface carbon accumulation that deactivates the catalyst, which is

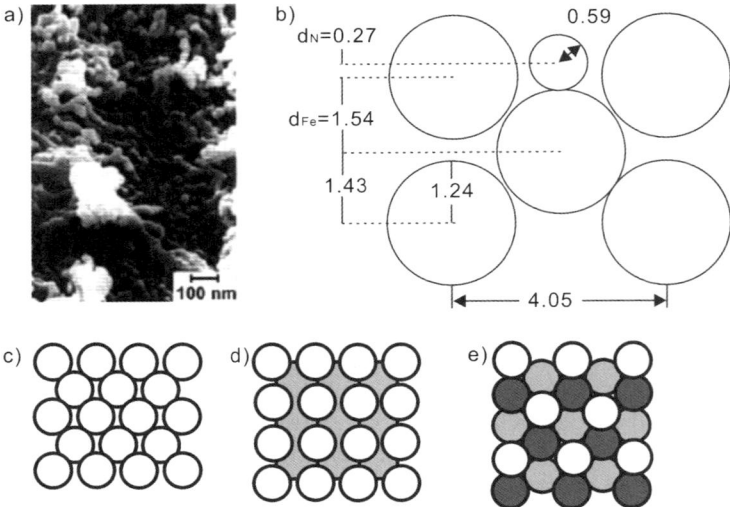

Figure 4.42 (a) Surface topography of an industrial ammonia synthesis iron oxide catalyst on alumina-K(O), the first NPs of industrial importance. The catalytically active spot contains 33.2% O, 30.1% Fe, 29.0% K, 6.7% Al, and 1.0% Ca. (b, c) Detailed and schematic view of the Fe(100) plane, where an N atom and NH find enough space between the Fe atoms. Chemisorption and activation occurs there, distances are given in Å = 0.1 nm. At first N_2 is split which is supported by the electrons from H_2 and H_{ad}. NH_{ad} is the first reduction product on the way to NH_3.[78] (d, e) the (110) and (111) surfaces provide not enough space for an N atom, if the N_2 molecule is adsorbed there, they may rearrange to (100) (see Figure 5.55).

avoided by more expensive cobalt catalysts.[78] Among all possible C_1 species, CH is identified to be the most dominant on the surface, which is in good agreement with the experimental observations that iron–cobalt catalysts are immune to carbon deposition, and possess similar catalytic properties to the conventional cobalt catalysts used in the Fischer–Tropsch synthesis.[79]

The most probable C–C bond propagation in the Fischer–Tropsch catalysis is polymerization of alkene (C=C=CH$_2$) and acetylene (CCCH$_3$) precursors. Hydrogen addition follows the site preference γ position $>\beta$ position $>\alpha$ position.

Approximately 75% of the surface of the catalyst is present as a carbide phase ($Fe_{2.23}C$), the remaining 25% being magnetite (Fe_3O_4). Magnetite is a cubic mineral with an inverse spinel structure. In a close packing of oxygen atoms, Fe^{3+} and Fe^{2+} ions are octahedral and Fe^{3+} tetrahedral sites. Fast electron hopping occurs between the two cations in the octahedral sites. The dominant lattice faces of magnetite are (111) and (110), but none of them has a neutral charge and the surface tends to reconstruct. The actual surface structures of magnetite and of spinel metal oxides are not known in sufficient detail.

The kinetically significant steps of the Fischer–Tropsch reaction are the dissociation of carbon monoxide, the partial hydrogenation of the CH_x species, the insertion of methylene, the reductive elimination toward n-alkanes and the elimination of β-hydride. Of all possible C_1 species, CH was identified to be the most dominant on the real catalyst surface as well as on the resulting kinetic potential energy surface.[79]

Iron oxide–iron nanoparticles entrapped in carbon nanotubes and active in Fischer–Tropsch catalysis showed a ratio of iron carbide (Fe_xC_y) to oxide (FeO) of 4.7 for the encapsulated iron catalyst in comparison to 2.4 for the iron catalyst dispersed on the outer walls. The yield of C_5 hydrocarbons over the encapsulated iron catalyst was twice that over iron catalyst outside the carbon nanotubes and more than six times that over free iron catalyst. The confinement effect thus gave C_5 rather than C_3 intermediates.[80]

"Skeletal" iron catalysts have been prepared by alkali leaching of an $Fe_{50}Al_{50}$ alloy and were solidified by spraying a 1600 K melt on to a high-speed water-cooled copper roller under argon. The catalyst surface area was $42\,m^2\,g^{-1}$, the pore size 16.7 nm. This iron–aluminium catalyst favoured the production of light alkanes, which was attributed to fast reactions on small magnetite nanoparticles and large numbers of separated iron carbide nanocrystals.[81,82]

4.8 A Brief Comparison of Carbon and Iron

Both carbon and iron are reducing agents. Both are easily oxized: carbon by molecular oxygen, iron by water. Nevertheless, carbon, as formed in large amounts by the dehydration of carbohydrates or rearrangement of lignine, has survived underground for many millions of years. Elemental graphite or coal, containing only a little oxygen and nitrogen, has endured because the resonance-stabilized benzene polymers survive attack by water, oxygen and bacteria. Even the tetrahedral diamond is practically inert on Earth, because it forms very hard crystals.

Iron could have made steel by combining with carbon, but it never managed to become "stainless" without combining with Cr(0) or similar metals. For terrestrial minerals, it is not oxygen that oxidizes Fe(0) and Cr(0), but the omnipresent water, which permits only the metal oxides. Human civilization then brought the reductive power of carbon to the iron oxides and stabilized Fe(0) by Cr(0) and other metals in alloys.

Elemental iron is extracted from oxides, on the scale of ~ 100 kg/year for each human being (~ 5–6×10^8 t in total). About 3.6×10^9 t/year of oil or 500 kg are used up for each human being, and 6.2×10^6 t/year of coal (a factor of 500 less). For comparison, $\sim 10^{11}$ t/year or 15 t per person of glucose is produced by photosynthesis.

The science of carbon and iron nanochemistry is dominated by dynamic stereochemistry and by interactions with water. The specific ways in which carbon nanotubes are rolled up dominate their colour, conductivity and surface interactions; the properties and applications of C–C and nano-relevant polymers are functions of the relative arrangements of hydrophobic and hydrophilic domains; the fluidity of non-covalent membranes depends on *cis* double bonds and determines their use in red blood cells, in the retina and in neuronal ion transport as well as for the formation and stabilization of all kinds of nanocrystals.

The stereochemistry of iron is of comparable complexity and beauty. The different types of iron magnetism allow for the long-lived and fast responses of computer memories; the size and shape of steel domains dominate its material science; the attachment of molecules to the iron (111) plane and its edges gives it catalytic activity.

Carbon and iron are the central elements of human life, and water and oxygen are the compounds of central importance in reactions with carbon and iron. The teaching of chemistry in schools and universities should start with these four. If their actions and interactions are understood, this basic knowledge can easily be extended to other elements and molecules.

References

1. S. V. Didziulis, K. D. Butcher and S. S. Perry, Properties of titanium carbide, vanadium carbide, and titanium nitride, *Inorg. Chem.*, 2003, **42**, 7766–7781.
2. A. Khan, Theoretical studies of the structure of Ti_8C_{12} clusters: existence of C_{12} cage structure surrounded by metal atoms, *J. Phys. Chem.*, 1993, **97**, 10937–10941.
3. H. K. D. H. Bhadeshia, The structure of cementite, www.msm.cam.ac.uk/phase trans/2003/Lattices/cementite.html.
4. H. Natter, M. Schmelzer, M.-S. Löffler, C. E. Krill, A. Fitch and R. Hempelmann, Grain-growth kinetics of nanocrystalline iron studied in situ by synchrotron real-time X-ray diffraction, *J. Phys. Chem. B*, 2000, **104**, 2467–2476.
5. D. Yuan, L. Ding, H. Chu, Y. Feng, T. P. McNicholas and J. Liu, Horizontally aligned single-walled carbon nanotube on quartz from a large variety of metal catalysts, *Nano Lett.*, 2008, **8**, 2576–2579.
6. S. M. Lundgren, K. Persson, G. Mueller, B. Kronberg, J. Clarke, M. Chtaib and P. Claesson, Unsaturated fatty acids in alkane solution: adsorption to steel surfaces, *Langmuir*, 2007, **23**, 10598–10602.
7. M. N. McCain, B. He, J. Sanati, Q. J. Wang and T. J. Marks, Aerosol-assisted chemical vapor deposition of lubricating MoS_2 films. Ferrous substrates and titanium film doping, *Chem. Mater.*, 2008, **20**, 5438–5443.
8. O. P. Khatri, S. Math, C. D. Bain and S. K. Biswas, Frictional response of a silane monolayer to sliding in a humid environment, *J. Phys. Chem. C*, 2007, **111**, 16339–16345.
9. H. Jiang, M. K. Jantan, S. Manolache, F. S. Denes and M. G. Lagally, Plasma-enhanced synthesis of thin fluoropolymer layers with low Raman and fluorescence backgrounds, *Langmuir*, 2008, **24**, 8672–8677.
10. M. Deya, A. R. Di Sarli, B. del Amo and R. Romagnoli, Performance of anticorrosive coatings containing tripolyphosphates in aggressive environments, *Ind. Eng. Chem. Res.*, 2008, **47**, 7038–7047.

11. K. D. Demadis, M. Papadaki, R. G. Raptis and H. Zhao, Corrugated, sheet-like architectures in layered alkaline-earth metal *R,S*-hydroxyphosphonoacetate frameworks: applications for anticorrosion protection of metal surfaces, *Chem. Mater.*, 2008, **20**, 4835–4846.
12. S. Park, D. Lee, C. Yu, K. Lee and K. Lee, Effect of silica particle size on performance of porous stainless-steel-supported silica membranes prepared by the DRFF and SRFF method, *Ind. Eng. Chem. Res.*, 2008, **47**, 6211–6215.
13. A. M. Mebel and D. Y. Hwang, Theoretical study of the reaction mechanism of Fe atoms with H_2O, H_2S, O_2 and H^+, *J. Phys. Chem. A*, 2001, **105**, 7460–7467.
14. D. Schroeder, Gaseous rust: thermochemistry of neutral and ionic iron oxides and hydroxides in the gas phase, *J. Phys. Chem. A*, 2008, **112**, 13215–13224.
15. R. O. Clark, Identification of rust on iron and steel, *Ind. Eng. Chem. Anal. Ed.*, 1943, 464–465, 736.
16. C. Ruby, C. A. Dhyay, A. Geahin, G. Nguema and J. M. R. Geanin, In situ redox flexibility of FeII-III oxyhydroxycarbonate green rust and fougerite, *Environ. Sci. Technol.*, 2006, **40**, 4696–4702.
17. M. C. F. Wander, K. M. Rosso and M. A. A. Schoonen, Structure and charge hopping dynamics in green rust, *J. Phys. Chem. C*, 2007, **111**, 11414–11423.
18. H. Park, C. D. Vecitis and M. R. Hoffmann, Solar-powered electrochemical oxidation of organic compounds coupled with the cathodic production of molecular hydrogen, *J. Phys. Chem. A*, 2008, **112**, 7616–7626.
19. S. N. Jampala, M. Sarmadi, E. B. Somers, A. C. L. Wong and F. S. Denes, Plasma-enhanced synthesis of bactericidal quaternary ammonium thin layers on stainless steel and cellulose surfaces, *Langmuir*, 2008, **24**, 8583–8591.
20. T. Karlsson, P. Persson, U. Skyllberg, C. M Morth and R. Giesler, Characterization of iron(III) in organic soils using extended X-ray absorption fine structure spectroscopy, *Environ. Sci. Technol.* 2008, **42**, 5449–5454.
21. W.-X. Zhang, Nanoscale iron particles for environmental remediation: An overview, *J. Nanopart. Res.*, 2003, **5**, 232–332.
22. J. T. Nurmi, P. G. Tratnyek, V. Visarathy, D. R. Baer, J. E. Amonetter, K. Pecher, C. Wang, J. C. Linehan, D. W. Matson, R. L. Penn and M. D. Driessen, Characterization and properties of metallic iron nanoparticles: spectroscopy, electrochemistry, and kinetics, *Environ. Sci. Technol.*, 2005, **39**, 1221–1230.
23. E. J Reardon, R. Fagan, J. L. Vogan and A. Przepiora, Anaerobic corrosion reaction kinetics of nanosized iron, *Environ. Sci. Technol.*, 2008, **42**, 2420–2425.
24. F. Alonso, I. P. Beletskaya and M. Yus, Metal-Mediated Reductive Hydrodehalogenation of Organic Halides, *Chem. Rev.*, 2002, **102**, 4009–4092.

25. B. Schrick, B. W. Hydutsky, J. L. Blough and T. E. Mallouk, Delivery vehicles for zerovalent metal NPs in soil and groundwater, *Chem. Mater.*, 2004, **16**, 2187–2193.
26. N. Saleh, T. Phenrat, K. Sirk, B. Dufour, J. Ok, T. Sarbu, K. Matyjaszewski, R. D. Tilton and G. V. Lowry, Adsorbed triblock copolymers deliver reactive iron NPs to the oil/water interface, *Nano Lett.*, 2005, **5**, 2489–94.
27. N. O. Nunez, R. Pozas, M. Puerto Morales, P. Tartaj, P. Bonville, A. R. Gonzalez-Elipe, A. Caballero, M. Ocana and C. J. Serna, Acicular metallic particles obtained from Al-doped goethite precursors, *Chem. Mater.*, 2003, **15**, 951–957.
28. V. Sudha and M. V. Sangaranarayanan, Underpotential deposition of metals: structural and thermodynamic considerations, *J. Phys. Chem. B*, 2002, **106**, 2699–2707.
29. S. Park, S. Kim, S. Lee, Z. G. Khim, K. Char and T. Hyeon, Synthesis and magnetic studies of uniform iron nanorods and nanospheres, *J. Am. Chem. Soc.*, 2000, **122**, 8581–8582.
30. M. Chen, T. Pica, Y.-B. Jiang, P. Li, K. Yano, J. P. Liu, A. K. Datye and H. Fan, Synthesis and self-assembly of fcc phase FePt nanorods, *J. Am. Chem Soc.*, 2007, **129**, 6348–6349.
31. L. Dai, T. Tu, S. You, W. Deng and X. Hou, Asymmetric catalysis with chiral ferrocene ligands, *Acc. Chem. Res.*, 2003, **36**, 659–667.
32. M. Ginzburg, J. Galloro, F. Jakle, K. Nicole Power-Billard, S. Yang, I. Sokolov, C. N. C. Lam, A. W. Neumann, I. Manners and G. A. Ozin, Layer-by-layer self-assembly of organic-organometallic polymer electrostatic superlattices using poly(ferrocenylsilanes), *Langmuir*, 2000, **16**, 9609–9614.
33. Y. Wang, N. Coombs, A. Turak, Z.-H. Lu, I. Manners and M. A. Winnik, Interfacial staining of a phase-separated block copolymer with ruthenium tetroxide, *Macromolecules*, 2007, **40**, 1594–1597.
34. X. Wang, K. Liu, A. C. Arsenault, D. A. Rider, G. A. Ozin, M. A. Winnik and I. Manners, Shell-crosslinked cylindrical polyisoprene-b-polyferrocenylsilane (PI-b-PFS) Block copolymer micelles: one-dimensional (1D) organometallic nanocylinders, *J. Am. Chem. Soc.*, 2007, **129**, 5630–5639.
35. B. R. Martin, D. J. Dermody, B. D. Reiss, M. Fang, L. A. Lyon, M. J. Natan and T. E. Mallouk, Sub-100 nm confinement of magnetic NPs using localized magnetic field gradients, *Adv. Mater.*, 1999, **11**, 1021–1025.
36. A. R. Urbach, J. C. Love, M. G. Prentiss and G. M. Whitesides, Sub-100 nm confinement of magnetic nanoparticles using localized magnetic field gradients, *J. Am. Chem. Soc.*, 2003, **125**, 12704–12705.
37. S. Laurent, D. Forge, M. Port, A. Roch, C. Robic, L. Van der Elst and R. N. Muller, Magnetic iron oxide nanoparticles: synthesis, stabilization, vectorization, physicochemical characterizations, and biological applications, *Chem. Rev.*, 2008, **108**, 2064–2110.

38. X. Li and W. Zhang, Iron nanoparticles: the core-shell structure and unique properties for Ni(II) sequestration, *Langmuir*, 2006, **22**, 4638–464.
39. A. Cabot, V. Puntes, E. Shevchenko, Y. Yin, L. Balcells, M. A. Marcus, S. M. Hughes and A. P. Alivisatos, Vacancy coalescence during oxidation of iron NPs, *J. Am. Chem. Soc.*, 2007, **129**, 10358–10360.
40. T. G. St. Pierre, J. Webb and S. Mann, Ferritin, in: *Biomineralization*, ed. S. Mann, J. Webb and R. J. P. Williams, VCH/Wiley, Weinheim, 1989, pp. 295–243.
41. J. Kanner and L. Doll, Ferritin in turkey muscle tissue: a source of catalytic iron ions for lipid peroxidation?, *J. Agric. Food Chem.*, 1991, **39**, 247–249.
42. J. M. Dominguez-Vera and E. Colacio, Nanoparticles of prussian blue ferritin: a new route for obtaining nanomaterials, *Inorg. Chem.*, 2003, **42**, 6983–6985.
43. Q. Tang, H. K. Moon, Y. Lee, S. M. Yoon, H. J. Song, H. Lim and H. C. Choi, Redox-mediated negative differential resistance behavior from metalloproteins connected through carbon nanotube nanogap electrodes, *J. Am. Chem. Soc.*, 2007, **129**, 11018–11019.
44. S. A. Steiner, T. F. Baumann, J. Kong, J. H. Satcher Jr and M. S. Dresselhaus, Iron-doped carbon aerogels: novel porous substrates for direct growth of carbon nanotubes, *Langmuir*, 2007, **23**, 5161–5166.
45. L. M. Bronstein, X. Huang, J. Retrum, A. Schmucker, M. Pink, B. D. Stein and B. Dragnea, Influence of iron oleate complex structure on iron oxide NP formation, *Chem. Mater.*, 2007, **19**, 3624–3632.
46. M. Lattuada and T. A. Hatton, Functionalization of monodisperse magnetic NPs, *Langmuir*, 2007, **23**, 2158–2168.
47. L. Hong, A. Cacciuto, E. Luijten and S. Granick, Clusters of charged Janus spheres, *Nano Lett.*, 2006, **6**, 2510–2514.
48. M. Lattuada and T. A. Hatton, Preparation and controlled self-assembly of Janus magnetic NPs, *J. Am. Chem. Soc.*, 2007, **129**, 12878–12889.
49. E. Taboada, E. Rodriguez, A. Roig, J. Or, A. Roch and R. N. Muller, Relaxometric and Magnetic characterization of ultrasmall iron oxide nanoparticles with high magnetization. evaluation as potential $T1$ magnetic resonance imaging contrast agents for molecular imaging, *Langmuir*, 2007, **23**, 4583–4588.
50. J. S. Kim, C. A. Valencia, R. Liu and W. Lin, Highly-efficient purification of native polyhistidine-tagged proteins by multivalent NTA-modified magnetic NPs, *Bioconjugate Chem.*, 2007, **18**, 333–341.
51. D. Horak, M. Babic, P. Jendelova, V. Herynek, M. Trchova, Z. Pientka, E. Pollert, T. Hajek and E. Sykova, D-Mannose-modified iron oxide NPs for stem cell labeling, *Bioconjugate Chem.*, 2007, **18**, 635–644.
52. D. J. Kim, Y. W. Yi and J. H. Kim, *In situ* monitoring of bindings between dasatinib and its target protein kinases using magnetic nanoparticles in live cells, *J. Am. Chem. Soc.*, 2008, **130**, 16466–16467.

53. M. V. Kovalenko, M. I. Bodnarchuk, R. T. Lechner, G. Hesser, F. Schaeffler and W. Heiss, Fatty acid salts as stabilizers in size- and shape-controlled nanocrystal synthesis: the case of inverse spinel iron oxide, *J. Am. Chem. Soc.*, 2007, **129**, 6352–6353.
54. J. H. Bang and K. S. Suslick, Sonochemical synthesis of nanosized hollow haematite, *J. Am. Chem. Soc.*, 2007, **129**, 2242–2243.
55. Z. Liu, D. Zhang, S. Han, C. LI, B. Lei, W. Lu, J. Fang and C. Zhou, Single crystalline magnetite nanotubes, *J. Am. Chem. Soc.*, 2005, **127**, 6–7.
56. J. Bachmann, J. Jing, M. Knez, S. Barth, H. Shen, S. Mathur, U. Gosele and K. Nielsch, Ordered iron oxide nanotube arrays of controlled geometry and tunable magnetism by atomic layer deposition, *J. Am. Chem. Soc.*, 2007, **129**, 9554–9555.
57. E. C. Walter, T. Beetz, M. Y. Sfeir, L. E. Brus and M. L. Steigerwald, Crystalline graphite from an organometallic solution-phase reaction, *J. Am. Chem. Soc.*, 2006, **128**, 15590–15591.
58. J. H. Bang and K. S. Suslick, Sonochemical synthesis of nanosized hollow haematite, *J. Am. Chem. Soc.*, 2007, **129**, 2242–2243.
59. J. Ha, T. P. Trainor, F. Farges and G. E. Brown, Jr., Interaction of aqueous Zn(II) with hematite nanoparticles and microparticles. Part 1. EXAFS Study of Zn(II) adsorption and precipitation, *Langmuir*, 2009, **25**, 5574–5585.
60. M. D. Shultz, J. U. Reveles, S. N. Khanna and E. Carpenter, Reactive nature of dopamine as a surface functionalization agent in iron oxide NPs, *J. Am. Chem. Soc.*, 2007, **129**, 2482–2487.
61. R.-R. Jorge, N. Zafeiropoulos, C. Serafinelli, R. Rojas-Reyna, B. Voit, E. Lopez Cabarcos and M. Stamm, Synthesis and characterization of thermosensitive PNIPAM microgels covered with superparamagnetic Fe_2O_3 NPs, *Langmuir*, 2007, **23**, 10280–10285.
62. X. Yu, C. Cao and X. An, Facile conversion of fe nanotube arrays to novel r-Fe_2O_3, nanoparticle nanotube arrays and their magnetic properties, *Chem. Mater.*, 2008, **20**, 1936–1940.
63. C. Tenailleau, M. Allix, J. B. Claridge, M. Hervieu, M. F. Thomas, J. P. Hirst and M. J. Rosseinsky, Modular construction of oxide structures:compositional control of transition metal coordination environments, *J. Am. Chem. Soc.*, 2008, **130**, 7570–7583 (and Supplementary Material).
64. R. Pelka, A. Pattek-Janczyk and W. Arabczyk, Studies of the oxidation of nanocrystalline iron with oxygen by means of TG, Mossbauer, and XRD methods, *J. Phys. Chem. C*, 2008, **112**, 13992–13996.
65. N. Andersson, R. W. Corkery and P. C. A. Alberius, One-pot synthesis of well ordered mesoporous magnetic carriers, *J. Mater. Chem.*, 2007, **17**, 2700–2705.
66. M. V. Kovalenko, M. I. Bodnarchuk, R. T. Lechner, G. Hesser, F. Schaffler and W. Heiss, Fatty acid salts as stabilizers in size- and shape-controlled nanocrystal synthesis: the case of inverse spinel iron oxide, *J. Am. Chem. Soc.*, 2007, **129**, 6352–6353.

67. K. Hayashi, R. Fujikawa, W. Sakamoto, M. Inoue and T. Yogo, Synthesis of highly transparent lithium ferrite nanoparticle/polymer hybrid self-standing films exhibiting Faraday rotation in the visible region, *J. Phys. Chem. C*, 2008, **112**, 14255–14261.
68. Y. Li, Q. Zhang, A. V. Nurmikko and S. Sun, Enhanced magnetooptical response dumbbell-like Ag-$CoFe_2O_4$ nanoparticle pairs, *Nano Lett.*, 2005, **5**, 1689–1692.
69. A. J. Devey, R. Grau-Crespo and N. H. de Leeuw, Combined density functional theory and interatomic potential study of the bulk and surface structures and properties of the iron sulfide mackinawite (FeS), *J. Phys. Chem. C*, 2008, **112**, 10960–10967.
70. W. Han and M. Gao, Investigations on iron sulfide nanosheets prepared via a single-source precursor approach, *Crystal Growth and Design*, 2008, **8**, 1023–1030.
71. A. A. Bharde, R. Y. Parikh, M. Baidakova, S. Jouen, B. Hannoyer, T. Enoki, B. L. V. Prasad, Y. S. Shouche, S. Ogale and M. Sastry, Bacteria-mediated precursor-dependent biosynthesis of superparamagnetic iron oxide and iron sulfide nanoparticles, *Langmuir*, 2008, **24**, 5787–5794.
72. I. Cesar, A. Kay, J. A. Gonzalez Martinez and M. Grätzel, Translucent thin film Fe_2O_3 photoanodes for efficient water splitting by sunlight: nanostructure-directing effect of Si-doping, *J. Am. Chem. Soc.*, 2006, **128**, 4582–4583.
73. A. Kay, I. Cesar and M. Grätzel, New benchmark for water photo-oxidation by nanostructured r-Fe_2O_3 films, *J. Am. Chem. Soc.*, 2006, **128**, 15714–15721.
74. T. Yamamoto, Y. Umemura, O. Sato and Y. Einaga, Observation of the anisotropic photoinduced magnetization effect in Co-Fe Prussian blue thin films fabricated by using clay Langmuir-Blodgett films as a template, *J. Am. Chem. Soc.*, 2005, **127**, 16065–16073.
75. D. Gimenez-Romero, J. Agrisuelas, J. J. Garcýa-Jareno, J. Gregori, C. Gabrielli, H. Perrot and F. Vicente, Electromechanical phase transition in hexacyanometallate nanostructure (Prussian blue), *J. Am. Chem. Soc.*, 2007, **129**, 7121–7126.
76. T. Rayment, R. Schlögl, J. M. Thomas and G. Ertl, Structure of the ammonia synthesis catalyst, *Nature*, 1985, **315**, 311–313.
77. P. L. Holland, Electronic structure and reactivity of three-coordinate iron complexes, *Acc. Chem. Res.*, 2008, **41**, 905–914.
78. J. M. H. Lo and T. Ziegler, A first-principle study of chain propagation steps in the Fischer-Tropsch synthesis on Fe(100), *J. Phys. Chem. C*, 2008, **112**, 13681–13691.
79. J. M. H. Lo and T. Ziegler, Chemisorption and reactivity of CHx (x) 0–4) on Fe-Co alloy surfaces, *J. Phys. Chem. C*, 2008, **112**, 13642–13649.
80. W. Chen, Z. Fan, X. Pan and X. Bao, Effect of confinement in carbon nanotubes on the activity Fischer-Tropsch iron catalyst, *J. Am. Chem. Soc.*, 2008, **130**, 9414–9419.

81. G. Lozano-Blanco, J. W. Thybaut, K. Surla, P. Galtier and G. B. Marin, Single-Event microkinetic model for Fischer-Tropsch synthesis on iron-based catalysts, *Ind. Eng. Chem. Res.*, 2008, **47**, 5879–5891.
82. J. G. Fan, B. N. Zong, X. X. Zhang, X. K. Meng, X. H. Mu, G. B. Yu, M. H. Qiao and K. N. Fan, Rapidly quenched skeletal Fe-based catalysts for Fischer-Tropsch synthesis, *Ind. Eng. Chem. Res.*, 2008, **47**, 5918–5923.

CHAPTER 5
Transition Metals

5.1 Introduction

The nanochemistry of iron has been treated in the previous chapter. The present chapter focuses on other transition metals with partly occupied d-orbitals.

All other transition elements are less abundant, less versatile and more expensive than iron, and often toxic. Their use instead of iron should always be justified, and attempts should be made to replace them by modifications of Fe(0) or iron oxides. The approximate cost of many transition elements is given in Table 5.1.

The practical perspective of cost is, of course, only dominant in commercial planning. In the science of nanochemistry it is an original property, making progress possible, that is most important. For example, if an Ag(0) nanocrystal fluoresces, we must first understand the particular case of silver before we look for generality.

5.2 Titanium, Zirconium, and Hafnium

5.2.1 Titanium

Titanium is the lightest of all transition metals, and silver-white in colour. It is relatively difficult to produce (see below) and expensive. Its hardness, lightness and stability against corrosion recommend it as a bulk material for bicycles, spectacle frames, watches, submarines, marine platforms and military aircraft.

RSC Nanoscience & Nanotechnology No. 10
Metallic and Molecular Interactions in Nanometer Layers, Pores and Particles: New Findings at the Yoctolitre Level
By Jürgen Fuhrhop, Tianyu Wang
© Jürgen Fuhrhop and Tianyu Wang 2010
Published by the Royal Society of Chemistry, www.rsc.org

Table 5.1 Cost of transition metals (pure powder or similar form) in €/100 g, in increasing order. Data from the 2008 Alfa Aesar catalogue.

Fe	Iron	8
Hg	Mercury	20
Pb	Lead	24
Mo	Molybdenum	25
Cu	Copper	30
Ni	Nickel	32
W	Tungsten	40
Cr	Chromium	45
Mn	Manganese	60
Ti	Titanium	60
Co	Cobalt	78
Nb	Niobium	100
Zr	Zirconium	180
Te	Tellurium	200
Ag	Silver	220
Ta	Tantalum	240
Hf	Hafnium	450
Pr	Praseodymium	500
Tl	Thallium	510
Ga	Gallium	540
V	Vanadium	560
La	Lanthanum	575
Nd	Neodymium	590
In	Indium	600
Gd	Gadolinium	660
Er	Erbium	680
Dy	Dysprosium	688
Ho	Holmium	690
Sm	Samarium	780
Y	Yttrium	800
Ru	Ruthenium	2080
Yb	Ytterbium	2350
Pt	Platinum	2720
Tb	Terbium	3500
Tm	Thulium	6000
Eu	Europium	6450
Au	Gold	7100
Pd	Palladium	7200
Ir	Iridium	8200
Os	Osmium	11650
Re	Rhenium	13300
Lu	Lutetium	14000
Rh	Rhodium	22000

Titania Pigments

Some 95% of titanium is, however, used in the form of titanium dioxide (TiO_2, titania). This white pigment, used in paints, has an extraordinary light scattering, hiding power and brightness, all due to its high refractive index ranging from 2.55 (anatase) to 2.9 (rutile). This is even higher than that of transparent

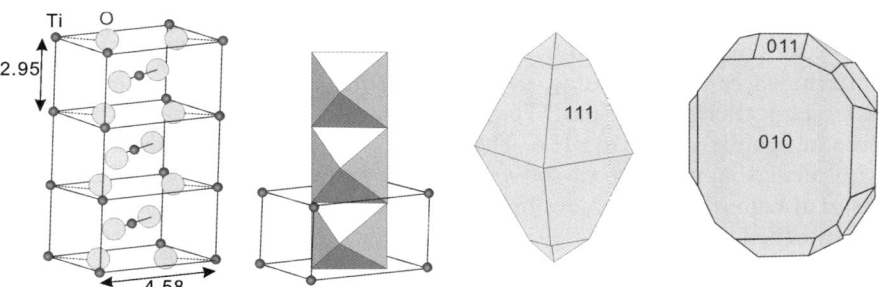

Figure 5.1 *Left to right:* Atomic arrangement of crystalline titania, the corresponding TiO_6 octahedrons of rutile, an anatase bipyramid and a brookite platelet.

diamond (2.42), where refraction of visible light causes the familiar sparkle. In rutile, each titanium atom sits in an oxygen octahedron, and the crystal structure is ditetragonal–bipyramidal. Anatase crystals are also ditetragonal–bipyramidal, but the crystal structure is more complex. Anatase converts to rutile at 915 °C. Brookite is rhombic bipyramidal and occurs in platelets (Figure 5.1).

Titania is an n-type semiconductor. With a small electric current the measured voltage ranges from 0 V in the absence of molecular oxygen and many defects in the crystal lattice to 5 V for perfectly crystalline titania. Some titanates, *e.g.* barium titanate ($BaTiO_3$), are ferroelectric because the counter-ion is so large (r_{Ba} = 156 pm) relative to Ti^{4+} (60 pm) that the latter rattles around in its octahedral hole. An electric field pulls the Ti^{4+} over to one side of the octahedron and thereby causes an enormous electric polarization of the whole crystal.

Pure rutile and anatase nanocrystals are commercially available as platelets and needles ranging in length from 7 to 20 nm. Simple hydrothermal treatment (210 °C; 10 M NaOH), microwave treatment and calcination at 450 °C converts the acicular nanocrystals to nanowires with typical diameters between 80 and 150 nm.[1]

The tetrachloride was converted to rutile, anatase and brookite nanoparticles by simply varying the pH of a boiling aqueous solution. The original pH of the $TiCl_4$ solution was ∼2.5; it was adjusted to 0 and 1 by adding HNO_3, and to 3 or 6 by $NH_3 \cdot H_2O$ (2.5%). The rutile nanoparticles occurred at pH 0, brookite nanoparticles at pH 1, and spherical and spindle anatase nanoparticles at pH 3 and pH 6. Centrifugation and drying produced the pure nanocrystals.[2]

A peroxotitanium complex was also prepared from $TiCl_4$ in ice water by addition of H_2O_2 (30%). The peroxotitanium complex is probably mononuclear, $Ti(O_2)OH^+$, at pH values <1; condenses to dinuclear $Ti_2O_5^{2+}$ between pH 1 and 3; and becomes electroneutral and polynuclear, $[Ti_2O_5(OH)_2]_n$ above pH 3. The pH manipulation of these peroxo complexes leads to a loss of molecular oxygen and formation of titania units, which then form the different nanocrystals. Pure, needle-like (= acicular, aspect ratio 10) rutile nanocrystals were directly formed only under highly acidic conditions or, less defined, from anatase nanocrystals at 915 °C. The brookite formed at pH 1 is readily distinguished by its (121) diffraction located at 30.6° in the X-ray diffraction pattern, where neither anatase nor rutile signals occur. Brookite is more stable in larger particles (11–35 nm). The anatase nanocrystals are round at pH 3 and become spindle-shaped at pH 6. The surface energy of the (101) plane increases with increasing pH. The crystal growth rate perpendicular to the (101) face is faster at high pH and leads to the observed spherical particles. Smaller particles are attached to large spindles at rising pH, which leads to less charge on the nanocrystals. The spindles then grow with time through an orientation-attachment mechanism.

At a fixed pH, the H_2O_2 concentration plays a role similar to that of the protons: it also destabilizes the nanocrystals. At pH 0, for example, increasing the [H_2O_2]/[Ti] ratio from 5 to 20 first caused the co-crystallization of anatase and brookite, then the less stable brookite disappeared followed by anatase, and finally, pure rutile came out (Figure 5.2).

Titanium has the largest, most diffuse 3d-orbitals of all the first-row transition metals. This extended space contains only two electrons and the titanium atom is therefore able to accept from and donate to large π-electron systems, such as the cyclooctatetraene (COT) ring. With this relatively large and flexible ring it forms $Ti_2(COT)_3$ with a planar ring on one side and a bent one on the other. Other COT metal complexes are less stable; they are typically active compounds, which may be decomposed to graphite and carbon rods. The diffuse titanium orbitals also bind four cylopentadiene units: two of them donate five electrons each, the other two only one each.

Bis(cyclooctatetraene)titanium ($Ti(COT)_2$) was prepared by the reduction of titanium butoxide with triethylaluminium and then slowly (within 48 h) oxidized to titania with dimethylsulfoxide (($CH_3)_2SO$) in apolar solvents at 120 °C. Precipitation of large and long-lived amorphous titania particles finally occurred and was arrested by the addition of coordinating bulky tributylphosphine/tributylphosphine oxide ligands to the anhydrous o-dichlorobenzene

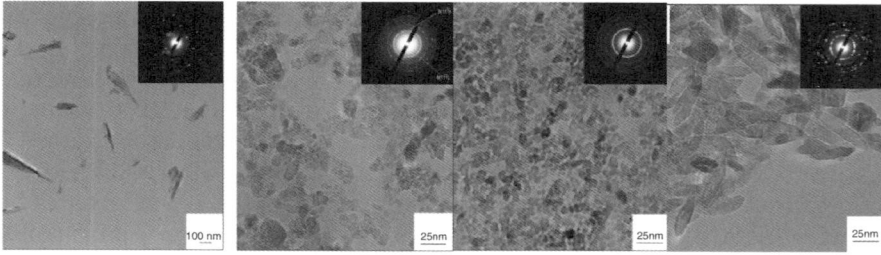

Figure 5.2 Electron micrographs of titania NCs as obtained by refluxing different peroxotitanium complex solutions at 100 °C for 48 h. The pH values of the solutions were (from left to right) pH = 0 to form acicular anatase, pH = 1 leading to brookite spheres, pH = 3 converting them to anatase spheres, and pH = 6 with anatase spindles as major products. © American Chemical Society.[1]

solution. The bulky ligands enforced the formation of small titania nanocrystals by chain termination (TiBu or TiOBu). The colour of the mixture changed slowly from the deep red of Ti(II) to the pale yellow of titania, but the mixture remained homogeneous. Addition of pentane precipitated a pale yellow solid, which could be redispersed in $CHCl_3$.

$$Ti(COT)_2 + 2OS(CH_3)_2 \rightarrow TiO_2 \downarrow + 2S(CH_3)_2 + 2COT$$

Commercial lithium-ion batteries usually depend on graphite anodes, which have an operating voltage close to lithium-ion reduction. Li^0 electroplating may occur. Titania has a redox potential of 1.5–1.8 V vs. the Li^+/Li redox couple, and therefore avoids lithium electroplating. Furthermore, lithium ion insertion/extraction in titania-based structures is accompanied by negligible lattice changes, which guarantees a long cycle life. However, the poor lithium ion and electronic conductivity of bulk titania structures limit the charge/discharge rate in comparison to bulk graphite. Titania nanostructures with large surface areas, in particular stable nanopores in titania, should help.

Such rutile nanoparticles with nanopores were made by hydrolysis and oxidation of titanium trichloride in the presence of sulfonic surfactants, which bind strongly to titania and stay in the nanocrystals (Figure 5.3). n-Decyl-1-sulfonate and $TiCl_3$ were stirred together in 1.0 M HCl and H_2O_2 was added to arrive at the peroxotitanate mentioned above. After 15 h at 60 °C a white mixture precipitated, was centrifuged, washed, dried at 60 °C overnight and calcined in air at 400 °C for 2 h to remove the sulfonate from the pores. The pore diameters ranged from 2.2 to 3.8 nm.[3]

Electrochemical experiments were carried out with a working electrode made of the calcined mesoporous rutile, carbon black, and poly(vinylidene fluoride) binder. This electrode was used in N-methylpyrrolidone mixed with 1 M $LiPF_6$ in ethyl carbonate/dimethyl carbonate as electrolyte. Voltages between 1 and 3 V vs. Li^+/Li were applied and up to 0.55 lithium ($Li_{0.55}TiO_2$, 185 mA h g^{-1})

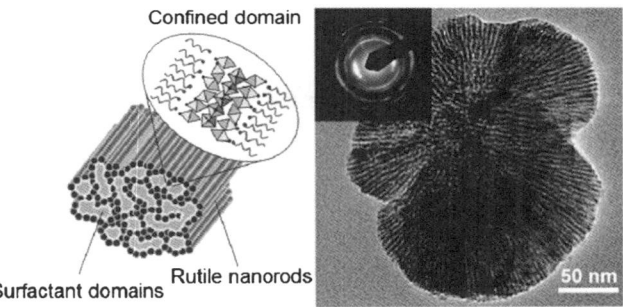

Figure 5.3 *Left:* Electron nanograph of rodlike rutile nanocrystals within a surfactant matrix, which became nanotubes after calcination at 420 °C. *Right:* An electron micrograph of a rutile electron diffraction pattern. © American Chemical Society.[3]

were charged and discharged reversibly hundreds of times without destruction of the pores and with excellent capacity retention.[3]

At high temperatures titania becomes orange-yellow. Golden TiO is obtained from Ti/TiO_2 at 1500 °C, and an amethyst-coloured Ti_2O_3 is obtained from titania with H_2 at 1000 °C.

Titania nanorods with an average diameter of 3–5 nm and a mean length up to 30–40 nm were synthesized by aminolysis of titanium oleate with oleyl amine. Its pronounced wettability drops dramatically under UV irradiation because photogenerated holes create oxygen vacancies at the semiconductor surface (Figure 5.4). These defects then promoted the dissociative adsorption of atmospheric water, which ultimately led to an increase of the surface hydroxylation. In the dark, molecular oxygen slowly replaced the UV-grafted hydroxyl groups.[4] Several semiconductor surfaces show such behaviour, but titania has been the most studied because of its photocatalytic activity and numerous applications, such as organic pollutant degradation, solar energy conversion, and the deactivation of microorganisms and malignant cells.

An electron paramagnetic resonance signal at $g = 1.979$ is characteristic of electrons trapped in tetrahedral Ti^{4+} sites and was found in Degussa P25, a commercial mixed-phase titania material consisting of anatase and rutile (Figure 5.5). The origin of this Ti^{4+} ion was traced back to the sintering process at 773 K, which produces small domains of rutile-like clusters instead of bulk rutile. These clusters occur at random at anatase–anatase nanoparticle contacts during the phase transformation and it is this intermediate phase that seems to be most active in the photooxidation of dyes on the titania catalysts.[5] Enhanced charge carrier mobility around the disturbed titanium site may be responsible for this effect, which was also found in layer structures of anatase and rutile, as well as for anatase on carbon nanotubes and sulfur-doped anatase. The mechanism of catalysis enhancement is presumably always the same.

The quantitative determination of molecular oxygen in an inert gas is often performed with semiconductors at temperatures between 600 and 1000 °C,

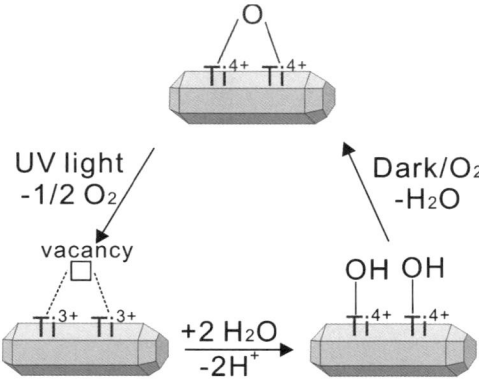

Figure 5.4 Model of the activation of the titania nanoparticle surface by UV light. © American Chemical Society.[4]

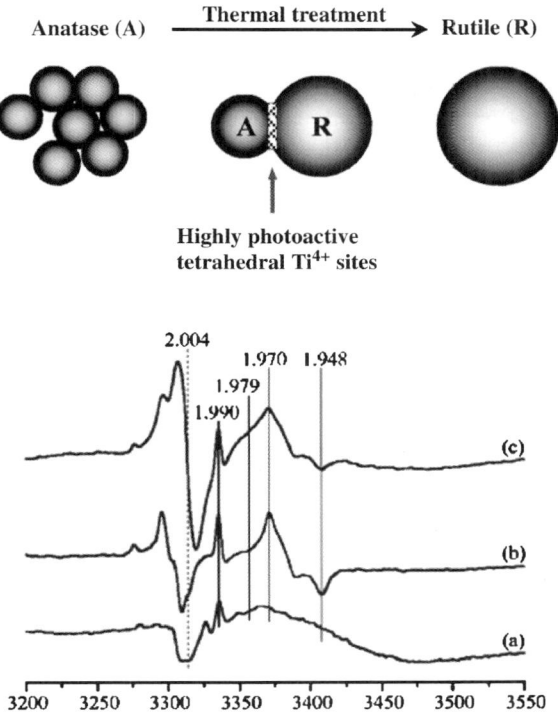

Figure 5.5 Model of the formation of tetrahedral Ti^{4+} at anatase–rutile interfaces. EPR spectra showing the corresponding $g = 1.979$ signal. © American Chemical Society.[5]

since the oxygen-ion-induced conductivity change of semiconductors, zinc sulfide for example, is very low at temperatures <600 °C. In the case of anatase nanocrystals, UV light may be used to increase the carrier density of the semiconductor. With 355 nm illumination at room temperature, four values of oxygen pressure (9.86, 9.28, 6.38 and 2.90 Torr) were measured rapidly (2 s) and reproducibly. The oxygen molecules were first physically adsorbed on the nanocrystal surface, then trapped electrons from the nanocrystals to form chemisorbed oxygen anions, lowered the carrier density in the anatase nanocrystal layer and produced measurable current drops (Figure 5.6). The fast response time is due to the large surface area of the small anatase nanocrystals.[6]

Fluorescing dyes, *e.g.* 8-hydroxyquinoline, sensitize titania for the absorption of visible light although the fluorescence is quenched at first. A colloid, however, showed no emission or transients of adsorbed dyes, which indicated that the excited state lifetime of the surface Ti(IV)–dye complex was less than the 11–25 ns that is typically observed for emissive metal oxide–dye assemblies. Visible light (wavelength >435 nm) also resulted in no photodecomposition of the dye after 17 h, because its diffusion-controlled photooxidation takes more than nanoseconds. Nevertheless, the dye modified the photochemistry of the

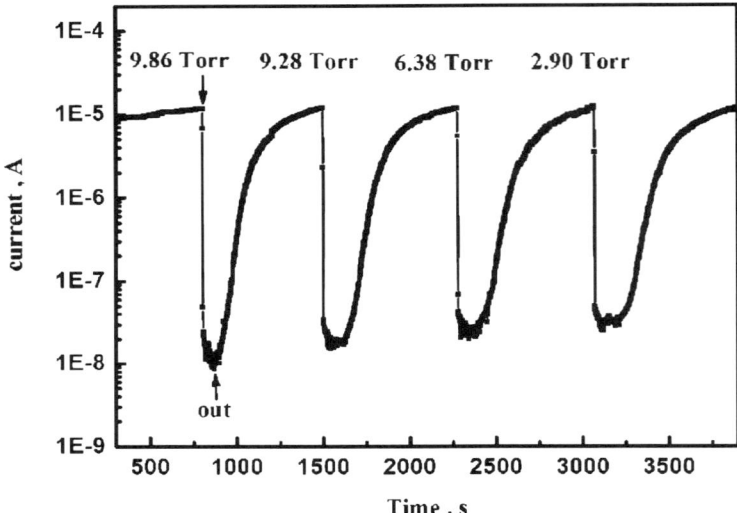

Figure 5.6 Current response of anatase nanocrystals to the different oxygen partial pressures indicated under 355 nm illumination at room temperature. The system was exposed constantly to oxygen at a pressure of 7.6×10^{-3} Torr. © American Chemical Society.[6]

titania powder: hydrogen evolved from water, which was photoreduced by ethylenediamine-tetraacetic acid as a sacrificial electron donor and $Pt(0)_{sol}$ as catalyst. The mechanism was clearly (1) excitation of the surface Ti(IV)–dye complex species and (2) electron injection from ethylenediamine. This system could be applied as the reducing part of a possible tandem cell for the photolysis of water by sunlight.[7]

Thin layers of titania are transparent to visible light, but absorb sunlight of wavelength <400 nm; they conduct electrons and have a chemically reactive surface.[8] In the popular scientific literature and advertisements, indoor photocatalysis is often recommended for air cleaning, and it is stated that pollutants are mineralized to carbon dioxide and water. Laboratory experiments showed an effect for nitrogen dioxide and formaldehyde, but volatile organic compounds and carbon monoxide were not degraded.[9,10] Wall paints with photocatalytic titania nanoparticles that work under indoor conditions have yet to be developed.

Commercial titania platelets have been supplemented by many other photoactive assemblies. For example, nanoparticle films with a thickness of 15 mm were fabricated by repeated dipping of a glass slide into a cationic polymer solution and then into a suspension of anionic titania nanoparticles, with some rinsing and drying steps. Anionic polymers with cationic titania nanoparticles were also applied, always with final drying at 150 °C, but no sintering.

Nanosheets of unilamellar titania ($Ti_{0.87}O_2$), 30 μm wide, were crystallized from a potassium molybdate (K_2MoO_4) melt, which served as a solvent for a K_2CO_3, Li_2CO_3, and TiO_2 mixture. The melt was cooled from 1473 to 1223 K, the molybdate decanted and the obtained crystals treated with a 0.5 M HCl solution for 3 days to be converted into an acidic titania sheet.

Interfacial charge transfer from the excited titania surface was also improved by phosphate treatment, in which titanium(IV)–isopropoxide in 2-propanol was simply precipitated with water and then treated with phosphoric acid. 10 nm nanoparticles prevailed in the product. The quantum efficiency of the UV photooxidation of ethanol to carbon dioxide was improved by a factor of 2 and the bandgap was lowered to 2.95 eV after calcination of the phosphate at 700 °C.[11]

Silica-coated silver nanoparticles in titania layers photoxidized methylene blue seven times faster than the titania surface alone (localized surface plasmon effect of silver), but it seems unlikely that silver will ever be used in wall paints.[12]

Biological applications of titania. Titania is also used as an absorber of UV light in sunscreen lotions. 10–20 nm nanoparticles are effective against long wavelength UV light and transparent between 400 and 700 nm. On the other hand, 20 nm titania nanoparticles are also photoactive and produce free radicals that might cause the complete destruction of skin cell DNA, even at low doses and in the absence of exposure to UV. Flow field-flow fractionation coupled with a plasma-atomic emission spectrometer were applied to determine the nanoparticle content of commercial lotions quickly and routinely.[13]

Microparticles from the Paris Metro are tested routinely for health risks by measurements of an increased expression of mRNA or proteins of MMP-2, -9 and -12 (p. 142) and haem oxygenase-1, both *in vitro* and *in vivo*. The Metro particles were quite active in all tests, pure industrial TiO_2 particles (150 nm) taken as a reference, were harmless.[14]

Bacteria attach to pharmaceutical polydimethylsiloxane (Dimeticon) surfaces through proteins. In Gram-negative bacteria these proteins assemble into pili, but in Gram-positive bacteria they do not. A thin titania film was attached to the polymer surface by liquid phase deposition from water and irradiated with UV light. Interfacial water, which was adsorbed to polar titania functional groups, now acted as a barrier and prevented bacterial adhesion. Titania-coated rubber gloves showed a greater antibacterial effect for Gram-negative than for Gram-positive bacteria (Figure 5.7).[15]

Photocatalysis. Commercial titania samples for photocatalysis (*e.g.* Degussa, Bayer, Earth Chemical) often contain 0.5% of Nb(V) as a p-dopant. Colloidal titania traps electrons in 30 ps and holes in 250 ns and releases the charges to adsorbants between nano- and milliseconds. Its surface charge is positive below pH 5 and negative above pH 9. The lifetime of surface-trapped electron–hole pairs is long enough to react with adsorbed molecules. Products are often formed before the back reaction takes place.

The photoelectrochemical conversion of UV light to electrical currents by anatase layers was discovered in 1969. Ever since, physicists and chemists have been trying to improve the quantum yield, to transfer the photocatalytic activity from the expensive UV light to visible solar radiation and to suppress the recombination of electron–hole pairs in titania. Even molecular nitrogen (N_2) was photoreduced to ammonia (NH_3) on "metal-like" titania surfaces.[16–18]

UV photochemistry is still used in the oxidation of harmful chlorocarbons to carbon dioxide. Multilayer coatings were generated *via* the alternating misting of a 10 wt% solution of 5–10 nm diameter titania nanoparticles made by hydrolysis of Ti(IV) isopropoxide ($Ti(OCH(CH_3)_2)$) and cationic poly(dimethyldiallylammonium chloride) applied layer by layer. Both layers had a thickness of $\sim$5–10 nm and the subphase was a 12.7 µm sheet of commercial, poly(vinylidene chloride). Chloroethyl ethyl sulfide ($Cl(CH_2)_2SH$) or its dimer rapidly permeated and dissolved in this coated foil, and irradiation with UV light gave carbon dioxide, water and acids (Figure 5.8). The oxidant was presumably the superoxide anion, O_2^-, which was formed together with hydroxyl radicals by the transfer of electrons and holes from the excited titania state to molecular oxygen.[19]

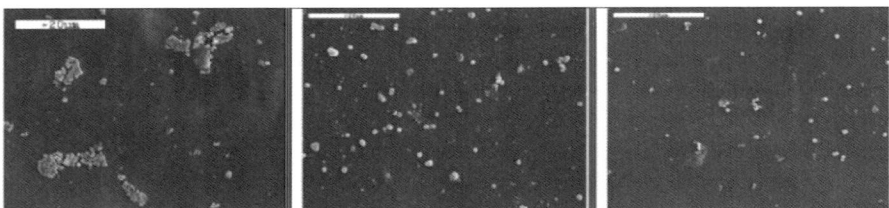

Figure 5.7 *Staphylococcus epidermis* bacteria grown on a naked polydimethylsiloxane surface, a thin titania coating and a UV-irradiated titania coating on the rubber material (area $\sim$100 µm^2 each). © American Chemical Society.[15]

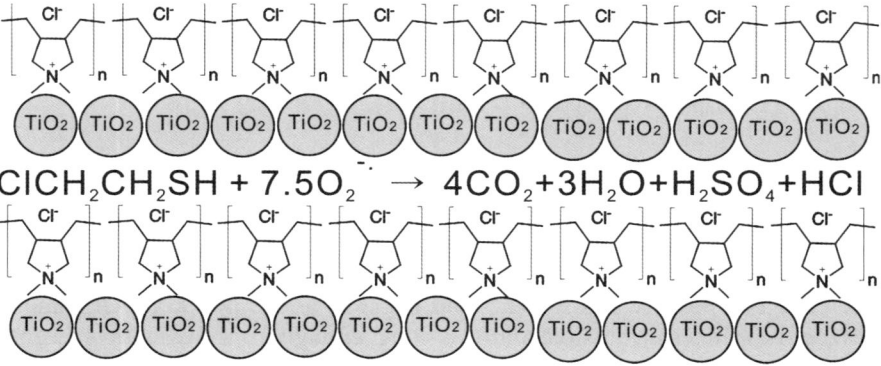

Figure 5.8 Model of titania nanoparticles between poly(dimethyldiallylammonium chloride) layers and the photooxidation of chloroethyl ethyl sulfide.[19]

Transition Metals

The photocatalytic activity of well-faceted titania nanocrystals was systematically studied in the mineralization of phenol by H_2O_2 under UV irradiation. It turned out that the photocatalytic activity increased with increasing crystallite size, suggesting that the electron–hole pair recombination plays an important role during the photodegradation of phenol to carbon dioxide and water. Relatively large prismatic rutile particles showed the highest activity, and a decrease the surface acidity by ammonia treatment lowered it. Phenol, H_2O_2 and OH^-, the main reactants, are thus most active on the nanocrystal surface at low pH.[20]

$$TiO_2 \xrightarrow{h.\nu} TiO_2 e^- + h^+$$

$$(OH_{ads}^- + h^+ \rightarrow OH^{\bullet}_{ads}; OH^{\bullet}_{ads} + C_6H_5OH_{abs} \rightarrow CO_2 + H_2O)$$

Fibres of poly(vinylpyrroHlidone) contain nitrogen chains which react directly with neighbouring precursors in layer deposition, *e.g.* titanium tetraisopropylate. Pure and crystalline anatase submicrotubes were obtained (Figure 5.9) with a tunable band gap energy of ~ 3.29 eV and well-developed porosity.[21] The poly(vinylpyrrolidone), dissolved in water, was electrospun at room temperature using a 1 mm flat-end needle, an 8 cm working distance to a grounded aluminum foil and voltages in the range 3–20 kV. The electrospun fibres were collected directly on the aluminium and used as templates. The titania thin films were deposited on to the non-woven fibre mats using titanium tetraisopropylate $(Ti(OiPr)_4)$ for the nanowire and argon as both a carrier and a purging gas. Pulsing times were 1.5 and 1.3 s for tetraisopropylate and water, and the total deposition of the titania thin films took 1000 cycles at 70 °C at 1×10^{-2} Torr or 1.3 Pa.

The visible light-induced cleavage of water into hydrogen and oxygen is the main target of photoelectrochemical research. Two photoelectrodes with different band gaps may be superimposed and coupled with a dark counter electrode to form a "tandem cell" (Figure 5.10). The two superimposed

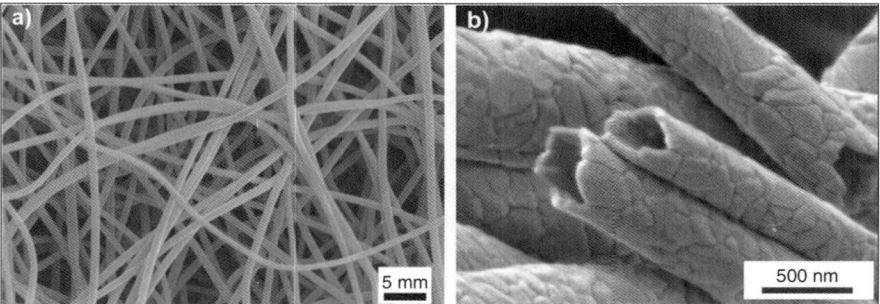

Figure 5.9 Electron micrographs of (a) titania tubes after calcination at 500 °C for 4 h and (b) a close-up image of broken parts of the tubes. © American Chemical Society.[21]

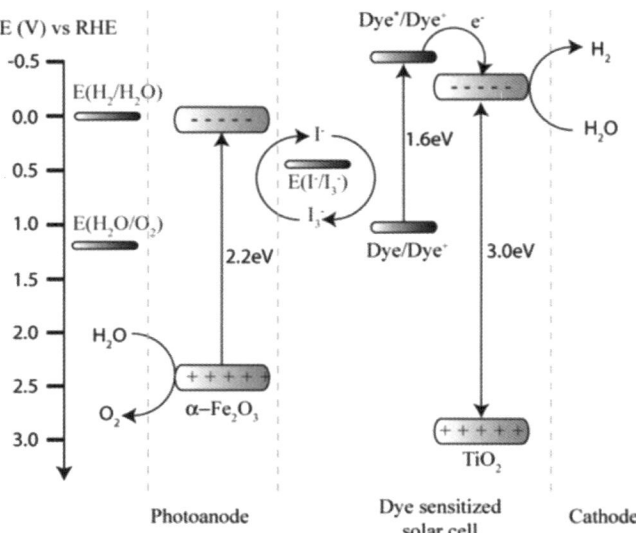

Figure 5.10 Model of a possible tandem solar cell for water photolysis. © American Chemical Society.[22,23]

photosystems should absorb complementary parts of the visible and near-IR range of the solar emission spectrum. Tungsten trioxide and haematite (γ-Fe_2O_3) are suitable semiconductor oxides to serve as oxygen-evolving anodes absorbing photons from the blue and UV parts of the solar emission (band gaps 2.2 and 2.6 eV respectively), but both have short charge carrier diffusion lengths and slow kinetics of water oxidation by the valence band holes. Titania is always used as the cathode for hydrogen evolution from protons.[22,23]

In the photolysis of water by sunlight the most difficult reaction is the formation of molecular oxygen. Ideally it should occur not only with yields of a few percent using near-UV light, but like plant photosynthesis, with visible sunlight and close to 100% efficiency. Furthermore, organic dyes such as chlorophyll bleach rapidly in the presence of reactive oxygen intermediates such as OH• radicals, and are only useful as sensitizers in the environmental destruction of organic halides in industrial waste water or in disinfection.

In order to achieve non-sensitized water oxidation with titania, the mechanism of the UV-initiated oxidation must first be known. It presumably occurs by the adsorption of a water molecule (Lewis base) to a surface-trapped hole (Lewis acid), an electrically neutral, triply coordinated oxygen atom, [Ti–O=Ti$_2$], on the surface of niobium-doped titania. Such oxygen vacancies and five-coordinated titanium atoms, sites with "too much" titanium with respect to oxygen, as well as bridging oxygen atoms, [Ti–O–Ti], are indeed the most frequent atoms on titania surfaces. The lifetime of both electrons and holes there becomes long enough to allow irreversible reactions with adsorbed electron acceptors on the surface to form molecular oxygen, *e.g.* from electron donors such as carbon compounds and water. The one oxygen atom that is coordinated to three Ti(IV) ions may be attacked by a water molecule and form an O–O bond. Polarization of the titania surface by water molecules will occur, combined with a crystal lattice relaxation, which is caused by the trapping of a one-electron deficiency produced by UV light. Two more holes and one additional water molecule then produce molecular oxygen and also begin to restore the original oxygen bridge.[24]

A commercial, (100)- and (110)-cut titania (rutile) wafer doped with Nb(V) provides another catalyst for dye-sensitized reduction. Atomically flat surfaces were obtained by immersion in 20% hydrogen fluoride for 10 min, washing with water and annealing at 600 °C. The water oxidation was not accompanied by any bond breaking at the surface crystal lattice of the catalytic nanoparticles, and the water oxidation catalysis did not lead to a surface roughening of the catalyst (Figure 5.11).

Support for the detailed mechanism above comes from photoluminescence bands around 810 and 840 nm for atomically flat (110) and (100) surfaces, which were assigned to radiative recombination transitions between the conduction-band electrons and surface-trapped holes formed at the triply coordinated oxygen atom. The intensity of photoluminescence and the concentration of surface-trapped holes decreased stepwise with increasing pH, sharply around pH 4, near the point of zero charge of titania ($\sim$pH 5.0), and then rapidly to zero near pH 13. Both observations are in accord with the assumption of a nucleophilic attack by water molecules at this site. Such a Lewis acid–base reaction may be supported by catalysts and opens up possibilities for the development of water oxidation induced by visible light. An electron-transfer type of mechanism, such as in photosynthesis or the photoreduction of protons, is much more difficult to manipulate and may therefore become a less popular subject for future research.

The major problem of titania anodes is their white colour: they react only with UV light, and the major part of visible sunlight is ineffective. The most

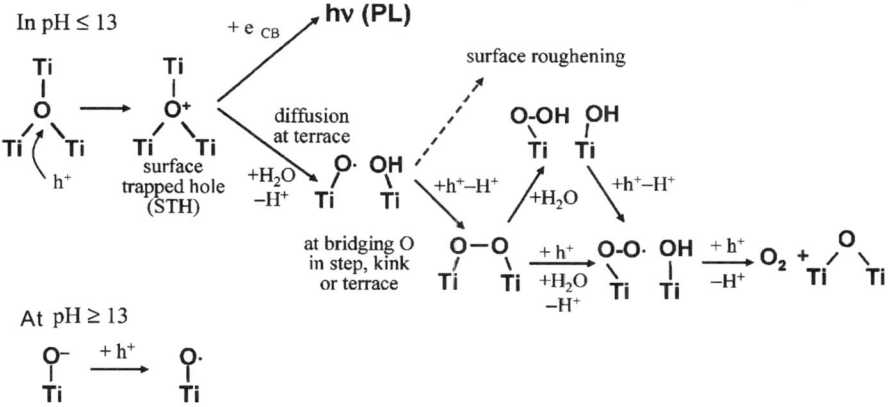

Figure 5.11 Photochemical production of electrons (e^-) and holes (h^+) from water, changes of the titania terraces and surface roughening by UV light. © American Chemical Society.[24]

direct way round this would be doping with nitrogenous species, which bring mobile electrons. Such nitrogenous species may either lie close to an oxygen ion or may substitute an oxygen in the lattice. At no other place was any nitrogenous species (either paramagnetic or diamagnetic) found to be stable by theoretical modelling of the titania matrix. The only well-defined colourized nitrogenous dopant seems to be the interstitial NO^{2-} dianion in an oxygen vacancy of the lattice. It absorbs blue light (437 nm), but can only be trapped in very low concentrations.[25]

Cobalt doping of titania with nitrogen and fluorine was achieved simply *via* a sol–gel procedure which consisted of mixing a solution of Ti(IV) isopropoxide in isopropanol with a solution of NH_4Cl and NH_4F. It was found that the incorporation of nitrogen was favoured in the presence of fluorine, and that the absorption of visible light and the photocatalytic activity under visible light irradiation were enhanced with respect to the singly doped materials (Figure 5.12). Density functional theory calculations showed a large stabilizing effect connected to the charge compensation between a p-type (N) and an n-type (F) dopant An internal charge transfer from the high-lying Ti^{3+} 3d to the low-lying N states reduces the overall energy cost of doping from ~ 5 to 3 eV, and no extra electrons by oxygen vacancy formation are necessary for doping. Fewer oxygen defects is probably the reason for the observed larger photostability and photocatalytic activity of the co-doped sample.[26]

An electrophoretic ink was formulated using light-scattering 250 nm titania particles coated with poly(styrene), poly(methacrylate) and poly(acrylate) anions (see Chapter 2, refs 101–105). The shell of crosslinked polystyrene prevented adsorption of the particles on the electrode surface and decreased the particle density in oil suspensions; 3-(trimethoxysilyl) propyl methacrylate encapsulates the polyanion and the latter provides the needed charge. The total polymer coating also lowers the density of the titania particles from 4.26 to 1.2, slowing down sedimentation.

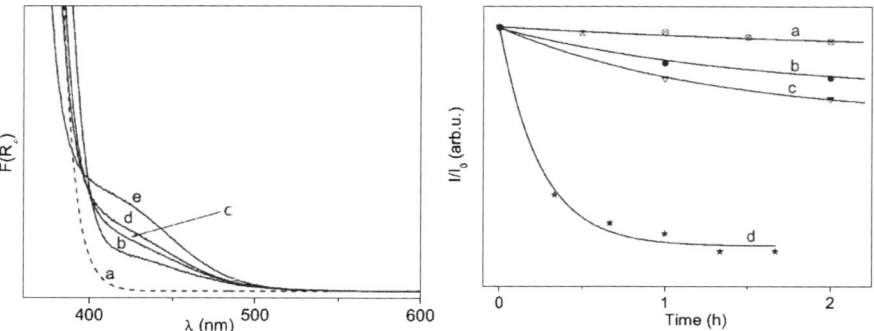

Figure 5.12 *Left:* UV–vis spectra of bare titania: (a) N/titania, (b) N:F = 1:1. *Right:* Methylene blue bleaching under visible light. (a) without catalyst, (b) titania, (c) N/titania; (d) N-F/titania, N:F = 1. © American Chemical Society.[26]

Transition Metals

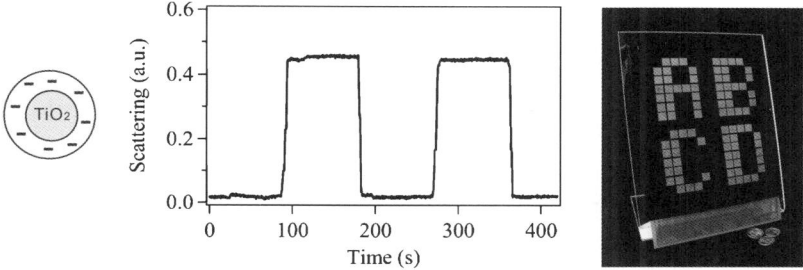

Figure 5.13 Optical response after an application of a voltage of 30 V to a dispersion of core–shell TiO_2-polystyrene particles (63 wt% polymer), prepared via precipitation polymerization in the presence of an anionic sulfonate monomer.[27] © American Chemical Society.

A 50 μm thick electrophoretic display panel using a white dispersion of these large titania–polymer particles in a black-dyed paraffin oil was black without an applied field or a negative voltage and showed a strong scattering effect when the particles were drawn to the electrode surface by a positive 30 V electric field (Figure 5.13).[27]

Coloured Fe_2O_3 films produced by ultrasonic spray pyrolysis may also be useful as oxygen-evolving photoanodes. The 200 nm thin films efficiently harvest visible light and offer very short distances between the photogenerated holes and the water interface. This allows for water oxidation by the holes although their diffusion length is only a few nanometres.[28]

The basic reaction of titania cathodes involves the formation of singly ionized molecular oxygen, O_2^-:

$$O_2(ads) + e \rightarrow O_2^-(ads)$$

This reaction has the following effects regarding the properties of oxygen on titania: the O_2^- anion is only weakly bound to titania and transformed to atomic O^-, which becomes irreversibly integrated into the titania surface lattice. The O^- ions then change the defect disorder, the related chemical potential of electrons and the reactivity of titania. Prolonged exposure to light led to an extremely slow removal of oxygen from its lattice sites.[29]

Energy conversion. Titania electrodes are not only used to reduce protons and oxidize OH^- in water-splitting tandem cells; they also convert solar energy into electricity. The oxidation potential of the sensitizer's excited state then determines the electron injection process into ruthenium polypyridyl complexes (11% solar-to-electric power conversion) or an asymmetrical bis-fluorene thiophene dye (−1.27 V; overall conversion efficiency of 8%; 91% photon-to-current conversion efficiency; stable for at least 1200 h) (Figure 5.14). The potential of these excited states was much more negative

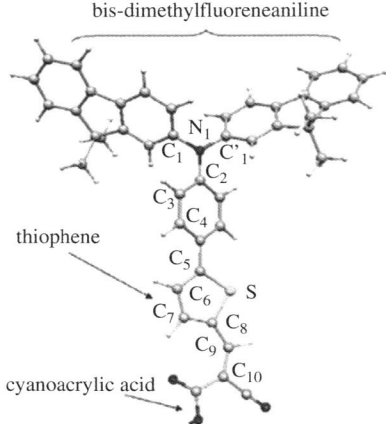

Figure 5.14 Molecular structure of the bis-fluorene dye for titania surfaces that convert light into electricity.[30]

than that of the titania conduction band, whereas the oxidation potentials of both sensitizers were more positive than that of the I^-/I_3^- redox couple (−0.4 V vs. NHE) in the electrolyte, which let the dye regeneration compete efficiently with recapture of the injected electrons by the dye cation radical.[30]

Both dyes, the ruthenium polymer and the bis-fluorene, were easy to prepare as surface layers, because both have flexible molecular skeletons, do not form ill-defined aggregates on the micron-thick titania film on glass, and are stable to sunlight for several months.

What about quantum dots (QDs) as sensitizers instead of dyes? Lead sulfide QDs have been tried, and it was found that electron affinity and ionization potential fit the energy levels of titania nanocrystals only if the diameter of the QD was <4.3 nm (Figure 5.15). Fluorescence transients of lead sulfide QDs coupled to titania nanoparticles were indeed consistent with electron transfer from excited lead sulfide QDs to titania nanoparticles: there was a 36 nm red shift when the QDs were combined with titania, and the QD fluorescence was quenched by >80% in the presence of the titania nanoparticles. The average lifetime of the lead sulfide–titania composite was estimated to be 0.7 μs, and the time constant of the assumed electron injection was 0.45 μs for 2.9 nm QDs. A Grätzel cell with 2.9 nm lead sulfide QDs infiltrated in 2–8 μm films of 25 nm titania nanoparticles deposited on fluorinated tin oxide electrodes in a I^-/I_3^- electrolyte gave a disappointing energy conversion efficiency of 0.28% for 800 nm light as compared to 11% visible light with the dye system described above.[31]

Connecting cadmium selenide QDs via Cd-SHCH$_2$CH$_2$COOTi bonds to films of titania nanoparticles was more efficient. Visible light excitation of the cadmium selenide QDs injected electrons into the titania nanocrystals and a titania–cadmium selenide photoanode in a photoelectrochemical cell gave a photon-to-charge carrier generation efficiency of 12%. The only trouble was that high light intensity led to a significant loss of electrons due to scattering

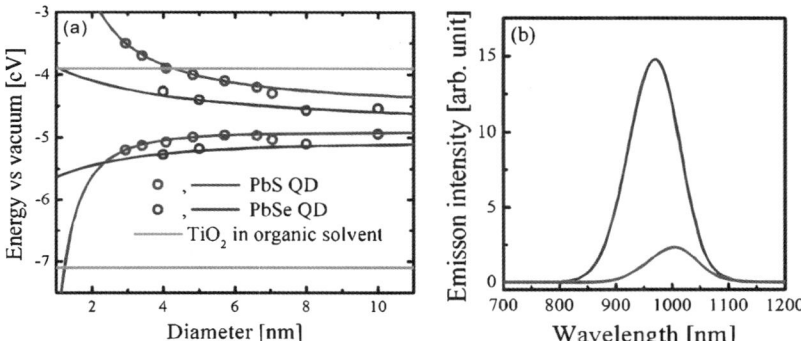

Figure 5.15 (a) Energy levels of various sizes of PbS and PbSe QDs and titania nanoparticles, as obtained from absorption spectra and cyclic voltammetry measurements. The energy levels for PbS and PbSe QDs were calculated. (b) Fluorescence spectra of PbS QDs before (*upper curve*) and after (*lower curve*) addition of 20 nm titania nanoparticles. © American Chemical Society.[31]

and internal charge recombinations at the titania/cadmium selenide interfaces.[32]

The nitrogen doping of titania nanoparticle films mentioned above causes a desirable red shift of the titania absorption into the visible region with an onset around 600 nm and a slightly enhanced photocurrent response relative to undoped titania films. The photon-to-current conversion efficiency was 6% at 400 nm for the titania/N-linker-cadmium selenide solar cells, and reached an astonishing 95% with 300 nm UV light in an Na_2S electrolyte.

Industrial processes. Titanium isopropylate (Ti(O*i*Pr)$_4$) exchanges its alcoholate groups stepwise against *tert*-butyl peroxide, chiral tartaric ethylester diol and an allylic alcohol and then oxidizes the double bond of the allylic alcohols stereoselectively to epoxides *via* allylic peroxide intermediates (Sharpless oxidation). In industry two hydroxyl groups of a zeolite pore are used instead of tartrate, because solid Ti(IV) silicate is much easier to separate on a large scale from the products than dissolved titanium tartrate. A titanium-ligated hydroperoxo intermediate has therefore been proposed as the oxygen-donating species in olefin epoxidations, but whenever such hydroperoxides were isolated, they were found to be inactive in olefin epoxidation. Modelling suggested that an asymmetric and labile titanium-peroxo species inserted in the zeolitic framework must at first be formed by protonation ([Si–O–Ti–O*OH$^-$] + H_3O^+ → [Si–OO*–Ti] + 2H_2O; Figure 5.16).[34]

The most important property of the octahedron made up of a central titanium with six oxide anions around it is the induced dipole moment that is produced if the titanium ion is pulled slightly out of the centre of the octahedron by interactions with neighbouring metal ions, *e.g.* in calcium or barium

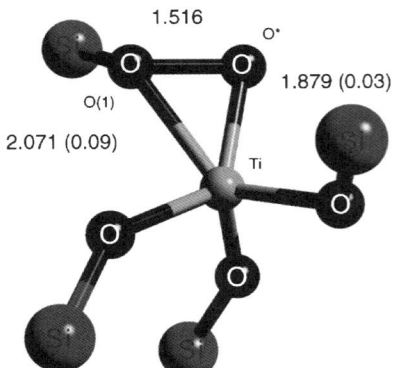

Figure 5.16 Simulated structure of the active oxygen atom O* in the zeolite-bound titanium peroxide. © American Chemical Society.[34]

perowskite. These dipole moments then do not level out within a crystal domain and the whole crystal is dominated by statistical compensations of the dipole moments. A strong electric field aligns the domains, however, and some of this alignment remains if the electric field is switched off. The dielectric constant (ε) of such a ferroelectric perowskite is ~ 1000, which means that the capacity of a reservoir condenser is increased by a factor of 1000. Some electronic equipment depends heavily on such ferroelectric compounds.

Viologen-modified anatase with periodic mesopores prepared from TiO_2–V^{2+} nanocrystals showed a fast colour change from light rose to dark blue at a potential of -0.62 V with a high contrast.[35] Bis(cyclooctatetraene)titanium reacted with dimethylsulfoxide in organic solvents at room temperature and in presence of trioctylphosphine to internally crystalline anatase nanoparticles.[36] Such nanocrystals could be assembled to molecularly thin monolayers, which assembled spontaneously to stacks of five or less such nanosheets showing 1 nm steps. This is much thinner than the unit cell dimension of anatase, extensive atomic diffusion is necessary for this surface transformation, which presumably leads to a mixed anatase–titania arrangement.[37]

Angle-dependent photoemission produces X-ray photoelectron spectra, which provided the first detailed insight into depth profiles of ultrathin multilayer structures. Titanium vapour deposited on to a C_{18} cadmium stearate monolayer supported on gold, silica, or platinum oxide (PtO_2) was studied as a first example, since titanium is widely used as an adhesion layer in organic thick-film metallization in microelectronics. After delaminating *in vacuo* both titania and reduced TiO_x species were found, with the relative proportion depending on oxygen availability. Oxygen is obtained during deposition from the ambient, organic film, and remarkably, from PtO_2 and silica substrates far below. On Au(0) substrates, up to 20% of the molecular monolayer formed titanium carbide, on silica substrates it dropped to $<15\%$, and on PtO_2 substrates to $<5\%$. The remainder was always titanium oxide (Figure 5.17).[38]

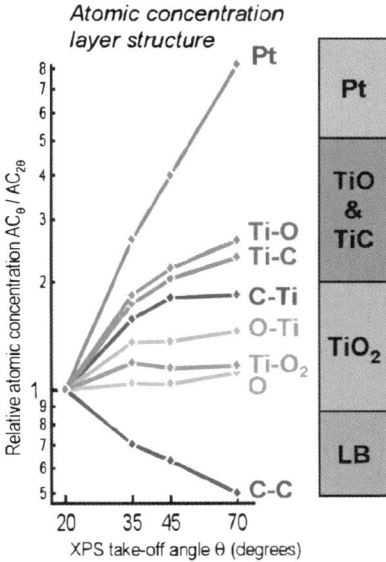

Figure 5.17 Quantification of atomic concentrations as a function of angle by angle-dependent X-ray photoelectron spectra. The cadmium-palmitate film is identified with the surface C–C carbon signature; the titanium oxide and titanium carbide are visible in the Ti(2p) spectra (Ti–O, Ti–C, Ti–O$_2$), the carbon C(1s) spectra (C–Ti) and the oxygen O(1s) spectra (O, O–Ti). The Pt(4f) signal is from a platinum capping layer. In particular, this atomic concentration pattern confirms that the TiO$_x$ and TiC reside above the titania in the fabricated structure. © American Chemical Society.[38]

5.2.2 Zirconium

Zr(IV)dioxide (ZrO$_2$, zirconia), dominates the nanochemistry of zirconium in the same way that titania does for Ti(0).

Organic solvents are more suitable than water for the formation of zirconia nanoparticles from zirconium salts or alcoholates, because some of these solvents have high boiling points, which are favourable for elimination reactions without proton catalysis. One prominent example is trioctylphosphine oxide (TOPO), which can be heated to 340 °C to produce uniform 4 nm zirconia nanoparticles from zirconium isopropylate and tetrachloride in multigram quantities:

$$\mathrm{Zr(OCH(CH_3)_2)_4 + ZrCl_4 \rightarrow 2ZrO_2 + 4(CH_3)_2CHCl}$$

Zirconia nanocrystals precipitated from Zr(IV) isopropoxide in benzyl alcohol within 4 days when heated in an autoclave to 240 °C and gave a turbid white suspension. After centrifugation the nanocrystals were dispersed by sonication in tetrahydrofuran, toluene, or pyridine containing 3-glycidoxy-, 3-amino- or 3-isocyanato-propyltriethoxysilane and coated with functional (–OH, NH$_2$ and

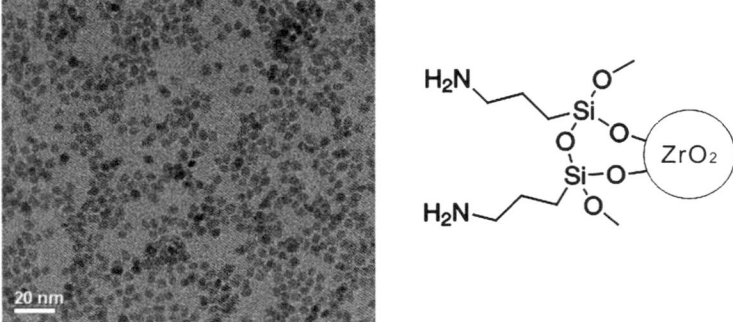

Figure 5.18 Electron micrographs of amine coated zirconia nanocrystals. © American Chemical Society.[39]

NCO) surfaces. After 12 h at 60 °C all three nanocrystals were silanated and were now soluble in the three solvents mentioned above (Figure 5.18).[39] Syntheses of metal oxide nanoparticles in non-aqueous sol–gel systems at low temperatures also led directly to white, highly crystalline materials without calcination. The organic groups attached to the surface of the nanocrystals prevent aggregation, and the zirconia core guarantees chemical inertness, thermal stability, high refractive index, and high hardness. The nanoparticles with vinyl surfaces were soluble in allylmalonic acid, which was polymerized in water and gave clear suspensions of the zirconia nanoparticles.[40]

In zirconia, a large proportion of the Zr(IV) ions may be replaced by Y(III) or other trivalent ions. A large number of oxygen gaps then remain unoccupied and the remaining oxygen anions move freely. If a porous Zr(IV)–O–Y(III) electrode is placed in a car engine between the exhaust gases and the outer air, it translates the concentration difference between the oxygen inside and outside the engine into a voltage, because the air side has many more oxide anions then the engine side (λ-sensor). A voltage of 200 mV corresponds to a "lean" mixture, where some CO is not burnt to CO_2. An output of 800 mV means a "rich" mixture, high in unburned fuel. The ideal stoichiometric point, with fully oxidized CO_2, is 450 mV. The lean–rich cycle of the λ-sensor is then used to adjust the fuel/air ratio to 1:14.7 by mass, although somewhat slowly. High-performance wideband sensors serve as an electronic feedback loop controlling the fuel pump.

Why are Zr(IV) ions replaceable by other ions so easily, and why are oxgen anions so mobile in incompletely filled crystal lattices? At first O^{2-} has a radius of 121 pm, Zr^{4+} of 73 pm and Y^{3+} of 116 pm. If an oxygen is missing, its neigbours can easily change places, because the small zirconium ion finds a lot of free space. This is also true for many other transition metal ions, *e.g.* Fe^{3+}, but only zirconia forms stable crystals with stoichiometries from $ZrO_{1.700}$ to $ZrO_{2.004}$. This means that up to 7% of the Zr is Zr(II), and up to 7% of the oxygen dianions are missing. It is thus understandable that other metal ions replace the Zr(II) and stabilize the oxygen defects.

Transition Metals

When zirconate layers were substituted simultaneously with short, rigid cyclohexylamine and long, flexible dodecylamine molecules, much empty space was left in the "intergallery region", which became porous (Figure 5.19).[41] Dispersion of the zirconate platelets by ultrasonication and dissolution of surface polymers became very fast and efficient.

Proteins intercalate spontaneously without any structural deformations into crystals of α-zirconium phosphate (Figure 5.20).[42] Such adducts are used as supported enzymes (lysozyme, chymotrypsin) and as sensors. Zirconium phosphate is a well-characterized layered material with hydrophilic hydroxyl functions present on its 2D lamellar surface (one hydroxyl group per $0.24\,\text{nm}^2$). One of the reasons that the crystal does not deform proteins is easy expansion into the defects which accommodates small as well as large guests. There must, however, also be plenty of hydration water for the phosphate protons and amino groups of the protein. The binding constants and pK values of the proteins correlate with the known pI values of the proteins, and electrostatic interactions determine the binding constants: neutral myoglobin with a pK of 7 binds weakly.

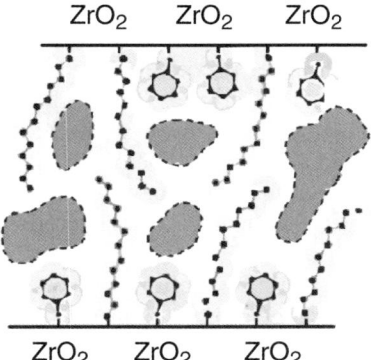

Figure 5.19 Cyclohexylamine- and dodecylamine-coated zirconia layers rapidly dissolve hydrophobic polymers symbolized by the dotted areas. © American Chemical Society.[41]

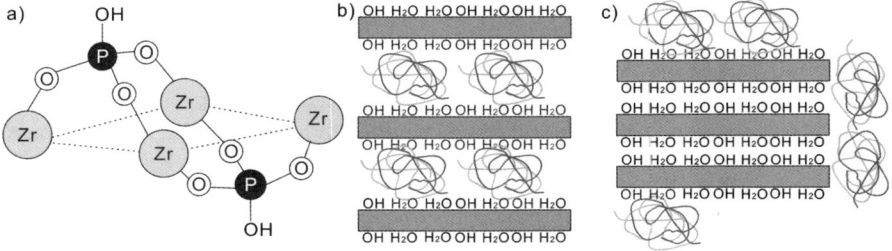

Figure 5.20 Zirconium phosphate layers bind proteins either in the galleries or on the outer surfaces. No protein deformation occured. © American Chemical Society.[42]

Zirconocenes with Zr(0) as a metallic centre, two π-donor (cyclopentadiene) and two σ-donor (H, Cl, OR, NR$_2$) ligands are applied in alkene polymerizations and ring enlargement with acetylene halides. Zirconia supported by tungstate and promoted with some iron and platinum is the most active catalyst for the skeletal isomerization of linear alkanes, because its surface hydroxyl groups are strong acids, heat stable and inert toward oxidising and reducing agents. Furthermore, zirconia is not toxic. Thin zirconia films or microcrystals for analytical purposes are usually prepared by electrodeposition of ZrOCl$_2$ at bare or functionalized gold electrodes.

Tungstated zirconia films become conductive in presence of ammonia gas, which is transformed to NH$_4^+$ on the acidic zirconia surface. NH$_4^+$ hopping between the acidic hydroxyl groups on the surface then leads to an electric current, which is measurable by tungstate–zirconia coated electrodes.

Zirconia has a strong affinity to the phosphoric group,[43] and nanoparticle layers on gold electrodes offer a large electrode surface area. A pair of redox peaks (E_{pa} 0.093 V and E_{pc} 0.037) and an irreversible reduction peak (E_{pc} −0.61 V) were observed with a methyl parathion in the potential range of −0.8 to +0.5 V. The irreversible reduction peak corresponded to the reduction of the nitro group to the hydroxylamine, and the reversible redox peaks were attributed to two-electron transfer processes. Concentrations down to 5 ng/mL were easily detectable with this electrode. One of the most important issues in the development of a chemical biosensor is the regeneration of the sensor surface. It was found that the anodic peak on the nanoparticle electrode disappeared completely after multiple scanning, indicating that the methyl parathion–zirconia complex dissociated. The electrode is self-cleaning, and the zirconia particles remained active.

Heterogeneous photocatalytic oxidation removes volatile organic compounds present in indoor air. Several companies offer air-conditioning systems which incorporate a titania photocatalyst and use the UV component of sunlight. In order to avoid the build-up of partially oxidized compounds and to achieve a better performance under solar illumination, other transition metals, in particular Sn^{4+}, were incorporated into Ti$_{1-x}$M$_x$O$_2$ phases. This led to a significant increase in the rate of photocatalytic oxidation of acetone or methylcyclohexane, or ammonia vapours to nitrogen.[44,45] Similarly, titania–zirconia nanoparticles, which introduce a higher surface acidity as well as "active defects" on the titania surface, improve the photocatalytic performance. All titania–zirconia materials were much more active at high relative humidity than pure titania. This was traced back to strong Lewis acidic sites on the catalyst surface. Similarly, titania daylight lamps are five times less efficient than UV lamps in the photocatalytic decomposition of methylcyclohexane to carbon dioxide and water.[44]

An aqueous solution of ZrO(NO$_3$)$_2$ 6H$_2$O was blended with *n*-heptane, Triton X-100 and the co-surfactant *n*-hexanol. Ti(OPri)$_4$ dissolved in 2-propanol was added to the obtained Zr(IV) microemulsion, and the mixture was stirred for 24 h. The titanium alkoxide hydrolysed and Zr(IV) was incorporated. The precipitate was centrifuged, rinsed with methanol, dried at 383 K for 24 h and calcined at 773 K for 3 h. The nanoparticle size ranged from 9.5 to 15.4 nm, the samples with higher zirconium content being smaller.

5.2.3 Hafnium

HfO_2 (hafnia) and zirconia are called "twin oxides", as the ionic radii of Hf^{4+} and Zr^{4+} are 78 Å and 79 Å respectively. Both oxides have large dielectric constants ($\varepsilon_{Hf}30$, $\varepsilon_{Zr}25$), high melting points (2758 and 2700 °C), and may replace silica as gate dielectrics to reduce the leakage current.

$HfCl_4$ and $Hf(OiPr)_4$ reacted at 360 °C in neat trioctylphosphine oxide to form hafnia nanoparticles with an average diameter of 5.5 nm. Uniform 4.3 nm $Hf_{0.45}Zr_{0.55}O_2$ nanocrystals were obtained from the reaction of $Hf(OiPr)_4$ with $ZrCl_4$.[46]

The hafnium sulfides HfS_2 and Hf_2S possess layered structures like graphite. Laser ablation gave no nanotubes (p. 57 ff), but only ill-defined multilayer onions.[47]

5.3 Vanadium, Niobium and Tantalum

5.3.1 Vanadium

Vanadium on titania catalyses the reduction of nitrogen monoxide (NO) to nitrogen. Bulk vanadium pentoxide (V_2O_5) is used as a catalyst for the $SO_2 \rightarrow SO_3$ conversion, and vanadium phosphorous oxide (*e.g.* $V(OH) PO_4$) catalyses the oxidation of butane (C_4H_{10}) to maleic anhydride ($C_4H_2O_3$).

The catalytic versatility of vanadium oxides is caused by partially filled d-orbitals; easy conversion of four oxidation states (II–V); extended d-orbitals away from the oxgen atoms, which can overlap with p-orbitals; shear planes, which facilitate the removal of oxygen from the crystal lattice; and an electronic surface state that allows simultaneous acceptance and donation of electrons to a substrate. This versatility may be enhanced by addition of other metal oxides, in particular molybdates.

Vanadium's uniquely verasatile one-electron transfers between the oxidation states –3 to +5, each state converting to the next by one-electron redox processes, permits a wide range of reaction conditions. The Lewis acidic nature of many vanadium and oxovanadium compounds, and the ability of oxovanadium compounds to functionalize substrates through oxygen transfer, are other key factors of vanadium catalysis.[48]

The thermal transformation of metal oxide nanoparticles with cyanamide or urea as nitrogen sources into metal nitrides highly depends on the size of the crystallites. In the case of titania, the upper size for complete transformation from the oxide to the nitride is <10 nm, whereas in the case of vanadium oxide, nanorods with sizes up to hundreds of nanometres react readily to the respective nitride. The degree of crystallinity plays a minor role for most of the metal oxides, but tantalum oxide must be an amorphous nanopowder.

Independent of the crystallite sizes of the oxidic precursor particles, which can be as large as 250 nm in the case of the VO_x nanorods, all of the obtained metal nitride nanoparticles are in the size range 3–20 nm, generally with a rather small size distribution.[49]

Uniform nanofibres of amorphous TiO_2/V_2O_5 mixtures were obtained by electrospinning a 2-propanol solution containing poly(vinylpyrrolidone) (PVP), acetic acid, and two sol–gel precursors, namely titanium tetraisoproxide, $Ti(OiPr)_4$ and vanadium oxytriisopropoxide, $VO(OiPr)_3$, followed by calcination in air at high temperatures. The compositions of the fibres were controlled by varying the ratio of the two alkoxides in the feeding solution. The addition of hexadecyltrimethylammonium bromide to the solution stabilized the spinning jet and improved the size uniformity of the fibres. V_2O_5 nanorods grew directly from electrospun nanofibres during calcination.[50]

A simple electrospinning procedure fabricates nanofibres of controllable hierarchical structure, e.g. of amorphous TiO_2/V_2O_5 nanofibres with a molar ratio of 1:1 (Ti:V), from simple solutions of metallalkoxides and PVP. It can be combined with calcination to change the morphology and phase structure of nanofibres. The fibres have a diameter of ~ 100 nm and are hundreds of microns long.[51]

In structural nanochemistry vanadyl ions are particularly useful in the analysis of molecular environments by electron paramagnetic resonance spectroscopy (Figure 5.21).[52] Three main reasons are: (1) the vanadyl ion VO^{2+} contains a single electron ($3d^1$), which produces an octet signal by electronnucleus coupling (^{51}V nuclear spin, $I = 7/2$, 99.8% natural abundance) and is centred at $g = 1.98$. The structure of this signal is sensitive to all asymmetries in the environment. (2) The V=O group is like a carbonyl group. It is polar, points in one direction, and adds protons, water and other nucleophiles. VO–ESR signals are extremely sensitive to all kinds of nucleophiles in the environment. (3) The "naked" metal side of vanadyl reacts with all kinds of hard ligands and is very different from the V=O side.

There are many crystal structures and perfectly analysed electron paramagnetic resonance (EPR) spectra in the literature for comparison. EPR spectroscopy of nanoparticle-bound vanadyl ions is therefore one of the most reliable and exact analytic methods of studying interactions of water and competitive ligands in otherwise ill-defined environments. This will be demonstrated here using hydrated and dehydrated silicalite nanoparticles as examples.

The vanadium silicalite nanoparticles were prepared by dissolving tetrapropylammoniumhydroxide in water, adding tetraethylsilicate and heating it to 70 °C. Vanadyl sulfate ($VOSO_4$) in water was added under vigorous stirring and the clear solution was aged at room temperature for 5 days. After dilution and acidification with 0.6 M HCl to pH 1, a clear solution of vanadium silicalite nanoparticles was obtained. SBA-15, a mesoporous hexagonal silica system with 3–15 nm pores, was added. Some of the vanadyl nanoparticles were thus entrapped by the silicate formed from the tetraester, and some migrated free into the SBA-15 network.

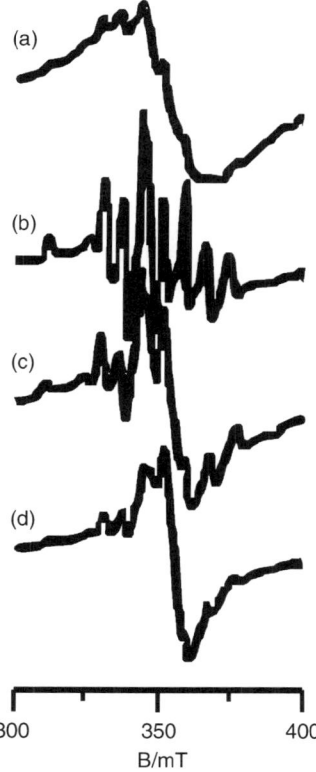

Figure 5.21 Electron spin resonance spectra of VO–silicalite nanoparticles: (a) freshly prepared; (b) dehydrated under vacuum at room temperature; (c) dehydrated under vacuum at 373 K; (d) sample rehydrated. © American Chemical Society.[52]

The ESR spectra were recorded at 10 K and showed a broadened signal for the VO^{2+} silicate and a poorly resolved octet for isolated VO^{2+} species. Dehydration led to a reduction of the broad signal, the isolated vanadyl sites becoming relatively stronger. Further dehydration of the sample at 373 K led to the vanishing of the broad absorption centred at $g = 1.98$, and a well-resolved octet characteristic of isolated VO^{2+} ions in square-pyramidal coordination. Rehydration and ageing re-established the old, broadened signal. The spectrum showed the typical VO signal and its broadening upon aggregation, nothing special about the nanoparticles or about stereoselective V=O interactions with the silicate pores or the amine solvent.

The molecular interactions of VO with the solvent (tetrapropylammonium hydroxide) and with silica were then analysed by a sophisticated pulsed (ps) NMR–EPR experiment called HYSCORE (= HYperfine Sublevel CORrelation). It correlates nuclear magnetic frequencies in one electron spin manifold with nuclear frequencies in another electron spin manifold by means of mixing

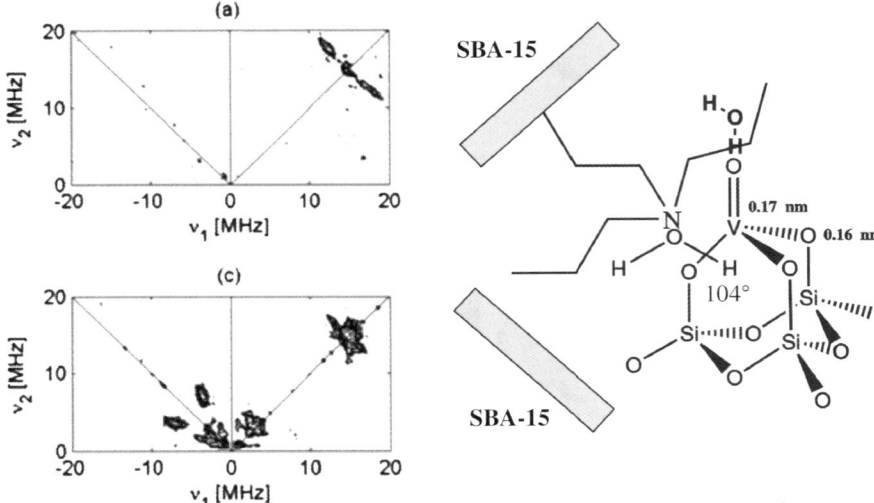

Figure 5.22 Electron-spin-echo (ESE) Detected electron-spin-resonance (ESR) hyperfine sublevel correlation (HYSCORE) spectra of the VO–silicalite nanoparticles in the SBA-15 network (observer position (346.5 mT) with impulse ô=176 ns). (a) fresh sample (b) dehydrated (c) model of the dehydrated/hydrated VO environments as calculated from the spectral data. © American Chemical Society.[52]

pulses. The combination of two NMR spectra of the probe under the influence of ESR evaluates the environment of VO in two dimensions and therefore makes it possible to determine the position of diamagnetic atoms, *e.g.* H, N and Si, with respect to the paramagnetic V=O vector.

The HYSCORE spectrum of the fresh VO–silicate nanoparticles was dominated by a pronounced ridge in the (+,+) quadrant, centred at ∼15 MHz (the ^{1}H Larmor frequency) with a width (T value) of ∼8.2 MHz. Large positive isotropic and dipolar components of the A^H matrix were found for the protons of water molecules lying in the VO^{2+} equatorial plane, with a_{iso} = 7–8.5 MHz and T = 4.7–5 MHz. Values of a_{iso} = 0 MHz and T = ∼4–5 MHz were characteristic of the protons of water molecules that coordinated perpendicularly to the V=O plane (Figure 5.22). The stereochemistry of the nanoparticle surface thus became accessible.

After dehydration, the proton ridge becomes shorter with a maximum extension of ∼5 MHz. The most relevant alteration occurs in the (−,+) quadrant where the features of a new, complex signal show up. The new spectrum is dominated by a pair of cross-peaks centred at about (−3.5, 6.9) and (−6.9, 3.5) MHz. These peaks are assigned to the double-quantum transitions arising from the hyperfine interaction of the unpaired electron of VO with a nitrogen nucleus bound at the equator of the V=O bond. The lack of combination peaks indicates one single nitrogen. No clear evidence could be obtained from the spectrum concerning the interaction of VO with the ^{29}Si nuclei because their signals overlapped with the nitrogen features.

Ageing of the sample was also characterized and showed quantitative changes in the V=O environment (not shown here). The result of the HYSCORE anaylsis is summarized in Figure 5.22. The information about the silicon interactions came mainly from IR spectra and indirect arguments.

5.3.2 Niobium

10–30 nm NbO_4 nanocrystals were made by similar procedures to titania nanocrystals, and have similar photocatalytic activities. They are thought to be probably more active under illumination with visible light.[53]

Niobium becomes superconducting at 5 K and is useful for the construction of nanosized superconducting quantum interference devices (SQUIDs). A SQUID is a superconducting ring that is interrupted by one or two thin layers of insulating, semiconducting or normally conducting material. This interruption must be so thin that superconducted electron pairs can tunnel through (Josephson contact). Currents flow in a SQUID only if the magnetic current is an integer multiple of the magnetic flux quantum, 2.07×10^{-15} V s. An applied magnetic field therefore allows the observation and quantification of the magnetization in small magnetic clusters in scanning SQUID microscopes. To make such a microscope, niobium bridges 80 nm × 50 nm × 150 nm with Nb–Ga semiconductor insulation were burnt into a 50 nm × 5 μm niobium lead on silica wafers with a focused ion beam (Figure 5.23). An applied magnetic field increased currents through the semiconducting Nb–Ga layer.[54]

Density functional theory predicts that mixed $Mo_{1-x}Nb_xS_2$ nanotubes should exhibit metallic character and that the density of states close to the Fermi level can be tuned by the degree of niobium doping. Mixed phase $Mo_{1-x}Nb_xS_2$ nanocrystals were prepared from $MoCl_5$ and H_2S vapours in a vertical reactor within a two-stage furnace at 250 and 850 °C in the absence of oxygen and water. Fullerene-type nanoparticles with 640 pm molybdenum sulfide layers were obtained. Elemental mapping by energy-filtered electron microscopy revealed that the niobium was distributed uniformly throughout the nanoparticles and in a very thin amorphous niobium oxide layer, seen as an outer envelope. Niobium oxide exists in three principal forms in the nanoparticles, Nb_2O_5, NbO_2, and NbO, but the oxide was amorphous in nature (Figure 5.24).[55] The Mo → Nb substitution leads to the formation of new states in the band gap of molybdenum sulfide nanoparticles and nanotubes (see p. 331 ff).

5.3.3 Tantalum

Tantalum was named after the mythological Greek figure Tantalus, whom the gods condemned to endure thirst while standing in a pool of water. The refusal of tantalum to dissolve in aqueous acids gives a similar impression. It is caused by a thin and very tight protective layer of tantalum oxide (Ta_2O_5) at the metal

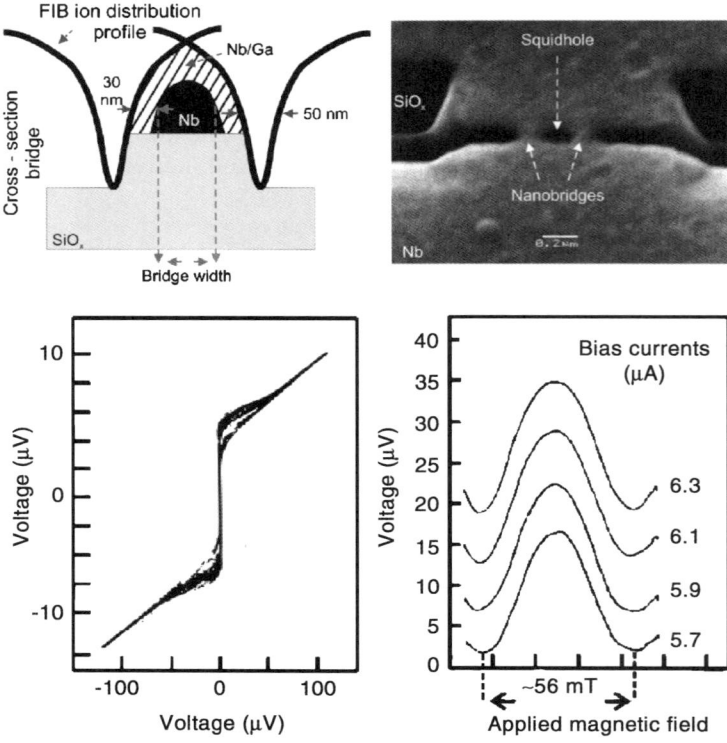

Figure 5.23 Model and scanning electron micrograph of a SQUID based on niobium nanobridges with designed widths of 80 nm and $I-V$ curve for different values of the applied magnetic field and the corresponding $V-H$ characteristics of the device. © American Chemical Society.[54]

surface. Consequently, tantalum and its oxide are used as protective coatings in chemical reactors, aircraft, biomedical implants, surgical instruments, and sensors.

Tantalum oxide has a very high refractive index and is used, like titania, in antireflective coatings for lenses and solar panels. As a piezoelectric material it is applied in surface acoustic wave devices such as band-pass filters and various types of mechanical sensors. Because of their very high dielectric constant and good compatibility with silicon, thin films of tantalum oxide are used in thin film transistors, ion sensors, and storage capacitors for dynamic random access memory. Tantalum oxide is also photocatalytically active in the photolysis of water to yield molecular hydrogen and oxygen. The first, much cited water splitting nanoparticle which used visible light and no dye was reported in 2001. It contained a mixture of semiconducting tantalum and indium oxides together with some nickel or ruthenium. The single phase $In_{1-x}Ni_xTaO_4$ layers were synthesized at 1100 °C from In_2O_3, Ta_2O_5 and NiO, impregnation with aqueous $Ni(NO_3)_2$ or $RuCl_3$ solution improved the photocatalytic activity. Visible light

Transition Metals

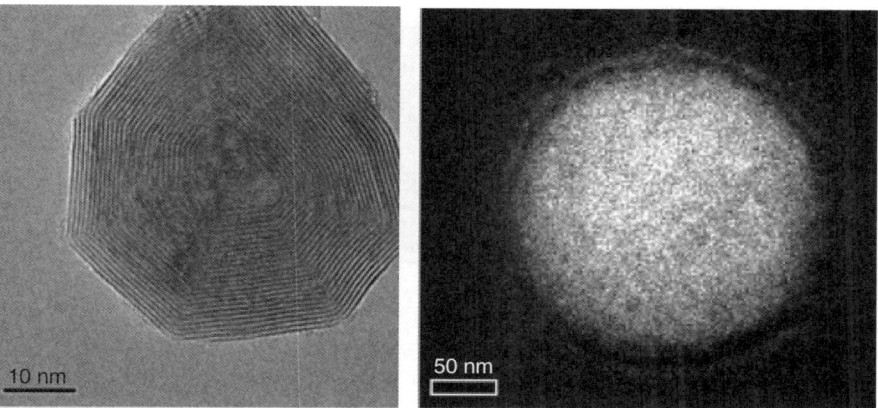

Figure 5.24 *Left:* Electron micrograph of a layered $Mo_{1-x}Nb_xS$ microcrystal. *Right:* Niobium map measured around the Nb L3 edge (2370–2470 eV) showing the homogeneous distribution. © American Chemical Society.[55]

came from a 300 W Xe arc lamp, the gases evolved were determined with a gas chromatograph. The rates of H_2 and O_2 evolution were about 16.6 and 8.3 mmol h^{-1}, and the quantum yield at 402 nm was estimated to be 0.66%.[56] Electrons presumably were collected on the Ru or Ni sites, which then produced hydrogen, the combined In–Ta oxides produced the holes for oxygen production.[56]

To produce tantalum oxide nanoparticles, tantalum pentachloride was dissolved in absolute alcohol, ammonia was added, and the precipitate was filtrated and calcined at 500 and at 700 °C for 5 h.

Tantalum oxide films with a controlled thickness of as little as 35 nm and up to > 100 nm are first grown on tantalum foil and then separated as sheets by the lift-off float-on technique. Water between the surfaces and the dissolution of metal oxide by hydrofluoric acid initiate the detachment of the film. During anodic oxide growth, the electrostatic field causes metal ions to enter the oxide film and travel to the electrolyte where they react to produce more oxide. Oxygen from the electrolyte also migrates through the oxide film to the metal interface, where more oxide is formed. A compact oxide film is formed and remains on the tantalum surface, with a final thickness determined by the anodization voltage.

Dimpled tantalum is ductile, high melting, chemically inert, and easily prepared. It can be safely used as a template or mould for nanostructure synthesis under extreme conditions, as demonstrated by a simple sputter coating and flame annealing procedure for gold nanoparticles.

Purely inorganic tantalum oxide films with a controlled thickness of 35–100 nm were fabricated from commercially available cold-rolled tantalum foils. The metal foil was anodized at 15 V in a 9:1 volumetric mixture of concentrated sulfuric acid and concentrated hydrofluoric acid. This "electropolishing" generated highly ordered dimples 30–50 nm in diameter and

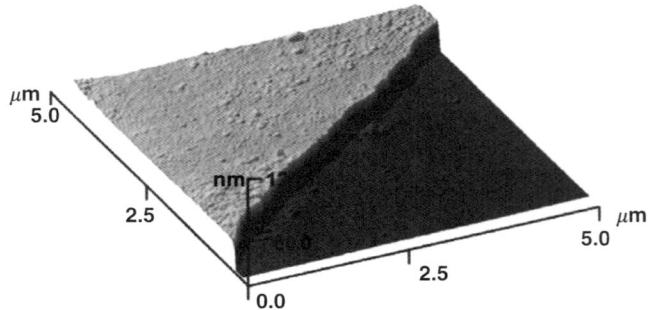

Figure 5.25 Atomic force micrograph of dimples in an "electropolished" (= anodized) Ta_2O_5 film. © American Chemical Society.[57]

~8–10 nm deep (Figure 5.25) a 5 min treatment with 0.5% HF converted the valleys to holes. The film was then anodized again in 1 M H_2SO_4 and 2 wt% HF at 20 V to weaken the adhesion of the anodic film to the underlying metal and peel it off.[57]

Tantalum nitride (Ta_3N_5) nanoparticles were made in a tube furnace with a current of ammonia gas:

$$3Ta_2O_5 + 10NH_3 \rightarrow 2Ta_3N_5 + 15H_2O$$

The band gap of 2.08 eV in tantalum nitride makes it suitable as a catalyst for reactions driven by visible light. For example, 18 nm nanoparticles catalysed the photooxidative bleaching of methylene blue much better than $TiO_{2-x}N_x$ nanoparticles of the same size.[58]

5.4 Chromium, Molybdenum, Tungsten and Manganese

5.4.1 Chromium

Chromium metal, Cr(0), occurs as bcc crystals; it is hard and brittle with a silver lustre. Chromium is mainly used in stainless steel (>13%), and as an electroplated protective coating for more reactive metals. It is inert to oxygen and water at room temperature, but forms trioxides at red heat.

Chromium dioxide (CrO_2) is one of the most remnent ferromagnetic materials and its nanoparticles are therefore used in magnetic tapes which reproduce high frequencies much better than iron tapes. It is also a strong reducing agent (Cr^{3+}/Cr^{2+}, $E° = -0.41$ V).

The most stable oxidation state of chromium is III. Its symmetrical t_{2g}-configuration leads to a large variety of metal complexes, which is only surpassed by Co(III), Mo(II) and W(II). Cr(VI)O_3 is strongly acidic and has an extraordinarily low melting point of 197 °C (MoO_3 and WO_3 melt at

795 and 1473 °C respectively). Its deep red crystals consist of corner-bridged tetrahedron chains.

Chromium metal catalyses the polymerisation of olefins. On silica it is selective for ethylene, rejects other olefins and prevents any co-polymerization. Together with iron it is also active in the reduction of nitrogen to ammonia. Cr_2O_3, often combined with Al_2O_3, catalyses oxidation, dehydrogenation and fluorination reactions.

Benzene is generally a poor ligand for metals, because its π-system is too rigid to be deformed. Nevertheless, chromium vapour and benzene yield directly dibenzene chromium ($Cr(C_6H_6)_2$) with planar benzene units in 60% yield. This can only be explained by the overwhelming stabilizing effect of the 18-electron configuration (3 × 6) which is provided by one chromium atom and two benzene molecules.

Electrodeposition of the chromium chloride ($CrCl_3$) electrolyte on silica surfaces led to spherical nanoparticles or homogeneous chromium films with and without hexagonal microrods on top. The granular chromium nanoparticles and the chromium film on silica consisted of a predominantly chromium metallic core and a Cr_2O_3 outer shell covered by surface CrO_3, whereas the hexagonal nanorods were made up of a metallic chromium core and a CrOOH shell.[59]

Polystyrene microspheres with a carboxylate surface (memobeads) were coated layer by layer, first with poly(allylamine hydrochloride) (PAH), followed by magnetic chromium dioxide (CrO_2), then PAH again to immobilize the CrO_2 nanoparticles in a plane, then poly(styrene sulfonate), PAH and poly(acrylic acid) (PAA) to stiffen the whole surface assembly. This allowed a well-defined sterochemistry of the magnetic nanoparticles. First the optimal fixation of the microparticles for encoding and coupling of captured patterns on the surface was attained, and secondly the microscopic readout of dot and bar codes became possible by perfect orientation of the micron-sized particles with the ferromagnetic CrO_2 nanoparticle assemblies (Figure 5.26). They became magnetized in an external magnetic field and remained so for a long periods of time even without the field. Encoding by an alignment was long-lived and no thermal movement occurred later.[60]

5.4.2 Molybdenum

Nitrogenase, the enzyme which reduces molecular nitrogen to ammonia, consists of two proteins. One of them, known as MoFe protein or molybdoferredoxin, is brown and and air sensitive and contains 2 molybdenum ions, 24–36 iron ions and about the same number of sulfur ions. Highly energetic, antibonding orbitals of N_2 presumably take up electrons from molybdenum sulfides.

As a wide band gap n-type semiconductors, molybdenum and tungsten trioxides, MoO_3 and WO_3, exhibit photochromic and electrochromic properties, with high colouration efficiency, long lifetime and fast response time. Display and sensor devices and the so-called "smart window" have been developed with

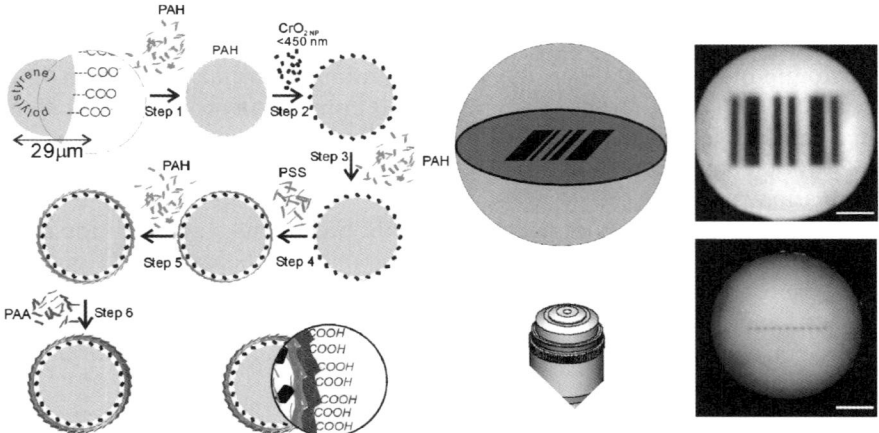

Figure 5.26 *Left:* Scheme of the layer by layer coating of carboxylated polystyrene microspheres with oppositely charged poly(allylamine hydrochloride; PAH) (PAA), negatively charged magnetic CrO_2 nanoparticles, and poly(styrenesulfonate, PSS). The ferromagnetic CrO_2 nanoparticles were localized between between two fluid PAH layers (see text). *Middle:* Observation of the magnetic nanoparticle pattern under the light microscope *Right:* Non-parallel and parallel oriented nanoparticle patterns. © American Chemical Society.[60]

these inorganic dyes, whose chromophores can be described as A_xMO_3 or as molybdenum and tungsten bronzes, where A is H, Li, Na, or K; M is Mo or W; and x is between 0 and 2. The injected electrons are trapped by some M^{6+} centres and colouration is attributed to the intervalence charge-transfer transition between M^{6+} and M^{5+}. Thin, transparent MoO_3 films were prepared by vacuum deposition on ITO.[61]

α-MoO_3 nanoplatelets were made from molybdenum isopropoxide, $(Mo(OC_3H_7)_5$, in 1-butanol. Simple ageing produced a clear blue gel, which was mixed with poly(ethyleneglycol) and transferred to a glass surface by dipping. The film was calcined at 500 °C which removed the polymer and left MoO_3 nanocrystals with a thickness of tens of nanometres (Figure 5.27).[62]

Alkylamines intercalate into layered molybdic acid, MoO_3. Short-chain amines $C_nH_{2n}NH_2$ ($n<9$) intercalated directly with full replacement of interlayer water molecules; long-chain amines required the previous intercalation of short-chain amines as "door openers" and allowed some water molecules to remain in voids. Subsequent acid treatment removed the short-chain amines. Intermediate chain length (*e.g.* $n=8$) gave compact "nanodisks"; long-chain amines remained in the interlamellar space. Salt formation between the molybdic acid layers is obviously only possible with single water-soluble amines; micelles do not enter the molybdic acid layers. They can only be integrated molecule by molecule into the preformed inner hydrocarbon layer. α,ω-Diamines were not tried.[62]

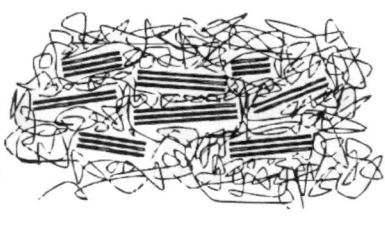

Figure 5.27 Scanning electron micrograph and model of MoO_3 nanoplatelets. The 3D growth of the nanocrystals was prevented by poly(ethyleneglycol) at 300 °C. © American Chemical Society.[61]

Heating of MoO_3 or WO_3 *in vacuo* or with or without some metal powder leads to intensely blue or violet deformed rutile structures containing MoO_2 or WO_2.

Hollow MoS_2 and MoO_3 nanospheres were prepared by sonochemical deposition of their suspensions on silica powder followed by hydrofluoric acid dissolution of the silica cores. MoS_2 is an extremely active catalyst for the hydrodesulfurization of thiophene, and the hollow MoS_2 nanoparticles were even more efficient. Surprisingly, upon heating the MoO_3 nanospheres rearranged to hollow cubic crystals of MoO_3, namely sharp-edged, truncated cubes with inner voids.[63]

Anisotropy is a prerequisite for the induction of long relaxation times of magnetization and superparamagnetic behaviour. poly(vinyl-pyrrolidone) was used to confine the growth of the coordination network and to stabilize the nanoparticles. 3 nm trimetallic $CuNiMo(CN)_8$ nanoparticles showed photo-induced superparamagnetism. Tiny MoO_2 crystallites were distributed uniformly among the polymer chains, but as the temperature increased above 300 °C, the MoO_2 nanoparticles transformed to MoO_3 layer-structured crystallites, which were still surrounded by polymer chains. The in-plane crystal growth of MoO_3 involved the formation of Mo–O bonds, but the inter-plane region was intercalated effectively by the poly(ethyleneglycol), and crystallization could not overcome the intercalation effect of this polymer at 300 °C. At 500 °C the polymer decomposed completely, and crystalline orthorhombic MoO_3 formed, still with an anisotropic geometry. Irradiation of these nanoparticles with 406–415 nm light converted the paramagnetic Cu(II) and Ni(II) ions by an electron transfer from Mo(IV) to Cu(II) to ferromagnetically coupled Mo(V)/Ni(II) pairs.[64]

In industry, molybdenum is commonly used in multifunctional catalysts for oxidation and, in particular, for desulfonations. Sulfur is removed from fossil feedstocks (0.1–5 wt%) by reduction to H_2S at 300–400 °C on the surface of molybdenum and tungsten sulfides. The desulfuration step is related to sulfur vacant sites, which are produced by hydrogen. Molybdenum sulfide catalysts also desulfurate thiophenes, which are otherwise very difficult to degrade. The hydrogen sulfide is later oxidized to sulfur with oxygen over cobalt/molybdenum catalysts.

Polycrystalline Mo/MoO_x nanoflakes were first made from $Mo(CO)_6$ powder in phenyl ether/oleic acid, slowly heated up to 310 °C to avoid sublimation of the molybdenum carbonyl, which started at 150 °C. The flakes were subsequently transformed to single-crystalline 2 nm MoO_3 and MoS_2 nanocrystals by in situ addition of oxygen or sulfur into the colloidal solution.[65]

Molybdene sulfide is a semiconductor and it was predicted theoretically that partial replacement of molybdenum by niobium would convert it to a metal by a shift of the Fermi level. This theory was experimentally verified in mixed Mo–Nb nanoparticles.

Molybdene sulfide is a quasi-2D compound, another candidate for a monolayer paper: all atoms within a layer are bound by strong covalent forces, but the layers are held together only by weak van der Waals forces. Foreign atoms or molecules therefore intercalate easily, or form closed cage structures (inorganic fullerenes) and inorganic nanotubes within the layers. Using density functional theory, the molybdene sulfide layers were modelled with some of the molybdenum replaced by niobium. This indicated that for niobium-substituted molybdene sulfide tubes the density of electronic states close to the Fermi level can be tuned over a wide range by the degree of niobium doping. This happens because the unit cell of semiconducting molybdene sulfide is similar to that of graphite, made up of two layers in a hexagonal arrangement, and each molybdenum atom is covalently bonded to six sulfur atoms in trigonal biprism coordination. A Mo→Nb substitution leads to a decreasing number of electrons in comparison with pure molybdene sulfide, so the Fermi level shifts and the density of states near the Fermi energy is increased. The substitution of niobium for molybdenum should therefore transform the semiconductor molybdene sulfide to a metal without a band gap.

The synthesis of such $Mo_{1-x}Nb_xS_2$ nanotubes was carried out starting from a $MoCl_5$: $NbCl_5$ precursor ratio of 55:1. The chlorides were first heated in an auxiliary furnace to 250 °C. A two-stage furnace setup was preset to 600 °C, and purged with N_2 and 5% H_2, mixed with the vapours of the metal chloride precursors. The temperature was raised to 850 °C and H_2S was introduced, diluted tenfold with N_2. The typical time period for formation of sulfide and nanotubes was 30 min. The black powder was then collected at the hot zone of the main furnace by means of a quartz wool filter. The control over the extent of niobium substitution in the nanoparticles was described as "rather limited".

Conducting atomic force microscopy (AFM) indicated a band gap of 1.0 eV for an ensemble of pure molybdene sulfide nanoparticles; for the bulk molybdene sulfide phase it is l.2 eV. The $Mo_{1-x}Nb_xS_2$ nanoparticles indeed exhibited a metallic character and no band gap; their resistance was 10 MΩ as opposed to 60 MΩ for the undoped MoS_2. Part of the remaining resistance was probably

Transition Metals

caused by an oxide layer in some of the particles and imperfect contact with the AFM tip.[66]

In water splitting, the first step in hydrogen evolution involves bonding of the proton to the catalyst, $H^+ + e^- + * \rightarrow H^*$, where * denotes a site on the catalyst's surface that is able to bind to hydrogen. The second step is the release of molecular hydrogen by one of two processes: either $2H^* \rightarrow H_2 + 2*$ or $H^+ + e^- + H^* \rightarrow H_2 + *$. Density functional theory can calculate the free energy of reversible atomic hydrogen bonding to the catalyst. It rules out metals that form strong bonds to atomic hydrogen, such as nickel and molybdenum, as good catalysts because the hydrogen release step will be too slow. Metals that do not bind to atomic hydrogen, *e.g.* gold, are also excluded, because the proton/electron-transfer step is not fast enough. A good catalyst will adsorb atomic hydrogen with a free energy close to that of the reactant or product.

Density functional theory calculations on the functional molybdenum sulfide centre of nitrogenase showed that the active FeMo cofactor site can bind hydrogen atoms only to the three equatorial sulfur ligands that comply with the $\Delta G°_H = 0$ requirement (Figure 5.28). Molybdenum sulfide nanocrystal edges have a similar structure to the FeMo cofactor: in both structures the sulfur atom that binds the hydrogen is doubly coordinated to metal atoms, either to molybdenum or to iron. Only the edges of molybdene sulfide are interesting in this context; its basal plane should be catalytically inactive. The first H atom that bonds to the edge is strongly bound, but at an H coverage >0.25, the differential free energy of adsorption is only 0.1 eV. Nanometre-sized

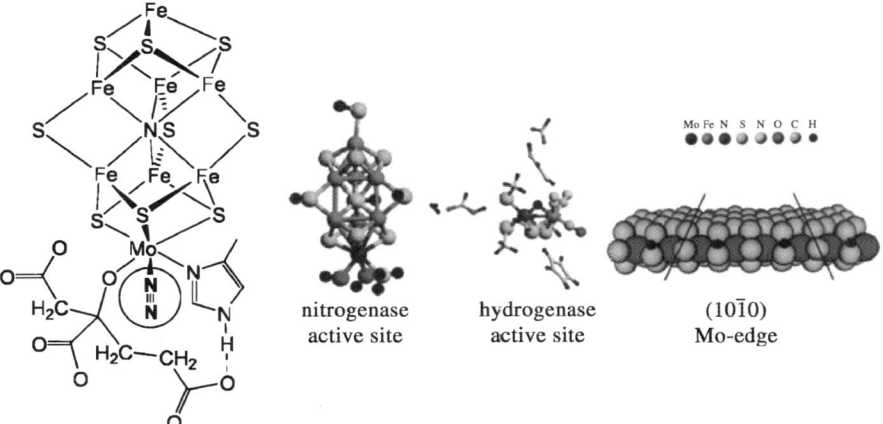

Figure 5.28 *Left:* Model of the nitrogenase FeMo cofactor with three hydrogen atoms bound to the equatorial sulphur atoms of Fe and N_2 to Mo. The carboxylate of the (*R*)-homocitrate has been substituted by N_2; three hydrogen atoms or protons were added to FeS. *Right:* MoS_2 slab with sulfur monomers present at the Mo edge. The coverage is 50%, *i.e.* hydrogen is bound at every second sulfur atom. The lines mark the dimension of the unit cell in the *x*-direction. © American Chemical Society.[67,68]

molybdene sulfide nanocrystallites supported for example on graphite, which is conducting but otherwise inert, should be good catalysts for hydrotreating, the hydrogenation of sulfur compounds in crude oil with water.[67,68]

4 × 1 nm molybdene sulfide nanocrystals nucleated along the graphitic steps and made a cathode for proton reduction at pH 0. Molybdene sulfide was also a promoter for the hydrogen evolution activity of NiS_x electrodes. Iron sulfides as biomimetic Haber–Bosch catalysis analogues were not investigated.

5.4.3 Tungsten

Next to carbon, tungsten has the second highest melting point of all elements (3400 °C), and it has the smallest expansion coefficient. Pure tungsten is soft, but small amounts of carbon or oxygen make it hard and brittle; with carbon, hardness values are between 4 and 8.5. At normal temperatures it is stable to water, oxygen and acids. Tungsten wires produce visible light at a temperature of 3250 °C. Chemically, tungsten resembles vanadium and molybdenum, but as a catalyst it is most active in alkyne polymerization and metathesis, a reversible cyclization of carbon chains with olefin ends, because it has a tendency to form carbenes.

Tungsten nanoclusters on graphite activated molecular nitrogen, catalysing the oxidation of almost inert nitrogen to dinitrogen oxide (N_2O) at 140 K. The observation that only wet nitrogen reacted indicated water as oxidant. Since no nitrogen atoms could be detected, it was assumed that N_2 itself added to a tungsten cluster and took up the water oxygen. Density functional calculations and spectra suggested a widening of the N_2 triple-bond of 109 pm to 125–129 pm by side-on addition to tungsten tetra-, penta- and hexamers, corresponding roughly to a double bond, which could indeed add water, if one also considers polarization by a metal (Figure 5.29).[69]

The oxidation of dibenzothiophene to the sulfone by hydrogen peroxide was catalysed by isolated tungsten atoms in tetrahedral sodium tungstate (Na_2WO_4), or in octahedral coordination on titania nanotube surfaces. The tetrahedral tungsten atoms were shown to be at least twice as active as the octahedral ones, but no reasons could be given.[70]

Nanopowders composed of molybdenum trioxide and tungsten trioxide nanoparticles were evaluated by various small-angle scattering techniques and

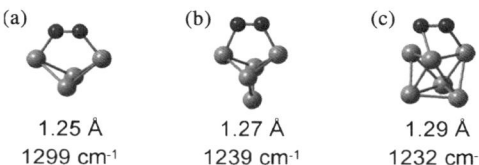

Figure 5.29 Density functional theory models of tungsten clusters with bound N_2 molecules and their calculated bond lengths (see text). © American Chemical Society.[69]

by scanning electron microscopy. Particle size distribution functions derived from both measurements and physical model calculations agreed well.[71]

Filled tungsten disulfide onions were obtained in a one-step reaction, hollow onions in a two-step procedure by chemical vapour deposition of tungsten hexacarbonyl (W(CO)$_6$). In contrast to the much better organized RuO$_2$ onions, the inner part of the tungsten sulfide onion is more amorphous. Orientation is triggered by incoming outer layers first and leads to the arrangement of inner layers later. Layer formation proceeds more smoothly in MoS$_2$ onions than with WS$_2$.[72,73]

A commercial heteropoly tungstic phosphoric acid, H$_3$PW$_{12}$O$_{40}$, with a polymeric oligoethyleneglycol-ammonium counterion, (CH$_3$)(C$_{18}$H$_{37}$)N$^+$[(CH$_2$CH$_2$O)$_n$H][(CH$_2$CH$_2$O)$_m$H] may be considered as a nanoparticle already and forms an anhydrous liquid by itself. It is of interest because it conducts protons by four orders of magnitude faster along the mobile oxygen chains than solids.[74]

5.4.4 Manganese

Manganese is the third most abundant transition metal on Earth, after iron and titanium, and is similar to iron in its chemical and physical behaviour. It is more variable, however, since its stable oxidation states range from −3 to +7. Manganese compounds are as widely used as redox catalysts as iron compounds. Chiral manganese Schiff bases, for example, are the best catalysts for the enantioselective epoxidation of non-substituted alkenes. Whenever Sharpless is not applicable (no allylic alcohol), try manganese.[75]

Hausmannite (Mn$_3$O$_4$) and rhodochrosite (MnCO$_3$) come to the surface by weathering of silicates and colloidal manganese oxides eventually float into the sea, where they sinter to manganese lumps. Steel always contains some manganese as sulfur catcher: MnS disappears in the slag.

Under water, manganese metal produces hydrogen. The metal burns in oxygen, nitrogen, chlorine and fluorine to yield Mn$_3$O$_4$, Mn$_3$N$_2$, MnCl$_2$, and MnF$_2$ + MnF$_3$ respectively. It also combines directly with boron, carbon, silicon, phosphorus, arsenic and sulfur. The oxidation number Mn^{2+} is most stable, because of its half-occupied d^5-configuration. Mn(IV)O$_2$ serves as a classical depolarizor in batteries, suppressing the evolution of hydrogen gas at the carbon electrode: MnO$_2$ + H$^+$ + e$^-$ → MnO(OH).

Mn is the typical metal for "molecular magnets", which are usually Mn(III) or Mn(III/IV) complexes (see refs. 75–77, and Chapter 1, ref. 42). Slow paramagnetic relaxation in isolated molecules is called single-molecule magnetism and crystalline Mn(III/IV) oxides, carboxylates and oximes can be deformed by their ligands in such a way that unpaired electrons with the same spin interact with each other without being allowed to form a p-electron pair (see Chapter 1.5 and Chapter 1, refs. 28–29,42,44).

The concept of molecular-based materials is being pursued in many directions. Instead of using solids consisting of extended lattices, such as in large oxide crystals, the goal is to make up a solid lattice of small molecular building blocks. The crucial points are (1) to arrange these chains in the solid state, so that there is a net ferromagnetic coupling between them (see Chapter 1, refs. 36,37) and (2) to achieve this not only at <5 K, but at room temperature. Three recent examples describe the state of the chemical art concerning molecular manganese magnets.

Molecular magnets gain their properties from the combination of a large ground-state spin quantum number (S), a magnetoanisotropy of the easy-axis type, and a negative zero-field splitting parameter (D) indicated by electron paramagnetic signal splitting. The expression "easy-axis" indicates here simply the favorite direction of spontaneous magnetization of ferromagnets, which is not only determined by the magnetic moment of the electron spin, but also influenced by magnetocrystal and shape anisotropies. The negative splitting constant D in the EPR spectrum is due to the single-electron property: an appreciable thermal energy is needed to populate higher energy levels. D is experimentally determined by the temperature dependence of the EPR signals. Stronger signals and splittings at higher temperature indicate a negative D.

The intermolecular interactions and long-range domain ordering of traditional magnets play only minor roles in molecular magnets. High-spin molecules with an easy-axis magnetic anisotropy show slow magnetic relaxation of the spin reorientation along the magnetic anisotropy axis, and at very low temperatures the spin flips not thermally but by quantum processes. These single-molecule magnets have a double minimum potential for the reversal of the magnetic moment and may lead to ultrasmall memory devices as well as to quantum computing. By combining different metal ions it should be possible to prepare molecules with large magnetic anisotropy and higher spin ground states due to ferro- or antiferromagnetic interactions.

Organic stereochemistry helps to convert Mn(III) low-spin complexes into high-spin ones without changing the metallic core. The primary reason that the oxime complexes are single-molecule magnets is the $S=6$ ground state, whereas the corresponding carboxylates (–COOH instead of –CNOH) are antiferromagnetically coupled with no magnetic moment imposing structural distortion on the molecule. Both oximate and carboxylate ligands usually lead to the same antiferromagnetic couplings, but if a displacement of a central O^{2-} ion out of the Mn_3 plane occurs, and the Mn–N–O–Mn bridge of the oxime shows a twisting reflected by an abnormal torsion angle, the strongly bonded, monoatomic O^{2-} bridge represents an excellent mediator of electron exchange interactions, providing the primary exchange pathway in metal oxide clusters, and paramagnetism occurs.

Since Mn(III) possesses both empty and partially occupied d-orbitals, the net observed coupling between two interacting metal ions is the sum of competing ferromagnetic and antiferromagnetic interactions. Overlap of π-symmetry magnetic orbitals is the primary contribution to the antiferromagnetic exchange between Mn(III) atoms, but the displaced oxygen prevents it. The ferromagnetic coupling of the parallel spins is hardly disturbed by the distortion, but the D value of the coupling constant is negative, which is unusual. This is again caused by the tilting of the Mn(III) anisotropy axes from the Mn_3

plane, which also ensures a significant molecular axial anisotropy. The easy axis now lies perpendicular to it and the single-ion z-axes become efficient ferromagnetic transfer axes (Figure 5.30).[76]

Dark red crystals of the [Mn(III)$_2$Ni(II)$_2$Cl$_2$(N-(2-hydroxybenzyl)-3-amino-1-propanol)$_2$] complex were easy to prepare by simple mixing of ligands and metal ions in 40% yield and contained an incomplete, face-sharing double cube with two Mn(III) and two Ni(II) ions with total $S=6$ spin ground state. The equatorial coordination planes of the Mn(III) ions with five O and one N ligand had a dihedral angle of 89.68° to each other. The D value of 7.05 T was negative because of the appreciable thermal energy needed to populate higher-energy M levels. Magnetization experiments were carried out for oriented crystals in an eicosane matrix at 0.55 K, and the data showed a step-like hysteresis loop, which proves that the molecule is a single-molecule magnet (Figure 5.31).[77]

The ferromagnetic intermetallic Mn$_2$Ga$_5$, a compound of the d-metal manganese and the p-metal gallium, has been synthesized and characterized by single-crystal X-ray diffraction and magnetic property measurements. The saturated magnetic moment of 2.71 μ_B was attributed to the high density of the states at the Fermi level, where the manganese d- and gallium p-orbitals are orthogonal by symmetry, giving net zero orbital interaction, no band dispersion and superdegenerate d-bands near the Fermi level, leading to the observed strong ferromagnetism. The manganese $3d_{xz}$- and $3d_{yz}$-orbitals provide non-bonding interactions with surrounding gallium p-orbitals close to the Fermi level, and this allows for many electrons here (Figure 5.32) (see Chapter 1, ref. 30).

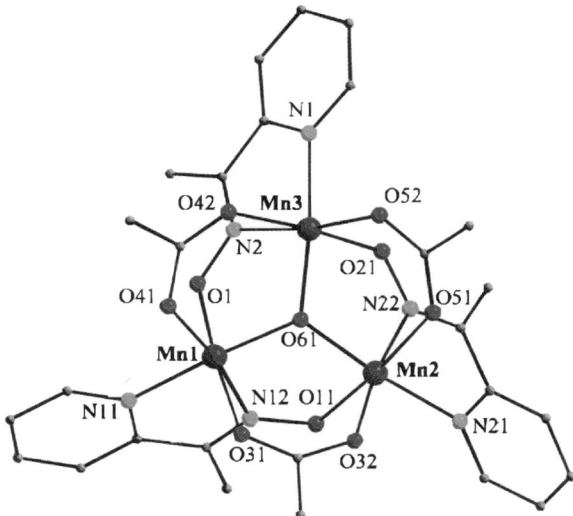

Figure 5.30 The molecular structure of the distorted (O1, O11 and O21) Mn(III)$_3$ trioxime molecular $S=6$ magnet.[76] See text for details. © American Chemical Society.

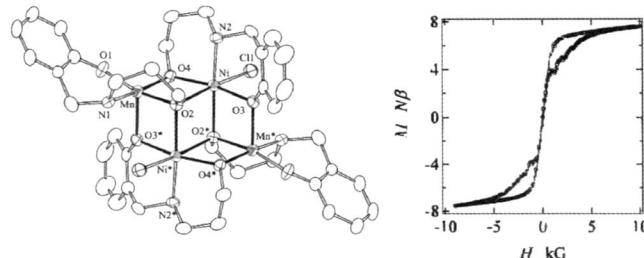

Figure 5.31 The $S=6$ Mn(III)$_2$Ni(II)$_2$ complex and the change of its magnetization M/Nb in an external magnetic field of changing polarization. © American Chemical Society.[77]

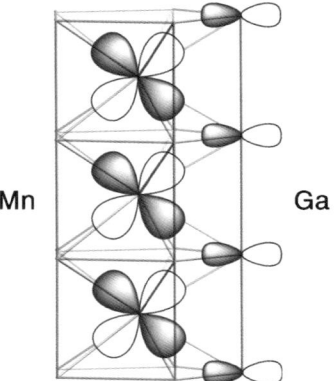

Figure 5.32 Arrangements of manganese 3d$_{xz}$ with respect to gallium p-orbitals. (Chapter 1, ref. 30) © American Chemical Society.

In nanocrystals the various oxidation states of manganese change by interaction with noble (He), oxidizing (O$_2$) or reducing (CH$_4$) gases at high temperatures without destroying the crystallinity. Temperature-dependent Raman spectra recorded in a stream of helium revealed an astonishing temperature-dependent, reversible solid-state redox reaction in an inert helium atmosphere. Figure 5.33a shows the spectrum of the original Mn(III)$_2$O$_3$ at room temperature, which remained essentially unchanged up to a temperature of 500 °C (Figure 5.33b), when a new band at 648 cm^{-1} belonging to Mn$_3$O$_4$ = Mn(II)OMn$_2$(III)O$_3$ nanocrystals appeared. This oxidation without an oxidant was fully reversible when the nanocrystal was cooled back to 25 °C again in the helium flow (Figure 5.33c). Essentially the same temperature-dependent spectra were obtained in a redox mixture of 0.5% CH$_4$ and 3.0% O$_2$ in helium, where methane combustion occurred. This time the 648 cm^{-1} band was already seen at 400 °C. Under the same experimental conditions bulk Mn$_2$O$_3$ did not show this reaction. The transition of MnO$_x$ in nanocrystals at elevated

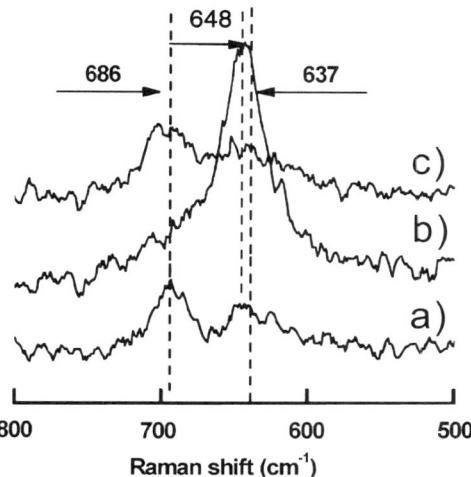

Figure 5.33 Raman spectra of Mn_2O_3 nanocrystals at (a) 25 °C, (b) 500 °C (Mn_3O_4; 648 cm^{-1}) and (c) upon cooling back to 25 °C. © American Chemical Society.[77]

temperatures is clearly due to the loss of lattice oxygen. In the methane–oxygen mixture the formation of the Mn_3O_4 species coincided with the onset of methane oxidation. Manganese oxide catalysis thus depends strongly on the oxygen mobility in the oxide lattice. This is the the primary factor for the activity of the manganese oxide nanocrystals, followed by their high surface area.[77]

Hausmannite (Mn_3O_4) nanocrystals with a faceted structure (20–40 nm) have been prepared by mixing aqueous solutions of manganese nitrate and hexamethylenetetramine at 20–80 °C. The temperature needed to reduce Mn_3O_4 to manganese oxide nanocrystals was higher with larger particle size. Upon re-oxidation, smaller manganese oxide nanoparticles formed an intermediate Mn_5O_8 phase (MnO → Mn_3O_4 → Mn_5O_8 → Mn_2O_3), while large ones gave Mn_2O_3 directly. Oxidation to MnO_2 required high temperatures.

Core–shell MnO–Mn_3O_4 nanoparticles (diameter 5–60 nm) were prepared from Mn(II)acetylacetonate, $Mn(CH_3CO)_2CH_2$ in 1,2-hexadecanediol–dibenzyl ether mixtures followed by the addition of oleylamine in molar ratios of either 2 or 20, leading to small or large nanoparticles. The slurry was mechanically stirred and heated to 150–290 °C. The amine reduced the inside oxides to antiferromagnetic MnO, oxygen caused a thin, highly anisotropic and ferromagnetic Mn_3O_4 shell.[78]

α-MnO_2 nanospheres (2 nm) and magnetic nanorods (aspect ratio 25) were made by air oxidation of $MnCl_2$ in variable sulfonate micellar templates under alkaline condition at room temperature.[79]

Metal–air batteries are light in weight and have a high energy density per weight unit, because air is used as one of the active materials. The air cathodes of hearing aids are composed of a mixture of carbon powder, oxygen reduction

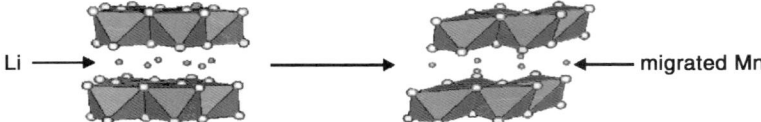

Figure 5.34 Model of the roughening of Li_2MnO_3 layers by acid treatment and manganese migration. © American Chemical Society.[81]

catalysts, and a binder. Manganese oxyhydroxide (MnOOH) has the highest electrocatalytic activity in oxygen reduction, which proceeds *via* a two-electron transfer followed by the MnOOH-catalysed disproportionation of peroxide to water and molecular oxygen.

MnOOH nanoparticles have been prepared by reduction of permanganate ($KMnO_4$) with sucrose at pH <1 accompanied by massive amounts of manganese(II) sulfate ($MnSO_4$). A dark brown slurry of an MnOOH phase was formed, without any disproportionation to Mn^{2+} and MnO_2, which usually occurs under acidic conditions. The high Mn^{2+} concentration suppressed it and favoured MnOOH nanoparticle formation. The nanoparticles were then integrated into the carbon cathode of a lithium–air battery and increased its energy content by 38%.[80] Treatment of this lithium manganate with 4 M H_2SO_4 at room temperature for 22 h yielded the protonated material $H[H_{0.18}Li_{0.15}Mn_{0.67}]O_{1.90}$. Some of the manganese had been removed from the layers by protons, migrated into the interlayer space and replaced the lithium ions there. The surface area became rougher and increased in area, which enhanced the catalytic activity in the reduction of oxygen (Figure 5.34).[81]

Spinel ferrite $MnFe_2O_4$ nanoparticles were synthesized on mesoporous silica gel as a hard template. The magnetic nanoparticles (<10 nm) were effective as a Fenton catalyst for the decomposition of hydrogen peroxide under neutral and basic conditions.[82]

$Yb_{14}MnSb_{11}$ nanocrystals are easily produced from the elements. Elementary ytterbium, manganese, antimony and tin were arranged in alumina crucibles in the ratio 14:6:11:86, sealed in quartz ampoules under argon and brought up to 1100 °C, when the tin melted. The mixture was then cooled to 700–800 °C and spun in a centrifuge at 6500 rpm to separate the solid products from the tin flux. Silver-coloured single-crystal ingots of $Yb_{14}MnSb_{11}$ remained. High-performance, high-temperature thermoelectric layers were made from the corresponding finely ground, polycrystalline powder, which was hot-pressed in high-density graphite dies to obtain macroscopic cylinders. Hot-pressed pellets of $Yb_{14}MnSb_{11}$ were used to analyse the electronic properties of the material. The resistivity increased linearly with increasing temperature. The thermopower of $Yb_{14}MnSb_{11}$ showed a positive increase with increasing temperature and reached a maximum of $+185\,\mu V/K$ at 1275 K. A low thermal conductivity 7–9 mW/cm K) may be due a less rigid structure; defects appeared not to play a role.[83]

High-temperature thermoelectric power generation occurred on graphite cylinders with a hot side of 1275 K and a cold side of 975 K. A thermoelectric

Transition Metals

efficiency of 4.3% was achieved, whereas commercially used p-SiGe achieves only 2.6%. Segmenting $Yb_{14}MnSb_{11}$ cylinders (1275 to 975 K) with $CeFe_4Sb_{12}$ (975 to 775 K) increased the efficiency to 7.3%. If the cold side of the device was lowered to 300 K, the thermoelectric efficiency was further enhanced to a value of 18.6% with the following p-segmentation:

$Yb_{14}MnSb_{11}/CeFe_4Sb_{12}/(GeTe)_{0.85}(AgSbTe_2)_{0.15}/(Bi,Sb)_2Te_3$.[83]

Manganese oxide nanoparticles (40 nm) were effectively internalized by cultured neuronal phenotype PC-12 cells, depleting dopamine and its metabolites, dihydroxyphenylacetic acid and homovanillic acid from aqueous buffer solutions. The nanoparticles were not toxic.[84]

Cadmium semiconductor nanocrystals with a direct band gap (p. 225ff) are toxic, however, and this limits the use of these nanoparticles emitting in the visible light region. Visible phosphorescence may also come from non-toxic Mn(II)-doped zinc sulfide nanoparticles, which have not yet been achieved. ZnSe/ZnMnS nanoparticles work, however, with an average quantum yield of 25%, and selenium is much less toxic than cadmium.[85]

5.5 Technetium and Rhenium

Technetium and rhenium provide much stronger metallic bonds than manganese, and the oxidation number II is rare. Re(III) clusters with M–M bonds are common. Whereas the permanganate ion (MnO_4^-) is a strong oxidant, TcO_4 and ReO_4 are weak oxidants.

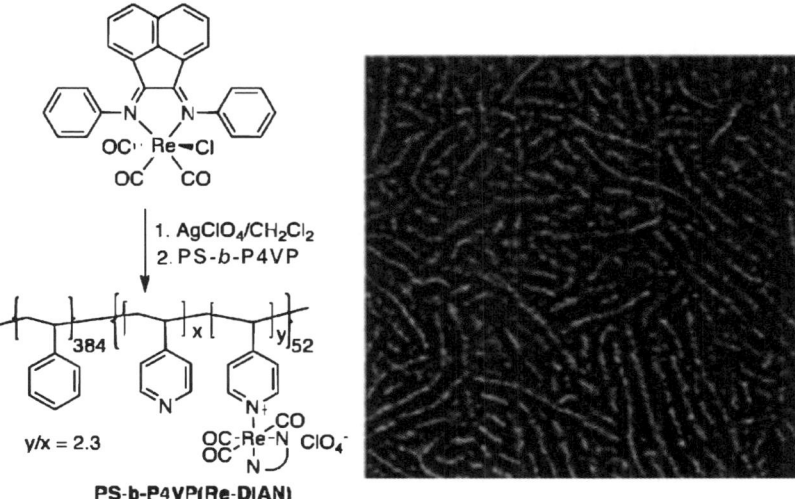

Figure 5.35 (a) Structures of the hydrophobic rhenium complex and its polymer adduct. (b) AFM image of the spin-coated layer of the polymer adduct. © American Chemical Society.[88]

The ^{99m}Tc isotope is a standard radiolabel for proteins. It has been used as ^{99m}Tc-pertechnetate to follow organic polymer–protein nanoparticles on their way through cell membranes or entrapped as dirhenium decacarbonyl [Re$_2$(CO)$_{10}$] in poly(-lactide) based nanoparticles in an oil-in-water emulsion.[86,87]

The Re(I)-(phenylimino)acenaphthene complex is a photosensitizer in photovoltaic devices and changes the morphology of poly(styrene)-poly(vinyl-4-pyrrolidone) from spherical to cylindrical (diameter 15–20 nm). The hydrophobic polymers wrap round the hydrophobic ligand (Figure 5.35).[88]

In rhenium disulfide (ReS$_2$), the tendency to form layered assemblies is even more pronounced and they "crystallize" to form onions with a hollow centre.[89] Core–shell nanoparticles were prepared by the reduction of gold or silver salts over ReO$_3$ nanoparticle seeds.[90]

5.6 Ruthenium and Osmium

5.6.1 Ruthenium

Ruthenium is the only metal other than iron that splits the nitrogen triple bond on a technologically large scale, supported by small graphite or magnesium oxide particles. The high price of ruthenium is balanced here by its higher tolerance to ammonia, which allows higher pressures and lower temperatures. Ruthenium black, usually prepared by reduction of ruthenium trichloride (RuCl$_3$) with formaldehyde in alkaline water, also catalyses hydrogenations of aromatics.

In homogeneous catalysis the Ru(0) 4d^{7}5s^1 configuration allows oxidation states from –2 in Ru(CO)$_4^{2-}$ (d^{10}) to +8 in RuO$_4$ (naked 4d and 5s). Ruthenium's high electron transfer capability and low redox potentials, as well as the extraordinary stability of ruthenium–carbon bonds, led to the development of Grubb's catalysts for ring-opening and ring-closing metathesis polymerizations.

Low melting and boiling points as well as binding enthalpies of Ru(0) organometallics indicate that the Ru(0) ($n-1$) d-electrons provide less polar carbon–metal bonds than the previous groups.

Fe(0) is by far the most reactive metal of the Fe–Ru–Os triad. Ru(0) and Os(0) are stable in air, although finely dispersed metallic osmium smells of OsO$_4$. Iron cannot occur with the oxidation state of VIII, Fe(VI) in (FeO$_4$)$_3$ being the maximum, whereas RuO$_4$ and OsO$_4$ are common compounds. Iron and steel need reductive protection by alloyed metals or by Fe$_2$O$_3$ coatings, where as metallic ruthenium and osmium are quite stable without an oxide surface layer.

Metallic Ru(0) nanoparticles were made from Ru(0) (η4-cyclooctadiene) (η6-cyclooctatriene) in tetrahydrofuran/methanol (19:1) by hydrogenation. The olefins were first reduced to the noncoordinating, saturated hydrocarbons and then expelled the metal atoms, which assembled to form nanocrystals (Figure 5.36). These ruthenium nanocrystals did not react with highly strained

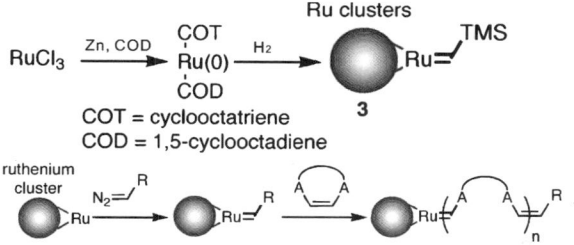

Figure 5.36 Ruthenium nanoparticle formation and metathetical growth of a polymer from a nanoparticle activated with a substituted diazoalkane. © American Chemical Society.[91]

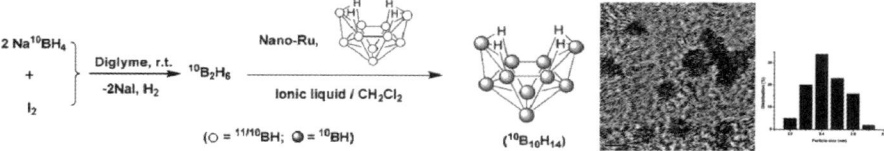

Figure 5.37 *Left:* Synthesis and molecular structure of the decaborane(14) ($B_{10}H_{14}$). *Right:* Electron micrograph and size histogram of the Ru(0) nanoparticles. © American Chemical Society.[92]

cyclic olefins at room temperature, but added trimethylsilyldiazomethane, $(CH_3)_3SiCHN_2$. The diazoalkane reacted first with the particle surface, released N_2 and formed a surface alkylidene, which was the active site for the ring-opening metathesis. As the polymers grew to surround the nanoparticles, phase separation occurred.[91]

Black Ru(0) nanoparticles were made by the hydrogenation of the tricyclopentadiene$_3$-dianion Ru$_2$(0) complex [CpRuCpRuCp]PF$_6$ in ethylene glycol at 180 °C and dissolved in the ionic liquid trihexyltetradecylphosphonium dodecylbenzenesulfonate. After removal of excess ethylene glycol a black sticky residue was obtained. The ruthenium nanoparticles were isolated by centrifugation of a dichloromethane solution. Catalytic $^{10}B/^{11}B$ isotope exchange took place within 6 h in a solution of decaborane(14) ($B_{10}H_{14}$), with the Ru(0) nanoparticles in the ionic liquid at 50 °C. The Ru(0) nanoparticles have diameters ~3 nm and are quite stable without any coating (Figure 5.37).[92]

Ru$_x$ and Ru$_x$Se$_y$ nanocrystals were prepared by thermolysis of Ru$_3$(CO)$_{12}$ and selenium in an organic solvent. In the ruthenium–selenium nanocrystals a crystalline 4 nm ruthenium core was covered by an amorphous, catalytically active "eggshell" of ruthenium oxide, selenide, selenite, and metal organic residues. Annealing *in vacuo* at 900 °C for 2 h produced hexagonal ruthenium dipyramids of up to 10 nm. The selenium, which had been bound covalently to ruthenium, was partially oxidized to $(SeO_3)^{2-}$, protected the ruthenium crystallites against oxidation and also provided ruthenium–selenium complexes on

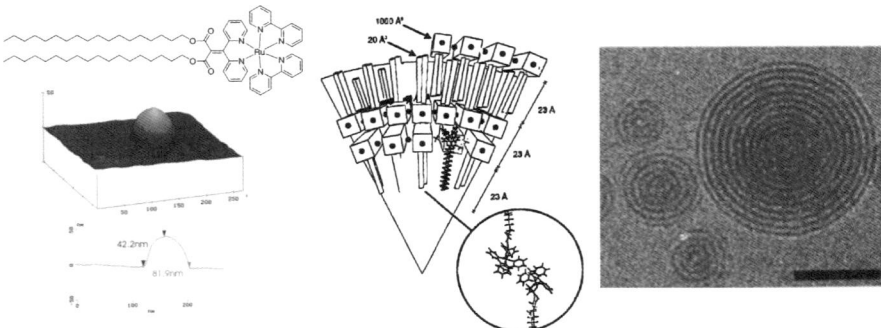

Figure 5.38 *Left:* Molecular structure of an amphiphilic tris-bipyridyl ruthenium hexafluorophosphate. *Centre:* Model of its multilayered micelle in water and the interdigitated bilayer. *Right:* Cryo-electron micrograph of the dense onion-type micelle without hollow centre, and an AFM image on a mica surface.[95]

the surface. The ruthenium–selenium nanocrystals were active in the catalytic electroreduction of O_2 to H_2O_2.[93] Thiol-stabilized Ru(0) nanocrystals were also described.[94]

The most stable and rigid micellar structure known was produced from a water-insoluble tris-bipyridyl ruthenium hexafluorophosphate salt. The head group of this amphiphile is so wide that even the double-chained hydrophobic blocks interdigitate. This double-chain packing rendered the assembly insoluble in water and enforced a tight packing down to the most curved centre of multilayered micelles with a negligibly small water content. This onion-like molecular assembly is probably the only known micelle that is isolable in solid form, and it was imaged with AFM tips on an aluminosilicate surface (Figure 5.38). The micelle was also fluorescing and photoactive.[95]

The broad visible absorption band of tris-bipyridyl Ru(II) and Os(II) complexes at 450–470 nm comes from a metal-to-ligand charge transfer. Both complexes also fluoresce, which is unique for transition metals with incomplete electron occupation of the d-shell. The d^5 ions behave here like the full shell d^{10} zirconium, but only Ru(II) and Os(II) show this behaviour, not Fe(II).

The luminescence behaviour of Ru(II) nanocrystals is also peculiar. The Ru(II)(bipyridyl)$_3$-dichloride dissolved in water emits at 620 nm (Figure 5.39). The amphiphilic hexafluorophosphaten was not soluble in water, but gave stable suspensions after ultrasonication. After 5 min of sonication the emission yield of the turbid dipersion amounted to 150% as compared to Ru(bipy)$_3$Cl$_2$. Cryotransmission electron microscopy did showed not the expected fluid micelles, but granular nanocrystals. After 45 min sonication, when the multi-layered structure of Figure 5.38 had finally formed, the fluorescence decreased to a constant value of $\sim 30\%$. In organic solution (chloroform, acetonitrile) the fatty acid diester did not luminesce like the chloride in water, but luminescence

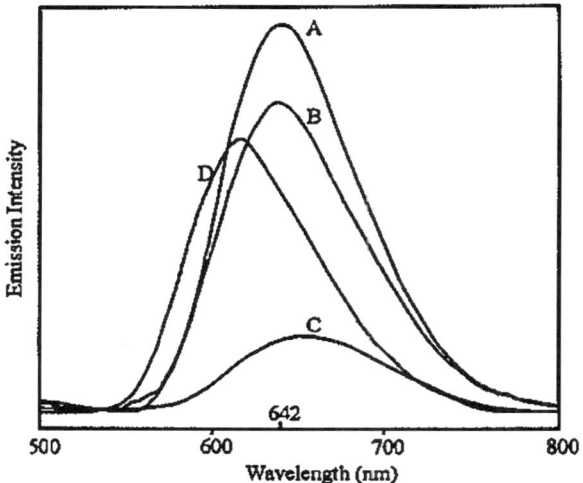

Figure 5.39 Relative luminescence intensity of aqueous suspensions of (A) Ru(bipy)$_3$Cl$_2$, (B) the nanocrystal of the double chain Ru(bipy)$_3$ (5μM) of Figure 5.38 after 5 min sonication at 50 °C, and (C) after 30 min (multilayered micelle). The visible absorption at 468 nm was identical for all four samples. © American Chemical Society.[96]

reappeared in frozen chloroform solutions at 77 K. Os(bipy)$_3$Cl$_2$ shows similar fluorescence to the ruthenium complex, but is shifted to the red by ∼100 nm.

poly(N-Isopropylacrylamide) was introduced on p. 116 as a thermosensitive polymer, which becomes reversibly dehydrated around 40 °C and may be precipitated reversibly at high temperatures and redissolved at lower temperatures. If such a polymer is connected with the tri(2,2′-bipyridine) ruthenium(II) bis(hexafluorophosphate), Ru(II/III)(bpy)$_3$, the hydrophilicity of the polymer can be enhanced by oxidizing Ru(II) to Ru(III). The oxidized polymer will then swell in water, and the reduced one will shrink. Furthermore, the solubility in water will be higher for the oxidized polymer micelle and in presence of NaCl it may still be soluble even when the reduced species flocculates. The microgels in the oxidized Ru(III) state flocculated at slightly higher temperature (34 °C) than the reduced Ru(II) state (32 °C).

In the presence of the oxidants sodium bromate (NaBrO$_3$), nitric acid (HNO$_3$) and malonic acid as proton donor and acceptor. gels were solubilized at 27 °C if enough of the Ru(III)(bpy)$_3$ complex was present. The swollen nanoparticles then contained excess charge and re-protonated the malonic acid, so the Ru(II)(bpy)$_3$ nanoparticles flocculated and precipitated. The low potential caused re-oxidation of the Ru(II) to Ru(III) and the nanoparticles solubilized. Flocculation made the dispersion turbid, and the redispersion transparent. Under optimized conditions this process was optically reversible and became "self-oscillating" (Figure 5.40).[96]

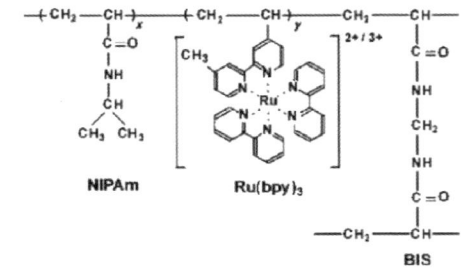

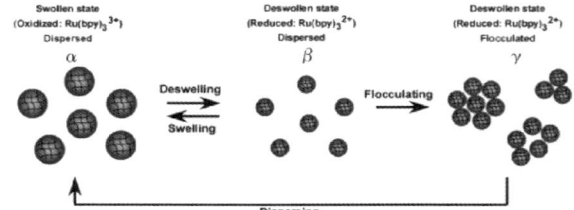

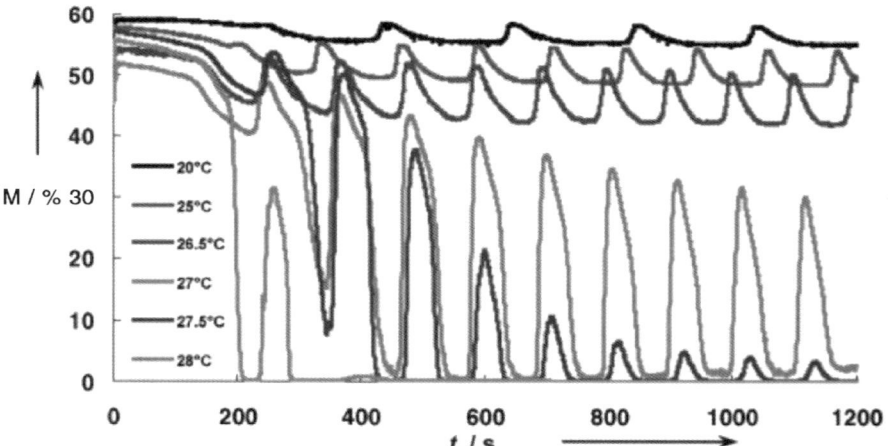

Figure 5.40 *Top*: Molecular structure of the (4-vinyl-4′-methyl-2,2′-bipyridine) bis(2,2′-bipyridine) Ru(II) bis(hexafluorophosphate) complex on the poly(*N*-isopropylacrylamide) cross-linked with *N,N*′-methylenebisacrylamide. *Middle*: Model of swollen Ru(III) and dehydrated Ru(II) nanoparticles and their flocculated clouds. *Bottom*: Self-oscillating transparency profiles of the dispersions in presence of malonic acid, $NaBrO_3$, and HNO_3 at 27 °C.[100]

5.6.2 Osmium

Two luminophores, tris(2,2′-bipyridyl)osmium(II)bis(hexafluorophosphate) (OsBpy) and tris(2,2′-bipyridyl)dichlororuthenium(II)hexahydrate (RuBpy), were simultaneously entrapped inside 10 nm silica nanoparticles coated with biotin or avidin at precisely controlled ratios. A single-wavelength excitation then produced dual emissions.[97]

If $Ru(bis\text{-}pyridine)_3^{2+}$ is connected by a rigid 4.2 nm poly(phenylene) wire to $Os(bis\text{-}pyridine)_3^{2+}$ its fluorescence is quenched because its excited electron is taken up by the osmium complex (redox quenching). An electron or an electron pair can only go from the reducing ruthenium to the oxidizing osmium; the system works as a rectifier.[98]

5.7 Cobalt, Rhodium and Iridium

5.7.1 Cobalt

Cobalt and nickel are both ferromagnetic, like iron (p. 28). They are perhaps best known as a team in the magnetic alloy known as alnico, containing aluminum, nickel and cobalt and used for permanent magnets that are 25 times stronger than normal steel magnets. Alnico magnets are fine-particle magnets like the sintered $Nd_2Fe_{14}B$ and $SmCo_5$ magnets, although the particles are much smaller.

Temperature-dependent measurements of the magnetic properties of 10 nm cobalt nanoparticles showed a rapid rise in magnetic moment at ~ 250 K during warming and an abrupt drop in magnetic moment at 234 K during cooling. These changes followed the melting and freezing of the 1,2-dichlorobenzene solvent (melting point 255–256 K or –18 °C). The rapid rise in magnetization was thus due to physical movement and rearrangement of the nanoparticle chains; sudden crystallization caused the abrupt drop in magnetic moment at 234 K. The cobalt nanoparticles in a colloidal solution also formed linear chains along the field lines of an external magnetic field. The chains remained after removal of the magnetic field, but lost linearity. At 298 K, the magnetic moment reaches its maximum value at 79.6 kA/m (1000 Oe) and is nearly constant for the strength of the external magnetic field values between 1000 and 50 000 Oe. Although the coercivity and remanence ratio are near zero at 298 K there are still interparticle interactions between the cobalt nanoparticles.

Hydrophobic Co(0) nanoparticles with a diameter of 10 ± 1.4 nm were prepared from $Co_2(CO)_8$ dissolved in 1,2-dichlorobenzene by addition of trioctylphosphine oxide and oleic acid. The black colloidal solution was extracted using an airtight syringe and stored in a glass vial under argon. The colloidal sample was stored in air and used as an aged sample for magnetic measurements. A dried sample was also prepared by placing the colloidal cobalt nanoparticles in a gelatin capsule, attracting the cobalt nanoparticles to the bottom using a hand-held magnet, extracting the solvent, and blowing the sample dry with argon.[99]

Similar polystyrene-coated cobalt nanoparticles produced 1D chains after dissolution in dodecanedioldimethacrylate, which were first oriented by application of a magnetic field and then fixated by polymerization of the solvent with UV light (Figure 5.40).[100]

Various coatings have been applied to Co(0) nanoparticles for protection against oxidation. The simplest is graphite, coming from the combustion of vinylphenoxyphthalonitrile at 700 °C. The cobalt-specific saturation magnetization value increased upon heating due to a growth in particle size and improved crystallinity.[101] Polymer-coated ferromagnetic nanoparticles were made by thermolysis of dicobalt octacarbonyl ($Co_2(CO)_8$) in the presence of end-functionalized polystyrenes bearing amine, phosphine oxide, or carboxylic acid ligands (p. 95).[102] Magnetic fields caused parallel alignment of the uniform nanoparticles (Figure 5.41). Stable colloidal dispersions of cobalt nanoparticles were also prepared by refluxing toluene solutions of dicobalt octacarbonyl in the presence of poly(dimethylsiloxane-b-(methylvinylsiloxane-co-methyl(2-trimethoxysilethyl)siloxane). The polymer-coated cobalt nanoparticles were subsequently heated at 700 °C to form siliceous shells, which protect them from oxidation and make them applicable in medicine.[103]

Trioctylphosphine oxide and octadecylamine in refluxing 1,2-dichlorobenzene at 455 K were used to stabilize cobalt nanoparticles formed from dicobalt octacarbonyl. The black colloidal solution contained ∼1% of the aminated cobalt nanoparticles and an electron micrograph of an aged sample showed 11 nm non-oxidized Co(0) nanocrystals after 70 days in the dichlorobenzene solvent (Figure 5.42).[104]

Colloidal Co(0) nanoparticles can also be formed from 0.5 M cobalt chloride under the same conditions as citrate Au(0) nanocrystals with aqueous sodium citrate solution, but sodium borohydride must be added, because Co(0) is a much stronger reducing agent than Au(0). Upon reaction addition of $HAuCl_4$ uniform hollow gold nanospheres were produced in aqueous solution as "sacrificial electroless deposition templates":

$$3Co(0) + 2Au(III) \rightarrow 3Co(II) + 2Au(0)$$

In the presence of a magnetic stir-plate these weakly aggregated Co(0/II) nanoparticles produced long, well-organized gold nanotubes (Figure 5.43).[105]

An Fe_2O_3 film on an SnO_2 water-splitting photoelectrode was treated with Co(II) nitrate ($Co(NO_3)_2$) yielding an 80 mV cathodic shift, and a rising photocurrent of 2.7 mA/cm^2 at 1.23 V. A $Co(OH)_2$ monolayer on a basic haematite surface had the same effect. It formed large cobalt oxide (Co(II, III, IV)O$_x$) clusters, which were electrocatalytically active in water oxidation to molecular oxygen. Valence-band holes created by visible light illumination of Fe_2O_3 are on Fe(IV) rather than on O^{2-} as in $2p^6$ TiO_2. Fe(IV) then oxidized slowly, but Co(IV) linked to strongly electrophilic oxo groups decayed within seconds and nucleophilic attack by hydroxide created an O–O single bond in the form of a peroxo intermediate. Molecular oxygen was released, accompanied by hydrogen transfer to a vicinal oxo group. Rehydroxylation of the liberated Co(II) coordination site by the electrolyte then closed the cycle (Figure 5.44). Hole trapping by surface Co(II) sites thus allowed accumulation of oxidation

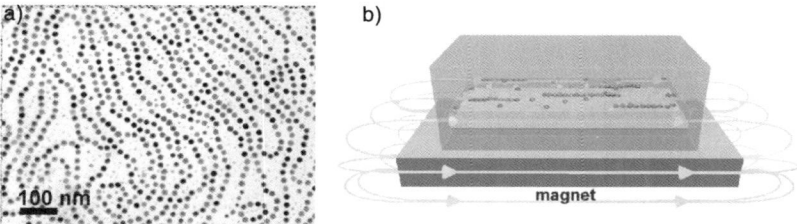

Figure 5.41 (a) Transmission electron micrograph of ferromagnetic poly(styrene) coated 19 nm cobalt nanoparticles drop-cast over a permanent magnet (b). © American Chemical Society.[102]

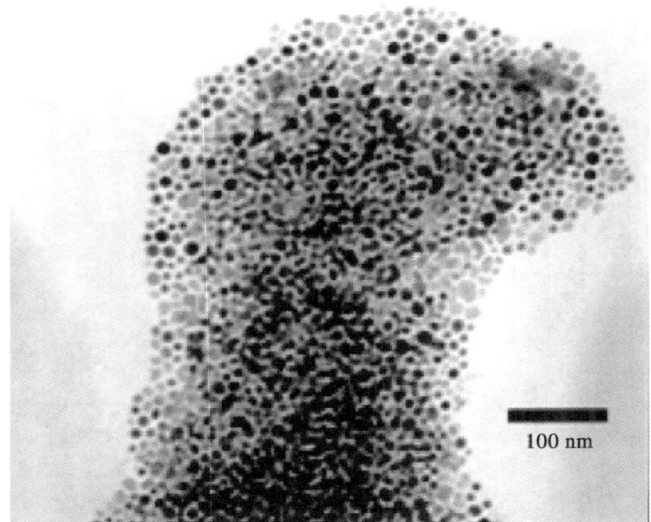

Figure 5.42 Electron microscopic characterization of 70-day-old cobalt nanoparticles in 1,2-dichlorobenzene. © American Chemical Society.[104]

equivalents without formation of highly energetic OH• radicals. Co(II, III, IV) deposition enhances the incident photon to current efficiency as a function of wavelength for the haematite electrode by a factor of 1.2 between 350 and 500 nm, and slightly more outside this region. At 1.23 V this electrode is much more efficient in water photooxidation than the best published Fe_2O_3 thin films or niobium-doped Fe_2O_3 single crystals.[106]

The visible spectra of low-spin Fe(II) and Co(III) complexes should be similar, because both metal ions are isoelectronic. In the case of Fe(II), however, spin-pairing needs a much stronger ligand field and ligands with such a field strength usually produce charge-transfer bands at low energy, which then cover the d–d-transitions.

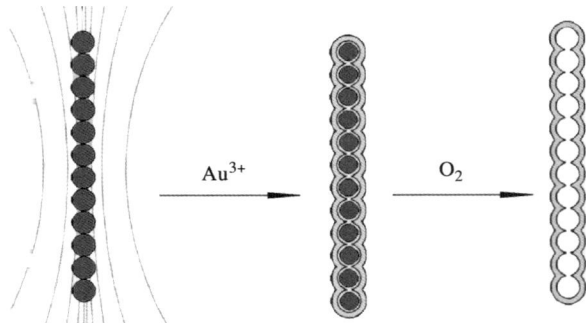

Figure 5.43 Alignment of Co(0/II) nanoparticles along a magnetic field, formation of a Au(0) coating and removal of Co(III) nanoparticles after oxidation with O_2. © American Chemical Society.[105]

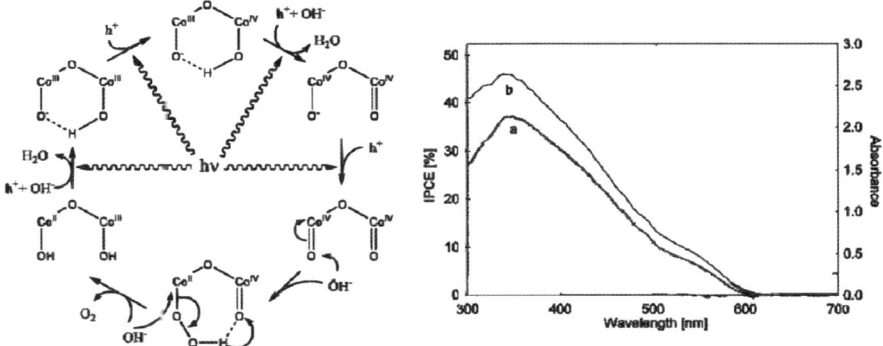

Figure 5.44 *Left:* Mechanistic proposal for water photooxidation at the cobalt-modified Fe_2O_3 surface: capture of four photogenerated holes by vicinal cobalt centres accompanied by deprotonation of surface hydroxyl groups creates highly electrophilic oxo groups on Co(IV). Nucleophilic hydroxide attack on one Co(IV)–O results in O–O bond, peroxo formation which decomposes to O_2 under hydrogen transfer to the second Co(IV)–O. Rehydroxylation of the dangling Co(II) coordination site by the electrolyte closes the cycle. The same mechanism can be envisaged with iron in place of cobalt. *Right:* Incident photon to current efficiency (=IPCE) spectrum of (a) the unmodified Fe_2O_3 (b) after cobalt deposition, both at 1.23 V against hydrogen. © American Chemical Society.[106]

The weak blue colour of Co(III) ions is of minor importance as a pigment for ceramics, but is often used to obtain white tones of slightly yellow materials containing Fe(III) impurities. Cobalt compounds are added to oil paints as catalysts for the oxidations and crosslinkings that cause the "drying" of the paints.

Transition Metals

The most important cobalt-catalysed process is the hydroformylation of alkenes:

$$RCH=CH_2 + CO + H_2 \rightarrow RCH_2CH_2CHO + RCH(CH_3)CHO$$

Common cobalt catalysts are $HCo(CO)_4$ (ligand free) and phosphine derivatives such as $HCo(CO)_3PR_3$. A π-complex is first formed, which rearranges to an alkylcarbonyl and is finally hydrogenated. The cobalt compounds catalyse not only the C–C bond formation, but also the final hydrogenation.

Co(II)/Co(0) on alumina, like iron, also catalyse the conversion of CO and H_2O to hydrocarbons and do not produce graphite (Fischer–Tropsch synthesis, p. 289). 15 wt% cobalt catalysts supported by γ-Al_2O_3 were prepared by mixing a freshly prepared alumina slurry (surface area ∼350 m g^{-1}, pore diameter ∼12.5 nm) and cobalt nitrate $(Co(NO_3)_2 \cdot 6H_2O)$ at a pH below the point of zero charge of γ-Al_2O_3. A repulsive interaction occurred between the supporting surface and the Co^{2+} ions and induced a non-homogeneous, possibly bimodal, cobalt particle size distribution with about one-half of the needles having a diameter of <4 nm and the rest being larger.

Three basic mechanisms have been proposed for Fischer–Tropsch syntheses: a carbide mechanism, a CO insertion mechanism, and a hydroxycarbene mechanism. For cobalt-catalysed syntheses the carbide mechanism, where chain growth proceeds *via* surface CH_2 insertion, has been widely accepted. This monomeric building block is formed by CO dissociation ($CO_{ads} \rightarrow C_{ads} + O_{ads}$) and hydrogenation of surface carbon atoms ($C_{ads} + H_{ads} \rightarrow CH_{ads}$; $CH_{ads} + H_{ads} \rightarrow CH_{2ads}$), while the surface oxygen atom is removed by the formation and desorption of water.[107,108] Hydrodesulfurization and hydrocracking are other typical activities of cobalt catalysts. Cobalt aluminophosphates (CoALPOs) with 0.38 nm pores catalyse the oxidation of alkanes by air.

The main problem with cobalt catalysts is their carcinogenicity. Rhodium functions similarly as a catalyst, but is harmless.

5.7.2 Rhodium

A convenient route to colloids of noble metals (osmium, rhodium, iridium, palladium, platinum, and gold) is reduction of the metal halides in alcohols in the presence of PVP. Small rhodium nanocrystal seeds were synthesized by the reduction of rhodium chloride ($RhCl_3$) in ethylene glycol in the presence of poly(vinylpyrrole). In ethylene glycol with poly(vinylpyrrolidone) (PVP) these seeds grew spontaneously to isotropic 3.6–5.5 nm cubic Rh(0) nanocrystals, which were precipitated with 2-propanol/hexane. A second addition of $RhCl_3$ to the Rh(0) seed sol led to the formation of larger particles (Figure 5.45).[109]

The nature of the rhodium nanoparticles produced depended on temperature. At 190 °C isotropic single-crystal 13 nm rhodium cubes were formed. Lowering the reaction temperature resulted in more anisotropic growth, giving single-crystalline rhodium cubes with horns at 140 °C and multipods at 90 °C. Below 90 °C no reduction of rhodium chloride occurred. The anisotropic geometry of

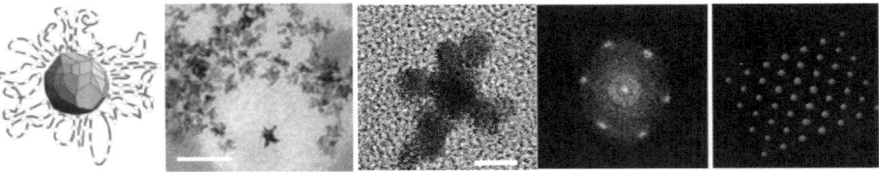

Figure 5.45 Model and electron microscopic image (bars = 50 nm and 4 nm), electron diffraction and lattice simulation of PVP-stabilized Rh(0) nanoparticles. © American Chemical Society.[109]

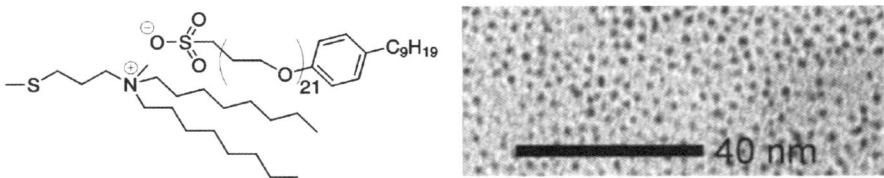

Figure 5.46 Molecular structure and electron micrograph of platinum droplets. Rhodium, Palladium and gold react similarly. © American Chemical Society.[110]

the particles arises either from seeded growth or oriented attachment, which favours the conservation of 111 surfaces at the expense of the others. In case of the rhodium nanocrystals, the seeded growth mechanism prevailed.

N,N-Dioctyl-N-(3-mercaptopropyl)-N-methylammonium bromide was ligated to the surface of rhodium, platinum, palladium and gold nanoparticles and an amphiphilic oligoethyleneglycol sulfonate was added at a bromide:sulfonate molar ratio of 1:1. The resulting metal nanoparticles were all viscous liquids at room temperature and the 2–10 nm droplets became solid at 16 °C[110] (Figure 5.46).

The pyrrolidone units of PVP form strong coordinating bonds with rhodium or platinum nanocrystals *via* amine-N and carbonyl-O donors, which bridge adjacent metal atoms on the nanocrystal surface, similar to the polymeric "platinum blue" and "rhodium blue" coordination compounds.[111]

The catalytic partial oxidation of methane with water is an exothermic process and produces hydrogen for fuel cells. Rhodium/alumina catalysts perform at temperatures between 600 and 1100 K, and CO is an important intermediate. The adsorption of CO on reduced rhodium frequently gives rise to Rh(I)(CO)$_2$ *gem*-dicarbonyl complexes, which are stabilized by a strong Rh-CO covalent σ-bond and a π-back bond. This reaction of CO with rhodium is supported by OH$^-$ ions, which disrupt only the very small Rh(0) particles or attack isolated Rh atoms:

$$Rh(0) + CO + OH^- \rightarrow O^{2-} + Rh(I)(CO)_2 + \tfrac{1}{2}H_2$$

Rh(0) nanoparticles give linear and bridging Rh(0)–CO complexes instead, which are less stable, because of a weak σ-bond. Carbon also deposits on

Transition Metals

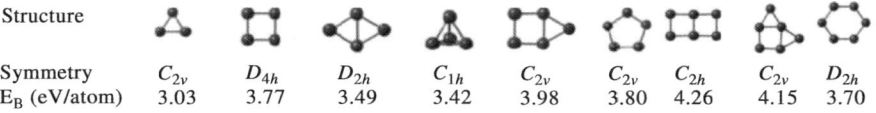

Structure									
Symmetry	C_{2v}	D_{4h}	D_{2h}	C_{1h}	C_{2v}	C_{2v}	C_{2h}	C_{2v}	D_{2h}
E_B (eV/atom)	3.03	3.77	3.49	3.42	3.98	3.80	4.26	4.15	3.70

Figure 5.47 Structure and binding energy (E_B) of some small iridium clusters (n = 3–6). © American Chemical Society.[113]

rhodium in the equilibrium mixture of the reducing agents methane, carbon monoxide and hydrogen with carbon dioxide and water:[112]

$$2CO \rightarrow CO_2 + C \text{ and } CH_4 \rightarrow 2H_2 + C$$

5.7.3 Iridium

Iridium is primarily known for its catalytic activity in company with other precious metals.[113] A process catalysed by iridium on its own is the carbonylation of methanol to form acetic acid, a liquid phase insertion of carbon monoxide into an iridium-bound methyl group and release of acetyliodide after addition of iodide.

Iridium has been shown to be very effective in the removal of pollutants such as NO_x from vehicle exhausts. The most stable adsorption state of NO towards any of the precious metal clusters was toward the iridium tetramers, square pyramidal and trigonal bipyramidal geometries. Density functional theory calculations assessed iridium clusters based on the binding energy per atom. Unlike platinum, palladium and rhodium with their preferred tetrahedron structures, for iridium the square configuration of atoms is most important (Figure 5.47).[113] Only ruthenium is more stable in the square configuration.

Many small metal clusters easily undergo coalescence, and electron microscopic analysis has shown liquid-like properties of surface atoms. Isomers with different arrangements of surface atoms all have the same energy, which is most typical for platinum. The small, rigid iridium cubes cause a localization of electrons, which produces high magnetic moments. Most of the 2–8 atom clusters have magnetic moments $>0.5\,\mu B$/atom, and many $>1.0\,\mu B$/atom. These substantial magnetic moments are also seen for a few of the larger structures.

5.8 Nickel, Palladium and Platinum

5.8.1 Nickel

Nickel is a ferromagnetic material with a small magnetocrystalline anisotropy energy. Only when it is grown as a nanowire might the anisotropy energy dominate the magnetism, and the nanowire will show high coercivity. Nickel nanowires 20 nm wide were fabricated by direct current electrodeposition in pores of anodic aluminum oxide templates (p. 182 ff) and were shown by electron microscopy to contain randomly arranged grains. Application of a magnetic field during nanowire growth allowed the grains to grow along the axis of the wire and enhanced the magnetism by shape anisotropy without any heat treatment. Nickel nanoparticles were prepared in air according to a modified polyol route using nickel chloride ($NiCl_2$) as precursor, sodium borohydride as reducing agent and PVP as coating. Fourier-transformed IR spectroscopy and XPS data showed an effective interaction between the Ni(0) nanoparticle surface and the carboxyl oxygen atoms of the polymer. Magnetic measurements indicated a single-domain superparamagnetism due to dipolar magnetic coupling between the nanoparticles (Figure 5.48).[114]

Nickel clusters were produced by the exploding wire method, silica shells *via* the sol–gel method. Metallic Ni(0) nanoparticles were encapsulated in submicron-sized spherical silica shells.[115]

Multicomponent metal oxide nanoparticles are most simply prepared by applying a three-solvent system to metal alkoxide mixtures: at first acetic acid acts as a solvent and complexing agent for the metal alkoxides; a strong aqueous acid such as HCl or HNO_3 then hydrophilizes the metal ions; and finally ethanol produces alkoxides, which allow combinations with structure-directing block copolymers.[116] Synthesis of mesoporous $NiO-2SiO_2-2ZrO_2$ nanoparticles, for example, started from a mixture of nickel nitrate ($Ni(NO_3)_2 \cdot 6H_2O$), tetraethylorthosilicate ($Si(OEt)_4$) and zircone butyrate ($Zr(OBu)_4$), all dissolved in acetic acid. HCl and Pluronic F12, the symmetrical triblock copolymer, poly(ethyleneoxide$_{106}$propyleneoxide$_{70}$ ethyleneoxide$_{106}$ (p. 70), dissolved in ethanol were added. The mixture was then stirred vigorously for 1 h, the ethanol was evaporated in a Petri dish and a transparent membrane was formed which was was dried, aged and calcined at 350 °C.

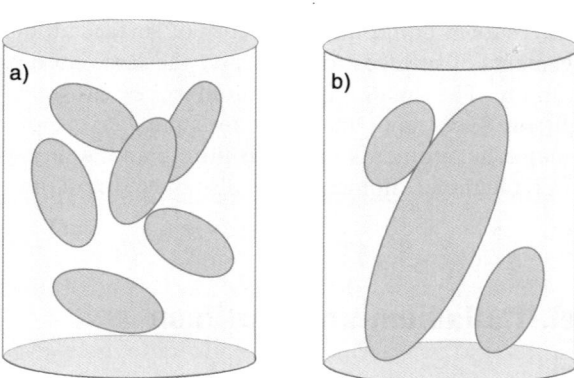

Figure 5.48 Schematic drawing of the grain size and orientation of a 20 nm wide nickel nanowire segment grown (a) without and (b) with an applied magnetic field.[114]

Mesoporous $NiO-2SiO_2-2ZrO_2-F12$ membranes were obtained directly in gram quantities.[117]

In catalysis the surfaces of Ni(0) nanocrystals are active in the dehydration of simple alcohols and the α-alkylation of ketones with primary alcohols. No added ligand or base is necessary.[118]

5.8.2 Palladium

Palladium is the only element with a d^{10} ground state and no s electrons of higher energy. Pd(0)–C bonds are as stable and easy to handle as Pd(I)–C and Pd(II)–C bonds. In the lab palladium is generally used on charcoal to hydrogenate olefins, and as $PdCl_2$ to form C–C bonds between aromatic compounds and alkenes (Heck reaction) or to oxidize arenes and alkenes. The commercial oxidation of ethylene to acetaldehyde with oxygen (Wacker process) uses Pd(II)/Cu(I) mixtures. Palladium black also dissolves molecular hydrogen and oxygen and splits both molecules to reactive atoms on the surface, because the Pd–H and Pd–O bonds are as strong as the H–H and O=O bonds but may easily migrate on the Pd(0) surface. Palladium is therefore a quite general catalyst: it is used industrially for the formation of C–C, C=O and C–H bonds.

$PdCl_2$ reacts with ethylene in water spontaneously to yield Pd(0) (palladium black) and acetaldehyde. The Pd(0) may then be reoxidized to Pd(II) with $CuCl_2$ and the formed Cu_2Cl_2 is reoxidized to $CuCl_2$ by O_2.

A polysiloxane matrix on Pd(II) nanoparticles prevented their agglomeration by steric and electrostatic forces and was stabilized with hydrido-poly(siloxane), which reduced the Pd(II) to Pd(0) and also crosslinked the polymer by hydrosilylation. These embedded Pd(0) nanoparticles then served (1) as catalyst and (2) as chromatographic stationary phase for multiphase reactions inside and outside of 250 μm fused silica capillaries at 200 °C with hydrogen as a reactive carrier gas. One educt library consisted of four unsaturated and functionalized compounds: 1-acetylcyclohexene, cyclohex-2-enone, nitrobenzene, and *trans*-methylpent-3-enoate. The compounds 1, 3, 4 and 5 did not react with H_2 and Pd(0) and were added simply as references. Capillary length, temperature concentration, etc were varied systematically and all synthetic and thermodynamic data of the four hydrogenation reactions were obtained within less than 1 h (Figure 5.49).[119] One result was that the adsorption of hydrogen on the palladium surface was fast, because the activation energy for this process was very low and because hydrogen easily passed through the polysiloxane matrix. The conversions and rate constants for the hydrogenation of cyclohex-2-enone, for example, varied in a broad range for the individual polysiloxane ratios. Higher conversions were observed for smaller palladium nanoparticles. Crosslinking simplified the dissolution of substrates, enhanced the accessibility of the palladium core and also prevented agglomeration of the palladium nanoparticles on heating.

When brown Na_2PdCl_4 powder and octylamine in 4-*t*-butyltoluene were sonicated, the brown colour of the Pd(II) salt faded and a colourless solution of Pd(II)-amine complexes formed. Refluxing at 192 °C gave black precipitates of

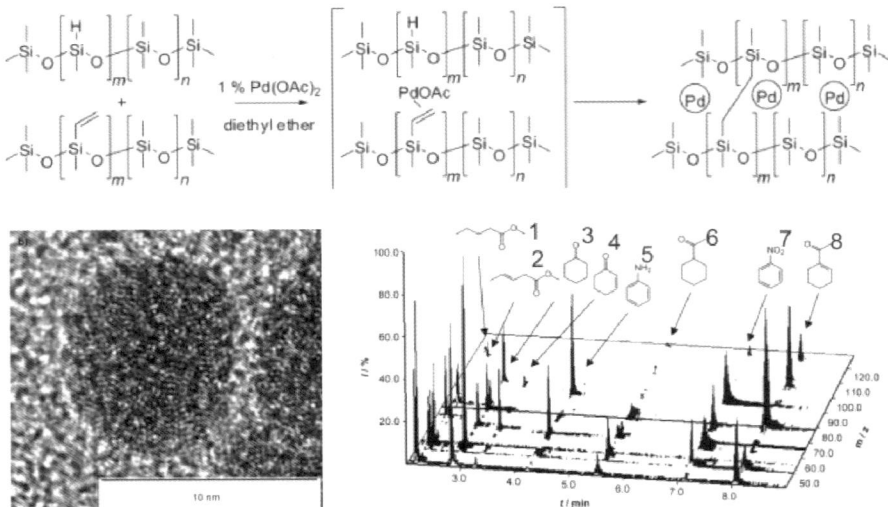

Figure 5.49 *Top:* Schematic mechanism of the formation of Pd(0) nanoparticles within methylated silicate layers. *Left:* Electron micrgraph of an 8 nm Pd(0) nanoparticle and its entrapment within a 250 μm fused silica capillary. *Right:* An example for the gas chromatographic separation and mass traces of the on-column hydrogenation of the four substrates 1, 2, 4 and 8 (see text).[119]

pure face centred cubic palladium nanocrystals.[120] Addition of dodecanethiol under the same conditions converted the crystallites to uniform nanoparticles with a diameter of 7.55±0.73 nm.

The functions of long-chain ligands are obviously often more complicated than simple chemical adsorption on the particle surface. The ligands may undergo decomposition and release certain components to form the nanoparticles. During heating, three main reactions occurred on the palladium nanoparticles. One was the amine-induced reduction of Pd(II) to Pd(0); the other was the formation of a tiara Pd(II) thiolate complex, $[Pd(SC_{12}H_{25})_2]_6$, and the third was the chemical adsorption and decomposition of Pd(II) thiolate species on the as-synthesized nanoparticles. After adsorption the thiolate species were no longer stable and a decomposition process occurred, which resulted in the sulfurization of Pd(0) nanoparticles to form a $(Pd_{16}S_7)_{14}$ core and a $(SC_{12}H_{25})_{57}$ shell (Figure 5.50).[120]

Classical hydrogen sensors rely on the change in resistance of thin palladium films upon exposure to hydrogen. More flexible hydrogen sensors were made from 75 μm poly(ethylene terephthalate) sheets covered with a thin layer of epoxy resin on a silicon wafer. The films of 1–2 nm single-wall carbon nanotubes had a density of ~14 tubes/μm^2 and were printed with an elastomeric polydimethylsiloxane stamp. Palladium nanoparticles on carbon nanotubes and polymer foil as well as the electrodes (2 nm titanium–50 nm palladium)

Transition Metals

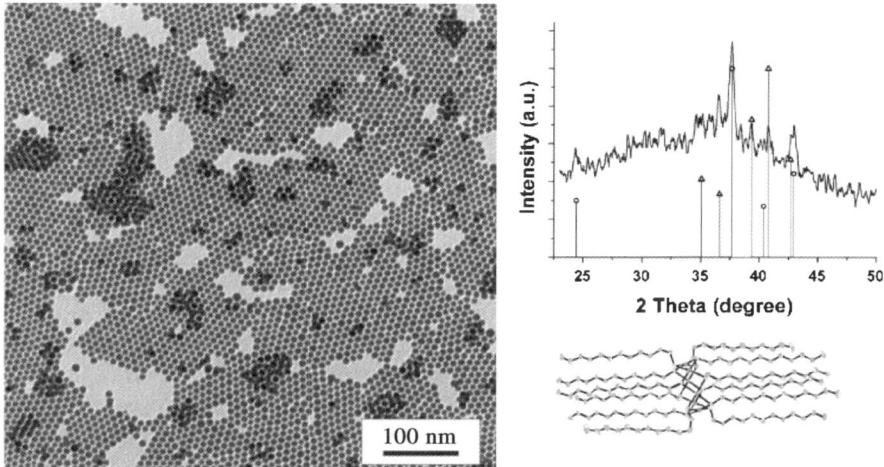

Figure 5.50 *Left*: Electron micrograph. *Upper right*: powder X-ray diffraction diagram of the palladium sample with octadecylthiolate. The standard pattern of $Pd_{16}S_7$ clusters are marked with O, the Pd_4S clusters with △. *Lower right*: A model of the nanoparticle constituents with the PdS core and the $SC_{12}H_{25}$ shell. © American Chemical Society.[120]

were added by chemical vapour deposition. Hydrogen molecules were detected by monitoring the change of resistance between the two titanium/palladium electrodes when the device was exposed to hydrogen. It worked well even when it was bent mechanically. The palladium nanoparticles carried ~40% of the current and were blocked when they adsorbed hydrogen (Figure 5.51).[121]

5.8.3 Platinum

Platinum is best known for hydrogen addition and abstraction of almost all functional groups of carbon compounds at low temperatures. Commercially the most important use is for hydrosilylation, where the Pt^{4+} state (H_2PtCl_6) is important as redox-active Lewis acid, which cannot be attained with the less expensive palladium. Nowadays the most powerful, but expensive, platinum catalysts for gas reactions involve molecular oxygen and/or carbon compounds. Both types of molecules are chemisorbed and the O–O or C–C bonds are weakened by hundreds of kJ/mol or even broken. Molecular fragments are formed with very low activation energy and may react with each other quickly. The reaction products usually do not bind to the Pt(0) metal surface; carbon dioxide and nitrogen, for example, are quickly released. Only sulfur compounds are bound so strongly to palladium and platinum that they poison the catalysts. Well-dispersed platinum nanoparticles catalyse fuel cell reactions, where they support anodic hydrogen or methanol oxidation as well as cathodic oxygen reduction.[122]

Monodisperse 8 nm platinum nanocubes were synthesized by dissolving Pt(II) (acetylacetonate)$_2$, oleic acid, and oleylamine in 1-octadecene at room temperature and heating the solution at 120 and 200 °C in the presence of trace amounts of Fe(CO)$_5$. Without Fe(CO)$_5$, which accumulates on the surface of the nanoparticles, the platinum nanoparticles were still formed, but the cubic shape as well as the size was less uniform (Figure 5.51). If oleylamine was used alone, spherical platinum nanoparticles prevailed, because stable Pt–NH$_2$R complexes hindered the nucleation and favoured amorphous structures. Oleic acid binds more weakly to platinum and facilitates the growth of cubes, but also favours aggregation of the nanoparticles.[123]

Platinum electrocatalysts for the reduction of molecular oxygen became 20-fold more active if platinum monolayers deposited on a Pd(111) single crystal or on carbon-supported palladium nanoparticles were mixed with another late transition metal (iridium, ruthenium, rhodium, rhenium or osmium). This superior activity and stability was caused by the lateral repulsion between the OH adsorbed on platinum and the OH or oxygen adsorbed on the neighbouring late transition metal atoms, which led to an extraordinary accessibility of free platinum atoms in the catalytic cycle.[124]

Dendrimer templating of platinum nanoparticles has been used to correlate catalyst composition and efficiency of the oxygen reduction reaction.[125] Dendrimer-encapsulated bimetallic nanoparticles in the <3 nm size range were thus obtained and were characterized prior to electrode immobilization. It was found that specific bimetallic compositions of platinum and palladium were more active in the catalytic reduction of oxygen than particles of identical size but composed exclusively of platinum.

Selective oxidation of carbon monoxide without affecting hydrogen has been achieved with platinum nanoparticles in mesoporous silica matrices. Infrared spectroscopy with ^{12}CO, ^{13}CO and ^{18}O$_2$ showed that molecular oxygen is not directly used for the carbon monoxide oxidation, but that the OH-oxygen of

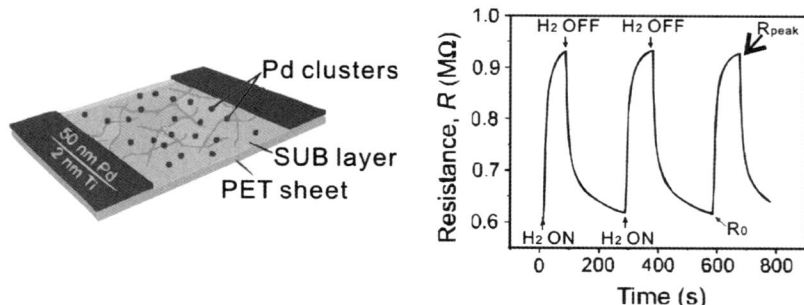

Figure 5.51 *Left*: Schematic picture of the flexible hydrogen sensor made of a poly(ethylene terephthalate) (=PET) sheet coated with a sublayer containing single-wall carbon nanotubes and palladium nanoparticles and thin metal electrodes. *Right*: Performance of the sensor for sensing hydrogen molecules in air at room temperature.[121]

mesoporous silica is incorporated into carbon dioxide, producing a vacant site on silicon. This is is then filled by oxygen. In catalysis matrices are often not innocent bystanders, but are directly involved in the chemistry; the only reason they do not disappear is that crystal lattices keep the reactive intermediates in place.[126]

Cyclohexadiene chemisorption on Pt(111) was irreversible and all the chemisorbed monolayer was dehydrogenated to form benzene upon heating. Further heating led to graphitic carbon. Platinum–tin surface alloys were made by evaporating 1–3 monolayers of tin on the clean Pt(111) crystal surface followed by annealing the sample to 1000 K for 10 s. Tin did not disturb the production of benzene, but completely eliminated its dehydrogenation to carbon ("coking resistant"). The rate-determining step in the benzene dehydrogenation was benzene desorption, which is much faster on the platinum–tin alloys and occurs at lower temperatures than on Pt(111). The effect is due to site-blocking by tin and to co-adsorbed hydrogen atoms coming from the dehydrogenation reaction. Benzene is thus quickly removed from the alloy surface and therefore not dehydrogenated. Only the noble metals form strong bonds reversibly with H, C and O.[127]

Cobalt–platinum magnetic barcode nanowires (p. 377 ff) were grown in a single electrolyte with a pH of 3.0 containing 0.3 M $CoSO_4$, 0.03 M K_2PtCl_6 and 0.485 M H_3BO_3 by using a pulsed-electrodeposition technique with alternative constant potentials of –1000 mV to deposit cobalt segments and – 350 mV to deposit platinum segments. The cobalt segments thus contained a small proportion of platinum, but the platinum segments were quite pure. The barcode nanowires were highly regular and uniform, with an average diameter of ∼65 nm, which corresponds closely to the Al_2O_3 nanochannel size. The magnetism of cobalt was strongest in the mixed alloy (Figure 5.52). Small segments of platinum had no effect on the cobalt magnetism, but larger segments destroyed it.[128]

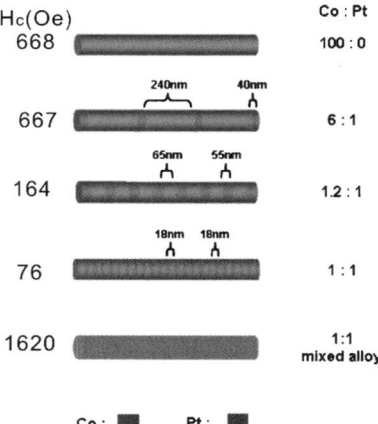

Figure 5.52 Schematic drawings of Co:Pt barcode rods and the cobalt magnetism. © American Chemical Society.[128]

The electrochemical reduction of a thin layer of platinum in dimethylformamide and in the presence of an alkali metal cation or a tetraalkylammonium salt produced small platinide grains covering the platinum surface. With NR_4BF_4 salts they appeared at the usual -1.6 V/SCE and were induced only after the electrogeneration of radical anions. The extent of the reaction was, however, limited by the conversion from the metallic conducting state to the insulating platinide salts. This limitation does not occur with nanoparticles, where all platinum is surface platinum. Negative redox states of platinum are thus formed and they are unstable in air and water. Dodecanethiol is immediately added and drives the nanoparticles from water into a hexane phase.[129]

Platinum nanoparticles on charcoal showed a hydrogen storage capacity of 1.2 wt% at 298 K and 10 MPa. The overall heats of adsorption (-23 to -15 kJ/mol) and the apparent activation energy for surface diffusion (7.6 kJ/mol) stabilized the hydrogen adduct. Desorption followed a slow spillover process to carbon.[130]

An aqueous phase synthesis of glucose, carboxymethylcellulose or octadecylamine stabilized platinum nanoparticles started with a reduction aqueous solutions of $H_2PtCl_6.6H_2O$ with sodium borohydride in presence of the polymer or D-glucose or the long-chain amine micelles under stirring at room temperature. The glucose and cellulose nanoparticles were stable in water; the dodecanethiol-coated platinum nanoparticles were dispersed in hexane or supercritical fluid carbon dioxide (Figure 5.53).[131]

Preferential oxidation of carbon monoxide over hydrogen is a practical way of purifying hydrogen for use in fuel cells. Platinum nanoparticles in mesoporous silica give unprecedented activity, selectivity, and durability in the preferential oxidation reaction below 353 K and it is the oxygen of mesoporous silica that is incorporated into carbon dioxide *via* OH groups without forming water (Figure 5.54).[132]

In the platinum-catalysed oxidation of carbon monoxide to carbon dioxide with oxygen strong rate oscillations are observed, when the speed of oxygen adsorption becomes rate limiting. This may occur because the oxygen sticking coefficient is structure sensitive on platinum surfaces and because certain Pt(0) surface phases, *e.g.* of Pt (110), are structurally unstable. Structural models of the Pt(110) surface planes are shown in Figure 5.54. The bulklike

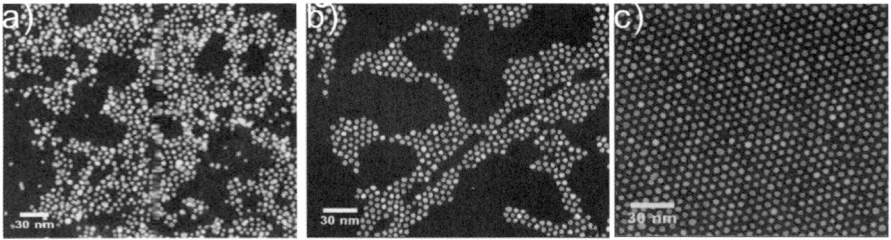

Figure 5.53 Transmission electron micrographs of images of Pt(0) nanoparticles stabilized with (a) glucose, (b) carboxymethylcellulose in water or (c) octadecylamine in hexane. © American Chemical Society.[131]

Transition Metals

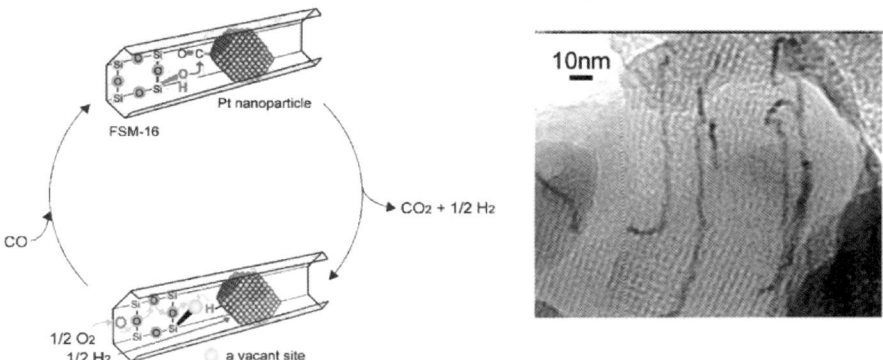

Figure 5.54 Proposed mechanism for the preferential oxidation of carbon monoxide over hydrogen with platinum nanocrystals in silica tubules and transmission electron micrograph of the platinum nanocrystals on FSM-16 mesoporous silica. © American Chemical Society.[132]

1×1 termination shows a hexagonal structure, the 1×2 phase a "missing row" geometry. The oscillation mechanism on this phase is also illustrated in Figure 5.55. On the carbon monoxide covered 1×1 phase of Pt(110) with all platinum rows present, the adsorption rate of oxygen, and hence the catalytic activity, will be high. As a consequence, more adsorbed carbon monoxide will be consumed by reaction and desorption than is supplied by adsorption, and its coverage decreases until, below a critical value, the surface will reconstruct into the 1×2 phase. This transition then causes a periodic switching between two states of different catalytic activity, the carbon monoxide binding energy switching between 37 and 27 kcal mol^{-1}.

The key observation leading to the model for oscillations on Pt(0)surfaces was that only unstable orientations, e.g (110) and (100) displayed oscillatory behaviour, whereas the stable Pt(111) surface exhibited bistability. The unstable structures again lead to very different space oscillations in p_{co}. For Pt(110) oscillations appear only in an extremely narrow range of carbon monoxide concentration, while Pt(100) displays oscillatory behaviour over a much larger concentration range (Figure 5.55). The simple geometry of a metal crystal thus gains an enormous stereochemical diversity if it is cut and if its surfaces participate in chemical reactions, *e.g.* in catalysis. Here lies one of the challenges of preparative nanocrystal chemistry: find the most active surfaces, synthesize the corresponding nanocrystals and find carriers and reaction conditions that stabilize the reaction system.[133–135]

5.9 Copper, Silver and Gold

Copper, silver, gold is the first triad after titanium, zirconium, hafnium in which all three free atoms have the same outer electron configuration in the ground state. Gold is the most electronegative of all metals. Its electronegativity value of 2.4 is the same as that of selenium, and close to those of carbon, sulfur

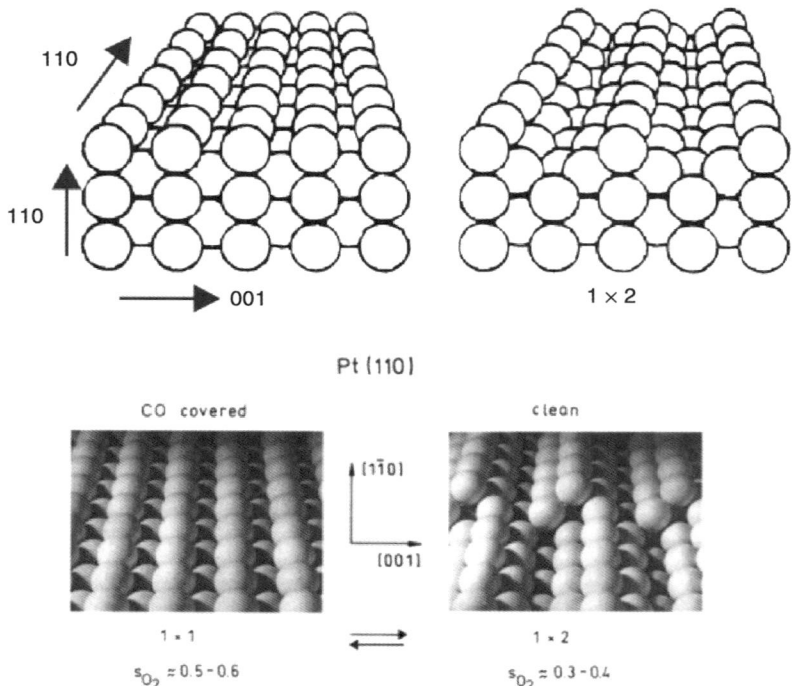

Figure 5.55 *Top:* Reconstructed and non-reconstructed surfaces for the (110) planes of platinum. *Bottom:* Ball model illustrating the same, CO-induced 1 × 1 t 1 × 2 surface phase transition of Pt(110). The different oxygen sticking coefficients of the two phases are responsible for rate oscillations during catalytic carbon monoxide oxidation. The model also demonstrates how the necessary mass transport of platinum atoms creates an atomic step on the surface. © American Chemical Society.[135]

and iodine (2.5). Their colours (copper reddish, silver white and gold yellow) and radiance are characteristic for these three "coin metals" and are caused by occupied d-levels whose energy is very close to the surface of the s–p conducting band (Fermi level) of the metals. Use in coinage also requires stability under air or high oxidation potentials: Cu^+/Cu +0.521 V, Ag^+/Ag +0.799 V, Au^+/Au +1.691 V. The reactivity decreases in the order copper > silver > gold. Copper and silver are, however, attacked by sulfur, but in gold it is only the Au(I) surface ions that are attacked. The relative softness or malleability of these metals is also a helpful property. One gram of gold gives 1.0 m² of gold foil, which is only 230 atoms thick.

5.9.1 Copper

The Cu(II) ion is smaller than the d^{10} Cu(I), but is more strongly hydrated with its two charges. The less stable d^9 ion is therefore more stable in water. In silver both ion radii are larger, the difference being insignificant; the d^{10} Ag(I) is therefore more stable. In gold the ionic diameter is much larger and the third electron is easy to remove. It prefers the oxidation number III. The oxides are Cu_2O (yellow or red), CuO (black), Ag_2O (dark brown) and $Au_2O_3 \cdot H_2O$ (black).

Due to its low electrical resistance and high electromigration resistance, copper has become the interconnect material of choice for printed circuit boards, integrated circuits, and multilayer sandwiches of giant magnetoresistive hard disk read heads. Gold or platinum plating then makes it more oxygen resistant in electronics. All three metals have a fcc structure and their respective lattice constants are 3.61, 4.08, and 3.92 Å. The identical crystal structure and comparable lattice constants stabilize multilayers and the different colours of the three metals allow the construction of variable surface appearances.

Copper nanoparticles oxidize readily; an oxygen-free environments is mandatory for working with them. Cu(0) nanoparticles 2.8±0.5 nm in diameter were prepared in the gas phase with a magnetron dc sputtering source. An oxygen-free high conductivity copper target was attached to the sputter head, which acted as the cathode. The anode cap with a central hole was placed 5 mm from the target and the discharge argon carried the sputtered copper out of the discharge into the condensation room, where the copper nanoparticles were collected on sodium chloride prisms and observed by a UV-vis spectrometer during the deposition process (Figure 5.56).[136]

The ^{64}Cu isotope is a positron emitter and copper nanoparticles with appropriate peptide coatings may be useful for positron emission tomography (PET).[137]

The spinel-type ferrites M(II)Fe$_2$O$_4$ have many industrial applications and by adjusting the chemistry of the M^{2+} cation their magnetic configurations and properties can be engineered. In $CuFe_2O_4$ nanoparticles all the copper ions occupy the octahedral sites and the iron ions are distributed between the tetrahedral and octahedral sites just as in the bulk material. They have been prepared by many procedures including electrochemical, reverse micelle, hydrothermal, mechanical milling, sol–gel, urea combustion leading to ash-like crystalline nanoparticles (see below) and precipitation in a polymer matrix.

Nanocrystalline $CuFe_2O_4$ and $CuFe_2O_4$–$xSnO_2$ were prepared by a urea–nitrate combustion method. For the $CuFe_2O_4$-SnO_2 nanocomposite preparation the stoichiometric quantities of $Cu(NO_3)_2 \cdot 6H_2O$, $Fe(NO_3)_3 \cdot 9H_2O$, $SnCl_4$, HNO_3, and $CO(NH_2)_2$ were dissolved in distilled water, refluxed at 110 °C with continuous stirring and the water was evaporated, until a highly viscous gel was obtained. The gel was ignited (urea and nitrate are present!) at 300 °C to evolve the unwanted gaseous products, resulting in the formation of desired product in the form of foamy powder, which was sintered at 1100 °C for 5 h to obtain an ultrafine powder.[138]

Thin-film solar cells are most promising if the semiconductor material has a band gap near 1 eV. Copper(I) sulfide is an indirect gap semiconductor with a bulk band gap of 1.21 eV and has been studied in combination with cadmium sulfide as electron acceptor, but copper diffusion and undesired doping of the

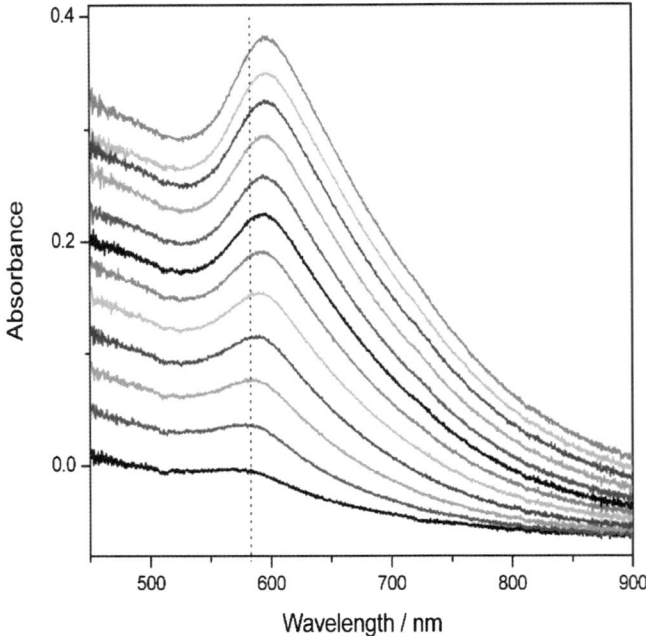

Figure 5.56 A series of transmission spectra of copper nanoparticles acquired *in situ* during deposition on to a sodium chloride plate. The lowest spectrum was acquired 1220 s after deposition was started, the second lowest 500 s later, and the rest every 400 s thereafter. For longer deposition times, there is an obvious shift of the 580 nm band (orange) to 640 nm (green). © American Chemical Society.[136]

cadmium sulfide layer led to long-term performance degradation. The mixed thin layers were replaced by separated layers of copper(I) sulfide nanocrystals and cadmium sulfide nanorods (p. 226 ff), in order to retard ion diffusion. The copper(I) sulfide nanocrystals were prepared from Cu(II) acetylacetonate and ammonium diethyldithiocarbamate in a mixture of dodecanethiol and oleic acid. They showed a broad absorption up to 1000 nm and photoluminescence at 1.32 eV. 300 nm spin-cast layers of copper(I) sulfide nanocrystals were then mixed with 100 nm layers of cadmium sulfide nanorods at 150 °C. The total roughness of the bilayer was 4 nm. The band offsets of both components fitted: all photogenerated electrons should have migrated readily to the cadmium sulfide, the holes to the copper(I) sulfide. A power conversion efficiency of 1.6% was observed with visible light. Only photons >700 nm showed no effect, although copper(I) sulfide absorbed up to 800 nm. A few copper ions had probably diffused a short way into the cadmium sulfide layer and blocked the electron transport slightly.[139] The barrier did not increase with time, however, as in conventional thin films.

5.9.2 Silver

Silver is a soft noble metal and reacts preferably with soft sulfur. The covalent S–S bond does not dissociate in water.

1–2% Na^+ ions relative to the concentration of COO^- groups in in aqueous solutions of long-chain sodium polyacrylates already induce a significant coil shrinking. If exposed to UV radiation, small Ag(0) nanoparticles formed within the shrunken anionic polyacrylate coils. The Ag(0) nanocrystals were characterized by enhanced light scattering and a characteristic plasmon absorption band around 410 nm. No coloured Ag(0) nanocrystal appeared, however, when the polyacrylate chains were short. Only if the coil size of the sodium poly(acrylate) chains exceeded a critical value was the internally achieved Ag^+ concentration was large enough to form colourized Ag(0)nanocrystals under UV light, which then generated solvated electrons or organic radicals with reducing properties. They cannot transfer an electron to Ag^+ ions in solution as the standard redox potential of –1.8 V of Ag^+/AgO is too negative. Positively charged silver clusters are, however, accessible. Six Ag^+ ions may be reduced per 100 eV of absorbed UV radiation energy.[140]

Metal-enhanced fluorescence of organic dyes occurs by the interactions of fluorophores with metallic nanoparticles. Fluorescence enhancement is accompanied by an increased photostability of the dye, decreased lifetime of excited states due to increased radiative decay, and increased transfer distances for fluorescence resonance energy transfer. Single molecule fluorescence spectroscopy is therefore a prime tool in single metal nanoparticle sensing.

The preparation of a fluorescent core-shell Ag@Si nanocomposites took three steps: (1) sodium citrate reduction of Ag(I) nitrate to form nanoparticles, (2) growth of a silica shell from tetraethoxysilane and (3) covalent linkage of the shell with fluorophores, *e.g.* Rhodamine 800. The diameter of the silver core was ∼130 nm and the thickness of the silica shell varied from 2 to 35 nm. The surface plasmon resonance peak for silver shifted from 620 to 720 nm as the thickness of the silica shell increased. The distance between the dye and the silver was always <10 nm and a 20-fold enhancement of the fluorescence emission was obtained for Rhodamine 800 by silver-enhanced fluorescence.[141]

13 nm Ag(0) nanocrystals are immobilized in silica pores (SBA-15) coated with an amine and a formaldehyde Schiff base up to 780 K. No obvious diffusion or evaporation occurred (Figure 5.57).

MCM-41, a mesoporous aluminosilicate (p. 89), was suspended in toluene and trimethylchlorosilane was added dropwise, followed by ethanol. Afterwards the mesoporous silica with the external $Si(CH_3)_3$ surface reacted internally with aminopropyl-triethoxyl silane at 353 K for 8 h and formaldehyde/ethanol/water was added. Addition of solver nitrate in dilute ammonia and calcination at 573 K produced metallic 7 nm silver nanoparticles inside the channels. At 773 K the nanoparticles increase in size from 7 nm to 17 nm, at 873 K to ∼34 nm. At 873 K the silver nanoparticles on the external surface evaporated after 10 min, while the mesopore-confined silver nanoparticles could still be observed (Figure 5.58).[142] An Ag(0) core–silica shell

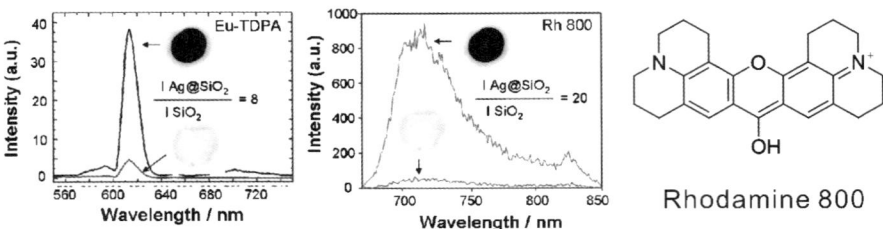

Figure 5.57 The fluorescence of a europium complex, Rhodamine 800, in solution (bottom spectrum) and attached to a silica shell of a silver nanocrystal (the peak at 720 nm). © American Chemical Society.[141]

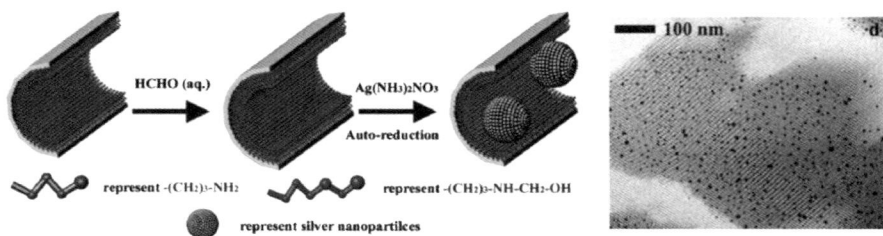

Figure 5.58 *Left:* The synthesis of Ag(0) nanocrystals within amine–Schiff base coated SBA-15 silica pores (diameter ∼20 nm) yielded immobile and non-evaporating particles at 770 K. *Right:* Electron micrograph of the nanocrystals in the silica pores. © American Chemical Society.[142]

nanoparticle was used to enhance the fluorescence of a lanthanide probe (a europium complex) eightfold, that of an organic dye 20-fold. The Ag(0) nanoparticle within the 10 nm thick silica shell yielded up to 20-fold fluorescence enhancements as compared to simple adsorption in silica nanoparticles.

Ag(0) nanoparticles were also prepared and chemically entrapped (at up to 770 K) in 20 nm silica pores. Without the confinement the Ag(0) nanoparticles rapidly grew fivefold and lost their crystallinity.[143]

Ag^+ ions were reduced with glucose and PVP under hydrothermal conditions in the temperature range 140–180 °C. Flexible Ag(0) nanowires with diameters of 16–24 nm and lengths of tens of microns were obtained (Figure 5.59). In addition, the wires exhibited a strong tendency to self-assemble upon drying. The nanowires were coated by carbon multilayers through carbonization by glucose, so the samples were described as Ag@C "nanocables". A strong UV absorption at 360 nm was traced back to plasmon resonances of a pentagonal nanocable coming from polygonal nanocrystals as nuclei, and the selective adsorption of PVP on specific crystal faces and carbonaceous growth units sticking preferentially to the edges of polygonal crystals with more dangling bonds.[144] Many other nanocrystals and wires have been coated accordingly with carbon. Trench patterns on a substrate that was coated with a 10–20 nm

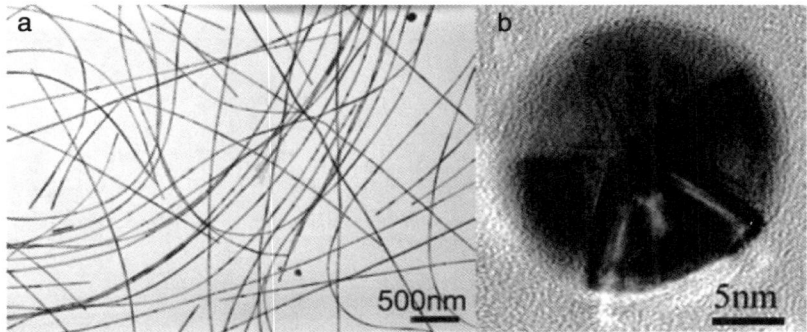

Figure 5.59 Electron micrographs of the pentagonal Ag(0) unit and the nanowires. © American Chemical Society.[145]

TiN film showed that the properties of the Ag(0) nanowires were controlled by grain orientation and grain-boundary organization, which were responsible for their anomalous strength: slip direction in these grains intersected with the twinning boundaries of the grains.[145]

If the size of a silver nanoparticle is similar to the electron Fermi wavelength (~0.5 nm for silver and gold), it displays strong single-electron excitations and emits fluorescence. Nanoparticles as large as the electron mean free path length, namely ~50 nm for silver and gold, do not fluoresce any longer. Collective excitations of electrons lead to plasmon resonance spectra only. Polycrystalline silver nanoparticles with grain sizes between 0.5 and 5 nm exhibit both bright red luminescence and a large enhancement effect on the Raman scattering signals of proximal molecules. The number of photons emitted from these nanoparticles exceeded that from quantum dots by two orders of magnitude, that of dye molecules by five orders of magnitude. Plasmons did not make major contributions to the luminescence, which likely arises from single-electron excitations between discrete energy states.[146] Again, small nanocrystals may act like molecules with defined energy levels, even if they are metals. The "electron gas" disappears abruptly; each electron is strictly localized.

The structural differences between the luminescent silver nanoparticles created by solid-phase synthesis and the non-luminescent silver particles of similar size (<20 nm) produced by solution-phase synthesis were analysed by electron microscopy. The non-luminescent nanoparticles typically showed crystalline domain sizes averaging 8 nm, while the luminescent nanoparticle domain sizes were in the 1–2 nm range. The small domains present in the nanoparticles likely result in discrete energy states that give rise to optical transitions; the subnanometre silver clusters and their junctions also give rise to a very large Raman enhancement effects.

Surface-enhanced Raman spectroscopy relies on strongly increased Raman signals generated by local electric fields near metallic gold and silver nanostructures. Maximum enhancement occurs in "hot spots", which are optimal geometries in the environment of the metal where maximum enhancement is

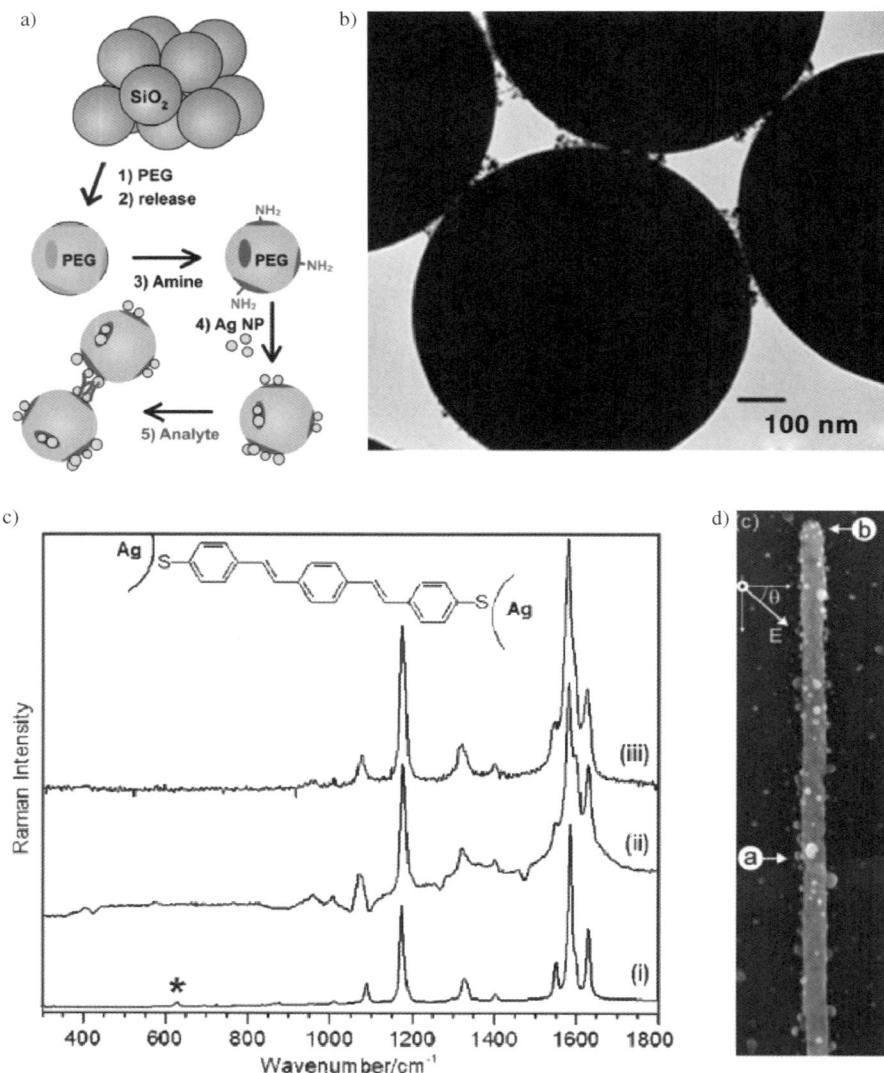

Figure 5.60 (a) The five-step process for the preparation of a (silica microsphere–silver nanocrystal)$_2$-(SH)$_2$-analyte bridge dimer for surface-enhanced Raman spectroscopy under a light microscope. (b) Electron micrograph of the nanocrystal bridges. (c) Raman spectrum of the bridges showing only the AgS part.[147] (d) Ag(0) with attached Ag(0) nanocrystals. Nanoparticles at position a were more efficient as Raman enhancers than those on the tip, position b. © American Chemical Society.[148]

Transition Metals

generated. Spectra can be obtained from a few single molecules on silver, if an optical microscope can locate them in a spot and if the laser beam of the Raman spectrometer can be focused on this spot. For this purpose some amine domains were first located on silica microspheres and used to bind much smaller silver nanocrystals.[147] The protein or organic probe for the Raman spectroscopy was then used as a crosslinker for two nanometre-sized silver nanocrystals on micrometre silica particles (Figure 5.60). Two connected silica particles were visible under the light microscope and the connecting bridge was examined with the Raman laser. Beautiful surface-enhanced Raman spectra were obtained from small disulfide molecules or proteins with SH groups in these obvious spots.[147] Ag(0) nanoparticles, which were attached on Ag(0) nanowires, (Figure 5.60(d)) were more efficient in the surface enhancement of Raman spectra at an interior location than at the spherical tip, which provided a less asymmetric environment.[148]

Photothermal surface-enhanced Raman spectroscopy with silver nanoparticles on cell membranes indicated that cell damage was most extensive in the vicinity of the nanoparticles. When the imaging power density was reduced by one order of magnitude, no detectable membrane damage was observed. Metal particles, laser irradiation and surface enhancement effects must obviously be applied very cautiously in living systems.[149]

Fluorescence and cyclization quantum yields of diarylethenes were strongly quenched by attached 3 nm silver nanoparticles, but the ring-opening of the non-fluorescent cyclization product was hardly suppressed. The fluorescing diarylethene transferred all its energy to the silver nanoparticles, but the non-fluorescent cyclization product had only a similar absorption spectrum, but no electronic interaction with Ag(0) nanocrystals (Figure 5.61).[150]

Poly(o-methoxyaniline) produced silver nanoparticles in aqueous silver nitrate through an interfacial redox process solution and then carried them into chloroform, since the nitrogen atoms of the quinoid poly(aniline) coordinated to adsorbed Ag^+ on the nanoparticle surface (Figure 5.62). The conductivity of the ordered silver nanoparticles in the polymer matrix was very low (10^{-11} S cm^{-1}), and the size of the silver nanoparticles varied with the polymer concentration between 9 and 20 nm.[151]

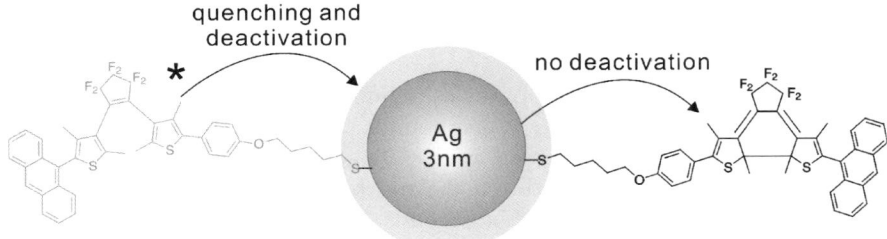

Figure 5.61 Only the left fluorescing dye interacts photochemically with silver nanocrystals; the right dye has the same colour, but is inert towards Ag(0). © American Chemical Society.[150]

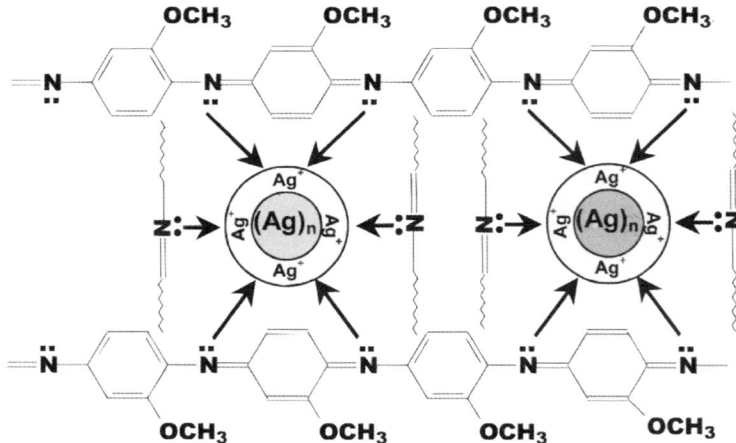

Figure 5.62 Stabilization of silver nanoparticles by poly(methoxy aniline) chains. © American Chemical Society.[151]

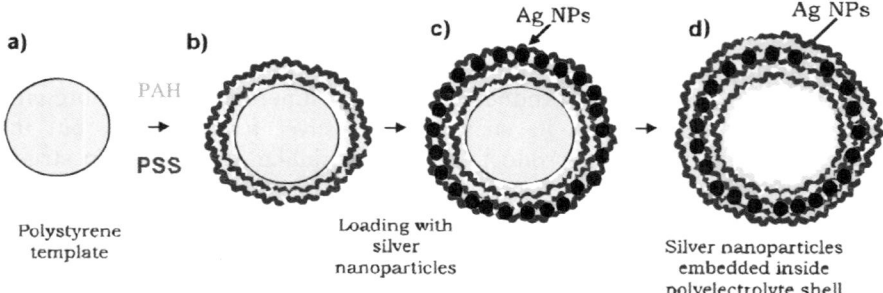

Figure 5.63 Fabrication of polyelectrolyte/silver microcapsules by the layer-by-layer technique using poly(sodium 4-styrenesulfonate)/poly(allylamine hydrochloride) and preformed silver nanoparticles which were adsorbed on to the multilayer. Laser radiation destroyed these assemblies and released compounds that were dissolved in the organic polymers or the aqueous centre. © American Chemical Society.[152]

The presence of Ag(0) nanoparticles significantly improves the efficiency of the release of encapsulated compounds by remote laser irradiation. Poly(styrene) spheres were coated with poly(allylamine) and poly(acrylate), silver nanoparticles were added and the poly(styrene) nucleus removed with tetrahydrofuran (Figure 5.63). Six seconds of 70 mW Nd:YAG FD laser irradiation destroyed the 10 μm capsules with 18 nm silver nanoparticles completely; for smaller 5 μm shells with 10 nm silver nanoparticles it took 22 s. A silver coat thus renders organic membrane structures extremely heat sensitive.[152]

5.9.3 Gold

Gold is a soft noble metal and reacts preferably with soft sulfur or sulfide anions. The covalent S–S bond does not dissociate in water.

Although Au_{55} clusters with a stoichiometry of $Au_{55}[P(C_6H_5)_3]_{12}C_{16}$ have often been described and analysed by high resolution microscopy (Figure 5.64), a careful analysis clearly proved heterogeneity in size and structure. Aggregation and formation of metallic gold from small clusters is a thermodynamically favored process, but uniform intermediates could not be isolated.[153–155]

A better defined gold nanocrystal, namely Au_{102} with 44 thiol caps was predicted by density functional theory calculations and realized by crystallization of $Au_{102}(SR)_{44}$ clusters, a "staple" motif. All 44 thiolate groups formed linear RS–Au–SR motifs on the cluster surface.[156] The density functional theory cacluations defined three planes for the gold atoms with two pentagons around a central gold atom and successive shells interacting with zero, one or two 4-mercapto-benzoic acid ligands, HSC_6H_5COOH, respectively (Figure 5.65).[157]

Larger, less defined soluble gold nanoparticles are usually made either by reduction of $AuCl_4^-$ ions with sodium borohydride in biphasic toluene–water systems and in the presence of alkylthiolates (Brust method) or by reduction with citrate at pH 9 in water (Figure 5.66). Both types of colloidal particles carry anions, citrate or sulfide, on their surface; both anions are soluble in the solvents surrounding the colloidal particles but do not separate from the gold surface. The Au(I) citrate surface is thus easily modified by change of pH or contact with different surfaces.[158]

Oleylamine-stabilized gold nanoparticles in chloroform rearranged to colloidal crystals when dodecanethiol was added. 60–80 nm non-rigid clusters of gold atoms formed in the first 60 s and direct nanoparticle aggregation, cluster aggregation, and heterogeneous aggregation then competed to produce a mixture of amorphous, single-crystalline, polycrystalline and core–shell type clusters. The typical mechanism of self assembly was the symmetrical

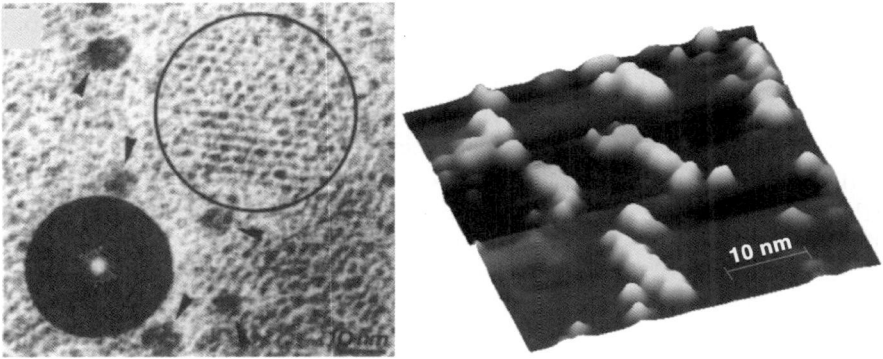

Figure 5.64 Transmission (*left*) and scanning electron micrographs (*right*) of Au_{55} clusters. © American Chemical Society.[154]

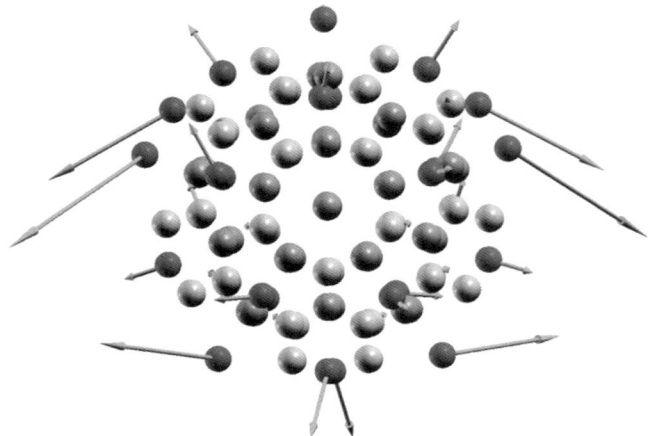

Figure 5.65 Arrangement of the gold atoms in $Au_{102}(MBA)_{44}$. From light to dark spots: zero, one and two 4-mercaptobenzoic acid ligands. Fewer than 70 atoms are visible. © American Chemical Society.[157]

aggregation of small clusters, *e.g.* $1+1 \rightarrow 2$, $2+2 \rightarrow 4$ *etc*. Face-centred cubic structures terminated by (100) and (111) planes appeared frequently, but there was no critical nucleation size for crystal growth. The liquid cluster intermediates rather grew until they reached a size of 35–200 nanoparticles, and only then slowly reordered to form crystals (Figure 5.67).[159]

The core sizes and ligand shell compositions of various kinds of small gold nanoparticles might be measured by mass spectrometry after tagging them with ionizable ligands, *e.g.* (ethylene glycol) thiolate ligands ($-S-(C_2H_4O)_5-CH_3$, –S–PEG). Poly(ethyleneglycol) is a good electrospray tag because of its ready coordination with protons and/or alkali metal ions and has made possible high-resolution ESI-MS measurements (Figure 5.68) as well as unambiguous assignments of chemical formulas to nanoparticles such as $Au_{38}(SC_2Ph)_{24}$.[160]

Galvanic replacement reactions between silver nanostructures and gold, platinum, or palladium compounds fabricated nanocages with single-crystal walls containing small pores formed by dealloying. Pores uniform in size and shape could be confined to the corners in Au(0)/Ag(0) nanocubes with "truncated" corners as obtained by thermal annealing of nanoparticles in which PVP capped only the 100 rather than the 111 facets. Only the poorly capped facets led to truncation of all corners and added $HAuCl_4$ attacked only these unprotected corner regions, the primary sites for the dissolution of Ag(0). Well-defined pores finally appeared at all corners of the nanobox (Figure 5.69).[161]

The Au–S– bond in gold nanoparticles as obtained by sodium borohydride reduction of Au(III) salts should be covalent, since the electronegativities of gold and sulfur are practically the same. Citrate is bound electrostatically by an Au(I) surface on Au(0) nanoparticles. X-ray photoelectronspectroscopy measurements of gold [Au(4f)] binding energies are always near to those of Au(0),

Figure 5.66 *From left to right:* transmission electron micrographs of citrate Au(I)/Au(0) nanocrystals prepared at pH 9 and pH 3; AFM images of citrate-depleted gold spheres in water as deposited on cystamine-coated Au(0) showing isolated spheres; highly oriented pyrolytic graphite (HOPG), where fusion took place; and on mica, where thick dendritic plates were formed by fusion.[158]

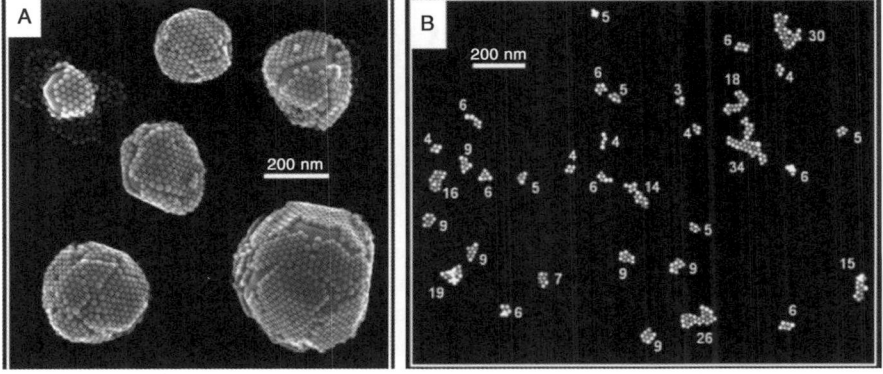

Figure 5.67 Growth of gold crystals from nanocrystals. © American Chemical Society.[159]

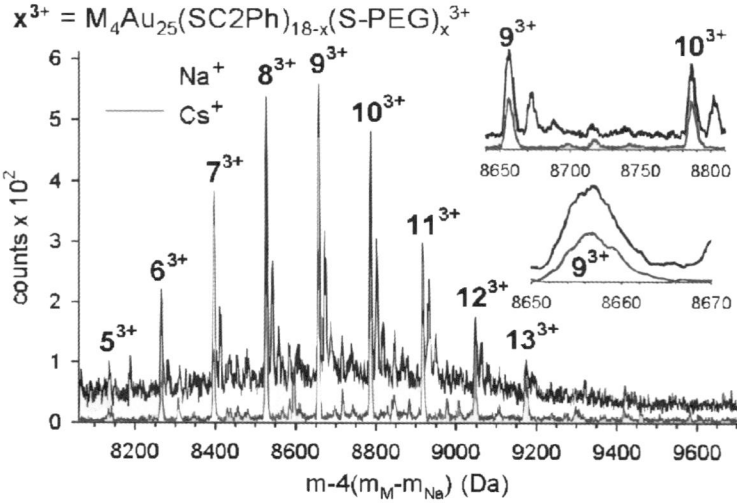

Figure 5.68 Mass spectrum of a modified Brust gold nanocrystal. © American Chemical Society.[160]

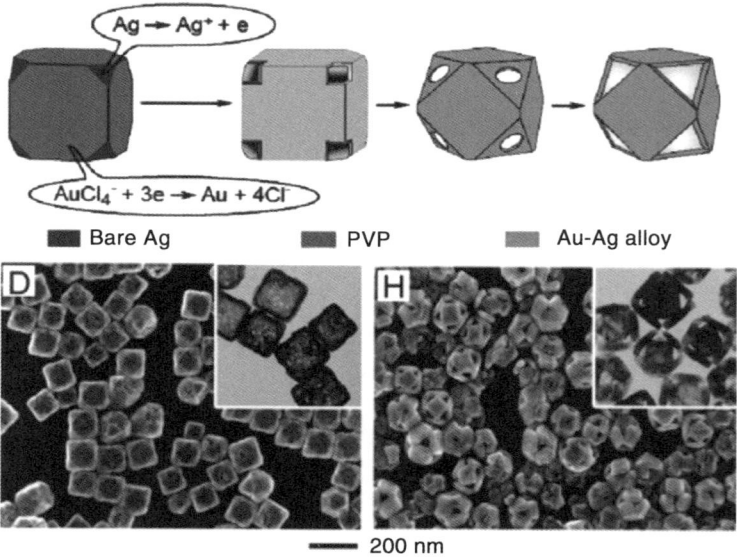

Figure 5.69 Au–Ag nanocages with pores at the corners. Models (*above*) and scanning electron micrographs (*below*). © American Chemical Society.[161]

whereas the S(2p) binding energies are close to those of RS–. Transmission electron micrographs show no sign of a C_{18} coating in 1.5–20 nm Brust gold nanoparticles and the crystalline areas extend up to the edges of the particles.

Transition Metals

The gold nanoparticles resulting from citrate reduction are somewhat larger, >20 nm. The soft gold droplet does not change its spherical appearance significantly on either graphite or mica surfaces, although most of its material is Au(0). Au(I)–Au(I) binding on the surface, rationalized by d^{10}–d^{10} interactions or the "aurophilic effect", presumably prevents the coalescence and/or crystallization of Au(0) within the nanoparticles. Citrate gold is long-lived in distilled water, where citrate anions should diffuse away from neutral gold surfaces: fusion occurs only if citrate removed. Two of the carboxyl groups can combine simultaneously to the Au(I) surface. The crystal structures of gold citrates are not known, because they disintegrate to form Au(0) and oxidation products of citric acid. On the surface of the gold colloid, however, the positive charge may be distributed over several gold atoms and the oxidation power may be too low.

Sodium borohydride decomposes citrate gold and its sulfide derivatives immediately, because it removes the Au(I) on the surface. Sulfide coatings are as stable as citrates in water and also in toluene. Lipoate-coated Au(0) nanoparticles aggregate in water upon protonation with hydrochloric acid, and dissolve again with sodium hydroxide (Figure 5.70). The gold nanoparticles are colourized by plasmon absorption of visible light: smaller nanoparticles are red, larger ones blue. The reversible acid–base titration changes the colour accordingly. Addition of sodium cyanate, which usually dissolves Au(0) and Au(I) as $Au(I)(CN)_4^{3-}$, has no effect on to sulfide-coated gold nanoparticles of low curvature (>20 nm). The coating must form a closed, impermeable monolayer, tightly bound to a more or less planar gold surface.[158] Tetrathiafulvalene (p. 92ff) is oxidized to the +1 state by citrate gold in the presence of pyridine.

The dipole of the keto group of acetone also attaches to citrate gold nanoparticles and renders the gold surface insensitive to pH changes and sodium borohydride. The dissolution of gold by acetone certainly starts with the

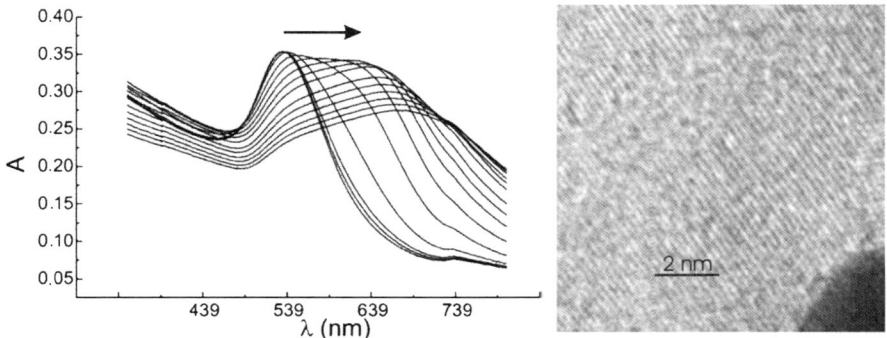

Figure 5.70 *Left:* Changes in the visible spectrum of an aqueous solution (NaOH, pH 11) of lipoate-coated gold nanoparticles upon titration with HCl. Back titrations with NaOH reverse the changes quantitatively (not shown). *Right:* Electron micrograph of a lipoate-coated Au(I)/Au(0) platelet. © American Chemical Society.[158]

formation of the μl-adsorbate, which readily redissolves in water/acetone. The basic keto oxygen atom may take up protons from water as well as Au(I) ions from the colloid's surface, or it may also add to the α-carbon atom as an enolate. Such structures are known from gold–ketone adducts and may contain condensation products of acetone, forming the viscous films that are detected on the surface of aged acetone–gold platelet probes by atomic force microscopy. The reduction of $AuCl_4^-$ by acetone takes many hours and presumably uses the enolate of acetone as the reducing agent. Water/acetone soluble platelets contain large crystalline areas and do not precipitate irreversibly. The inversion of the dipole by treatment with hydrochloric acid does not destroy it. Its broad absorption in the near IR around 750 nm is the ultimate long-wavelength plasmon band. The platelets can be stabilized by a covalent lipoic acid coat instead of electroneutral acetone adsorption (Figure 5.71).[158]

Metal surfaces quench the fluorescence of adsorbed quantitatively by electron transfer to the excited state. This effect may, however, be cancelled by atomic roughness of the metal surface. An octaanionic porphyrin carboxylate, for example, was bound to the Au(0/I) surface of Au(0) electrodes at pH 12 and stayed there after repeated washing with 10^{-2} M KOH. The fluorescence on sputtered gold surfaces amounted to 10% of the intensity observed on an organic subphase or silica gel (Figure 5.72). This rest fluorescence was presumably caused by fluorescence enhancement through atomic Au(0) islands on the smooth surface.[162] Gold nanoflares were, however, also used as partial

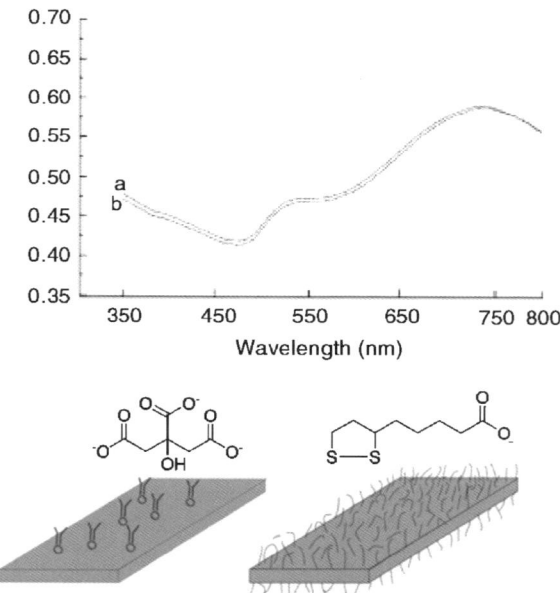

Figure 5.71 *Left:* Visible spectrum of acetone-coated Au(0)/Au(I) platelets. *Right:* Model of the replacement of acetone by lipoate. © American Chemical Society.[158]

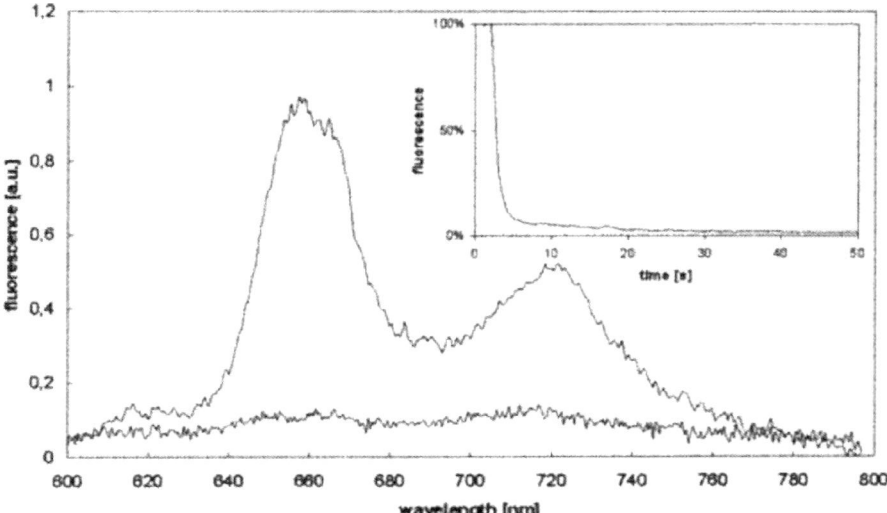

Figure 5.72 Fluorescence of a porphyrin directly bound to the Au(I) surface by COO⁻ substituents.[162] Time course of the fluorescence quenching after additon of a manganese (II) porphyrin leading to the lower spectrum. © American Chemical Society.

fluoresence quenchers in transfection experiments with living cells.[163] The actual effect of metallic surfaces on the electro- and photochemistry of adsorbed compounds should always be tested experimentally – predictions and theoretical arguments are quantitatively fortuitous.

Coated Au(0) nanoparticles are weakly magnetic. Measured values for the magnetic moments of magnetic gold range from 0.002 to to 0.3 µB; all of them reflect the generation of "holes", presumably Au(I), in the 5d energy level as a consequence of different bonds among the surface gold atoms and different organic groups. In the 0.3 µB case, 100% of the surface gold atoms was covered with alkanethiolates and the sulfur/metal ratio was 0.3, indicating that each S atom would be linked to three surface gold atoms. The S atoms would then lean into the hollow formed by adjacent atoms at the surface inducing σ- and π-bonding orbitals and cause the observed partial quenching of the magnetic orbits of the metal atoms at the surface of the nanoparticles. Ag(0) and Cu(0) nanoparticles showed a similar "chemical magnetism".[164]

Metal nanorods can be prepared within the nanochannels of a nanoporous alumina sheet (p. 182 ff), which work as reaction vessels and matrices. Multimetal rods with nanometre-long segments of platinum, silver and gold were produced by first evaporating a thin metal film, e.g. gold, on to one side of the perforated alumina sheet and using it as an electrode. The metallized alumina sheet was then plunged into salt solutions of metals for a limited time period and a reducing potential produced metallic nanorods on top of the electrode and within the pores. Electrolysis time and metal ion concentration determined the length of the rod, and its width corresponded roughly to the radius of the pore. Gold, platinum and silver segments of any length and any order were thus combined in barcoded nanorods. Metal-specific reactions may also introduce different fluorescent dyes on gold, platinum or silver segments "stereoselectively". Isonitrile, for example, binds indiscriminately to gold and platinum; thiols have a greater affinity for gold than butylisonitrile and thus displace it from the gold segments, but not from platinum. The histidine units of proteins bind to iron/nickel segments, but not to gold.[165]

Commercial nanobarcodes are usually 300 nm in diameter, and are also produced by sequential electrodeposition of metals such as gold, silver, palladium, nickel, or platinum in alumina template membranes. Gold and silver are particularly attractive because of the large difference in reflectivity between these metals under blue illumination, which makes the barcodes readable by optical microscopy. Brand protection in the retail market and multiplexed bioanalysis are feasible when different patterns are located in different environments and changes of their conductivities are registered regularly.

Biomolecules such as antibodies or DNA are attached to the barcodes in aqueous buffer where silver rapidly degrades by oxidation through dissolved oxygen. Trisodium citrate is commonly used to form Ag(0) nanoparticles from Ag^+. With the barcode wires it was indeed found that citrate ions adsorbed to the silver surface *via* their carboxylate moieties, competed effectively for binding sites with various oxygen species and itself reacted with oxygen species adsorbed to the Ag(0). Any Ag^+ that did form was re-reduced by the excess citrate. Addition of 40 mM citrate to the buffer markedly slowed down Ag(0) oxidation and kept the barcodes readable for several months (Figure 5.73).[166]

We end this section with a couple of examples of gold nanoparticle aggregates. Such aggregations occur with each type of metal and metal oxide nanoparticle, but are best analysed with the easily accessible and spectroscopically observable gold particles.

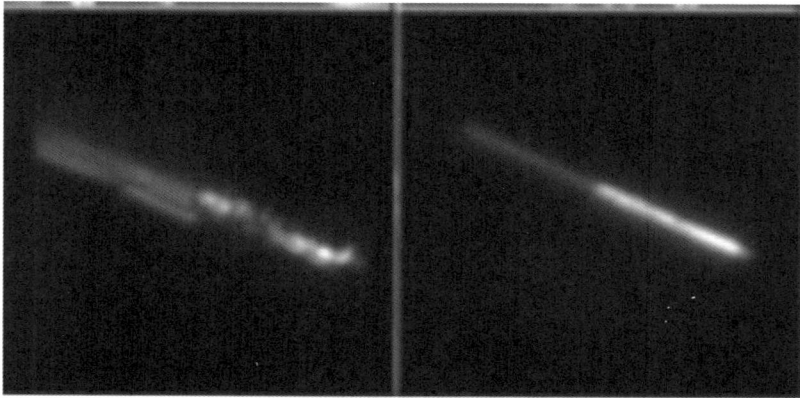

Figure 5.73 Nanowires (patterned 000/111) after storage in non-reducing buffer (left) or citrate buffer (right) after 17 days. © American Chemical Society.[166]

The stability of mixtures of oppositely charged ions and nanoparticles is fundamentally different at molecular and colloidal scales. Ion pairs of salts can remain stable in solution until reaching a certain concentration or ratio, but charged microparticles precipitate continuously due to residual van der Waals forces and poor solvation of large aggregates. When "ionic-like" gold nanoparticles were mixed with N,N,N-trimethyl(11-mercaptoundecyl)ammonium chloride and mercaptoundecanoic acid they aggregated upon mixing without any visible precipitation. Only when charge neutrality was attained did a sudden precipitation occur (Figure 5.74). Mixtures of oppositely charged nanoparticles of various sizes (3–11 nm) and relative charges always precipitate sharply at the point of overall electroneutrality. This remains true if the surface charges are diluted by electroneutral components, e.g. 11-mercaptoundecanol $(HS(CH_2)_{10}OH)$.[166] Gold nanoparticle functionalization with ammonium coatings also promoted the association and ligation of peptide fragments by complementary electrostatic interactions, e.g. with the phosphate groups of DNA.[167,168]

Three-layer magnetic haematite Fe_3O_4 nanoparticles with a reactive Au(0) surface were produced from rough, aminated silica particles with a diameter of ~150 nm as templates (Figure 5.75). The electrostatic fixation of negatively

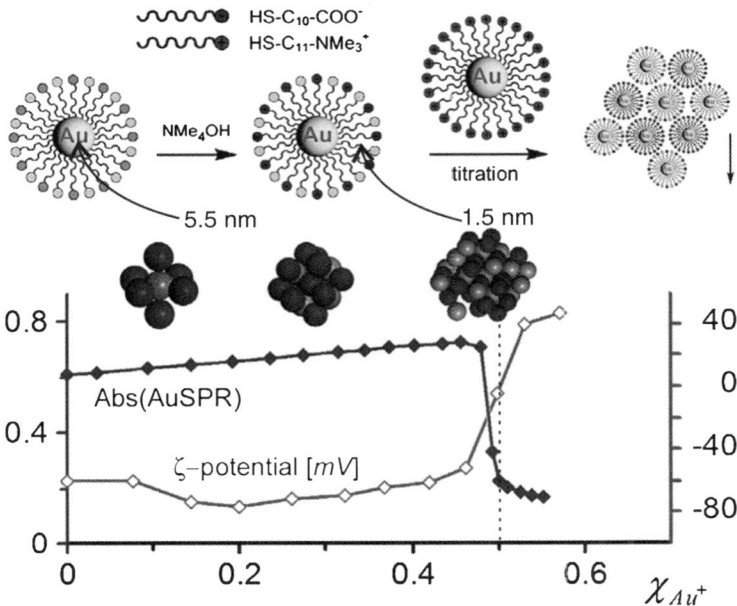

Figure 5.74 The mixture of Au(0) nanoparticles with electropositive ammonium and negative carboxylate coatings precipitates at the point of electroneutrality. Before this point, the charge of the excess nanoparticles keeps the agglomerates dissolved. © American Chemical Society.[166]

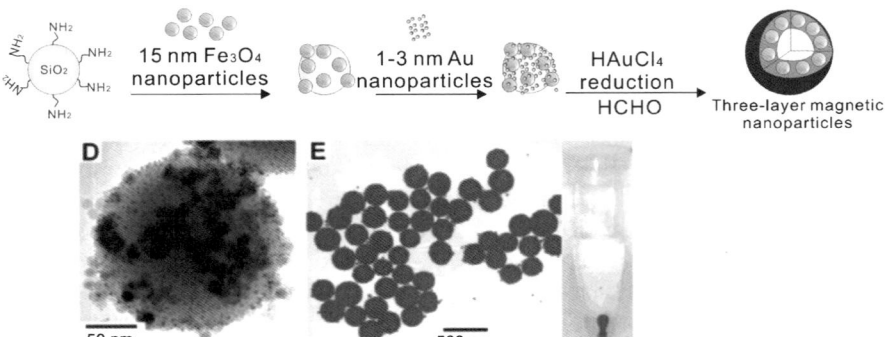

Figure 5.75 Composite nanoparticle made of an aminated silica core (100 nm), attached Fe_3O_4 nanoparticles (15 nm, magnetic) and gold nanoparticles everywhere (<3 nm). A final silicate coating stabilized the nanoparticle. © American Chemical Society.[168]

charged 15 ± 1 nm superparamagnetic Fe_3O_4 on the ammonium layer surface and subsequent reduction of added $HAuCl_4$ with formaldehyde produced a composite magnetic colloid with a broad plasmon band above 600 nm which appeared red in colour due to light scattering. The particles were magnetic and when exposed to a bar magnet they moved 30 times faster in water than the unmodified Fe_3O_4 nanoparticles. Upon further hybridization with complementary linking oligonucleotides the particles aggregated and separated again by raising the temperature above the melting point of the duplex DNA linkers. This process was accompanied by changes in the UV-vis spectra. A silica coating stabilized the Fe_3O_4–gold nanoparticle assembly.[168]

Gold–platinum nanoparticles catalyse oxidations in water and at room temperature better than platinum or gold surfaces alone. Central problems of catalytic nanoparticles, however, are coagulation and handling; "stabilization" through thiol alkanes, in particular, may destroy the nanoparticles and/or alter the catalytic properties profoundly ("catalyst poisoning"). Spherical polyelectrolyte brushes made of poly(2-aminoethylmethacrylate), however, allowed the addition of $[PtCl_6]^{2-}$ ions to gold nanoparticles leading to gold–platinum alloys. This prevented aggregation and established a high catalytic activity of the gold–platinum alloy in the oxidation of alcohols in water. The oxidation was carried out with pure oxygen and 1 mmol of the alcohol in water containing the basic catalyst potassium carbonate (K_2CO_3). The yields of aldehydes and ketones were close to quantitative.[169]

Gold–TCOOH and gold–TAsp particles bind to a large surface spanning the front face of cytochrome c, much as cytochrome c oxidase does. Simple Coulombics lead to diffuse binding. In contrast, gold-TPhe binds to a smaller surface near Lys72, which bears a close resemblance to the surface recognized by cytochrome c peroxidase.[170]

co-Poly(styrene)poly(vinylpyrrolidine) (p. 112ff) was used as a binding centre for $HAuCl_4$ or $Cd(OAc)_2$ and small Au(0) and cadmium sulfide

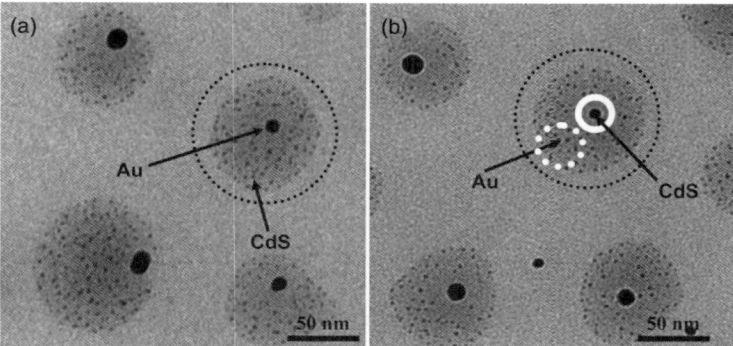

Figure 5.76 Electron micrographs of an embedded gold nanoparticle surrounded by smaller CdS nanoparticles (*left*) synthesized by a two-step reaction starting with HAuCl$_4$ solid salt and dilute HCl (pH 5) and of a bigger core-embedded CdS nanoparticle surrounded by smaller gold nanoparticles (*right*) synthesized from solid Cd(OAc)$_2$ dissolved in dilute HCl (pH 5). N$_2$H$_4$ and Na$_2$S in water (30 wt%) were used as reductants. © American Chemical Society.[171]

nanoparticles were prepared on the block copolymer surface with hydrazine (N$_2$H$_4$) as reducing agent for Au(III) and sodium sulfide (Na$_2$S) for the precipitation of cadmium sulfide. The size of the Au(0) and Cd(II)S nanoparticles was exclusively determined by the amount of metal within the polymer micelle. Toluene was then added and the micelle structure changed to an inverted one with a polar Au(0) or cadmium sulfide poly(vinylpyrrolidine) core and a poly(styrene) shell. The maximum absorptions indicating the surface plasmon resonance of gold nanoparticles as well as the optical transition of the first excitation state of cadmium sulfide nanoparticles were clearly shown at 545 and 410 nm, respectively (Figure 5.76).[171]

5.10 Scandium, Yttrium and the Lanthanides

Scandium and yttrium are the first two elements of the third group of the periodic system. The 15 elements with atomic numbers 57–70 are called the lanthanoids. These elements have the s,p,d electronic configuration of xenon plus two electrons in the 6s^2 orbital. Lanthanum, cerium, gadolinium and lutetium also add one electron to the 5d orbital.

All lanthanoids have the oxidation number +3 in their oxides and salts, which means that they have empty 6s and 5d orbitals, and only fill up the 4f orbital from f^0 (lanthanum) to f^{14} (lutetium). The 4f electrons are too close to the nucleus to be chemically accessible. Exceptions are Ce(IV) and Pr(IV), which are at the beginning of the series and have high energy f-orbitals, and Tb(IV), which is stabilized by a half-occupied f-orbital (4f^7). Eu(II) (4f^7), which reduces water, and Tb(II) (4f^{14}) are also stabilized by half-filled and fully filled f-orbitals, respectively. The metallic radii are all ~180 pm. The heavy lanthanoid trications are weaker bases and precipitate from aqeous solutions earlier than the lighter elements upon addition of base. The general chemical properties are uniformly close to those of calcium; the filling of the f orbitals has little chemical relevance.

The ionic radius of eightfold coordinated +3 ions decreases with increasing atomic number from 116 (^{57}La) to 98 (^{71}Lu). La(III) is much less hydrated than Lu(III). Redox potentials are all close to -2.3 V. Heavy lanthanoids are eluted first in the chromatography of 2-hydroxybutyrates ($CH_3CH_2CHOHCOO^-$), because of their smaller radius and tighter binding to the carboxylate. f–f Transitions of electrons produce much narrower absorption bands than d–d transitions. The f-orbitals have smaller radii than the occupied 5s and 5p-orbitals, and f electrons are shielded and hardly changed by ligand fields.

The colour and luminescence of lanthanoid compounds are often bright and attractive. Sunglasses, for example, are often colourized with Pr(III) (green), Nd(III) (pink) or Ho(III) (yellow). The luminescence of lanthanoids becomes an important source of domestic lighting. Electrons from electrodes (cathodic luminescence), UV light (photoluminescence) and electric fields (electroluminescence) are used to excite f electrons of lanthanoid metals, which are integrated into alumina, borate or other oxide and sulfide matrices. A combination of $(Ba,Eu(II))MgAl_{10}O_{17}$ for blue light, $(Ce, Gd, Tb)MgB_5O_{10}$ for green and $(Y,EuIII)_2O_3$ for red produces the white light of fluorescent lamps. A mixed oxide sulfide of yttrium (Y_2O_2S) doped with some Eu(III) gives the red light on television screens.

Partially occupied f-orbitals also produce very strong magnetic fields with very little electricity, and lanthanoids are used in compact high-strength magnets. The most common types of rare earth magnets are samarium–cobalt and neodymium–iron–boron (NIB) magnets (see p. 385). Such magnets are a prerequisite for the construction of headphones and the small electric motors used in computers, cars and personal stereos.

5.10.1 Scandium

The scandium atom is extremely electron deficient and binds several non-cyclopentadienyl ligands. Scandium triflate, $Sc(CF_3COO-)_3$, for example, is a useful Lewis acid in organic synthesis and the polar Sc–C bond is especially susceptible to insert alkenes, such as ethylene, styrene or norbornene.

^{45}Sc, which is 100% naturally abundant, possesses a gyromagnetic ratio close to that of ^{13}C, a nuclear spin of $I = 7/2$, a moderate nuclear quadrupole moment and a relative receptivity of 1780 in comparison to ^{13}C. The chemical shift range is approximately 250 ppm and ^{1}H–^{45}Sc dipolar coupling can be observed in some instances. While the presence of these interactions complicates solid-state NMR spectra, they also provide information about molecular structure and dynamics. Scandium is therefore an excellent nucleus for NMR experimentation in solution and in the solid state.[172]

5.10.2 Yttrium

The classical use of yttrium is the neodymium–yttrium–aluminium–granate or Nd:YAG laser. A "granate" is a cubic silicate mineral with a stoichiometry X(II)$_3$Z(III)$_2$(SiO$_4$)$_3$ (X = Ca, Mg, Fe(II), Mn; Z = Al(III), Fe(III), Ti, V, Cr), but it is a misleading name for yttrium aluminium oxide (Y$_3$Al$_5$O$_{12}$), the basis of YAG. In highly yttrium-doped alumina most of the yttrium appears as a Y$_3$Al$_5$O$_{12}$ phase in which the Y(III) is located on the octahedral Al(III) sites, where the larger Y(III) ion induces a distortion of the hexagonal oxygen ion lattice in the neighbourhood of Y (Y(III) is 208 pm in diameter, Al(III) 135 pm).

The growth of porous yttrium/aluminum oxides was started by the formation of a layered phase of Y(III)/Al(III) with a long-chain alkylsufate as counterion, e.g sodium dodecylsulfate at pH 5.7–6.1. The homogeneous precipitation of mixed yttrium/aluminium oxide layers was then initiated by added urea and heating. The urea decomposed within 20 h at 60–80 °C to ammonium hydroxide and raised the pH to 8. Hexagonal oxides (YO$_6$ and AlO$_6$) are formed. The cell dimension of the yttrium-based system is little dependent on the alkyl chain length of the sulfate and it was converted into an ordered porous material with a specific surface area of 662–797 m^{2}g^{-1} by the anion exchange of the long-chain alkyl sulfate by acetate ions, in contrast to pure aluminium hexagonal mesophases, which collapsed. The much higher stability of the yttrium/aluminium hexagonal phase upon acetate treatment was attributed to the increased ratio of 6-coordinate to 4-coordinate aluminium polyhedra, which is enforced by the larger Y^{3+} cation.[173]

If 1% of neodyme was added as a dopant to the crystal lattice, the Nd-YAG crystal emitted hot, coherent near-IR radiation (1064 nm; weaker bands near 940, 1120, 1320 and 1440 nm). Such neodyme-YAG lasers are useful in engraving, etching, or marking of a variety of metals and plastics. Frequency-doubled Nd:YAG lasers (wavelength 532 nm) are used in medicine for eye surgery and the treatment of minor vascular defects on the face and legs, and for cosmetic prcedures such as hair removal. Pulsed Nd:YAG surgical lasers typically have an optical switch in the laser cavity that waits for a maximum population inversion of the neodymium ions before it opens. The light wave then depopulates the excited neodymium, achieving 20 MW, 10 ns pulses.

Lanthanide-doped nanoparticles are a tempting alternative to dye molecules (p. 64 ff, 79, 123, 130 ff), liquid crystals (p. 38) and quantum dots (p. 39) in colour displays and/or light bulbs. They do not bleach, are not toxic and have low emission band widths of 10 to 20 nm. f–f transitions also provide multiline spectra, which are particularly useful for multicolour labelling.

Yttrium vanadate (YVO$_4$) doped with lanthanoid and phosphate ions, Y(P$_{0.75}$V$_{0.25}$)O$_4$ dissolved in metal-chelating PVP, for example, is excited by UV light and the excitation energy will readily migrate through vanadate groups to a quenching site and dissipate non-radiatively. Owing to the strong coupling between the electrons, the thermally activated energy migration is so efficient that the host YVO$_4$ essentially did not show any visible emission at room temperature. Eu^{3+}, Dy^{3+} and Sm^{3+} activators, however, emitted light at 618, 576 and 604 nm, respectively.[174] The introduction of phosphate into the YVO_4 lattice hampered the energy transfer. The nanoparticles showed even stronger emission from the $[VO_4]^{3-}$ groups and further addition of lanthanoid dopants yielded a dual emission from the host and the dopant. Selective fine-tuning of the emission colour from deep blue to green, red and yellow was achieved.

5.10.3 Lanthanides

Lanthanum

Magnesium and calcium ions are not only present in the phosphates of DNA, bones and teeth, but also in several enzymes connected with phosphorylation. Here they can be replaced by trivalent lanthanide ions, from La(III) through Lu(III), which have ionic radii ranging from slightly greater to slightly smaller than that of Ca(II). This is usually unproblematic; the enzymes remain active, since Ln(III) ions exhibit the same preference for oxygen donor ligands as Ca(II), have coordination numbers generally >6, and show little stereochemical preference in their coordination chemistry. Lanthanides go everywhere, where they find enough oxygen ligands.

Proteins, with their charged groups and counterions, are usually able to adjust to the gain of a single positive charge without significant structural consequences. With the exceptions of La(III) ($4f^0$) and Lu(III) ($4f^{14}$), all of the Ln(III) ions are open-shell, paramagnetic species. These various $4f^n$ configurations give rise to low-lying excited states, which cause visible absorption. Individual orbitals are split by the sizable spin-orbit coupling interaction and by the ligand field into many closely spaced components and several lanthanide ions are able to relax by the emission of photons, luminescing in solution at room temperature. Luminescence can be measured spectrometrically with high sensitivity; a lanthanide in a protein is therefore a valuable indicator. The emission properties of different lanthanide ions can also take up electrons from different neighbouring amino acids. Excitation in the UV region hits the aromatic amino acids (phenylalanine, tyrosine, tryptophan), which transfer their energy to the Tb(III) ion. Eu(III) bound to nucleic acids is especially sensitive to the fluorescent nucleotide 4-thiouridine as an energy donor.[175]

Amine-coated nanoparticles doped with lanthanide oxide luminescent nanoparticles were coupled to a protein labelled with an organic fluorophore. Their stepwise photobleaching as well as their initial emission allowed the counting of the numbers of proteins coupled to single nanoparticles.[176]

Polymers should not be in direct contact with the lanthanides because their vibrational and electronic transitions quench f–f transitions. Small amounts of the Ln^{3+} ions doped into LaF_3 nanoparticles are more appropriate, and the nanoparticles are then solubilized ("stabilized") by amphiphilic ammonium di-*n*-octadecyldithiophosphate ($[NH_4]$–$[S_2P(OC_{18}H_{37})_2]$) ligands.

Solid state ^{19}F, ^{31}P, ^{45}Sc and ^{139}La NMR showed (1) that the lanthanum trifluoride (LaF_3) nanoparticles remained crystalline in presence of lanthanum, ytterbium and scandium surface ligands; (2) that doping with paramagnetic Yb^{3+} ions leads to severe broadening of ^{19}F powder patterns and only moderate broadening of ^{139}La powder patterns, while no significant paramagnetically induced shifts are observed, indicating that the paramagnetic ions are uniformly distributed in the lanthanum positions, (3) the Sc^{3+} ions are completely incorporated into the extended lanthanum trifluoride structure, as demonstrated by the ^{45}Sc NMR data. ^{31}P resonances finally indicate partial hydrolysis of the dithiophosphate and suggest that the La^{3+} surface interacts more strongly with the O-containing head groups.[177]

Air-sphere silica inverse opals with a photonic stop band at 551 nm were combined with lanthanum trifluoride nanoparticles doped with photoluminescent Tb^{3+}, emitting at 542 nm ($^5D_4 \rightarrow {}^7F_4$ transition). This embedding of the luminescence in the centre of the photonic stop band increased the lifetime of the luminescence by 40%.[178]

Cerium

Iron alloys containing 50% of cerium and 25% of lanthanum (ferrocerium) are used as ignition metals. Their powdered abrasion products self-ignite in air and inflame hydrocarbon vapours (as in the familiar cigarette lighter). The calcium-like reactivity is also helpful in steel alloys, where cerium and other lanthanides remove sulfur and other gases in the process of steel formation.

Cerium dioxide (ceria, CeO_2), the most important dioxide in the series, prevents the formation of tar in self-cleaning ovens and is an important catalyst for the combustion of soot from diesel engine exhaust.

Ceria nanoparticles freshly prepared by hydrolysis of alkoxides exhibited Ce^{3+}–O–Ce^{4+}-type defects predominantly on the surface, as evidenced by a 650 nm reflectance band and ESR signals. In nanocrystalline ceria the variation in the signal with increasing reduction temperatures indicates the migration of subsurface oxygen anions, which cause a sudden disappearance of the EPR signal of the Ce^{3+} ions. These ions promote the dehydrogenation of ethylbenzene to styrene by N_2O at 598 K.[179]

Biological particles usually have a negatively charged surface. Cell membranes and nucleic acids carry phosphate anions on their surface, and proteins usually have excessive glutamate and aspartate and produce nanoparticles with a positive zeta (ζ) potential. This is the electrokinetic potential of colloidal particles, is the potential difference between the dispersion medium and the stationary layer of fluid attached to the dispersed particles, usually bulk and hydration water.[180] A value of ±25 mV separates instable particles with low surface charges and a high tendency to coagulate and precipitate from electrically stable particles with highly charged surfaces (25–60 mV). The zeta potential cannot be measured directly, but is calculated from the mobility of particles in an electric field. Ceria nanoparticles having positive zeta potential were found to adsorb more bovine serum albumin while the samples with negative zeta potential showed little or no protein adsorption. Cellular uptake was stronger for negatively charged nanoparticles, however, because they were not adsorbed on the membrane surface.[181]

Neodymium

Magnets based of the $Nd_2Fe_{14}B$ type are powerful permanent magnets at room temperature. They are produced by milling a sintered $Nd_{16}Fe_{76}B_8$ alloy, which adsorbs large quantities of hydrogen gas (0.4 wt%) at room temperature. The volume of the powder thereby increases and favours the formation of fine powders. After heating, Fe, Fe_2B and $NdH_{2.9}$ are formed at first, vacuum annealing at 800–1000 °C decomposes the hydride to neodymium metal and highly coercive (p. 29) $Nd_2Fe_{14}B$ is formed. The sintered magnets are micro- and nanometre ferromagnetic particles in a nonmagnetic matrix. They contain three basic phases: $Nd_2Fe_{14}B$ (Φ-phase), $Nd_1Fe_4B_4$ (η-phase) and a neodymium-rich phase called n. Rearrangements of the domain walls of these phases and the magnetic anisotropy of the iron sublattice in $Nd_2Fe_{14}B$ crystals lead to high-anisotropy fields. This anisotropy originates from the parallel coupling of the light rare earth metal neodymium with the iron sublattice moment. Heavy lanthanoid metals, *e.g.* Y, couple antiparallel with iron and magnetization is lower.

The experimental findings that the coercivity increases even in relatively low magnetic fields, that magnetizing fields can be substantially lower than the induced coercivity and that the temperature dependence of the coercive field is perfectly described by a nucleation model indicate that the high coercivity of $Nd_2Fe_{14}B$ magnets is not caused by a pinning effect. Materials for nucleation are pressed under an applied magnetic field in order to align the magnetization axes of particles and are then sintered at high temperatures to compact the magnet to full density. A wall pinning effect, on the other hand, occurs if the domain wall energy is different in different positions of discontinuous domains and is trapped in regions of low energy. "Unpinning" or formation of larger domains then occurs only in very large magnetic fields.[182,183]

Neodymium magnets are very strong, with grades from N24 to N54. The number represents the magnetic energy product, in megagauss oersteds (1 MG Oe = 7.958 T A/m = 7.958 J/m^3). The most powerful grades lose their magnetism at a Curie temperature of 80 °C. High-temperature grades will operate at up to 230 °C, but their strength is close to the lower limit. In 2008 1 kg of neodymium magnets cost ~100 €, such a magnet holds up 1300 times its own mass. They have replaced marginally weaker and significantly more heat-resistant samarium-cobalt magnets in most applications, due mainly to their lower cost. Neodyme–iron–boron magnets ($Nd_2Fe_{14}B$) are used in the read-write heads of computer hard disk drives.

For Nd:YAG lasers see yttrium, p. 382.

Samarium

Samarium–cobalt magnet alloys $SmCo_5$ (N16–N25) and Sm_2Co_{17} (N20–N32) contain ~25% of samarium and can be used at higher temperatures. Samarium cobalt is manufactured by sintering and inherent cracks are to be expected, favouring pinning and unpinning mechanisms for the formation of large magnetic domains (see previous section).

Europium

Uniform 30 nm silica-coated luminescent Eu(III) nanoparticles emit luminescence with a long lifetime of 770 μs when excited by UV light (absorption 336 nm, emission 615 nm) and are stable against photobleaching.[184]

Gadolinium

Gadolinium oxide nanoparticles with a polysiloxane shell containing poly(ethyleneglycol) (PEG) outside and organic dyes inside have been used as positive contrast agents in magnetic resonance imaging (MRI) and in fluorescence imaging in cancerogenic rats. The hybrid luminescent particles were able to circulate in the blood vessels for at least 3 h without aggregation. No irreversible uptake was observed in the very thin capillaries of the lung. The absence of luminescent particles in the liver showed a low recognition by the immune system, and the particles were naturally eliminated by renal excretion. This metabolically safe behaviour is attributable to the small size of the particles but also to the PEG corona since the injection of similar particles devoid of PEG led to accumulation both in the liver and in the lung.

The gadolinium oxide core induced an enhancement of the positive contrast of MRI as compared to widely used contrast agents in clinical MRI, whereas the fluorescence imaging results from the presence of organic dyes in the polysiloxane shell. The data revealed that the longitudinal relaxivity per particle increased with the core size ($3.3 < 5.2 < 8.9$ nm). As ^{157}Gd (natural abundance 20%) possesses a high neutron capture cross-section (66 times higher than ^{10}B), these particles also appear to be a good alternative to boron compounds for neutron-capture therapy. This alternative therapy would consist of destroying the tumour, loaded with hybrid Gd_2O_3 nanoparticles, by energetic particles resulting from an interaction between the internalized biocompatible nanoparticles and a harmless thermal neutron beam.

$NaGdF_4$ nanoparticles doped with cerium and terbium dissolved in nonpolar solvents and showed green luminescence under UV excitation.[185,186]

Terbium

Three-dimensional photonic structures with a complete photonic band gap in the visible and near-IR region, such as inverse opal structures, are attractive candidates for the realization of 3D photoluminescence. Tb^{3+} ions embedded in inverse opals and coupled to lanthanum trifluoride nanoparticles, in order to obtain efficient shielding of the terbium ions from the quenching effects of the environment, gave a record photoluminescence lifetime of 4.0 ms at 546 nm.[187]

Turnip yellow mosaic virus is an icosahedral plant virus with an average diameter of 28 nm and can be isolated inexpensively in gram quantities from turnip or Chinese cabbage (Figure 5.77). The virus consists of a single-stranded

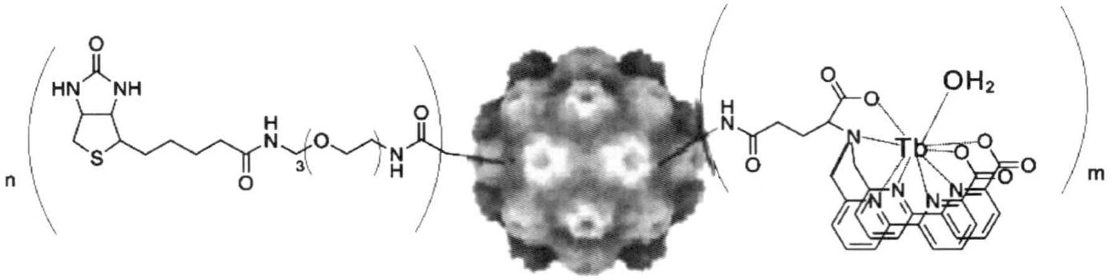

Figure 5.77 Model of the labelled turnip yellow mosaic virus and the biotin-Tb^{3+} labels. It reacts with AlexaFluor488-avidin dye in solution (not shown). © American Chemical Society.[188,189]

RNA with a high molecular weight of 1.9 MDa and 180 chemically identical protein subunits of 20 kDa each – another 3.6 MDa. Three carboxyl groups on the surface of the protein are accessible to amino or carbodiimide groups of small dye molecules and fluorescein signals increase linearly over a wide range of dye/virion ratios, with negligible quenching even at dye loadings up to ~40 dye/virion, corresponding to a local dye concentration of 4.6 mM considering that the virus occupies a space of 1.4×10^{-23} m^3. Several molecules of a terbium complex were also fixated to this bionanoparticle and produced a 308 nm luminescent signal under UV light. Several biotin groups served as ligands for the fluorescing protein derivative AlexaFluor488-avidin, which was used as the energy receptor for the terbium signals.

The critical radius for Förster resonant energy transfer (FRET, R_o) from terbium to fluorescein isothiocyanate, corresponding to the distance at which energy transfer and spontaneous decay of the acceptor are equally probable, was determined as 4.1 nm. A clear communication between the fixed signalling and receptive sites was established experimentally and traced back to the long lifetime of both excited states, that of the lanthanide energy donor and that of the fluorescence dye on the protein. It was also found that the average luminescence lifetime of the terbium decreased during titration with Alexa-Fluor488-avidin and other fluorescent dyes. Quantitative evaluation of the data suggested that only one biotin group per virus was present.[188,189]

References

1. C.-C. Chung, T. W. Chung and T. C.-K. Yang, Rapid synthesis of titania nanowires by microwave-assisted hydrothermal treatments, *Ind. Eng. Chem. Res.*, 2008, **47**, 2301–2307.
2. Y. Zhang, L. Wu, Q. Zeng and J. Zhi, An approach for controllable synthesis of different-phase titanium dioxide nanocomposites with peroxotitanium complex as precursor, *J. Phys. Chem. C*, 2008, **112**, 16457–16462.
3. D. Wang, D. Choi, Z. Yang, V. V. Viswanathan, Z. Nie, C. Wang, Y. Song, J. Zhang and J. Liu, Synthesis and Li-ion insertion properties

of highly crystalline mesoporous rutile TiO_2, *Chem. Mater.*, 2008, **20**, 3435–3442.
4. G. Caputo, C. Nobile, T. Kipp, L. Blasi, V. Grillo, E. Carlino, L. Manna, R. Cingolani, P. D. Cozzoli and A. Athanassiou, Reversible wettability changes in colloidal TiO_2 nanorod thin-film coatings under selective UV laser irradiation, *J. Phys. Chem. C*, 2008, **112**, 701–714.
5. G. Li, N. M. Dimitrijevic, L. Chen, J. M. Nichols, T. Rajh and K. A. Gray, The important role of tetrahedral Ti^{4+} sites in the phase transformation and photocatalytic activity of TiO_2 Nanocomposites, *J. Am. Chem. Soc.*, 2008, **130**, 5402–5403.
6. H. S. Jung, J.-K. Lee, J. Lee, B. S. Kang, Q. Jia, M. Nastasi, J. H. Noh, C. M. Cho and S. H. Yoon, mobility enhanced photoactivity in sol-gel grown epitaxial anatase TiO_2 films, *Langmuir*, 2008, **24**, 2695–2698.
7. Y. Yao, G. Li, S. Ciston, R. M. Lueptow and K. A. Gray, Photoreactive TiO_2/carbon nanotube composites: synthesis and reactivity, *Environ. Sci. Technol.*, 2008, **42**, 4952–4957.
8. P. Periyat, S. C. Pillai, D. E. McCormack, J. Colreavy and S. J. Hinder, Improved high-temperature stability and sun-light-driven photocatalytic activity of sulfur-doped anatase TiO_2, *J. Phys. Chem. C*, 2008, **112**, 7 644–7652.
9. P. Wang, T. Xie, L. Peng, H. Li, T. Wu, S. Pang and D. Wang, Water-assisted synthesis of anatase TiO_2 nanocrystals: mechanism and sensing properties to oxygen at room temperature, *J. Phys. Chem. C*, 2008, **112**, 6648–6652.
10. T. Althammer and F. Fuhrmann, Photocatalytic surface reactions on indoor wall paint, *Environ. Sci. Technol.*, 2007, **41**, 6573–6578.
11. L. Korosi, S. Papp, I. Bertoti and I. Dekany, Surface and bulk composition, structure, and photocatalytic activity of phosphate-modified TiO_2, *Chem. Mater.*, 2007, **19**, 4811–4819.
12. K. Awazu, M. Fujimaki, C. Rockstuhl, J. Tominaga, H. Murakami, Y. Ohki, N. Yoshida and T. Watanabe, A Plasmonic photocatalyst consisting of silver nanoparticles embedded in titanium dioxide, *J. Am. Chem. Soc.*, 2008, **130**, 1676–1680.
13. C. Contado and A. Pagnoni, TiO_2 in commercial sunscreen lotion: flow field-flow fractionation and ICP-AES together for size analysis, *Anal. Chem.*, 2008, **80**, 7594–7608.
14. R. Bachoual, J. Boczkowski, D. Goven, N. Amara, L. Tabet, D. On, V. Leçon-Malas, M. Aubier and S. Lanone, Biological effects of particles from the Paris subway system, *Chem. Res. Toxicol.*, 2007, **20**, 1426–1433.
15. O. Girshevitz, Y. Nitzan and C. N. Sukenik, Solution-deposited amorphous titanium dioxide on silicone rubber: a conformal, crack-free antibacterial coating, *Chem. Mater.*, 2008, **20**, 1390–1396.
16. A. G. Agrios, l. Cesar, P. Comte, M. K. Nazeeruddin and M. Grätzel, Nanostructured composite films for dye-sensitized solar cells by electrostatic layer-by-layer deposition, *Chem. Mater.*, 2006, **18**, 5395–5397.

17. K. Fukuda, Y. Ebina, T. Shibata, T. Aizawa, I. Nakai and T. Sasaki, Unusual crystallization behaviors of anatase nanocrystallites from a molecularly thin titania nanosheet and its stacked forms: increase in nucleation temperature and oriented growth, *J. Am. Chem. Soc.*, 2007, **129**, 202–209.
18. M. Vettraino, M. Trudeau, A. Y. H. Lo, R. W. Schurko and D. Antonelli, Room-temperature ammonia formation from dinitrogen on a reduced mesoporous TiO_2 surface with metallic properties, *J. Am. Chem. Soc.*, 2002, **124**, 9567–9573.
19. K. C. Krogman, N. S. Zacharia, D. M. Grillo and P. T. Hammond, Photocatalytic layer-by-layer coatings for degradation of acutely toxic agents, *Chem. Mater.*, 2008, **20**, 1924–1930.
20. A. Testino, I. R. Bellobono, V. Buscaglia, C. Canevali, M. D'Arienzo, S. Polizzi, R. Scotti and F. Morazzoni, Optimizing the photocatalytic properties of hydrothermal TiO_2 by the control of phase composition and particle morphology. a systematic approach, *J. Am. Chem. Soc.*, 2007, **129**, 3564–3575.
21. G.-M. Kim, S.-M. Lee, G. H. Michler, H. Roggendorf, U. Gosele and M. Knez, Nanostructured pure anatase titania tubes replicated from electrospun polymer fiber templates by atomic layer deposition, *Chem. Mater.*, 2008, **20**, 3085–3091.
22. V. H. Houlding and M. Grätzel, Photochemical H_2 generation by visible light. Sensitization of TiO_2 particles by surface complexation with 8-hydroxyquinoline, *J. Am. Chem. Soc.*, 1983, **105**, 5695–5696.
23. A. Duret and M. Grätzel, Visible light-induced water oxidation on mesoscopic Fe_2O_3 films made by ultrasonic spray pyrolysis, *J. Phys. Chem. B*, 2005, **109**, 17184–17191.
24. A. Imanishi, T. Okamura, N. Ohashi, R. Nakamura and Y. Nakato, Mechanism of Water photooxidation reaction at atomically flat TiO_2 (rutile) (110) and (100) surfaces: dependence on solution pH, *J. Am. Chem. Soc.*, 2007, **129**, 11569–11578.
25. S. Livraghi, M. R. Chierotti, E. Giamello, G. Magnacca, M. C. Paganini, G. Cappelletti and C. L. Bianchi, Nitrogen-doped titanium dioxide active in photocatalytic reactions with visible light: a multi-technique characterization of differently prepared materials, *J. Phys. Chem. C*, 2008, **112**, 17244–17252.
26. C. Di Valentin, E. Finazzi, G. Pacchioni, S. Livraghi, A. M. Czoska, M. C. Paganini and E. Giamello, Density functional theory and electron paramagnetic resonance study on the effect of N-F codoping of TiO_2, *Chem. Mater.*, 2008, **20**, 3706–3714.
27. M. P. L. Werts, M. Badila, C. Brochon, A. Hébraud and G. Hadziioannou, Titanium dioxide-polymer core–shell particles dispersions as electronic inks for electrophoretic displays, *Chem. Mater.*, 2008, **20**, 1292–1298.
28. A. Duret and M. Gratzel, Visible Light-induced water oxidation on mesoscopic Fe_2O_3 films made by ultrasonic spray pyrolysis, *J. Phys. Chem. B*, 2005, **109**, 17184–17191.

29. J. Nowotny, T. Bak, L. R. Sheppard and M. K. Nowotny, Reactivity of titanium dioxide with oxygen at room temperature and the related charge transfer, *J. Am. Chem. Soc.*, 2008, **130**, 9984–9993.
30. S. Kim, J. K. Lee, S. O. Kang, J. Ko, J.-H. Yum, S. Fantacci, F. De Angeli, D. Di Censo, K. Nazeeruddin and M. Grätzel, Molecular engineering of organic sensitizers for solar cell applications, *J. Am. Chem. Soc.*, 2006, **128**, 16701–16707.
31. B. Hyun, Y. Zhong, A. C. Bartnik, L. Sun, H. D. Abruna, F. W. Wiese, J. D. Godreau, J. R. Matthews, T. M. Leslie and N. F. Borelli, Electron injection from colloidal PbS quantum dots into titanium dioxide nanoparticles, *ACS nano*, 2008, **2**, 2206–2212.
32. I. Robel, V. Subramanian, M. Kuno and P. V. Kamat, Quantum dot solar cells. harvesting light energy with CdSe nanocrystals molecularly linked to mesoscopic TiO_2 films, *J. Am. Chem. Soc.*, 2006, **128**, 2385–2393.
33. T. Lopez-Luke, A. Wolcott, L. Xu, S. Chen, Z. Wen, J. Li, E. De La Rosa and J. Z. Zhang, Nitrogen-doped and CdSe quantum-dot-sensitized nanocrystalline TiO_2 films for solar energy conversion applications, *J. Phys. Chem. C*, 2008, **112**, 1282–1292.
34. E. Spano, G. Tabacchi, A. Gamba and E. Fois, On the role of Ti(IV) as a Lewis acid in the chemistry of titanium zeolites: formation, structure, reactivity, and aging of Ti-peroxo oxidizing intermediates. A first principles study, *J. Phys. Chem. B*, 2006, **110**, 21651–21661.
35. S. Y. Choi, M. Mamak, N. Coombs, N. Chopra and G. A. Ozin, Electrochromic performance of viologen-modified periodic mesoporous nanocrystalline anatase electrodes, *Nano Lett.*, 2004, **4**, 1231–1235.
36. J. Tang, F. Redl, Y. Zhu, T. Siegrist, L. E. Brus and M. L. Steigerwald, An organometallic synthesis of TiO_2 NPs, *Nano Lett.*, 2005, **5**, 543–548.
37. K. Fukuda, Y. Ebina, T. Shibata, T. Aizawa, I. Nakai and T. Sasaki, Unusual crystallization behaviors of anatase nanocrystallites from a molecularly thin titania nanosheet and its stacked forms: increase in nucleation temperature and oriented growth, *J. Am. Chem. Soc.*, 2007, **129**, 202–209.
38. J. J. Blackstock, C. L. Donley, W. F. Stickle, D. A. A. Ohlberg, J. J. Yang, D. R. Stewart and R. S. Williams, Oxide and carbide formation at titanium/organic monolayer interfaces, *J. Am. Chem. Soc.*, 2008, **130**, 4041–4047.
39. K. Luo, S. Zhou, L. Wu and G. Gu, Dispersion and functionalization of nonaqueous synthesized zirconia nanocrystals via attachment of silane coupling agents, *Langmuir*, 2008, **24**, 11497–11505.
40. S. Zhou, G. Garnweitner, M. Niederberger and M. Antonietti, Dispersion behavior of zirconia nanocrystals and their surface functionalization with vinyl group-containing ligands, *Langmuir*, 2007, **23**, 9178–9187.
41. W. J. Boo, L. Sun, J. Liu, A. Clearfield and H.-J. Sue, Effective intercalation and exfoliation of nanoplatelets in epoxy via creation of porous pathways, *J. Phys. Chem. C*, 2007, **111**, 10377–10381.

42. C. V. Kumar and A. Chaudhari, Proteins immobilized at the galleries of layered R-zirconium phosphate: structure and activity, *J. Am. Chem. Soc.*, 2000, **122**, 830–837.
43. G. Liu and Y. Lin, Electrochemical sensor for organophosphate pesticides and nerve agents using zirconia NPs as selective sorbents, *Anal. Chem.*, 2005, **77**, 5894–5901.
44. M. D. Hernandez-Alonso, J. M. Coronado, B. Bachiller-Baeza, M. Fernandez-Garcia and J. Soria, Influence of structural and surface characteristics of $Ti_{1-x}Zr_xO_2$ nanoparticles on the photocatalytic degradation of methylcyclohexane in the gas phase, *Chem. Mater.*, 2007, **19**, 4283–4291.
45. A. Satsuma, K. Shimizu, K. Kashiwagi, T. Endo, H. Nishiyama, S. Kakimoto, S. Sugaya and H. Yokoi, Ammonia sensing mechanism of tungstated-zirconia thick film sensor, *J. Phys. Chem. C*, 2007, **111**, 12080–12085.
46. J. Tang, J. Fabbri, R. D. Robinson, Y. Zhu, I. P. Herman, M. L. Steigerwald and L. E. Brus, Solid-solution NPs: use of a nonhydrolytic sol-gel synthesis to prepare HfO_2 and $Hf_xZr_{1-x}O_2$ nanocrystals, *Chem. Mater.*, 2004, **16**, 1336–1342.
47. M. Nath, C. N. R. Rao, R. Popovitz-Biro, A. Albu-Yaron and R. Tenne, NPs produced by laser ablation of HfS_3 in liquid medium: inorganic fullerene-like structures of Hf_2S, *Chem. Mater.*, 2004, **16**, 2238–2243.
48. T. Hirao, Vanadium in modern organic synthesis, *Chem. Rev.*, 1997, **97**, 2707–2724.
49. J. Buha, I. Djerdj, M. Antonietti and M. Niederberger, Thermal transformation of metal oxide NPs into nanocrystalline metal nitrides using cyanamide and urea as nitrogen source, *Chem. Mater.*, 2007, **19**, 3499–3505.
50. K. L. Knappenberger Jr., C. E. Jones Jr., M. A. Sobhy, I. Iordanov, J. Sofo and A. W. Castleman Jr., Anion photoelectron spectroscopy and density functional investigation of vanadium carbide clusters, *J. Phys. Chem. A*, 2006, **110**, 12814–12821.
51. R. Ostermann, D. Li, Y. Yin, J. T. McCann and Y. Xia, V_2O_5 nanorods on TiO_2 nanofibers: a new class of hierarchical nanostructures enabled by electrospinning and calcination, *Nano Lett.*, 2006, **6**, 1297–1302.
52. M. Chiesa, V. Meynen, S. Van Doorslaer, P. Cool and E. F. Vansant, Vanadium silicalite-1 NPs deposition onto the mesoporous walls of SBA-15. Mechanistic insights from a combined EPR and Raman study, *J. Am. Chem. Soc.*, 2006, **128**, 8955–8963.
53. L. Zhang, I. Djerdj, M. Cao, M. Antonietti and M. Niederberger, Nonaqueous sol-gel synthesis of a nanocrystalline $InNbO_4$ visible-light photocatalyst, *Adv. Mater.*, 2007, **19**, 2083–2086.
54. A. G. P. Troeman, H. Derking, B. Borger, J. Pleikies, D. Veldhuis and H. Hilgenkampen, NanoSQUIDs based on niobium constrictions, *Nano Lett.*, 2007, **7**, 2152–2156.

55. F. L. Deepak, H. Cohen, S. Cohen, V. Feldman, R. Popovitz-Biro, O. Azulay, O. Millo and R. Tenne, Fullerene-like (IF) NbxMo$_{1-x}$S$_2$ nanoparticles, *J. Am. Chem. Soc.*, 2007, **129**, 12549–12562.
56. Z. Zou, J. Ye, K. Sayama and H. Arakawa, Direct splitting of water under visible light irradiation with an oxide semiconductor photocatalyst, *Nature*, 2001, **414**, 625–627.
57. S. Singh, M. T. Greiner and P. Kruse, Robust inorganic membranes from detachable ultrathin tantalum oxide films, *Nano Lett.*, 2007, **7**, 2676–2683.
58. Q. Zhang and L. Gao, Ta$_3$N$_5$ NPs with enhanced photocatalytic efficiency under visible light irradiation, *Langmuir*, 2004, **20**, 9821–9827.
59. L. Y. Zhao, A. C. Siu, L. J. Pariag, Z. H. He and K. T. Leung, Electrochemical deposition of chromium core-shell nanostructures on H-Si(100): evolution of spherical nanoparticles to uniform thin film without and with atop hexagonal microrods, *J. Phys. Chem. C*, 2007, **111**, 14621–14624.
60. S. Derveaux, B. G. De Geest, C. Roelant, K. Braeckmans, J. Demeester and S. C. De Smedt, Multifunctional layer-by-layer coating of digitally encoded microparticles, *Langmuir*, 2007, **23**, 10272–10279.
61. Y. A. Yang, Y. W. Cao, B. H. Loo and J. N. Yao, Microstructures of electrochromic MoO$_3$ thin films coloured by injection of different cations, *J. Phys. Chem. B*, 1998, **102**, 9392–9396.
62. M. I. Shukoor, H. A. Therese, L. Gorgishvili, G. Glasser, U. Kolb and W. Tremel, From layered molybdic acid to lower-dimensional nanostructures by intercalation of amines under ambient conditions, *Chem. Mater.*, 2006, **18**, 2144–2151.
63. N. A. Dhas and K. S. Suslick, Sonochemical preparation of hollow nanospheres and hollow nanocrystals, *J. Am. Chem. Soc.*, 2005, **127**, 2368–2369.
64. D. Brinzei, L. Catala, C. Mathoniere, W. Wernsdorfer, A. Gloter, O. Stephan and T. Mallah, Photoinduced superparamagnetism in trimetallic coordination NPs, *J. Am. Chem. Soc.*, 2007, **129**, 3778–3779.
65. J. C. Park and H. Song, Synthesis of polycrystalline Mo/MoOx nanoflakes and their transformation to MoO$_3$ and MoS$_2$ nanoparticles, *Chem. Mater.*, 2007, **19**, 2706–2708.
66. F. L. Deepak, H. Cohen, S. Cohen, Y. Feldman, R. Popovitz-Biro, D. Azulay, O. Millo and R. Tenne, Fullerene-like (IF) Nb$_x$Mo$_{1-x}$S$_2$ nanoparticles, *J. Am. Chem. Soc.*, 2007, **129**, 12549–12562.
67. B. Hinnemann, P. G. Moses, J. Bonde, K. P. Jørgensen, J. H. Nielsen, S. Horch, I. Chorkendorff and J. K. Nørskov, Biomimetic hydrogen evolution: MoS$_2$ nanoparticles as catalyst for hydrogen evolution, *J. Am. Chem. Soc.*, 2005, **127**, 5308–5309.
68. L. C. Seefeldt, I. G. Dance and D. R. Dean, Substrate interactions with nitrogenase: Fe versus Mo, *Biochemistry*, 2004, **43**, 1401–1409.
69. W. Yamaguchi and J. Murakami, Low-temperature formation of nitrous oxide from dinitrogen, mediated by supported tungsten nanoclusters, *J. Am. Chem. Soc.*, 2007, **129**, 6102–6103.

70. M. A. Cortes-Jacome, M. Morales, C. Angeles Chavez, L. F. Ramirez-Verduzco, E. Lopez-Salinas and J. A. Toledo-Antonio, WOx/TiO$_2$ Catalysts via titania nanotubes for the oxidation of dibenzothiophene, *Chem. Mater.*, 2007, **19**, 6605–6614.
71. J. T. Mang, R. P. Hjelm, S. F. Son, P. D. Peterson and B. S. Jorgensen, Characterization of components of nano-energetics by small-angle scattering techniques, *J. Mater. Res.*, 2007, **22**, 1907–1920.
72. N. Zink, J. Pansiot, J. Kieffer, H. A. Therese, M. Panthoefer, F. Rocker, U. Kolb and W. Tremel, Selective synthesis of hollow and filled fullerene-like (IF) WS2 nanoparticles via metal-organic chemical vapor deposition, *Chem. Mater.*, 2007, **19**, 6391–6400.
73. N. Zink, H. A. Therese, J. Pansiot, A. Yella, F. Banhart and W. Tremel, In situ heating TEM study of onion-like WS$_2$ and MoS$_2$ nanostructures obtained via MOCVD, *Chem. Mater.*, 2008, **20**, 65–71.
74. A. B. Bourlinos, K. Raman, R. Herrera, Q. Zhang, L. A. Archer and E. P. Giannelis, A liquid derivative of 12-tungstophosphoric acid with unusually high conductivity, *J. Am. Chem. Soc.*, 2004, **126**, 15358–15359.
75. a) V. L. Pecoraro, M. J. Baldwin, A. Gelsco, Interaction of manganese with dioxygen and its reduced derivatives, *Chem. Rev.*, 1994, **94**, 807–826; b) G. C. Dismukes, Manganese enzymes, *Chem. Rev.*, 1996, **96**, 2909–2926; c) K. K. Yachandra, K. Sauer, M. P. Klein, Manganese clusters in photosynthesis, *Chem. Rev.*, 1996, **96**, 2927–2950; d) M. Sono, M. P. Roach, E. D. Coulter, J. H. Jawson, Heme-containing oxygenases, *Chem. Rev.*, 1996, **96**, 2841–2888.
76. J. S. Garitaonandia, M. Insausti, E. Goikolea, M. Suzuki, J. D. Cashion, N. Kawamura, H. Ohsawa, I. T. C. Stamatatos, D. Foguet-Albiol, S. C. Lee, C. C. Stoumpos, C. P. Raptopoulou, A. Terzis, W. Wernsdorfer, S. O. Hill, S. P. Perlepes and G. Christou, "Switching on" the properties of single-molecule magnetism in triangular manganese(III) complexes, *J. Am. Chem. Soc.*, 2007, **129**, 9484–9499.
77. H. Oshio, M. Nihei, S. Koizumi, T. Shiga, H. Nojiri, M. Nakano, N. Shirakawa and M. Akatsu, A heterometal single-molecule magnet of [Mn(III)$_2$Ni(II)$_2$Cl$_2$(salpa)$_2$], *J. Am. Chem. Soc.*, 2005, **127**, 4568–4569.
78. G. Salazar-Alvarez, J. Sort, S. Surinach, M. D. Baro and J. Nogues, Synthesis and size-dependent exchange bias in inverted core-shell MnO/Mn$_3$O$_4$ nanoparticles, *J. Am. Chem. Soc.*, 2007, **129**, 9102–9108.
79. S. Jana, S. Basu, S. Pande, S. K. Ghosh and T. Pal, Shape-selective synthesis, magnetic properties, and catalytic activity of single crystalline α-MnO$_2$ nanoparticles, *J. Phys. Chem. C*, 2007, **111**, 16272–16277.
80. V. M. B. Crisostomo, J. K. Ngala, S. Alia, A. Dobley, C. Morein, C. Chen, X. Shen and S. L. Suib, New synthetic route, characterization, and electrocatalytic activity of nanosized manganite, *Chem. Mater.*, 2007, **19**, 1832–1839.

81. J. K. Ngala, S. Alia, A. Dobley, V. M. B. Crisostomo and S. L. Suib, Characterization and electrocatalytic behavior of layered Li_2MnO_3 and its acid-treated form, *Chem. Mater.*, 2007, **19**, 229–234.
82. T. Valdes-Solis, P. Valle-Vigon, S. Alvarez, G. Marban and A. B. Fuertes, Manganese ferrite nanoparticles synthesized through a nanocasting route as a highly active Fenton catalyst, *Catal. Comm.*, 2007, **8**, 2037–2042.
83. S. R. Brown, S. M. Kauzlarich, F. Gascoin and G. J. Snyder, $Yb_{14}MnSb_{11}$: New high efficiency thermoelectric material for power generation, *Chem. Mater.*, 2006, **18**, 1873–1877.
84. S. M. Hussain, A. K. Javorina, A. M. Schrand, H. M. Duhart, S. F. Ali and J. J. Schlager, The interaction of manganese nanoparticles with PC-12 cells induces dopamine depletion, *Toxicol. Sci.*, 2006, **92**, 456–463.
85. S. Wang, B. R. Jarrett, S. M. Kauzlarich and A. Y. Louie, Core/shell quantum dots with high relaxivity and photoluminescence for multimodality imaging, *J. Am. Chem. Soc.*, 2007, **129**, 3848–3856.
86. Y. Wang, K. Nakamura, X. Liu, N. Kitamura, A. Kubo and D. J. Hantowich, Simplified preparation via streptavidin of antisense oligomers/carriers nanoparticles showing improved cellular delivery in culture, *Bioconj. Chem.*, 2007, **18**, 1338–1343.
87. M. Hamoudeh, H. Salim, D. Barbos, C. Paunoiu and H. Fessi, Preparation and characterization of radioactive dirhenium decacarbonyl-loaded PLLA nanoparticles for radionuclide intra-tumoral therapy, *Eur. J. Pharm. Biopharm.*, 2007, **67**, 597–611.
88. K. W. Cheng and W. K. Chan, Morphology of rhenium complex-containing polystyrene-block-poly(4-vinylpyridine) and its use as self-assembly templates for nanoparticles, *Langmuir*, 2005, **21**, 5247–5250.
89. A. Yella, H. A. Therese, N. Zink, M. Panthofer and W. Tremel, Large scale MOCVD synthesis of hollow ReS_2 nanoparticles with nested fullerene-like structure, *Chem. Mater.*, 2008, **20**, 3587–3593.
90. S. Ghosh, K. Biswas and C. N. R. Rao, Core-shell nanoparticles based on an oxide metal: $ReO_2@Au$ (Ag) and $ReO_3@Si$ (TiO_2), *J. Materials Chem.*, 2007, **17**, 2412–2417.
91. F. Ren, A. K. Feldman, M. Carnes, M. Steigerwald and C. Nuckolls, Polymer growth by functionalized ruthenium nanoparticles, *Macromolecules*, 2007, **40**, 8151–8155.
92. Z. Yinghuai, E. Widjaja, S. L. P. Sia, W. Zhan, K. Carpenter, J. A. Maguire, N. S. Hosmane and M. F. Hawthorne, Electron micrograph and histogram of the Ru NCs and their catalytic action in B-isotope exchange. Ruthenium(0) nanoparticle-catalyzed isotope exchange between ^{10}B and ^{11}B nuclei in decaborane(14), *J. Am. Chem. Soc.*, 2007, **129**, 6507–6512.
93. S. Fiechter, I. Dorbandt, P. Bogdanoff, G. Zehl, H. Schulenburg, H. Tributsch, M. Bron, J. Radnik and M. Fieber-Erdmann, Surface modified ruthenium nanoparticles: structural investigation and surface analysis of a novel catalyst for oxygen reduction, *J. Phys. Chem. C*, 2007, **111**, 477–487.

94. T. Tsukatani and H. Fujihara, New method for facile synthesis of amphiphilic thiol-stabilized ruthenium nanoparticles and their redox-active ruthenium nanocomposite, *Langmuir*, 2005, **21**, 12093–12095.
95. C. Draeger, C. Bottcher, C. Messerschmidt, A. Schulz, L. Ruhlmann, U. Siggel, L. Hammarstrom, H. Berglund-Baudin and J.-H. Fuhrhop, Isolable and fluorescent mesoscopic micelles made of an amphiphilic derivative of tris-bipyridyl ruthenium hexafluorophosphate, *Langmuir*, 2000, **16**, 2068–2077.
96. D. Suzuki, T. Sakai and R. Yoshida, Self-flocculating/self-dispersing oscillation of microgels, *Angew. Chem.*, 2008, **120**, 931–934.
97. L. Wang, W. Yang and W. Tan, Dual-luminophore-doped silica nanoparticles for multiplexed signaling, *Nano Lett.*, 2005, **5**, 37–43.
98. S. Welter, N. Salluce, A. Benet.ti, N. Rot, P. Belser, P. Sonar, A. C. Grimsdale, K. Mullen, M. Lutz, A. L. Spek and L. De Cola, Rodlike bimetallic ruthenium and osmium complexes bridged by phenylene spacers. synthesis, electrochemistry, and photophysics, *Inorg. Chem.*, 2005, **44**, 4707–4716.
99. G. Cheng, C. L. Dennis, R. D. Shull and A. R. Hight Walker, Influence of the colloidal environment on the magnetic behavior of cobalt nanoparticles, *Langmuir*, 2007, **23**, 11740–11746.
100. J. J. Benkoski, S. E. Bowles, B. D. Korth, R. L. Jones, J. F. Douglas, A. Karim and J. Pyun, Field induced formation of mesoscopic polymer chains from functional ferromagnetic colloids, *J. Am. Chem.. Soc.*, 2007, **129**, 6291–629.
101. M. A. Zalich, V. V. Baranauskas, J. S. Riffle, M. Saunders and T. G. St. Pierre, Structural and magnetic properties of oxidatively stable cobalt nanoparticles encapsulated in graphite shells, *Chem. Mater.*, 2006, **18**, 2648–2655.
102. P. Y. Keng, I. Shim, B. D. Korth, J. F. Douglas and J. Pyun, Synthesis and self-assembly of polymer-coated ferromagnetic nanoparticles, *ACS Nano*, 2007, **1**, 279–292.
103. M. A. Zalich, M. L. Vadala, J. S. Riffle, M. Saunders and T. G. St. Pierre, Structural and magnetic properties of cobalt nanoparticles encased in siliceous shells, *Chem. Mater.*, 2007, **19**, 6597–6604.
104. G. Cheng, C. L. Dennis, R. D. Shull and A. R. Hight Walker, Influence of the colloidal environment on the magnetic behavior of cobalt nanoparticles, *Langmuir*, 2007, **23**, 11740–11746.
105. A. M. Schwartzberg, T. Y. Olson, C. E. Talley and J. Z. Zhang, Gold nanotubes synthesized via magnetic alignment of cobalt nanoparticles as templates, *J. Phys. Chem. C*, 2007, **111**, 16080–16082.
106. A. Kay, I. Cesar and M. Grätzel, New benchmark for water photo-oxidation by nanostructured Fe_2O_3 films, *J. Am. Chem. Soc.*, 2006, **128**, 15714–15721.
107. J. W. Bae, Y.-J. Lee, J.-Y. Park and K.-W. Jun, Influence of pH of the impregnation solution on the catalytic properties of Co/γ-alumina for Fischer-Tropsch synthesis, *Energy Fuels*, 2008, **22**, 2885–2891.

108. C.-F. Huo, Y.-W. Li, J. Wang and H. Jiao, Formation of CHx Species from CO dissociation on double-stepped Co(0001): exploring Fischer-Tropsch mechanism, *J. Phys. Chem. C*, 2008, **112**, 14108–14116.
109. J. D. Hoefelmeyer, K. Niesz, G. A. Somorjai and T. D. Tilley, Radial anisotropic growth of rhodium nanoparticles, *Nano Lett.*, 2005, **5**, 435–438.
110. S. C. Warren, M. J. Banholzer, L. S. Slaughter, E. P. Giannelis, F. J. DiSalvo and U. B. Wiesner, Generalized route to metal nanoparticles with liquid behavior, *J. Am. Chem. Soc.*, 2006, **128**, 12074–12075.
111. Y. Borodko, S. M. Humphrey, T. D. Tilley, H. Frei and G. A. Somorjai, Charge-transfer interaction of poly(vinylpyrrolidone) with platinum and rhodium nanoparticles, *J. Phys. Chem. C*, 2007, **111**, 6288–6295.
112. E. Finocchio, G. Busca, P. Forzatti, G. Groppi and A. Beretta, State of Supported rhodium nanoparticles for methane catalytic partial oxidation (CPO): FT-IR studies, *Langmuir*, 2007, **23**, 10419–10428.
113. T. Pawluk, Y. Hirata and L. Wang, Studies of iridium nanoparticles using density functional theory calculations, *J. Phys. Chem. B*, 2005, **109**, 20817–2082.
114. F. Tian, J. Zhu and D. Wei, Fabrication and magnetism of radial-easy-magnetized Ni nanowire arrays, *J. Phys. Chem. C*, 2007, **111**, 12669–12672.
115. G. G. Couto, J. J. Klein, W. H. Schreiner, D. H. Mosca, A. J. A. de Oliveira and A. J. G. Zarbin, Nickel nanoparticles obtained by a modified polyol process: synthesis, characterization, and magnetic properties, *J. Coll. Interface Sci.*, 2007, **311**, 461–468.
116. B. Borak, S. Laskowski, O. Heczko, A. Aaltonen, A. Baszczuk, M. Jasiorski, O. Soderberg, B. Mazurek, M. Oja, S.-P. Hannula and K. Maruszewski, Submicron-sized hybrid Si-Ni powders produced with the sol-gel method, *Materials Sci.*, 2007, **25**, 167–175.
117. J. Fan, S. W. Boettcher and G. D. Stucky, Nanoparticle assembly of ordered multicomponent mesostructured metal oxides via a versatile sol-gel process, *Chem. Mater.*, 2006, **18**, 6391–6396.
118. F. Alonso, P. Riente and M. Yus, The α-alkylation of methyl ketones with primary alcohols promoted by nickel nanoparticles under mild and ligandless conditions, *Synlett*, 2007, **12**, 1877–1880.
119. O. Trapp, S. K. Weber, S. Bauch, T. Bäcker, W. Hofstadt, B. Spliethoff, High-throughput kinetic study of hydrogenation over palladium nanoparticles: combination of reaction and analysis, *Chem. Eur. J.*, 2008, **14**, 4657–4666.
120. Z. Yang, K. J. Klabunde and C. M. Sorensen, From monodisperse sulfurized palladium nanoparticles to tiara Pd(II) thiolate clusters: influence of thiol ligand on thermal treatment of a palladium(II)-amine system, *J. Phys. Chem. C*, 2007, **111**, 18143–18147.
121. Y. Sun and H. H. Wang, High-performance, flexible hydrogen sensors that use carbon nanotubes decorated with palladium nanoparticles, *Adv. Mat.*, 2007, **19**, 2818–2823.

122. A. Dedieu, Theoretical studies in palladium and platinum molecular chemistry, *Chem. Rev.*, 2000, **100**, 543–600.
123. C. Wang, H. Daimon, Y. Lee, J. Kim and S. Sun, Synthesis of monodisperse Pt nanocubes and their enhanced catalysis for oxygen reduction, *J. Am. Chem. Soc.*, 2007, **129**, 6974–6975.
124. J. Zhang, M. B. Vukmirovic, K. Sasaki, A. U. Nilekar, M. Mavrikakis, A. Manos and R. Radoslav, Mixed-metal Pt monolayer electrocatalysts for enhanced oxygen reduction kinetics, *J. Am. Chem. Soc.*, 2005, **127**, 12480–12481.
125. H. Ye and R. M. Crooks, Effect of elemental composition of PtPd bimetallic nanoparticles containing an average of 180 atoms on the kinetics of the electrochemical oxygen reduction reaction, *J. Am. Chem. Soc.*, 2007, **129**, 3627–3633.
126. A. Fukuoka, J. I. Kimura, T. Oshio, Y. Sakamoto and M. Ichikawa, Preferential oxidation of carbon monoxide catalyzed by platinum nanoparticles in mesoporous silica, *J. Am. Chem. Soc.*, 2007, **129**, 10120–10125.
127. J. W. Peck and B. E. Koel, Selective dehydrogenation of 1,3-cyclohexadiene on ordered Sn/Pt(111) surface alloys, *J. Am. Chem. Soc.*, 1996, **118**, 2708–2717.
128. J. Choi, S. J. Oh, H. Ju and J. Cheon, Supporting information on the massive fabrication of free standing 1-dimensional Co/Pt nanostructures and modulation of ferromagnetism via a programmable barcode layer effect, *Nano Lett.*, 2005, **5**, 2179–2183.
129. J. Ghilane, M. Guilloux-Viry, C. Lagrost, J. Simonet and P. Hapiot, Reactivity of platinum metal with organic radical anions from metal to negative oxidation states, *J. Am. Chem. Soc.*, 2007, **129**, 6654–6661.
130. Y. Li and R. T. Yang, Hydrogen storage on platinum nanoparticles doped on superactivated carbon, *J. Phys. Chem. C*, 2007, **111**, 11086–11094.
131. J. Liu, J. Sutton and C. B. Roberts, Synthesis and extraction of monodisperse sodium carboxymethylcellulose-stabilized platinum nanoparticles for the self-assembly of ordered arrays, *J. Phys. Chem. C.*, 2007, **111**, 11566–11576.
132. A. Fukuoka, J.-I. Kimura, T. Oshio, Y. Sakamoto and M. Ichikawa, Preferential oxidation of carbon monoxide catalyzed by platinum nanoparticles in mesoporous silica, *J. Am. Chem. Soc.*, 2007, **129**, 10120–10125.
133. P. J. Chupas, K. W. Chapman, G. Jennings, P. L. Lee and C. P. Grey, Watching nanoparticles grow: the mechanism and kinetics for the formation of TiO_2-supported platinum nanoparticles, *J. Am. Chem. Soc.*, 2007, **129**, 13822–13824.
134. J. Kua and W. A. Goddard, Chemisorption of organics on platinum. 1. The interstitial electron model, *J. Phys. Chem. B*, 1998, **102**, 9481–9491.
135. B. V. Bergeron, K. C. White, J. L. Boehme, A. H. Gelb and P. B. Joshi, Variable absorbance and emittance devices for thermal control, *J. Phys. Chem. C*, 2008, **112**, 832–838.

136. D. B. Pedersen and S. Wang, Surface plasmon resonance spectra of 2.8±0.5 nm Cu NPs in both near and far fields, *J. Phys. Chem. C*, 2007, **111**, 17493–17499.
137. G. Sun, J. Xu, A. Hagooly, R. Rossin, Z. Li, D. A. Moore, C. J. Hawker, M. J. Welch and K. L. Wooley, Strategies for optimized radiolabeling of nanoparticles for in vivo PET imaging, *Adv. Mater.*, 2007, **19**, 3157–3162.
138. V. Krishnan, R. Kalai Selvan, C. O. Augustin, A. Gedanken and H. Bertagnolli, EXAFS and XANES investigations of $CuFe_2O_4$ nanoparticles and $CuFe_2O_4$-MO_2 (M) Sn,Ce) nanocomposites, *J. Phys. Chem. C*, 2007, **111**, 16724–16733.
139. Y. Wu, C. Wadia, W. Ma, B. Sadtler and A. P. Alivisatos, Synthesis and photovoltaic application of copper(I) sulfide nanocrystals, *Nano Lett.*, 2008, **8**, 2551–2555.
140. K. Huber, T. Witte, J. Hollmann and S. Keuker-Baumann, Controlled formation of Ag nanoparticles by means of long-chain sodium polyacrylates in dilute solution, *J. Am. Chem. Soc.*, 2007, **129**, 1089–1094.
141. K. Aslan, M. Wu, J. R. Lakowicz and C. D. Geddes, Fluorescent core-shell Ag@Si nanocomposites for metal-enhanced fluorescence and single nanoparticle sensing platforms, *J. Am. Chem. Soc.*, 2007, **129**, 1524–1525.
142. J. Sun, D. Ma, H. Zhang, X. Liu, X. Han, X. Bao, G. Weinberg, N. Pfander and D. Su, Toward monodispersed silver nanoparticles with unusual thermal stability, *J. Am. Chem. Soc.*, 2006, **128**, 15756–15764.
143. X. M. Sun and Y. D. Li, Cylindrical silver nanowires: preparation, structure, and optical properties, *Adv. Mater.*, 2005, **17**, 2626–2630.
144. X. Wang, Q. Peng and Y. Li, Interface-mediated growth of monodispersed nanostructures, *Acc. Chem. Res.*, 2007, **40**, 635–643.
145. B. Wu, A. Heidelberg, J. J. Boland, J. E. Sade, X. M. Sun and Y. D. Li, Microstructure-hardened silver nanowires, *Nano Lett.*, 2006, **6**, 468–472.
146. J. Zheng, J. Ding, B. Tian, Z. Lin Wang and X. Zhuang, Luminescent and Raman active silver nanoparticles with polycrystalline structure, *J. Am. Chem. Soc.*. 2008, **130**, 10472–10473.
147. G. Braun, I. Pavel, A. R. Morrill, D. S. Seferos, G. C. Bazan, N. O. Reich and M. Moskovits, Chemically patterned microspheres for controlled nanoparticle assembly in the construction of SERS hot spots, *J. Am. Chem. Soc.*, 2007, **129**, 7760–7761.
148. S. J. Lee, J. M. Baik and M. Moskovits, Polarization-dependent surface-enhanced raman scattering from a silver-nanoparticle-decorated single silver nanowire, *Nano Lett.*, 2008, **8**, 3244–3247.
149. Q. Hu, L. L. Tay, M. Noestheden and J. P. Pezacki, Mammalian cell surface imaging with nitrile-functionalized nanoprobes: biophysical characterization of aggregation and polarization anisotropy in SERS imaging, *J. Am. Chem. Soc.*, 2007, **129**, 14–15.
150. H. Yamaguchi and K. Matsuda, Masahiro, Excited-state behavior of a fluorescent and photochromic diarylethene on silver nanoparticles, *J. Phys. Chem. C*, 2007, **111**, 3853–3862.

151. A. Dawn, P. Mukherjee and A. K. Nandi, Preparation of size-controlled, highly populated, stable, and nearly monodispersed Ag nanoparticles in an organic medium from a simple interfacial redox process using a conducting polymer, *Langmuir*, 2007, **23**, 5231–5237.
152. D. Radziuk, D. G. Shchukin, A. Skirtach, H. Mohwald and G. Sukhorukov, Synthesis of silver nanoparticles for remote opening of polyelectrolyte microcapsules, *Langmuir*, 2007, **23**, 4612–4617.
153. W. Vogel, B. Rosner and B. Tesche, Structural investigations of Au55 organometallic complexes by X-ray powder diffraction and transmission electron microscopy, *J. Phys. Chem.*, 1993, **97**, 11611–11616.
154. D. H. Rapoport, W. Vogel, H. Colfen and R. Schlögl, Ligand-Stabilized metal clusters: reinvestigation of the structure of "$Au_{55}[P(C_6H_5)_3]_{12}Cl_6$", *J. Phys. Chem. B*, 1997, **101**, 4175–4183.
155. S. Hoeppener, L. Chi and H. Fuchs, Formation of Au_{55} strands on a molecular template at the solid-liquid interface, *Nano Lett.*, 2002, **2**, 459–463.
156. P. D. Jadzinsky, G. Calero, C. J. Ackerson, D. A. Bushnell and R. D. Kornberg, Structure of a Thiol monolayer-protected gold nanoparticle at 1.1 Å resolution. *Science*, 2007, **318**, 430–433.
157. Y. Li, G. Galli and F. Gygi, Electronic structure of thiolate-covered gold nanoparticles: $Au_{102}(MBA)_{44}$, *ACS Nano*, 2008, **2**, 1896–1902.
158. G. Li, M. Lauer, A. Schulz, C. Boettcher, F. Li and J.-H. Fuhrhop, Spherical and planar gold(0) nanoparticles with a rigid gold(I)-anion or a fluid gold(0)-acetone surface, *Langmuir*, 2003, **19**, 6483–6491.
159. O. C. Compton and F. E. Osterloh, Evolution of size and shape in the colloidal crystallization of gold nanoparticles, *J. Am. Chem. Soc.*, 2007, **129**, 7793–7798.
160. J. B. Tracy, G. Kalyuzhny, M. C. Crowe, R. Balasubramanian, J.-P. Choi and R. W. Murray, Poly(ethylene glycol) ligands for high-resolution nanoparticle mass spectrometry, *J. Am. Chem. Soc.*, 2007, **129**, 6706–6707.
161. J. Chen, J. M. McLellan, A. Siekkinen, Y. Xiong, Z.-Y. Li and Y. Xia, Facile synthesis of gold-silver nanocages with controllable pores on the surface, *J. Am. Chem. Soc.*, 2006, **128**, 14776–14777.
162. W. Fudickar, J. Zimmermann, L. Ruhlmann, J. Schneider, B. Roder, U. Siggel and J.-H. Fuhrhop, Fluorescence quenching and size selective heterodimerization of a porphyrin adsorbed to gold and embedded in rigid membrane gaps, *J. Am. Chem. Soc.*, 1999, **121**, 9539–9545.
163. D. S. Seferos, D. A. Giljohann, H. D. Hill, A. E. Prigodich and C. A. Mirkin, Nano-flares: probes for transfection and mRNA detection in living cells, *J. Am. Chem. Soc.*, 2007, **129**, 15477–15479.
164. J. S. Garitaonandia, M. Insausti, E. Goikolea, M. Suzuki, J. D. Cashion, N. Kawamura, H. Ohsawa, I. Gil de Muro, K. Suzuki, F. Plazaola and T. Rojo, Chemically Induced Permanent Magnetism in Au, Ag, and Cu nanoparticles: localization of the magnetism by element selective techniques, *Nano Lett.*, 2008, **8**, 661–667.

165. R. L. Stoermer, J. A. Sioss and C. D. Keating, Stabilization of silver metal in citrate buffer: barcoded nanowires and their bioconjugates, *Chem. Mater.*, 2005, **17**, 4356–4361.
166. (a) A. M. Kalsin, B. Kowalczyk, S. K. Smoukov, R. Klajn and B. A. Grzybowski, Ionic-like behavior of oppositely charged nanoparticles, *J. Am. Chem. Soc.*, 2006, **128**, 15046–15047; (b) A. M. Kalsin, B. Kowalczyk, P. Wesson, M. Paszewski and B. A. Grzybowski, Studying the thermodynamics of surface reactions on nanoparticles by electrostatic titrations, *J. Am. Chem. Soc.*, 2007, **129**, 6665–6666.
167. Y. Fillon, A. Verma, P. Ghosh, D. Ernenwein, V. M. Rotello and J. Chmielewski, Peptide ligation catalyzed by functionalized gold nanoparticles, *J. Am. Chem. Soc..*, 2007, **129**, 6676–6677.
168. S. I. Stoeva, F. Huo, J.-S. Lee and C. A. Mirkin, Three-layer composite magnetic nanoparticle probes for DNA, *J. Am. Chem. Soc.*, 2005, **127**, 15362–15363.
169. M. Schrinner, S. Proch, Y. Mei, R. Kempe, N. Miyajima and M. Ballauff, Stable bimetallic gold–platinum nanoparticles immobilized on spherical polyelectrolyte brushes: synthesis, characterization, and application for the oxidation of alcohols, *Adv. Mater.*, 2008, **20**, 1928–1933.
170. H. Bayraktar, C.-C. You, V. M. Rotello and M. J. Knapp, Facial control of nanoparticle binding to cytochrome c, *J. Am. Chem. Soc.*, 2007, **129**, 2732–2733.
171. H. D. Koh, N. G. Kang and J.-S. Lee, Location control of Au/CdS nanoparticles in block copolymer micelles, *Langmuir*, 2007, **23**, 11425–11429.
172. A. J. Rossini and R. W. Schurko, Experimental and theoretical studies of ^{45}Sc NMR interactions in solids, *J. Am. Chem. Soc.*, 2006, **128**, 10391–10402.
173. M. Yada, M. Ohya, K. Ohe, M. Machida and T. Kijima, Porous yttrium aluminum oxide templated by alkyl sulfate assemblies, *Langmuir*, 2000, **16**, 1535–1541.
174. F. Wang, X. Xue and Xi. Liu, Multicolour tuning of (Ln, P)-doped YVO$_4$ nanoparticles by single-wavelength excitation, *Angew. Chem.*, 2008, **120**, 920–923.
175. W. DeW. Horrocks Jr. and D. R. Sudnick, Lanthanide ion luminescence probes of the structure of biological macromolecules, *Acc. Chem. Res.*, 1981, **14**, 384–392.
176. D. Casanova, D. Giaume, M. Moreau, J. L. Martin, T. Gacoin, J. P. Boilot and A. Alexandrou, Counting the number of proteins coupled to single nanoparticles, *J. Am. Chem. Soc.*, 2007, **129**, 12592–12593.
177. A. Y. H. Lo, V. Sudarsan, S. Sivakumar, F. van Veggel and R. W. Schurko, Multinuclear solid-state NMR spectroscopy of doped lanthanum fluoride nanoparticles, *J. Am. Chem.. Soc.*, 2007, **129**, 4687–4700.
178. M. Aloshyna, S. Sivakumar, M. Venkataramanan, A. G. Brolo and F. C. J. M. van Veggel, Significant suppression of spontaneous emission in Si photonic crystals made with Tb^{3+}-doped LaF$_3$ nanoparticles, *J. Phys. Chem. C*, 2007, **111**, 4047–4051.

179. B. Murugan and A. V. Ramaswamy, Defect-site promoted surface reorganization in nanocrystalline ceria for the low-temperature activation of ethylbenzene, *J. Am. Chem. Soc.*, 2007, **129**, 3062–3063.
180. L. Qi, J.-P. Chapel, J.-C. Castaign, J. Fresnais and J.-F. Berret, Stability and adsorption properties of electrostatic complexes: design of hybrid nanostructures for coating applications, *Langmuir*, 2007, **23**, 11996–11998.
181. S. Patil, A. Sandberg, E. Heckert, W. Self and S. Seal, Protein adsorption and cellular uptake of cerium oxide nanoparticles as a function of zeta potential, *Biomaterials*, 2007, **28**, 4600–4607.
182. J. Schoenes, Magneto-optical properties of metals, alloys and compounds, in *Materials Science and Technology*, ed. R. W. Cahn, P. Haasen and E. J. Kramer, Wiley-VCH, Weinheim, 2005, Vol. 3a, pp. 147–256.
183. K. H. J. Buschow, Permanent magnet materials, in *Materials Science and Technology*, ed. R. W. Cahn, P. Haasen and E. J. Kramer, Wiley-VCH, Weinheim, 2005, Vol. **38**, 452–528.
184. M. Tan, Z. Ye, G. Wang and J. Yuan, DEG diethylene glycol preparation and time-resolved fluorometric application of luminescent europium nanoparticles, *Chem. Mater.* 2004, **16**, 2494–2498.
185. J. C. Boyer, J. Gagnon, L. Cuccia and J. S. Capobianco, Synthesis, characterization, and spectroscopy of $NaGdF_4$: $Ce^{3+},Tb^{3+}/NaYF_4$ core/shell nanoparticles, *Chem. Mater.*, 2007, **19**, 3358–3360.
186. J.-L. Bridot, A. C. Faure, S. Laurent, C. Riviere, C. Billotey, B. Hiba, M. Janier, V. Josserand, J. L. Coll, R. van der Elst, R. Muller, S. Roux, P. Perriat and O. Tillement, Hybrid gadolinium oxide nanoparticles: multimodal contrast agents for in vivo imaging, *J. Am. Chem. Soc.*, 2007, **129**, 5076–5084.
187. M. Aloshyna, S. Sivakumar, M. Venkataramanan, A. G. Brolo and F. C. J. M. van Veggel, Significant suppression of spontaneous emission in Si photonic crystals made with Tb^{3+}-doped LaF_3 nanoparticles, *J. Phys. Chem. C*, 2007, **111**, 4047–4051.
188. H. N. Barnhill, S. Claudel-Gillet, R. Ziessel, L. J. Charbonniere and Q. Wang, Prototype protein assembly as scaffold for time-resolved fluoroimmuno assays, *J. Am. Chem. Soc.*, 2007, **129**, 7799–7806.
189. H. N. Barnhill, R. Reuther, P. L. Ferguson, T. Dreher and Q. Wang, Turnip yellow mosaic virus as a chemoaddressable bionanoparticle, *Biocon. Chem.*, 2007, **18**, 85228–85229.

Subject Index

acetal cross-linking 94
acicular–crystals (needles) 139, 165, 176, 250, 269
actuator 23
adhesion carbon tubes 63
AFM (atomic force microscopy) tips 64 ff
aldehydes 94
AlNiCo magnets 29 ff
alumina (aluminium oxide) 181 ff
alumina layers 13
alumina porous sheets 182 ff
alumina–cellulose 136
aluminium 162, 181 ff
aluminium enrichment 263
aluminium nitride 207
aluminogermanate tubes 197
aluminosilicate MCM 41 89, 365 ff
Alzheimer 181 ff
amide bonds 10 ff
aminated silica particles 90 ff, 181, 194
amines 88
amines–molybdenum trioxide 330
amines–zirconate layers 319
aminophosphonate 228
ammonium–carboxylate gold particles 379
ammonium binding to cells 260
anatase (titanium dioxide) 301 ff, 304 ff
anion–potassium–sodium 167
anodization 183
antiferromagnetism 26
antimony 213
antimony telluride 199
arsenic 212 ff

arsenides 162
atomic force microscopy (AFM) 5, 14, 64, 205, 285, 341, 371
atomically flat mirrors 200
atomically flat silica particles 10, 81, 194
atomically flat surface layers 13 ff
atomically flat titania 311
atom-transfer radical polymerization (ATRP) 105 ff
ATRP (atom-transfer radical polymerization) 105 ff
austenite (a steel) 248 ff

ballistic electron transport 60 ff, 68 ff
band gap 39 ff, 204, 207, 223
barcode rods 19, 144, 146, 183, 359, 377 ff
barcode rod–DNA 146
barcode rod cobalt–platinum 359
barcode rods gold–silver 377 ff
barium 176 ff
barium titanate 176 ff, 301
biotin 388
bipolarons 22, 77, 124 ff
bismuth 213 ff
bismuth telluride 218
bits 25
block copolymers 105–118
Bohr radius 39
bonding anisotropy 185
bones 174 ff, 220 ff
borates 179
boron 162, 179 ff
boron carbide 180 ff

boron conjugated polymer 179
boron doping 189
boron ionization potential 179
boron isotopes 180
boron luminescence 179
boron neutron capture 180
boron nitride 202 ff
boron nitride aggregates 207
boron nitride tubes 206 ff
brookite (titanium dioxide) 301 ff
bungarotoxin (protein) lanthanides 141 ff
butterfly wings 176
byte 31 ff

cadmium 221 ff, 225 ff
cadmium chalconides 39 ff
cadmium particle–DNA mismatch 145 ff
cadmium selenide 40 ff, 215 ff, 228, 314
cadmium stearate layer 316
cadmium sulfide 215, 226
cadmium telluride 217 ff, 229 ff
cages 14, 266, 374
calcium 174 ff
calcium carbonate 174 ff
calcium silicate 245
calixarene 14 ff
calixarene–dodecylsulfate 131 ff
cancer treatment 180 ff
carbon arc synthesis 73
carbon monoxide–platinum 358 ff, 361
carbon monoxide 318
carbon monoxide oxidation 44 ff
carbon nitride 207 ff
carbon rods 302
carbon tube adhesion 63
carbon tube chirality 64
carbon tube electric motor 72
carbon tube solubility 63
carbon tubes 22, 53, 56, 60, 63 ff, 69, 71, 75
carbon tubes AFM tips 64, 65
carbon tubes doping 66 ff
carbon tubes fluorination 63
carbon tubes from ferritin 69 ff
carbon tubes fullerenes 65
carbon tubes in alumina pores 71
carbon tubes photoluminescence 66 ff

carbon tubes porphyrin adsorption 64
carbon tubes Raman spectra 68
carbon tubes soluble 75
carbon–iron comparison 291 ff
carbonyl sodium 169 ff
carboxylate–metal ion 95
catalysis 43 ff
catalysis chromium 329
catalysis iron 289 ff
catalysis molybdene sulfide 333
catalysis nickel 355
catalysis platinum 357 ff
cavitands 14 ff
cellobiose 13
cellobiose–maltose adsorption 11
cellulose 136 ff, 285 ff
cellulose acetate particles 138
cellulose–platinum 360
cellulose–titania 136
cellulose–alumina 136
cellulose–iron oxide particles 138 ff
cellulose–polystyrene 136 ff
cellulose–silver particles 136 ff
cementite (iron carbide) 247 ff
ceria (cerium dioxide) electron
    paramagnetic resonance 385
ceria particle zeta potential 383
ceria–proteins 385
cerium 385
cerium ignition metal particles 385
channels 190 ff
chemical vapor deposition (CVD) 84 ff
chirality 64
chlorine *vs* oxygen oxidation 26
chlorocarbon Fe(0) reduction 261 ff
chlorosilanes 195
chromium 328 ff
chromium dioxide 328 ff
chromium dioxide magnets 29 ff, 330
chromium trioxide 328 ff
circular dichroism 78 ff (polyenes),
    90 (chiral soaps)
citrate gold 91, 372 ff
clay (aluminium silicate) 13, 192 ff
click coupling 68
clusters iridium 353

Subject Index

clusters nickel 354
cobalt 347–351
cobalt catalysts 351
cobalt ferrite 285
cobalt ferromagnetism 26, 346
cobalt(0) particle rows 350
cobalt(0) particles 348 ff
coercivity 19
collagen 174 ff
compact disk burning 36 ff
computer memories and switches 31 ff
condenser 316
conducting AFM 332
conducting bands 34
conducting polyaniline 123 ff
conducting polymers 122–130
conducting polymers doping 125
conductivity 29, 60 ff, 68 ff, 122, 306
copolymer–glucose ester 136
copolymer micelles 108 ff
copolymers 105–118
copolymers surface attached 107
copper 362–364
copper ferrites 363
copper particles 363 ff
copper positron emission 107 ff
corundum (red aluminate) 181
cost main group elements 161
cost transition metals 300
Coulomb blockade 62 ff
covalent carbon compounds 53
crystal planes 6 ff
cubes iron oxide 280
cubes rhodium 351
Curie temperature 27, 34, 250
CVD (chemical vapor deposition) 84 ff
cyclohexadiene–platinum 359
cyclooctatetraene 302 ff
cytochrome P 450 271 ff

dangling bonds 55, 185
decaborane–ruthenium particle 343
dendrimer chains 121
dendrimeric polyglycerol 121 ff
dendrimers 119 ff, 176
dendrimers polyene 79 ff

density functional theory (DFT) 44, 46 ff, 325, 333 ff
detonation diamonds 54
DFT (density functional theory) 44, 46, ff
diacetylene cross-linking 80
diagnostic iron oxide particles 278 ff
diamonds 53 ff
diesel soot combustion 385
dimethylmercury 232
DNA 227
DNA beacon fluorescence 143 ff
DNA synthesis on gold particles 143
DNA–barcodes 378
DNA–gold silver barcode wire 146
DNA–lanthanum 385
DNA–mismatch cadmium particle 145 ff
dodecylsulfate 133
dopamine 13, 282 ff
DVD/CD polyenes 78 ff

easy axis magnets 336
electrets 110
electric conductivity 21 ff
electric motor 72
electrical conductivity carbon tubes 59 ff
electroluminescence 37 ff
electroluminescence polyenes 80 ff
electron confinement 41
electron micrographs 36, 61, 65, 70 ff, 85, 89, 97, 108, 112, 128, 138 ff, 141, 146, 303, 309, 373 ff
electron paramagnetic resonance (EPR) 305 (titania), 322 ff (VO), 385 (Ce)
electron spin 24
electron spin echo 324
electronegativity 44
electronic hardness 44
electronic spectra carbon tubes 59
electrophoresis titania 313
electroplating 20 ff
energy levels carbon tubes 62 ff
enzyme entrapment 319
europium 387

fatty acid curds 89 ff, 95
fatty acid fibre formation 96

Fermi energy 27 ff
Fermi level 34, 124 ff
Fermi sphere 27 ff
ferritin 69, 79 ff
ferroelectric 301
ferromagnetism 26, 28 ff
fluorescence 35, 143, 187, 304, 344, 346, 365, 388
fluorescence quenching 12 (porphyrins yoctowells), 42 (Ag)
fluorinated alcohols 146
fluorination carbon tubes 63
fluorine 220 ff
fluormethine layers 147
fluoroapatite–gelatin 177
fluorocarbon hydrocarbon chains 147
fluorographenes 147
f-orbital magnetism 25
force field simulation 169
formic acid decomposition 43 ff
Förster (or fluorescence) resonance energy transfer (FRET) 388
four point conductivity probe 126 ff
fractal 176
FRET (Förster or fluorescence resonance energy transfer) 388
friction 119, 175
fuel cell rhodium 352, 357, 359
fullerene 65 ff
fulvalene (bicyclic tetrasulfide) 131

gadolinium 387
gallium 184 ff
gallium arsenide 39, 162, 184, 212
gallium arsenide wire 186
gallium nitride 38, 208 ff
gallium selenide 184 ff, 216
gallium selenide wires 185
gas chromatography 356
gas separation 255
germania (germanium oxide) 198
germanium 162, 196 ff
germanium crystals 196 ff
germanium nitride 208
germanium octamer 198
germanium telluride 199, 219

germanium telluride wires 199 ff
giant magnetoresistance 32 ff
goethite (iron(III) oxide) 8, 263
gold 371–381
gold cluster 5, 371 ff
gold in silica pores 20
gold nanoparticles 185
gold particles 371 ff
gold particles–single strand DNAs 145 ff
gold particles–cytochrome c 380
gold plasmons 41
gold rods 377 ff
gold sulfides 372 ff
gold–lipoate 376
gold–cadmium sulfide 380 ff
gold–DNA 143
gold–platinum particles 380
gold–porphyrin 377
gold–silver barcode wire–DNA 146
gold–silver cages 374
grafting 107, 119
gramicidine 169 ff
graphene chemical vapor deposition (CVD) 84 ff
graphene hydrazine addition 84
graphene lithium intercalation 85
graphene oxidation 83 ff
graphene sheets 53 ff, 56 ff
graphenes fluorinated 147
graphite 53 ff, 56
graphite electrode 165
green rust 257 ff
greigite (iron sulfide) 286
guanine 226 ff

Haber–Bosch ammonia production 289 ff
haematite (iron (III) oxide) 12, 245, 269 ff
haematite–gold 370 ff
haematite particle magnetism 34
haemoglobin 271 ff
hafnium 321
hard disk $(CH)_x$ and $(CN)_x$ surfaces 73 ff
hard disk 13, 31, 73 ff, 269
hard magnets 250 ff, 269
hausmannite (manganese oxide) 335, 339
head slider 23

Subject Index

heat conductivity 35 ff
heat treatment 35 ff
helium reflection 200
histidine 228
hole diffusion 259
human civilization 246
Hume–Rothery phase 6 ff
hyaline cartilage 176
hydrazine 84
hydrocarbons 72 ff
hydrodesulfurization 214 ff
hydrogen 162 ff
hydrogen isotope separation 205 ff
hydrogen sensor platinum 358
hydrogen sensors palladium 356 ff
hydrogen storage platinum 163 ff, 360
hydrogenation palladium catalysis 355
hydrophobic effect 98
hydrosulfides–gold 91
hydrothermal synthesis 191
hydroxyapatite 174 ff
hyperfine sublevel correlation spectroscopy (HYSCORE) 323 ff
HYSCORE (hyperfine sublevel correlation spectroscopy) 323 ff

ignition metal particles cerium 385
indium 22, 186 ff
indium arsenide 212 ff
indium phosphide 39, 211
infrared spectra 23 (titania),125 ff (polyaniline), 188 (ITO), 218 (GdTe), 354 (Ni–CO)
inter–intralayer bonding 185
inverse micelles 99 ff
ion conductance 22, 166 ff
ionic carbides 55
iridium 353
iridium clusters 353
iridium–nitrogen monoxide 353
iron 7, 245 ff
iron carbide 247 ff
iron carbonyl 270 ff
iron oxidation 255 ff
iron oxide–carbon aerogel 274 ff
iron oxide catalysis 289 ff

iron oxide particles–cellulose 138 ff
iron oxides 8 ff, 245, 269 ff
iron phosphonate coating 255
iron polyphosphate coating 254 ff
iron sulfide sheets 286
iron water addition 255 ff
iron wires 19, 266 ff
iron–silicon ceramic 267
iron(III) hopping 258 ff
ITO (indium tin oxide) 162, 186 ff
ITO reactivity 187 ff

Janus effect 9 ff
Janus particles 195

Kapton (block copolymer) 108 ff

Langmuir–Blodgett monolayers 10
lanthanide–protein 141 ff
lanthanide light emitters 38
lanthanides 381 ff, 384 ff
lanthanum 384
lanthanum trifluoride particles opals 385
lanthanum–DNA 385
lanthanum–proteins 384, 385
laponite (magnesium silicate) 192
laser irradiation 199
laser vaporization 247
laser-burnt grooves 13, 36, 40
layers 34
layers of silicon 189 ff
lead 200 ff
lead selenide 163, 217, 315
lead sulfide 215, 314 ff
lead telluride 219
LEDs (light emitting diodes) 37 ff
light emitting diodes (LEDs) 37 ff
lime (calcium oxide) light 37, 175
limestone (calcium carbonate) 245
limonite (iron(III) oxide) 245
lipoate–gold 91 ff, 376
lithium 161
lithium battery 165
lithium ferrite–cellulose 285 ff
lithium intercalation 85
lithium ion conductivity 303

lithium manganate particles 340
lithium nitride 164
lithium stearate 164 ff
living polymerization 118 ff
long chain amines 223
long chain carboxylic acids 89 ff, 96 ff, 252
long chain phosphines 223
long chain tartaric amide fibres 139 ff
lubricants 253
luminescence 37 ff, 160, 189, 382, 384 ff
luminescence lanthanide 382, 384 ff
luminescent silicon 189

magic thicknesses 13
magnesium 174
magnetic iron–cobalt–nickel rods 268
magnetic moment 24 ff
magnetism 24 ff
magnetism and reactivity 25
magnetism chromium dioxide 330
magnetism gold 377
magnetism iridium clusters 353
magnetism manganese 336 ff
magnetism neodyme–iron–boron (NIB) 382
magnetism neodyme 386
magnetism nickel 353 ff
magnetism samarium–cobalt 386 ff
magnetite Janus particles 276 ff
magnetite (iron (II/III) oxide) 8, 142, 245, 269 ff, 274 ff, 276 ff
magnetite particles–proteins 142
magnetization 24
magnetization indium 186
main group elements 160
manganese–gallium magnet 338
manganese–nickel magnets 338
manganese trioxim magnet 337
manganese–zinc sulfide 341
manganese(III) hydroxy oxide particles 339
manganese 335–341
manganese dioxide particles 339
manganese ferrite 32
manganese ferrite catalysts 340

manganese gallate 34, 337 ff
manganese metal–air batteries 339 ff
manganese molecular magnets 335 ff
manganese oxide–neuronal cells 341
manganese particle redox reactions helium 338
manganese production 335
manganese(III) oxide Raman spectra 339
mass spectrum gold particles 374
mercury 163, 221 ff, 231
mercury electrosensibilization 232
metal clusters 32
metal hexamers 343
metalloporphyrin ordering 134 ff
metal-organic frameworks (MOFs) 103 ff
methacrylate coating 263
micelles 96 ff
micelles copolymer 108
microwave treatment 301
mineralization 175
Mn(III)–porphyrin quenching 12
MOFs (metal–organic frameworks) 103 ff
molecular ordering in yoctowells 134 ff
molybdene–niobium sulfide tubes 332
molybdene 329 ff
molybdene desulfonation catalysis 331
molybdene oxidation catalysis 331 ff
molybdene phosphide 212
molybdene sulfide on steel 253
molybdene sulfide layers 332 ff
molybdene trioxide–amines 330
molybdene trioxide–polyethyleneglycol 331
molybdene trioxide disks 330
molybdene trioxide platelets 331
molybdene(V)–nickel(II) 331
molybdic acid–amine 330
monolayer vesicles 101 ff
Moessbauer spectra 285

n-doping 189
nanometer smoothness 191
nanometre 1 ff
nanoparticle generalities 17 ff
nanowells silicon 189
negative-differential resistor 273 ff

Subject Index

neodyme–iron–boron (NIB) magnets 382
neodyme 383
neodyme magnets 386
neodyme–YAG (yttrium aluminium granate) laser 383
neodymium 386
neuronal cells–manganese oxide 341
neurotransmitter–fatty acid fibers 89 ff
neutron capture 180
neutron diffraction 285
nickel 353 ff
nickel catalysis 355
nickel clusters 354
nickel magnetism 353 ff
nickel phosphide 210
nickel(II) reduction by iron, 270
niobium–titania 310
niobium 325
nitrides 162, 202 ff
nitrogen doping titania 312, 315
nitrogen monoxide–iridium 353
nitrogen oxidation tungsten catalysis 334
nitrogenase 329, 333
nitroxide fragmentation 117 ff
NMR aminophosphonates 178
NMR aminophosphonates in bones 178
NMR bones 179
NMR docosahexadienoic acid 97
NMR lanthanides 384 ff
NMR polyenes 77
NMR scandium 382
NMR T2 contrast magnetite 277

Oersted 29, 34
opals 176, 187, 385
organic carbon 52 ff
oscillating catalysis 44 ff, 361 ff
oscillating flocculation 347 ff
oscillating transparency 348
osmium 346
osmium rectifier 346
oxygen–platinum 358
oxygen defects 318
oxygen determination 303 ff
oxygen molecules 26

p-doping 189
p–n junction 22 ff
palladium–silicate particles 356
palladium 355 ff
palladium hydrogen sensors 356 ff
palladium hydrogenation catalysis 355
palladium sulfide particles 356
palladium tetrachlorinates 355
paramagnetism 25 ff
particle appearance 18 ff
pearlite (a steel) 248 ff
permanent magnetism 27 ff, 34, 250
peroxotitanium 301 ff
phenalenyl radical 179
phenol reactions with iron oxides 259 ff
phosphate 22, 100 ff, 166 ff
phosphides 162 ff, 209 ff
phosphonium head groups 99 ff
phosphoric acid 183
phosphoric *vs* oxalic acid 183
phosphorous doping 189
photocatalysis 307 ff, 312, 319 ff, 326
photochemical disinfection 310
photocopier 215, 217
photoluminescence 59, 223 ff, 311, 387
photonic crystals 176
photooxidation 221, 308 ff
piezoelectric materials 23, 205, 326
plasma deposition 254
plasmon absorption 41, 91 ff, 136, 375 ff
platinide grains 360
platinum 357–361
platinum catalysis 44 ff, 357 ff
platinum hydrogen sensor 358
platinum hydrogen storage 360
platinum particles 358
platinum–carbon monoxide 358 ff, 361
platinum–cellulose 360
platinum–cyclohexadiene 359
platinum–oxygen 358
platinum–silicate 360 ff
poisons 201
polarons 22, 76 ff, 125
polymethyl methacrylate tubes 184
poly-N-isopropylacrylamide 117
polyacetylene 76 ff, 122

polyacrylic acid 112 ff
polyallylamine–polystyrenecarboxylate 330
polyaniline 22, 123 ff, 187 ff
polyaniline–silver 369 ff
polyaniline–Prussian blue 130
polyaniline X-ray photoelectron spectrum (XPS) 126 ff
polyaniline–horse radish peroxidase 129
polycarbonate 20
polyene dendrimers 79 ff
polyene electric conductivity 75 ff
polyenes 21, 74 ff, 78 ff
polyenes electroluminescence 80 ff
polyethyleneglycol–protein repulsion 86, 187
polyethyleneglycol 174, 228
polyethyleneimine–protein 89
poly-N-vinyl-caprolactam 115 ff
polyphenylene 122
polypyrrole 122
polyradical paramagnetism 26
polystyrene–cellulose 136 ff
polystyrene sulfonate–silver 369 ff
polythiophene 122 ff
polyvinylpyrrolidone 112 ff, 115, 309, 352, 381
poreous polycarbonate 20
pores 19
pores silicate fibres 19, (see also zeolites)
pores voltage gated 102 ff
pores zeolite 195 ff
porous alumina sheets 19 ff, 182 ff
porous aluminosilicate MCM 41 89, 365 ff
porous charcoal 59
porous Fe(0) 261
porous nickel–zirconium oxide silicate 354
porous steel 255
porous titania 302, 316
porphyrin–gold 377
positron emitter copper 363
potassium 162, 166 ff
potassium molybdate melt 306
potassium pore 169 ff
potassium transport 14

prebiotic synthesis 72 ff
price silicon 190
price transition metals 300
pristine 254
protein–magnetite particle 142
protein rods and wires 140 ff
proteins–ceria 385
proteins–lanthanum 385
pulsed laser 35

quantum confinement 189
quantum dots 39, 99, 190, 211 ff, 225 ff

Raman spectra 68 (C tubes), 339 (Mn), 367 ff (Ag)
Rayleigh instabilities 184
REDOR (Rotational Echo Double Resonance) 176
refractive index 176, 197, 326
resistance carbon tubes 61 ff, 68 ff
rewritable laser disks 198, 219
rhenium complex–copolymer rods 341
rhenium disulfide onions 342
rhodium 351 ff
rhodium cubes 351
rhodium fuell cells 352
rhodochrosite (manganese carbonate) 335
RNA 388
rods 20, 265 ff
room air cleaning 319
rust characterization 257
ruthenium 342–345
ruthenium bipyridyl onions 344
ruthenium bipyridyl 344 ff
ruthenium fluorescence 344
ruthenium–carbon 342
ruthenium–selenium 343 ff
ruthenium particle–decaborane 343
ruthenium tetroxide oxidation 267 ff, 342
rutile (titanium dioxide) 301 ff, 304 ff

Sabatiers volcano curve 43 ff
samarium 386 ff
samarium–cobalt magnets 386 ff
sapphire (blue aluminate) 181
saturated magnetization 27

Subject Index

scandium 382
scandium nitride 208
selenides 162
selenium 215 ff
semiconducting silicon 189 ff
semiconducting titania 301
semiconductor wires 217 ff
semiconductors 22 ff, 160
semiconductors serine–phosphate 102
short-range forces 169
silica entrapped polyethylene 194
silica sols 193
silicate surface 10, 13 ff, 81, 194
silicon 162, 189 ff
silicon dioxide (quartz) 190 ff
silicon layers 185
silicon monoxide 190 ff
silicon nitride 208 ff
silicon semiconductors 22
silver 365–370
silver fluorescence 365, 367
silver germanium 197
silver layer 13
silver metal fluorescence 42 ff
silver particle laser membrane destruction 370
silver particles 197, 307, 365
silver particles–cellulose 136 ff
silver selenide 215 ff
silver–polyaniline 369 ff
silver–carbon cables 366 ff
silver–sulfide particles 368 ff
single molecule fluorescence 365
single strand DNAs–gold particle 145 ff
slow diffusion 11
smart window molybdenates 329 ff
sodium 162, 166 ff
sodium NMR 169 ff
sodium pore 173
sodium potassium phosphate 168 ff
sodium sulfide electrolyte 315
soft 34, 250 ff
soft magnets 250 ff, 269
soil and iron 260
solar cells copper(I) sulfide–cadmium sulfide 363

solar energy conversion 313 ff
solitons 22, 75 ff
sphere–cube conversions 264 ff
spin coated 254
spin configuration boron nitride 204
spin splitting 27 ff
spontaneous spin splitting 28 ff
SQUID (superconducting quantum interference device) 326
stable polymeric monolayers 120 ff
steel 246 ff
steel domains 249
steel formation 246 ff
steel grains 249
steel scratches and carbon tubes 251 ff
Stoner criterion 28
strontium random access memory 33
sulfides–zinc oxide 92
sulfides 162
sulfonate head groups 99 ff
sulfur 214 ff
superparamagnetic behavior 34
surface attached copolymers 107
surface plasmon resonance 186

tandem cell 309 ff
tantalum 325 ff
tantalum dimples 327 ff
tantalum nitride 328
tantalum oxide 327
TAT fluorescing dye 180 ff
technetium 341
technetium radiolabel 342
teeth 178
tellurium 163, 217 ff
temperature sensitive microgels 115 ff
terbium 387 ff
tetrathiafulvalene 92 ff
thallium 188 ff
thermal conductivity 35 ff, 174
thermal conductivity carbon tubes 62
thermoelectric wires 160
thermoelectric yttrium manganese antimonide crystals 340
thermoelectrics 35 ff, 218
thulium NMR 169 ff

tiatania fluoride 221
tin 162, 200 ff
tin oxide 187
tin telluride 200 ff, 220
titania 23
titania anodes 311
titania atomically flat 311
titania charge separation 311
titania electrochemistry 303
titania electrophoresis 313
titania nitrogen doping 312
titania photocatalysis 312
titania photocurrent 305 ff
titania pigments 300 ff
titania rods 304
titania sheets 306
titania transparency 306
titania tubes 309
titania vanadium oxide wires 322
titania wall paints 306
titania wires 309
titania–cellulose 136
titania–niobium 310
titanium 299 ff
titanium carbide 247
titanium monoxide 304
titanium–dye 305
toluene–water fingers 5 ff
trans-cyclohexane 13
tribology 253 ff
tributylphosphine 302
trioctylphosphine 99
tubes 284
tubes iron oxide 280 ff
tubes titania 309
tungstate catalyzed thiophene oxidation 334
tungstate zirconia layers 320
tungsten 334 ff
tungsten carbide layer 55
tungsten disulfide onions 335
tungsten hexacarbonyl vapor deposition 335
tungsten particle catalysis nitrogen oxidation 334
tungsten trioxide 310
tungsten–graphite 334

tungstic phosphoric acid–ammonium-polyethyleneglycol 335
turnip virus–terbium 387
tyrosine 13

underpotential sodium deposition 264 ff
undulations 183
urea decomposition 383
UV/vis/near IR spectra 41 (ZnO), 43 (Ag), 59 ff (C tubes), 135, 377 (porphyrins), 315 (PbS), 350 (Fe oxide), 364 (Cu), 375 ff (Au)

vanadium 321 ff
vanadium electron paramagnetic resonance 322 ff
vanadium oxide titania wires 322
vanadium oxides 321 ff
vanadyl oxygen transfer 321
vesicles 101 ff
viologen (bipyridinium) 130 ff
viologen layer 316
viologen–fulvalene 131 ff
voltage gated channels 102 ff
voltammetry 59

wall pinning 386
water 3 ff
water photolysis 208, 259, 286 ff
water adsorption 187
water cluster 4
water compressibility 9 ff
water diffusion 11, 14
water diffusion carbon tubes 62
water luminescence 38 ff
water pentamer 4
water repellent surfaces 192
water splitting cobalt(II) 349 ff
water–ethanol contraction 83
water–polyethyleneglycol 87
wires 19, 280 ff, 309, 322
wurtzite (zinc oxide) 223

xerography 215, 217
XPS 316
X-ray photoelectron spectrum (XPS) 126 ff

yoctolitre 1 ff
yoctowell 10 ff, 134 ff
yoctowells–edge amphiphiles 134 ff
yttrium 318, 383
yttrium manganese antimonide
    crystals thermoelectric 340 ff
yttrium vanadate 383 ff
yttrium–aluminium 383

zeolites 14, 24, 195 ff
zeolites–dyes 22 ff
zeta potential ceria particles 383

zinc 221 ff
zinc blende 7 ff, 196
zinc cadmium selenide 163
zinc ITO 187
zinc oxide 39, 163, 222 ff
zinc selenide 215
zinc sulfide 228
zirconate layers–amines 319
zirconia (zirconium dioxide) 317 ff
zirconium–phosphate 101, 319 ff
zirconium 317 ff
zirconocenes 320